Chapman & Hall/CRC Mathematical and Computational Biology Series

Handbook of Chemoinformatics Algorithms

Edited by

Jean-Loup Faulon
Andreas Bender

CRC Press
Taylor & Francis Group
Boca Raton London New York

CRC Press is an imprint of the
Taylor & Francis Group an **informa** business
A CHAPMAN & HALL BOOK

CHAPMAN & HALL/CRC
Mathematical and Computational Biology Series

Aims and scope:
This series aims to capture new developments and summarize what is known over the entire spectrum of mathematical and computational biology and medicine. It seeks to encourage the integration of mathematical, statistical, and computational methods into biology by publishing a broad range of textbooks, reference works, and handbooks. The titles included in the series are meant to appeal to students, researchers, and professionals in the mathematical, statistical and computational sciences, fundamental biology and bioengineering, as well as interdisciplinary researchers involved in the field. The inclusion of concrete examples and applications, and programming techniques and examples, is highly encouraged.

Series Editors

Proposals for the series should be submitted to one of the series editors above or directly to:
CRC Press, Taylor & Francis Group
4th, Floor, Albert House
1-4 Singer Street
London EC2A 4BQ
UK

Published Titles

Algorithms in Bioinformatics: A Practical Introduction
Wing-Kin Sung

Bioinformatics: A Practical Approach
Shui Qing Ye

Biological Sequence Analysis Using the SeqAn C++ Library
Andreas Gogol-Döring and Knut Reinert

Cancer Modelling and Simulation
Luigi Preziosi

Cell Mechanics: From Single Scale-Based Models to Multiscale Modeling
Arnaud Chauvière, Luigi Preziosi, and Claude Verdier

Combinatorial Pattern Matching Algorithms in Computational Biology Using Perl and R
Gabriel Valiente

Computational Biology: A Statistical Mechanics Perspective
Ralf Blossey

Computational Neuroscience: A Comprehensive Approach
Jianfeng Feng

Data Analysis Tools for DNA Microarrays
Sorin Draghici

Differential Equations and Mathematical Biology, Second Edition
D.S. Jones, M.J. Plank, and B.D. Sleeman

Engineering Genetic Circuits
Chris J. Myers

Exactly Solvable Models of Biological Invasion
Sergei V. Petrovskii and Bai-Lian Li

Gene Expression Studies Using Affymetrix Microarrays
Hinrich Göhlmann and Willem Talloen

Glycome Informatics: Methods and Applications
Kiyoko F. Aoki-Kinoshita

Handbook of Chemoinformatics Algrithms
Jean-Loup Faulon and Andreas Bender

Handbook of Hidden Markov Models in Bioinformatics
Martin Gollery

Introduction to Bioinformatics
Anna Tramontano

An Introduction to Systems Biology: Design Principles of Biological Circuits
Uri Alon

Kinetic Modelling in Systems Biology
Oleg Demin and Igor Goryanin

Knowledge Discovery in Proteomics
Igor Jurisica and Dennis Wigle

Meta-analysis and Combining Information in Genetics and Genomics
Rudy Guerra and Darlene R. Goldstein

Modeling and Simulation of Capsules and Biological Cells
C. Pozrikidis

Niche Modeling: Predictions from Statistical Distributions
David Stockwell

Normal Mode Analysis: Theory and Applications to Biological and Chemical Systems
Qiang Cui and Ivet Bahar

Optimal Control Applied to Biological Models
Suzanne Lenhart and John T. Workman

Pattern Discovery in Bioinformatics: Theory & Algorithms
Laxmi Parida

Python for Bioinformatics
Sebastian Bassi

Spatial Ecology
Stephen Cantrell, Chris Cosner, and Shigui Ruan

Spatiotemporal Patterns in Ecology and Epidemiology: Theory, Models, and Simulation
Horst Malchow, Sergei V. Petrovskii, and Ezio Venturino

Stochastic Modelling for Systems Biology
Darren J. Wilkinson

Structural Bioinformatics: An Algorithmic Approach
Forbes J. Burkowski

The Ten Most Wanted Solutions in Protein Bioinformatics
Anna Tramontano

Chapman & Hall/CRC
Taylor & Francis Group
6000 Broken Sound Parkway NW, Suite 300
Boca Raton, FL 33487-2742

Printed in the United States of America on acid-free paper
10 9 8 7 6 5 4 3 2 1

International Standard Book Number: 978-1-4200-8292-0 (Hardback)

Library of Congress Cataloging-in-Publication Data

Handbook of chemoinformatics algorithms / editors, Jean-Loup Faulon, Andreas Bender.
p. cm. -- (Chapman & Hall/CRC mathematical and computational biology series)
Includes bibliographical references and index.
ISBN 978-1-4200-8292-0 (hardcover : alk. paper)
1. Cheminformatics--Handbooks, manuals, etc. 2. Algorithms. 3. Graph theory. I. Faulon, Jean-Loup. II. Bender, Andreas, 1976- III. Title. IV. Series.

QD39.3.E46H357 2010
542'.85--dc22 2010005452

Visit the Taylor & Francis Web site at
http://www.taylorandfrancis.com

and the CRC Press Web site at
http://www.crcpress.com

Contents

Preface

The field of handling chemical information electronically—known as Chemoinformatics or Cheminformatics—has received a boost in recent decades, in line with the advent of tremendous computer power. Originating in the 1960s in both academic and industrial settings (and termed by its current name only from around 1998), chemoinformatics applications are today commonplace in every pharmaceutical company. Also, various academic laboratories in Europe, the United States, and Asia confer both undergraduate and graduate degrees in the field.

But still, there is a long way to go. While resembling its sibling, bioinformatics, both by name and also (partially) algorithmically, the chemoinformatics field developed in a very different manner right from the onset. While large amounts of biological information—sequence information, structural information, and more recently also phenotypic information such as metabolomics data—found their way straight into the public domain, large-scale chemical information was until very recently the domain of private companies. Hence, public tools to handle chemical structures were scarce for a very long time, while essential bioinformatics tools such as those for aligning sequences or viewing protein structures were available at no cost to anyone interested in the area. More recently—luckily—this situation changed significantly, with major life science data providers such as the NCBI, the EBI, and many others also making large-scale chemical data publicly available.

However, there is another aspect, apart from the actual data, that is crucial for a scientific field to flourish—and that is the proper documentation of techniques and methods, and, in the case of informatics sciences, the proper documentation of algorithms. In the bioinformatics field, and in line with a tremendous amount of open access data and tools available, algorithms were documented extensively in reference books. In the chemoinformatics field, however, a book of this type is missing until now. This is what the editors, with the help of expert contributors in the field, are attempting to remedy—to provide an overview of some of the most common chemoinformatics algorithms in a single place.

The book is divided into 15 chapters. Chapter 1 presents a historical perspective of the applications of algorithms and graph theory to chemical problems. Algorithms to store and retrieve two-dimensional chemical structures are presented in Chapter 2, and three-dimensional representations of chemicals are discussed in Chapter 3. Molecular descriptors, which are widely used in virtual screening and structure–activity/property predictions, are presented in Chapter 4. Chapter 5 presents virtual screening methods from a ligand perspective and from a structure perspective including docking methods. Chapters 6 and 7 are dedicated to quantitative structure–activity relationships (QSAR). QSAR modeling workflow and methods to prepare the data are presented in Chapter 6, while the development and validation of QSAR models are discussed in Chapter 7. Chapter 8 introduces algorithms to enumerate and sample chemical structures, with applications in combinatorial libraries design. Chapters 9 and 10 are dedicated to computer-aided molecular design: from a ligand perspective in Chapter 9, where inverse-QSAR methods are reviewed, and from a structure perspective in Chapter 10, where *de novo* design algorithms are presented. Chapter 11 covers reaction network generation, with applications in synthesis design and biological network inference. Closing the strictly chemoinformatics chapters, Chapter 12 provides a review of Open Source software and database technologies dedicated to the field. The remaining chapters (13–15) present techniques developed in the context of bioinformatics and computational biology and their potential applications to chemical problems. Chapter 13 discusses possible applications of sequence alignment algorithms to tree-like structures such as glycans. Chapter 14 presents classical machine learning algorithms that can be used for both bioinformatics and chemoinformatics problems. Chapter 15 introduces a systems biology approach to study the kinetics of metabolic networks.

While our book covers many aspects of chemoinformatics, our attempt is ambitious—and it is probably impossible to provide a complete overview of "all" chemoinformatics algorithms in one place. Hence, in this work we present a selection of algorithms from the areas the editors deemed most relevant in practice and hope that this work will be helpful as a reference work for people working in the field.

MATLAB® and Simulink® are registered trademarks of The Math Works, Inc. For product information, please contact:

The Math Works, Inc.
3 Apple Hill Drive
Natick, MA 01760-2098, USA
Tel: 508 647 7000
Fax: 508-647-7001
E-mail: info@mathworks.com
Web: www.mathworks.com

Jean-Loup Faulon, *Paris, France*
Andreas Bender, *Leiden, the Netherlands*

Acknowledgments

The editors would like to first thank Robert B. Stern from the Taylor & Francis Group for giving them an opportunity to compile, for the first time, an overview of chemoinformatics algorithms. They also thank the authors for assembling expert materials covering many algorithmic aspects of chemoinformatics. Jean-Loup Faulon would like to acknowledge the interest and encouragement provided by Genopole's Epigenomics program and the University of Evry, France, to edit and coauthor chapters in this book.

The authors of Chapter 2 would like to thank Ovidiu Ivanciuc for providing relevant literature references. They also acknowledge the permission to reprint Algorithm 2.1 [Dittmar et al. *J. Chem. Inf. Comput. Sci.*, 17(3): 186–192, 1977. Copyright (1977) American Chemical Society]. Milind Misra acknowledges funding provided by Sandia National Laboratories, a multiprogram laboratory operated by Sandia Corporation, a Lockheed Martin Company, for the United States Department of Energy's National Nuclear Security Administration under contract DE-AC04-94AL85000.

Markus Meringer would like to thank Emma Schymanski for carefully proofreading Chapter 8.

Shawn Martin would like to acknowledge funding (to write Chapter 14) provided by Sandia National Laboratories, a multiprogram laboratory operated by Sandia Corporation, a Lockheed Martin Company, for the United States Department of Energy's National Nuclear Security Administration under contract DE-AC04-94AL85000.

Finally, Fangping Mu, Amy L. Bauer, James R. Faeder, and William S. Hlavacek acknowledge funding support (to write Chapter 15) provided in part by the NIH, under grants GM080216 and CA132629, and by the DOE, under contract DE-AC52-06NA25396. They also thank P.J. Unkefer, C.J. Unkefer, and R.F. Williams for helpful discussions.

Contributors

Tatsuya Akutsu
Institute for Chemical Research
Kyoto University
Uji, Japan

Amy L. Bauer
Theoretical Biology and Biophysics Group
Theoretical Division, Los Alamos National Laboratory
Los Alamos, New Mexico

Pablo Carbonell
Institute of Systems and Synthetic Biology
University of Evry
Evry, France

Robert D. Clark
Biochemical Infometrics and School of Informatics
Indiana University
Bloomington, Indiana

James R. Faeder
Department of Computational Biology
University of Pittsburgh School of Medicine
Pittsburgh, Pennsylvania

Jean-Loup Faulon
Department of Biology
University of Evry
Evry, France

Nikolas Fechner
Department of Computer Architecture
University of Tübingen
Tübingen, Germany

Alexander Golbraikh
Division of Medicinal Chemistry and Natural Products
University of North Carolina
Chapel Hill, North Carolina

Rajarshi Guha
NIH Chemical Genomics Center
Rockville, Maryland

Georg Hinselmann
Department of Computer Architecture
University of Tübingen
Tübingen, Germany

William S. Hlavacek
Theoretical Biology and Biophysics Group
Theoretical Division, Los Alamos National Laboratory
Los Alamos, New Mexico

Ovidiu Ivanciuc
Department of Biochemistry and Molecular Biology
University of Texas Medical Branch
Galveston, Texas

Shawn Martin
Sandia National Laboratories
Albuquerque, New Mexico

Markus Meringer
German Aerospace Center (DLR)
Oberpfaffenhofen, Germany

Milind Misra
Sandia National Laboratories
Albuquerque, New Mexico

Fangping Mu
Theoretical Biology and Biophysics Group
Theoretical Division, Los Alamos National Laboratory
Los Alamos, New Mexico

Diana C. Roe
Department of Biosystems Research
Sandia National Laboratories
Livermore, California

Alexander Tropsha
Division of Medicinal Chemistry and Natural Products
University of North Carolina
Chapel Hill, North Carolina

Donald P. Visco, Jr.
Department of Chemical Engineering
Tennessee Technological University
Cookeville, Tennessee

Jörg Kurt Wegner
Integrative Chem-/Bio-Informatics
Tibotec (Johnson & Johnson)
Mechelen, Belgium

Egon L. Willighagen
Department of Pharmaceutical Biosciences
Uppsala University
Uppsala, Sweden

1 Representing Two-Dimensional (2D) Chemical Structures with Molecular Graphs

Ovidiu Ivanciuc

CONTENTS

1.1 INTRODUCTION

Graphs are used as an efficient abstraction and approximation for diverse chemical systems, such as chemical compounds, ensembles of molecules, molecular fragments, polymers, chemical reactions, reaction mechanisms, and isomerization pathways. Obviously, the complexity of chemical systems is significantly reduced whenever they are modeled as graphs. For example, when a chemical compound is represented as a molecular graph, the geometry information is neglected, and only the atom connectivity information is retained. In order to be valuable, the graph representation of a chemical system must retain all important features of the investigated system and has to offer qualitative or quantitative conclusions in agreement with those provided by more sophisticated methods. All chemical systems that are successfully modeled as graphs have a common characteristic, namely they are composed of elements that interact between them, and these interactions are instrumental in explaining a property of interest of that chemical system. The elements in a system are represented as graph vertices, and the interactions between these elements are represented as graph edges. In a chemical graph, vertices may represent various elements of a chemical system, such as atomic or molecular orbitals, electrons, atoms, groups of atoms, molecules, and isomers. The interaction between these elements, which are represented as graph edges, may be chemical bonds, nonbonded interactions, reaction steps, formal connections between groups of atoms, or formal transformations of functional groups. The chapter continues with an overview of elements of graph theory that are important in chemoinformatics and in depicting two-dimensional (2D) chemical structures. Section 1.3 presents the most important types of chemical and molecular graphs, and Section 1.4 reviews the representation of molecules containing heteroatoms and multiple bonds with weighted graphs and molecular matrices.

1.2 ELEMENTS OF GRAPH THEORY

This section presents the basic definitions, notations, and examples of graph theory relevant to chemoinformatics. Graph theory applications in physics, electronics, chemistry, biology, medicinal chemistry, economics, or information sciences are mainly the effect of the seminal book *Graph Theory* of Harary [1]. Several other books represent essential readings for an in-depth overview of the theoretical basis of graph theory: *Graphs and Hypergraphs* by Berge [2]; *Graphs and Digraphs* by Behzad, Chartrand, and Lesniak-Foster [3]; *Distance in Graphs* by Buckley and Harary [4]; *Graph Theory Applications* by Foulds [5]; *Introduction to Graph Theory* by West [6]; *Graph Theory* by Diestel [7]; and *Topics in Algebraic Graph Theory* by Beineke and Wilson [8]. The spectral theory of graphs investigates the properties of the spectra (eigenvalues) of graph matrices, and has applications in complex networks, spectral embedding of multivariate data, graph drawing, calculation of topological indices, topological quantum chemistry, and aromaticity. The major textbook in the spectral theory of graphs is *Spectra of Graphs. Theory and Applications* by Cvetković, Doob, and Sachs [9]. An influential book on graph spectra applications in the quantum chemistry of conjugated systems and aromaticity is *Topological Approach to the Chemistry of Conjugated Molecules* by Graovac, Gutman, and Trinajstić [10]. Advanced topics of topological aromaticity are treated in *Kekulé Structures in Benzenoid Hydrocarbons* by Cyvin and Gutman [11]; *Introduction to the Theory of Benzenoid Hydrocarbons* by Gutman and Cyvin [12]; *Advances in the Theory of Benzenoid Hydrocarbons* by Gutman and Cyvin [13]; *Theory of Coronoid Hydrocarbons* by Cyvin, Brunvoll, and Cyvin [14]; and *Molecular Orbital Calculations Using Chemical Graph Theory* by Dias [15]. The graph theoretical foundation for the enumeration of chemical isomers is presented in several books: *Graphical Enumeration* by Harary and Palmer [16]; *Combinatorial Enumeration of Groups, Graphs, and Chemical Compounds* by Pólya and Read [17]; and *Symmetry and Combinatorial Enumeration in Chemistry* by Fujita [18]. A comprehensive history of graph theory can be found in the book *Graph Theory 1736–1936* by Biggs, Lloyd, and Wilson [19].

The first edited book on chemical graphs is *Chemical Applications of Graph Theory* by Balaban [20]. Several comprehensive textbooks on chemical graphs are available, such as *Chemical Graph Theory* by Trinajstić [21], *Mathematical Concepts in Organic Chemistry* by Gutman and Polansky [22], and *Handbook of Chemoinformatics* by Gasteiger [23]. Applications of topological indices in quantitative structure–activity relationships (QSAR) are presented in *Molecular Connectivity in Chemistry and Drug Research* by Kier and Hall [24], *Molecular Connectivity in Structure–Activity Analysis* by Kier and Hall [25], *Molecular Structure Description. The Electrotopological State* by Kier and Hall [26], *Information Theoretic Indices for Characterization of Chemical Structure* by Bonchev [27], and *Topological Indices and Related Descriptors in QSAR and QSPR* by Devillers and Balaban [28]. A comprehensive text on reaction graphs is *Chemical Reaction Networks. A Graph-Theoretical Approach* by Temkin, Zeigarnik, and Bonchev [29], and a graph-theoretical approach to organic reactions is detailed in *Synthon Model of Organic Chemistry and Synthesis Design* by Koča et al. [30]. Graph algorithms for drug design are presented in *Logical and Combinatorial Algorithms for Drug Design* by Golender and Rozenblit [31]. Graph theory concepts relevant to chemoinformatics are introduced in this section, together with examples of graphs and graph matrices.

1.2.1 Graphs

A *graph* $G(V, E)$ is an ordered pair consisting of a vertex set $V(G)$ and an edge set $E(G)$. Each element $\{i, j\} \in E$ (where $i, j \in V$) is said to be an edge joining vertices i and j. Because each edge is defined by an unordered pair of vertices from V, the edge from vertex i to vertex j is identical with the edge from vertex j to vertex I, $\{i,j\} = \{j, i\}$. The number of vertices N defines the order of the graph and is equal to the number of elements in $V(G)$, $N = |V(G)|$, and the number of edges M is equal to the number of elements in $E(G)$, $M = |E(G)|$. Several examples of graphs relevant to chemistry are shown in Graphs **1.1** through **1.5**.

Vertices and edges in a graph may be labeled. A vertex with the label i is indicated here as v_i. An edge may be denoted by indicating the two vertices that define that edge. For example, the edge connecting vertices v_i and v_j may be denoted by e_{ij}, $e_{i,j}$, $\{i,j\}$, or v_iv_j. Usually, graph vertices are labeled from 1 to N, $V(G) = \{v_1, v_2, \ldots, v_N\}$, and graph edges are labeled from 1 to M, $E(G) = \{e_1, e_2, \ldots, e_M\}$. There is no special rule in labeling graphs, and a graph with N vertices may be labeled in $N!$ different ways.

1.1

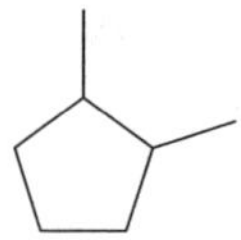

1.2

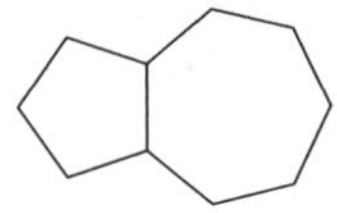

1.3

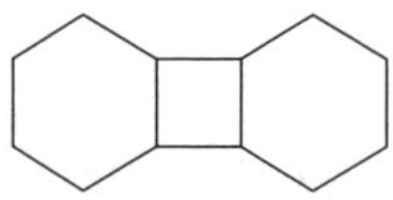

1.4

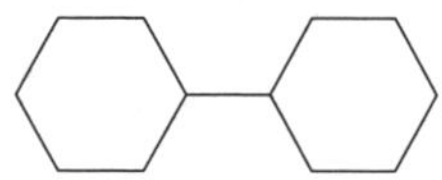

1.5

A *graph invariant* is a number, sequence of numbers, or matrix computed from the graph topology (information contained in the V and E sets) that is not dependent on the graph labeling (the graph invariant has the same value for all $N!$ different labelings of the graph). Two obvious graph invariants are the number of vertices N and the number of edges M. Other invariants of molecular graphs are topological indices, which are used as structural descriptors in quantitative structure–property relationships (QSPR), QSAR, and virtual screening of chemical libraries (cf. Chapters 4 and 5).

Graphs that have no more than one edge joining any pair of vertices are also called *simple graphs*. A *multigraph* is a graph in which two vertices may be connected by more than one edge. A *multiedge* of multiplicity m is a set of m edges that connects the same pair of distinct vertices. A *loop* $e_{ii} \in E$ is an edge joining a vertex v_i with itself. A loopgraph is a graph containing one or more vertices with loops.

Simple graphs cannot capture the complexity of real life systems, such as electrical circuits, transportation networks, production planning, kinetic networks, metabolic networks, or chemical structures. In such cases it is convenient to attach weights to vertices or loops, weights that may represent current intensity, voltage, distance, time, material flux, reaction rate, bond type, or atom type. A graph $G(V, E, w)$ is a *weighted graph* if there exists a function $w : E \rightarrow R$ (where R is the set of real numbers), which assigns a real number, called weight, to each edge of E. Graph **1.6** has all edge weights equal to 2, whereas in Graph **1.7** the edge weights alternate between 1 and 2. In the loopgraph **1.8** all edges have the weight 1 and the loop has the weight 2. Alkanes and cycloalkanes are represented as molecular graphs with all edges having a weight equal to 1, whereas chemical compounds containing heteroatoms or multiple bonds are represented as vertex- or edge-weighted molecular graphs. Section 1.4 reviews in detail the representation of chemical compounds with weighted graphs.

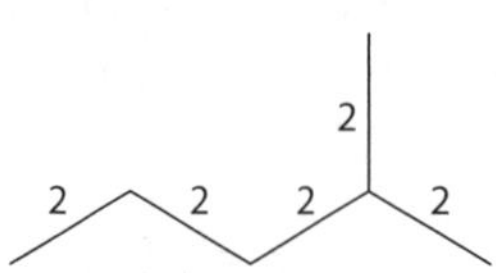

1.6

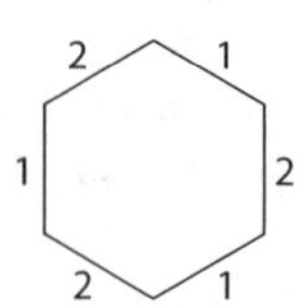

1.7

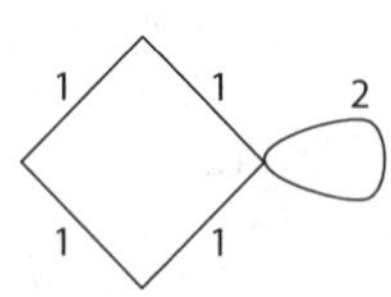

1.8

In many graph models, such as those of kinetic, metabolic, or electrical networks, it is useful to give each edge a direction or orientation. The graphs used to model such oriented systems are termed *directed graphs* or *digraphs*. A graph $D(V, A)$ is an ordered pair consisting of two sets $V(D)$ and $A(D)$, where the vertex set V is finite and nonempty and the arc set A is a set of ordered pairs of distinct elements from V. Graphs **1.9** through **1.12** are several examples of digraphs. A comprehensive overview of reaction graphs is presented by Balaban [32], and graph models for networks of chemical reactions are reviewed by Temkin et al. [29].

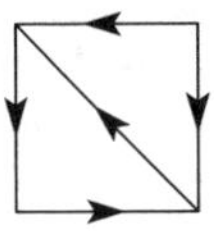
1.9

1.10

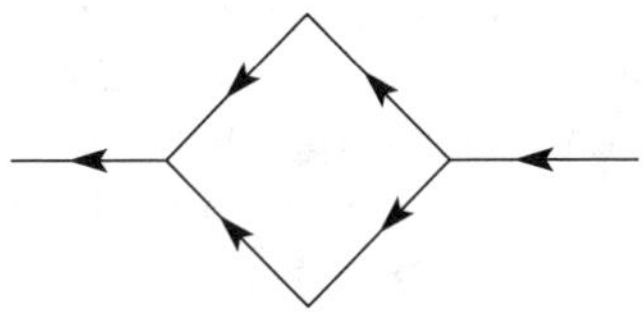
1.11

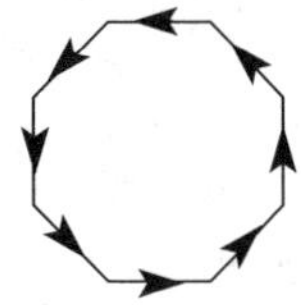
1.12

1.2.2 Adjacency, Walks, Paths, and Distances

Two vertices v_i and v_j of a graph G are adjacent (or neighbors) if there is an edge e_{ij} joining them. The two adjacent vertices v_i and v_j are said to be incident to the edge e_{ij}. The neighborhood of a vertex v_i is represented by the set of all vertices adjacent to v_i. Two distinct edges of G are adjacent if they have a vertex in common.

The degree of a vertex v_i, denoted by deg_i, is equal to the number of vertices adjacent to vertex v_i. The set of degree values for all vertices in a graph gives the vector Deg(G) whose ith element represents the degree of the vertex v_i. In a weighted graph $G(V, E, w)$, the valency of a vertex v_i, $\text{val}(w, G)_i$, is defined as the sum of the weights of all edges e_{ij} incident with vertex v_i [33,34]. The set of valencies for all vertices in a graph forms the vector Val(w, G) whose ith element represents the valency of the vertex v_i. From the definition of degree and valency it is obvious that in simple, nonweighted graphs, the degree of a vertex v_i, deg_i, is identical to the valency of that vertex, val_i. Consider the simple labeled graph **1.13**. A simple count of the neighbors for each vertex in **1.13** gives the degree vector $\text{Deg}(\mathbf{1.13}) = \{2, 2, 3, 2, 2, 3, 2\}$. The second example considers a weighted graph with the labeling given in Graph **1.14** and with the edge weights indicated in **1.15**. The degree vector of **1.14** is $\text{Deg}(\mathbf{1.14}) = \{2, 3, 2, 3, 2, 2, 3, 2, 3, 2\}$, and the valency vector is $\text{Val}(\mathbf{1.14}) = \{1.5, 4, 3, 4, 1.5, 1.5, 4, 3, 4, 1.5\}$. Both degree and valency are *graph invariants*, because their numerical values are independent of the graph labeling.

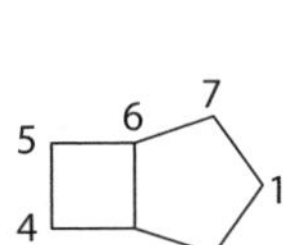

1.13

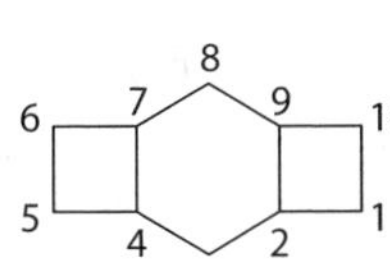

1.14

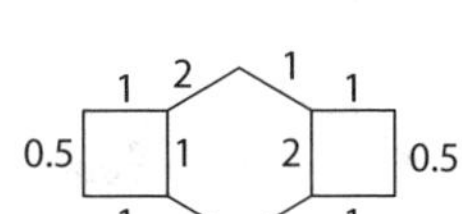

1.15

A *walk* W in a graph G is a sequence of vertices and edges $W(G) = \{v_a, e_{ab}, v_b, e_{bc}, v_c, e_{cd}, v_d, e_{de}, v_e, \ldots, v_i, e_{ij}, v_j, \ldots, v_m, e_{mn}, v_n\}$ beginning and ending with vertices, in which two consecutive vertices v_i and v_j are adjacent, and each edge e_{ij} is incident with the two vertices v_i and v_j preceding and following it, respectively. A walk may also be defined as a sequence of vertices $W(G) = \{v_a, v_b, \ldots, v_n\}$ in which two consecutive vertices v_i and v_{i+1} are adjacent. Similarly, a walk may be defined as a sequence of edges $W(G) = \{e_{ab}, e_{bc}, \ldots, e_{mn}\}$ in which two consecutive edges e_{ij} and e_{jk} are adjacent. In a walk any edge of the graph may appear more than once. The length of a walk is equal to the total number of edges that define the walk. A walk in which the initial and the terminal vertices coincide is called a closed walk. A walk in which the initial and the terminal vertices are different is called an open walk. A *trail* is a walk in which no edge is repeated. A certain vertex may appear more than once in a trail, if the trail intersects itself. A *path P* is a walk in which all vertices (and thus necessarily all edges) are distinct. The length of a path in a graph is equal to the number of edges along the path.

A *graph cycle* or *circuit* is a closed walk in which all vertices are distinct, with the exception of the initial and terminal vertices that coincide. In Graph **1.16** there are three cycles: $C_1(\mathbf{1.16}) = \{v_1, v_2, v_5, v_1\}$, with length three; $C_2(\mathbf{1.16}) = \{v_1, v_2, v_3, v_4, v_5, v_1\}$, with length five; and $C_3(\mathbf{1.16}) = \{v_2, v_3, v_4, v_5, v_2\}$, with length four. In Graph **1.17** there are three cycles of length five: $C_1(\mathbf{1.17}) = \{v_1, v_2, v_5, v_6, v_3, v_1\}$, $C_2(\mathbf{1.17}) = \{v_2, v_4, v_7, v_8, v_5, v_2\}$, and $C_3(\mathbf{1.17}) = \{v_7, v_9, v_{11}, v_{10}, v_8, v_7\}$.

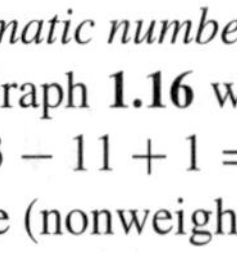

1.16

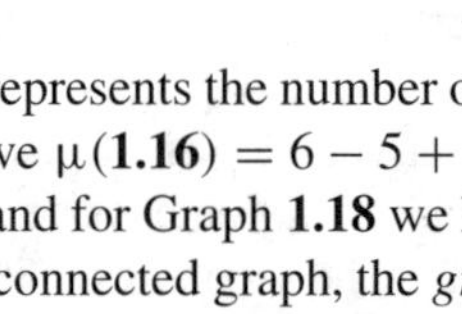

1.17

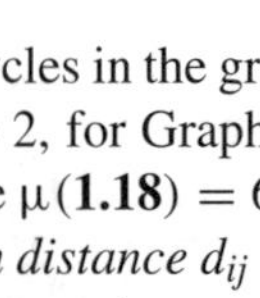

1.18

The *cyclomatic number* μ represents the number of cycles in the graph, $\mu = M - N + 1$. For Graph **1.16** we have $\mu(\mathbf{1.16}) = 6 - 5 + 1 = 2$, for Graph **1.17** we have $\mu(\mathbf{1.17}) = 13 - 11 + 1 = 3$, and for Graph **1.18** we have $\mu(\mathbf{1.18}) = 6 - 6 + 1 = 1$.

In a simple (nonweighted) connected graph, the *graph distance* d_{ij} between a pair of vertices v_i and v_j is equal to the length of the shortest path connecting the two vertices (i.e., the number of edges of the shortest path). The distance between two adjacent vertices is 1. The graph distance satisfies the properties of a metric:

a. The distance from a vertex v_i to itself is zero:

$$d_{ii} = 0, \quad \text{for all } v_i \in V(G). \tag{1.1}$$

b. The distance between two distinct vertices v_i and v_j is larger than 0:

$$d_{ij} > 0, \quad \text{for all } v_i, v_j \in V(G). \tag{1.2}$$

c. The distance between two distinct vertices v_i and v_j is equal to the distance on the inverse path, from v_j and v_i:

$$d_{ij} = d_{ji}, \quad \text{for all } v_i, v_j \in V(G). \tag{1.3}$$

d. The graph distance satisfies the triangle inequality:

$$d_{ik} + d_{kj} \geq d_{ij}, \quad \text{for all } v_i, v_j, v_k \in V(G). \tag{1.4}$$

The *eccentricity* $\text{ecc}(v_i)$ of a vertex v_i is the maximum distance from the vertex v_i to any other vertex v_j in graph G, that is, max [35] for all $v_j \in V(G)$. The *diameter* **diam**(G) of a graph G is the maximum eccentricity. If the graph G has cycles, then the *girth* of G is the length of a shortest cycle, and the *circumference* is the length of a longest cycle.

A graph G may be transformed into a series of *subgraphs* of G by deleting one or more of its vertices, or by deleting one or more of its edges. If $V(G')$ is a subset of $V(G)$, $V(G') \subseteq V(G)$, and $E(G')$ is a subset of $E(G)$, $E(G') \subseteq E(G)$, then the subgraph $G' = (V(G'), E(G'))$ is a subgraph of the graph $G = (V(G), E(G))$. A subgraph $G - v_i$ is obtained by deleting from G the vertex v_i and all its incident edges. A subgraph $G - e_{ij}$ is obtained by deleting from G the edge e_{ij}. Graph **1.19** has four subgraphs of the type $G - v_i$, **1.20** through **1.23**, which are obtained by deleting, in turn, one vertex and all its incident edges from Graph **1.19**.

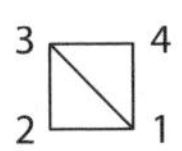

1.19

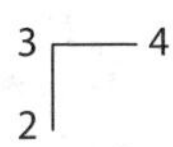

1.20

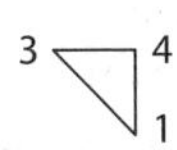

1.21

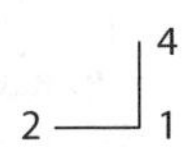

1.22

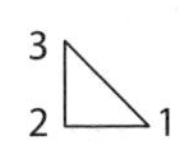

1.23

1.2.3 Special Graphs

A *tree*, or an *acyclic graph*, is a connected graph that has no cycles (the cyclomatic number $\mu = 0$). Alternative definitions for a tree are the following: a tree is a connected graph with N vertices and $N-1$ edges; a tree is a graph with no cycles, N vertices, and $N-1$ edges. A graph that contains as components only trees is a *forest*. A k-tree is a tree with the maximum degree k. Alkanes are usually represented as 4-trees. A *rooted tree* is a tree in which one vertex (the root vertex) is distinct from the other ones.

A graph with the property that every vertex has the same degree is called a *regular graph*. A graph G is called a k-regular graph or a regular graph of degree k if every vertex from G has the degree k. A ring R_N with N vertices is a 2-regular graph with N vertices, that is, a graph with all vertices of degree 2. The cycloalkanes cyclopropane, cyclobutane, cyclopentane, cyclohexane, cycloheptane, and cyclooctane are examples of 2-regular graphs. The 3-regular graphs, or *cubic graphs*, **1.24** through **1.27**, represent as molecular graphs the polycyclic hydrocarbons triprismane, tetraprismane (cubane), pentaprismane, and hexaprismane, respectively. Fullerenes are also represented as cubic graphs.

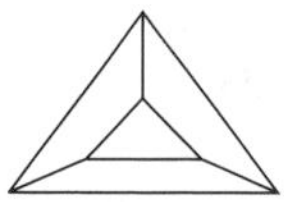

1.24

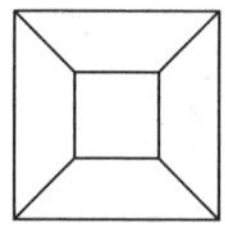

1.25

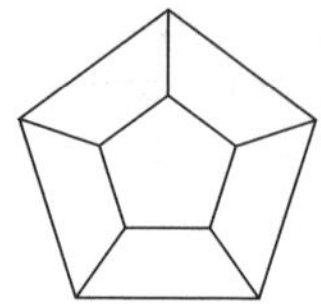

1.26

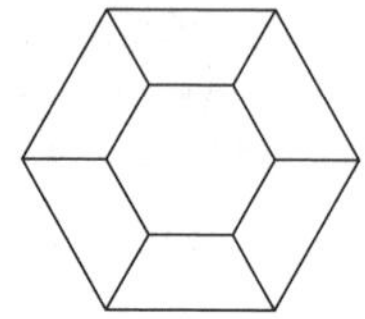

1.27

1.2.4 Graph Matrices

A graph is completely determined by indicating its adjacency relationships or its incidence relationships. However, the algebraic properties of graphs are easier studied by representing a graph as a matrix, such as adjacency matrix, incidence matrix, cycle matrix, path matrix, Laplacian matrix, distance matrix, and detour matrix. Graph matrices of chemical systems are used to investigate the spectral properties of molecular graphs [9], to apply the Hückel molecular orbitals method to conjugated molecules [10], to compute various topological indices for QSAR models [36,37], and to study the topology of biological networks [38]. In presenting graph matrices we consider only labeled, connected, simple graphs.

1.2.4.1 Adjacency Matrix

The *adjacency matrix* $\mathbf{A}(G)$ of a vertex labeled graph G with N vertices is a square $N \times N$ symmetric matrix in which $[\mathbf{A}]_{ij} = 1$ if vertex v_i is adjacent to vertex v_j and $[\mathbf{A}]_{ij} = 0$ otherwise. The adjacency matrix is symmetric, with all elements on the main diagonal equal to zero. The sum of entries over row i or column i in $\mathbf{A}(G)$ is the degree of vertex v_i, $\deg_i$. As an example we consider Graph **1.28** labeled from 1 to 8 and its adjacency matrix **A(1.28)**.

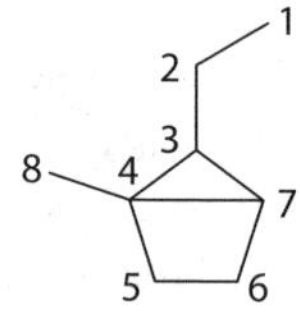

1.28

$$\mathbf{A}(\mathbf{1.28}) = \begin{array}{c|cccccccc|} & 1 & 2 & 3 & 4 & 5 & 6 & 7 & 8 \\ 1 & 0 & 1 & 0 & 0 & 0 & 0 & 0 & 0 \\ 2 & 1 & 0 & 1 & 0 & 0 & 0 & 0 & 0 \\ 3 & 0 & 1 & 0 & 1 & 0 & 0 & 1 & 0 \\ 4 & 0 & 0 & 1 & 0 & 1 & 0 & 1 & 1 \\ 5 & 0 & 0 & 0 & 1 & 0 & 1 & 0 & 0 \\ 6 & 0 & 0 & 0 & 0 & 1 & 0 & 1 & 0 \\ 7 & 0 & 0 & 1 & 1 & 0 & 1 & 0 & 0 \\ 8 & 0 & 0 & 0 & 1 & 0 & 0 & 0 & 0 \end{array}$$

From the definition of the adjacency matrix, it follows that if $[\mathbf{A}]_{ij} = 1$ then there is a walk of length one between vertices v_i and v_j. Higher powers of the adjacency matrix can be used to count the number of closed or open walks of a certain length between two vertices. The element $[\mathbf{A}^k]_{ij}$ of the kth power of the adjacency matrix $\mathbf{A}$ is the number of walks of length k between vertices v_i and v_j [1]. If $i = j$ then the element $[\mathbf{A}^k]_{ii}$ is the number of closed walks of length k that start and end at the same vertex v_i. Similarly, when $i \neq j$, the element $[\mathbf{A}^k]_{ij}$ is the number of open walks of length k starting from vertex v_i and ending at vertex v_j. Because the kth power of the adjacency matrix is symmetric, it follows that the number of walks of length k from v_i to v_j is equal to the number of walks of length k from v_j to v_i, that is, $[\mathbf{A}^k]_{ij} = [\mathbf{A}^k]_{ji}$. $\mathbf{A}^k$ matrices can also be used to determine the distances between vertices in simple graphs. If in a sequence of $\mathbf{A}^k$ matrices all elements $[\mathbf{A}^{k-1}]_{ij} = 0$ and $[\mathbf{A}^k]_{ij} \neq 0$, it follows that the distance between vertices v_i and v_j is k (the two vertices are separated by k edges). A general procedure for computing graph distances, which can be applied to general graphs, is presented in the section on the distance matrix.

Randić suggested the use of the closed walk counts of different lengths originating from a vertex to describe the environment of that vertex [39]. He defined the closed walk atomic code of the vertex v_i, CWAC$_i$, as the sequence $\{[\mathbf{A}^1]_{ii}, [\mathbf{A}^2]_{ii}, \ldots, [\mathbf{A}^k]_{ii}, \ldots, [\mathbf{A}^N]_{ii}\}$. The count of closed walks is also related to the graph spectrum and spectral moments. The complete set of graph eigenvalues $x_1, x_2, \ldots, x_N$ of the adjacency matrix $\mathbf{A}(G)$ forms the spectrum of a graph G, Sp($\mathbf{A}$,G) = $\{x_i, i = 1, 2, \ldots, N\}$. The kth spectral moment of $\mathbf{A}(G)$, SM($\mathbf{A}$, G)$_k$, is defined as the sum of the kth power of Sp($\mathbf{A}$, G). Finally, the sum of the diagonal elements of $\mathbf{A}^k$ (the trace of the kth power of the adjacency matrix which is equal to the count of closed walks of length k) equals SM($\mathbf{A}$, G)$_k$. Spectral moments represent a powerful theoretical tool in correlating structural features with various properties of chemical systems. Burdett used spectral moments to estimate the electronic properties of solids [40,41]. Spectral moments of conjugated compounds are correlated with the presence of certain subgraphs [42–44], thus making possible the calculation of the resonance energy per electron (REPE) from subgraph contributions [42]. A similar approach was proposed by Schmalz, Živković, and Klein for the decomposition of the π-electron energy of conjugated acyclic hydrocarbons in terms of various substructures [45].

1.2.4.2 Laplacian Matrix

Consider a simple graph G with N vertices and M edges, and its adjacency matrix $\mathbf{A}(G)$. We define the diagonal matrix $\mathbf{DEG}(G)$ with the diagonal elements $[\mathbf{DEG}]_{ii} = \deg_i$ (the degree of vertex v_i) and with the nondiagonal elements $[\mathbf{DEG}]_{ij} = 0, i \neq j$. The *Laplacian matrix* of the simple graph G, $\mathbf{L}(G)$, is the difference between $\mathbf{DEG}$ and $\mathbf{A}$ [46–48]:

$$\mathbf{L}(G) = \mathbf{DEG}(G) - \mathbf{A}(G). \tag{1.5}$$

The most significant chemoinformatics applications of the Laplacian matrix are in computing topological indices [48,49], defining the resistance distance matrix [50], and interpolating QSAR models based on molecular networks [51–54].

1.2.4.3 Distance Matrix

The *distance matrix* $\mathbf{D}(G)$ of a simple graph G with N vertices is a square $N \times N$ symmetric matrix in which $[\mathbf{D}]_{ij} = d_{ij}$, where d_{ij} is the distance between vertices v_i and v_j, that is, the length of the shortest path that connects vertices v_i and v_j [1,4]. The distance matrix is symmetric, with all elements on the main diagonal equal to zero. Applications of the distance matrix to chemical graphs may be found in several reviews [37,55]. As an example we consider Graph **1.29** labeled from 1 to 9 and its distance matrix **D(1.29)**.

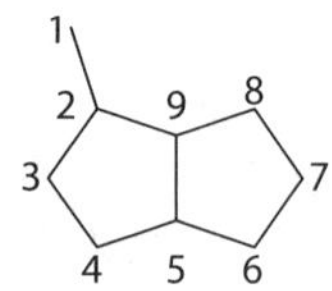

1.29

$$
\mathbf{D}(\mathbf{1.29}) = \begin{array}{c|ccccccccc|}
 & 1 & 2 & 3 & 4 & 5 & 6 & 7 & 8 & 9 \\
1 & 0 & 1 & 2 & 3 & 3 & 4 & 4 & 3 & 2 \\
2 & 1 & 0 & 1 & 2 & 2 & 3 & 3 & 2 & 1 \\
3 & 2 & 1 & 0 & 1 & 2 & 3 & 4 & 3 & 2 \\
4 & 3 & 2 & 1 & 0 & 1 & 2 & 3 & 3 & 2 \\
5 & 3 & 2 & 2 & 1 & 0 & 1 & 2 & 2 & 1 \\
6 & 4 & 3 & 3 & 2 & 1 & 0 & 1 & 2 & 2 \\
7 & 4 & 3 & 4 & 3 & 2 & 1 & 0 & 1 & 2 \\
8 & 3 & 2 & 3 & 3 & 2 & 2 & 1 & 0 & 1 \\
9 & 2 & 1 & 2 & 2 & 1 & 2 & 2 & 1 & 0
\end{array}
$$

In a simple graph, the distances between one vertex and all other vertices may be computed with the algorithm proposed by Dijkstra [35], which may also be applied to graphs with non-negative edge weights. Unlike the Dijkstra algorithm, the Floyd–Warshall algorithm [56,57] may be applied to graphs that have some edges with negative weights, as long as all cycle weights are non-negative.

ALGORITHM 1.1 FLOYD–WARSHALL

01. Consider the labeled, weighted graph G with N vertices, M edges, the vertex set $V(G)$, the edge set $E(G)$, and with a weight w_{ij} for each edge $e_{ij} \in E(G)$.
02. Define the cost matrix ${}^1\mathrm{Co} = {}^1\mathrm{Co}(G)$ of the labeled graph G as the square $N \times N$ symmetric matrix in which $[{}^1\mathrm{Co}]_{ii} = 0, [{}^1\mathrm{Co}]_{ij} = w_{ij}$ if $e_{ij} \in E(G)$, and $[{}^1\mathrm{Co}]_{ij} = \infty$ otherwise.
03. For each $k \in \{1, 2, \ldots, N\}$ do

```
04.     For each i ∈ {1,2,...,N} do
05.        For each j ∈ {1,2,...,N} do
06.          Update the cost matrix ^kCo:
             [Co]_ij = min{[^(k-1)Co]_ij, [^(k-1)Co]_ik + [^(k-1)Co]_kj}
07. End do
08. D = ^NCo
```

Step 06 in the Floyd–Warshall algorithm is based on the triangle inequality mentioned in Equation 1.4. If a graph contains cycles with negative weights, then the cost matrix Co has some negative numbers on the main diagonal. If $\mathbf{Co}_{ii} < 0$, then the vertex v_i belongs to at least one cycle with negative weight. The distance matrix is used to compute many important topological indices, such as Wiener index W [58], Balaban index J [59,60], Kier–Hall electrotopological indices [26,61], information theory indices [62], and molecular path code indices [63]. The distance matrix is the source of several molecular matrices [37,64], namely the reciprocal distance matrix [65], the distance-valency matrix [33], the distance complement matrix [66], the reverse Wiener matrix [67], the distance-path matrix [68,69], and the Szeged matrix [70,71]. These distance-related molecular matrices are used to compute topological indices and related graph descriptors for QSPR and QSAR.

1.3 CHEMICAL AND MOLECULAR GRAPHS

Chemical compounds are usually represented as molecular graphs, that is, nondirected, connected graphs in which vertices correspond to atoms and edges represent covalent bonds between atoms. The molecular graph model of the chemical structure emphasizes the chemical bonding pattern of atoms, whereas the molecular geometry is neglected. Among other applications, molecular graphs are used in chemoinformatics systems, chemical databases, design of combinatorial libraries, reaction databases, computer-assisted structure elucidation, molecular design of novel chemicals, and computer-assisted organic synthesis. Molecular graphs are the basis for computing the structural descriptors used in QSPR and QSAR models to predict physical, chemical, biological, or toxicological properties. The molecular graph representation of chemical structure reflects mainly the connectivity of the atoms and is less suitable for modeling those properties that are determined mostly by molecular geometry, conformation, or stereochemistry.

1.3.1 Molecular Graphs

A chemical structure may be represented by a large number of different molecular graphs, depending on the translation rules for depicting atoms and chemical bonds. The translation rules, that is, "atom → vertex" and "bond → edge," should preserve the features of the molecular structure that are relevant for the scope of the modeling, for example, database search, reaction representation, molecular design, or property prediction. Cayley introduced the concept of molecular graphs in 1874, as "plerograms" and "kenograms," in which graph edges correspond to covalent bonds [72]. In a plerogram all atoms (including hydrogen atoms) are represented as vertices, whereas in a kenogram only non-hydrogen atoms are represented, because the hydrogen atoms can be reconstructed from the skeleton of a molecule. In modern terminology a plerogram is a hydrogen-included molecular graph, and a kenogram is a hydrogen-excluded molecular graph (called also hydrogen-depleted or hydrogen-suppressed molecular graph).

Using different rules for converting a chemical structure into a molecular graph, methylcyclopropane can be represented by Graphs **1.30**, **1.31**, and **1.32**. Graph **1.30** is a hydrogen-included molecular graph with labeled vertices, Graph **1.31** is a hydrogen-included molecular graph in which hydrogen and carbon atoms are not differentiated, and Graph **1.32** is a hydrogen-excluded molecular graph.

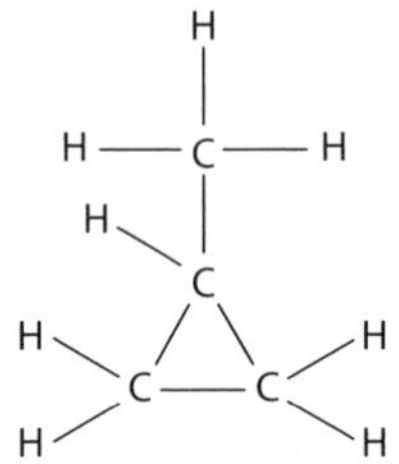

1.30

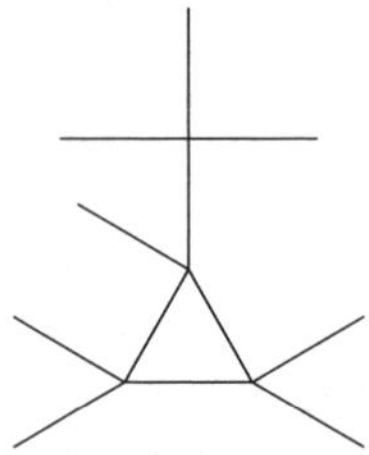

1.31

1.32

The usual graph representation of an organic chemical compound is as a nondirected, connected multigraph in which vertices correspond to non-hydrogen atoms and edges represent covalent bonds between non-hydrogen atoms. For hydrocarbons, the vertices in the molecular graph represent carbon atoms. Using this convention, alkanes are represented as 4-trees, that is, acyclic graphs with the maximum degree 4. Several studies compared structural descriptors (topological indices) computed from hydrogen-included and hydrogen-excluded molecular graphs of alkanes, and found that the topological indices are correlated [73,74]. These results support the preponderant use of hydrogen-excluded molecular graphs. To accommodate the presence of heteroatoms, a molecular graph has vertex labels corresponding to the atomic symbol of the heteroatoms, as shown for 2-methyl-1-bromobutane **1.33** (molecular graph **1.34**) and for ethyl *tert*-butyl ether **1.35** (molecular graph **1.36**).

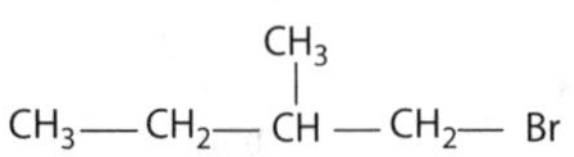

1.33

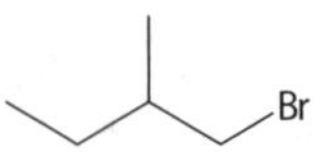

1.34

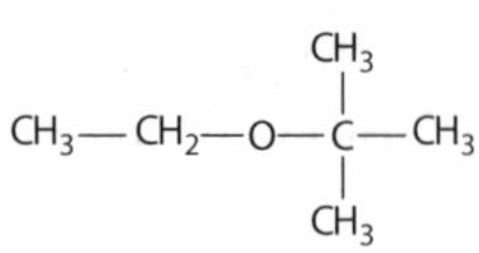

1.35

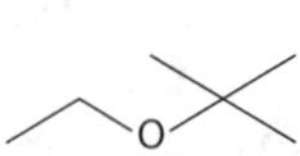

1.36

Multiple bonds are represented as multiedges, as shown for 1,4-dibromo-2-butene **1.37** (molecular graph **1.38**). Conjugated systems may be represented with the usual pattern of alternating double and single bonds, or with two lines, one continuous and the second broken, as shown for the aromatic system of benzyl chloride **1.39** (molecular graph **1.40**). The differences between these two representations of conjugated systems are significant when computing topological indices that have special parameters for aromatic bonds, and in chemical database registration, search, and retrieval.

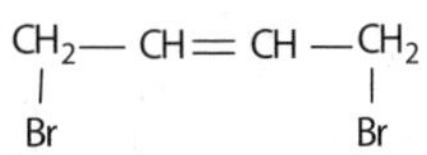

1.37

1.38

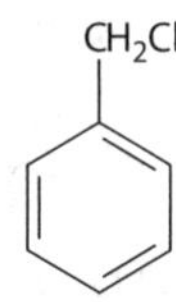

1.39

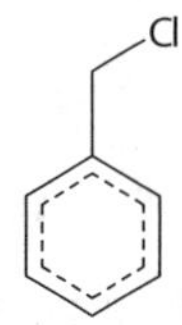

1.40

1.3.2 Molecular Pseudograph

There are a multitude of molecular graph models, each one developed with a specific set of rules, and fit for particular applications, such as structure elucidation, chemical synthesis design, or structure–property relationships. Koča et al. defined a mathematical model of organic synthesis design based on the graph theory formalism [30]. In this model, a chemical compound is represented by a molecular pseudograph (or general graph, containing multiedges and loops) $G(V, E, L, \varphi, \upsilon)$, where V is a vertex set, E is an edge set, L is a loop set, and φ is a mapping of the vertex set into the vocabulary υ of vertex labels. A single bond is represented by an edge, a double bond is represented by a multiedge of double multiplicity, and a triple bond is represented by a multiedge of triple multiplicity. A pair of free, unshared valence electrons of an atom is represented as a loop. Nitrogen is represented by a vertex with a loop, oxygen is represented by a vertex with two loops, whereas a halogen atom is represented by a vertex with three loops, as shown for 2-bromopropanoic acid **1.41** (molecular graph **1.42**) and for morpholine **1.43** (molecular graph **1.44**).

1.3.3 Molecular Graph of Atomic Orbitals

Toropov introduced the molecular graph of atomic orbitals (GAO) as a source of structural descriptors for QSPR and QSAR [75–77]. GAO is based on the hydrogen-included molecular graphs, in which each atom is substituted by the corresponding set

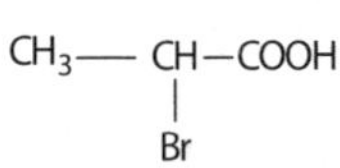

1.41

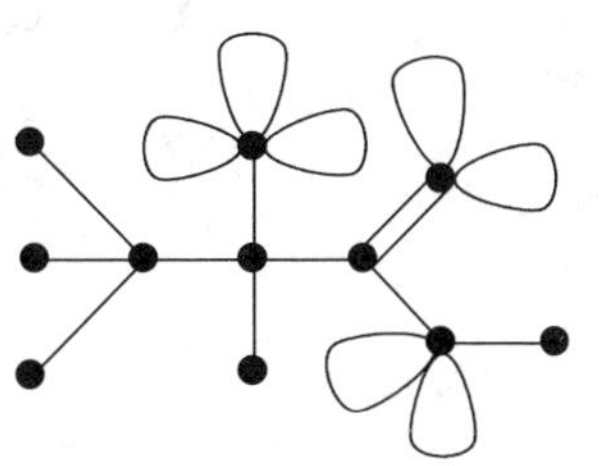

1.42

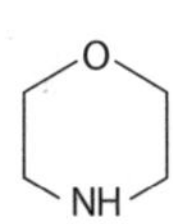

1.43

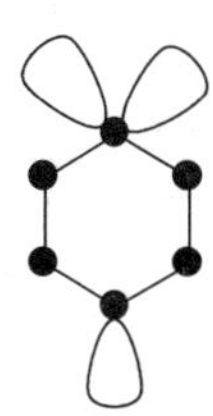

1.44

of atomic orbitals: H, $1s^1$; C, $1s^2$, $2s^2$, $2p^2$; N, $1s^2$, $2s^2$, $2p^3$; O, $1s^2$, $2s^2$, $2p^4$; F, $1s^2$, $2s^2$, $2p^5$; S, $1s^2$, $2s^2$, $2p^6$, $3s^2$, $3p^4$; Cl, $1s^2$, $2s^2$, $2p^6$, $3s^2$, $3p^5$; Br, $1s^2$, $2s^2$, $2p^6$, $3s^2$, $3p^6$, $3d^{10}$, $4s^2$, $4p^5$. Using this convention, C is represented in GAO by three vertices, Cl is represented by five vertices, and Br is represented by eight vertices. A covalent bond between atoms i and j is represented in GAO by $n_i \times n_j$ edges between the n_i atomic orbitals of atom i and the n_j atomic orbitals of atom j. As example we show the GAO of fluorobenzene (Figure 1.1). Another example of atomic orbitals graphs are the molecular graphs proposed by Pogliani, based on the hydrogen-excluded pseudograph augmented with information regarding the inner-core electrons [78–82].

1.3.4 Markush Structures

A major branch of chemoinformatics is represented by the development of efficient algorithms for the computer storage and retrieval of generic chemical structures. Using special topological representations, generic chemical structures encode into a single chemical graph an entire family of structurally related compounds. Among the different generic chemical structure representations, Markush structures have a special place because of their use in representing generic structures in patents. In a 1925

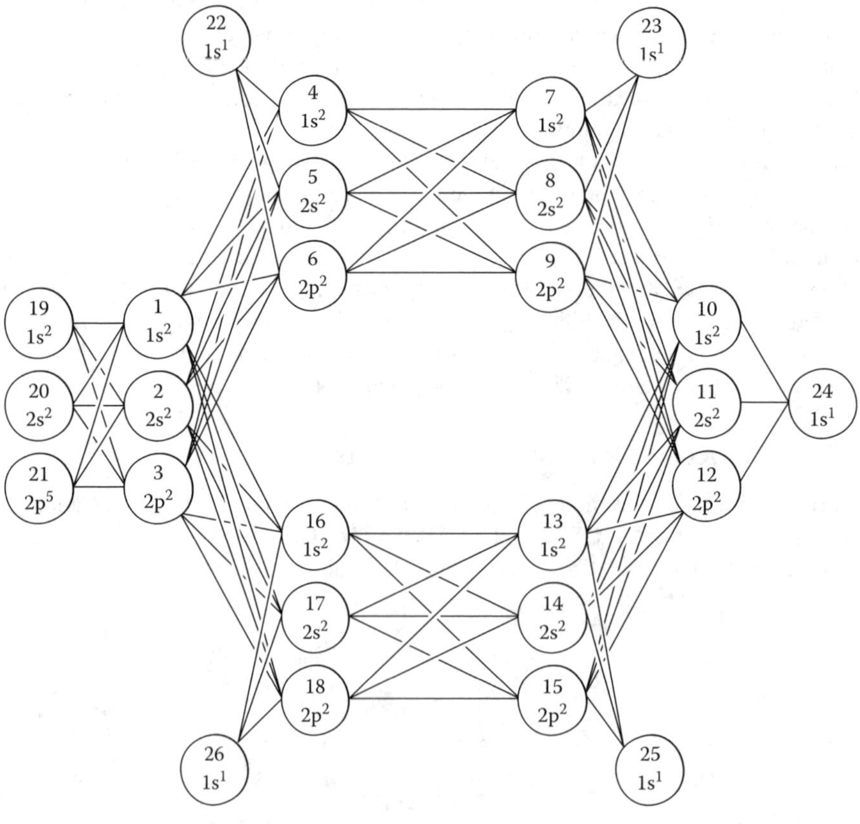

FIGURE 1.1 GAO of fluorobenzene.

court case Eugene Markush put forward such structures, which were later accepted in patent claims by the US Patent Office. Several approaches for the implementation of Markush structures are in use [83]. Among them, the Chemical Abstracts Service [84,85] and the Questel.Orbit [86] systems are more prominent. Markush structures **1.45** through **1.47** represent several examples of generic chemical structures.

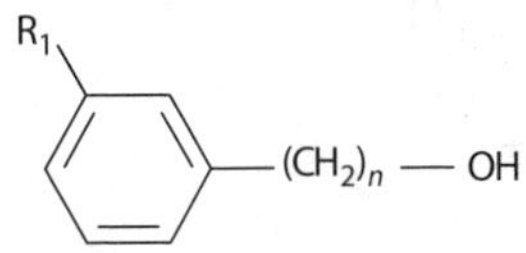

1.45

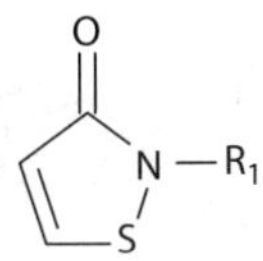

1.46

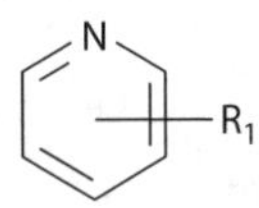

1.47

The Sheffield University group led by Lynch [87,88] developed graph representations for generic chemical structures, together with the GENSAL language [89] that is used to encode patent information into a computer-readable form [90]. The system developed by Lynch is a comprehensive collection of algorithms and procedures for the utilization of generic chemical structures: connection table representation [91], generation of fragment descriptors [92–94], computer interpreter for GENSAL [95,96], substructure search algorithm [97], reduced chemical graphs [98,99], algorithm to find the extended set of smallest rings [100], chemical ring identification [101], chemical graph search [102,103], and atom-level structure matching [104].

1.3.5 Reduced Graph Model

A more abstract representation of chemical structures is achieved with reduced graphs, in which each vertex represents a group of connected atoms, and an edge links two such vertices if in the original molecule there is a bond between an atom within one group and an atom in the second group [98,99]. A vertex in a reduced graph may represent a ring system, aromatic rings, aliphatic rings, or functional groups. There are several systems to transform a molecule into a reduced graph, by highlighting and grouping together different substructures in a chemical compound. We demonstrate here four types of reduced graphs that start from the same molecular graph and end up with different simplified representations.

Type 1. Vertices in the reduced graph correspond to ring systems (R) and connected acyclic components (Ac). The ring system R from compound **1.48** (shown inside a circle) corresponds to the central vertex in the reduced graph **1.49**.

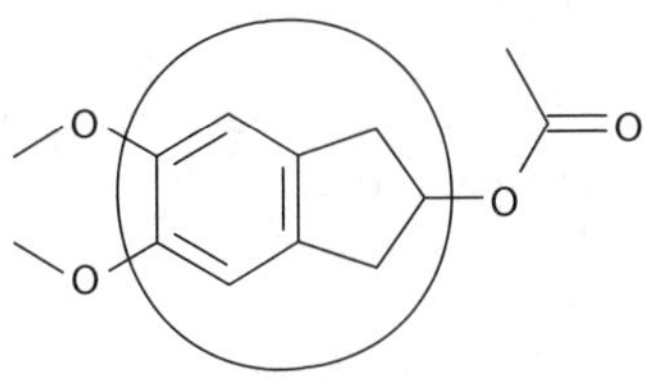

1.48

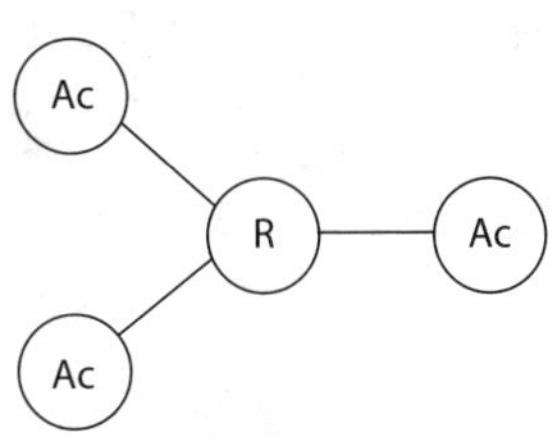

1.49

Type 2. Vertices in the reduced graph correspond to connected carbon components (C) and connected heteroatom components (H). Each heteroatom component in **1.50** is depicted inside an ellipse, and the corresponding reduced graph is shown in **1.51**.

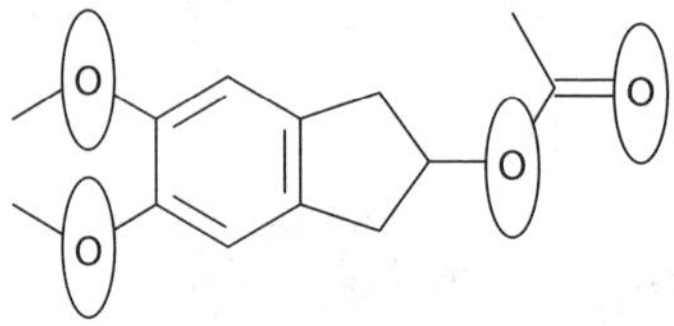

1.50

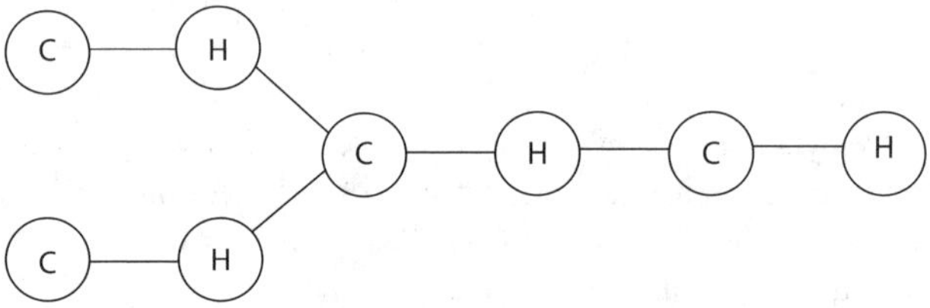

1.51

Type 3. Vertices in the reduced graph correspond to aromatic rings (Ar), aliphatic rings (R), and functional groups (F). Each functional group from molecular graph **1.52** is depicted inside a circle, with the final reduced graph depicted in **1.53**.

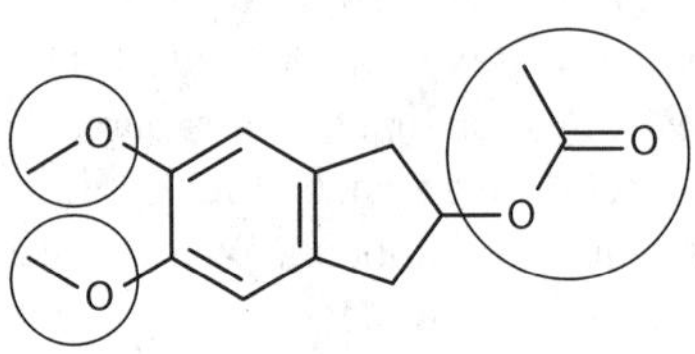

1.52

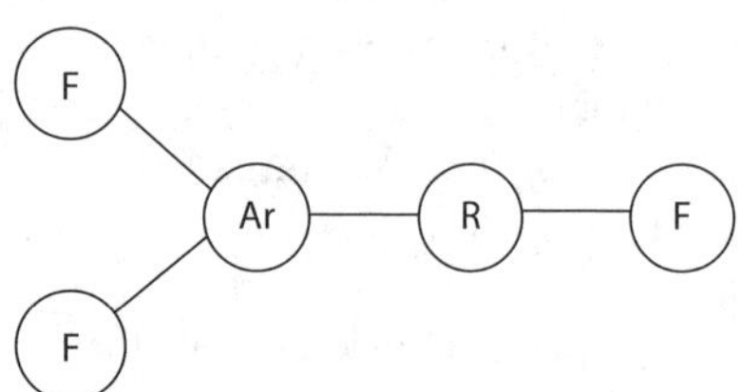

1.53

Type 4. Vertices in the reduced graph correspond to aromatic rings (Ar), functional groups (F), and linking groups (L). Each functional group from molecular graph **1.54** is depicted inside a circle, and the linker group is shown inside an ellipse. The corresponding reduced graph **1.55** has the same topology as reduced graph **1.53**, but with a different fragment type for the vertex between vertices labeled Ar and F.

When using a reduced graph to screen chemical libraries, different molecules may generate the same reduced graph, thus clustering together chemicals that have the same topological distribution of various types of subgraphs. The value of this approach is given by the fact that chemicals with similar bioactivities are translated into identical or highly similar reduced graphs. Several experiments show that reduced

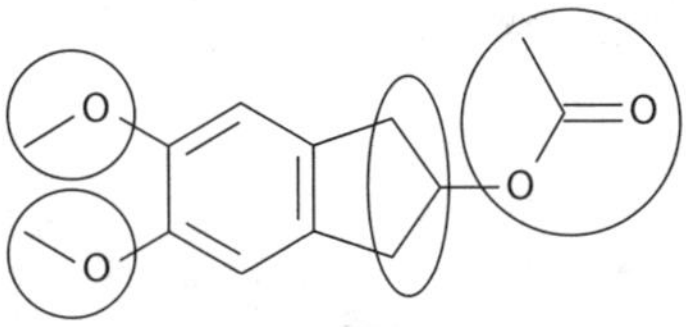

1.54

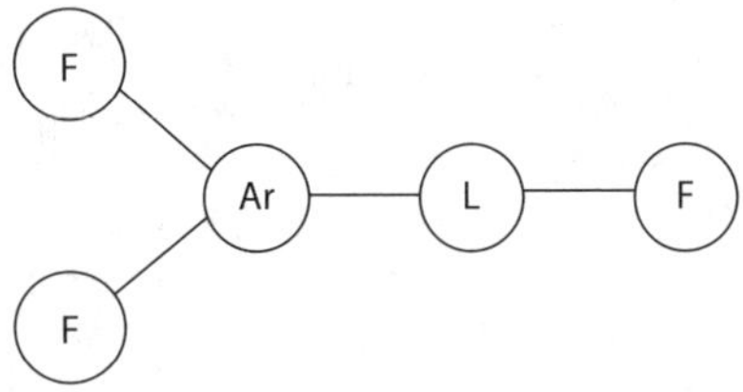

1.55

graphs may identify bioactive compounds that are missed with a fingerprint similarity search [105–108]. As expected, across a large spectrum of bioactivities, there is no definite advantage of using only reduced graphs, but these studies demonstrate the complementary nature of reduced graph similarity compared to fingerprint similarity.

1.3.6 Molecule Superposition Graphs

The molecular alignment of chemicals in a QSAR dataset is a characteristic of three-dimensional (3D) QSAR models. Similarly, the topological information encoded into the molecular graph may be used to obtain a 2D alignment of all molecules in a QSAR dataset. Such a molecule superposition graph, which is obtained from structurally related compounds by superposing the molecules according to a set of rules, may be considered as a supermolecule with the property that any molecule in the QSAR dataset is its subgraph. An early 2D alignment model is represented by the DARC (description, acquisition, retrieval, correlation) system, which applies the supermolecule approach by considering that molecules are composed of a common skeleton and a variable collection of substituents [109–114]. The contribution of the variable part of the structure to the overall property value of a molecule is determined by regression analysis to predict various physical, chemical, and biological properties.

An example of a DARC supermolecule is demonstrated for the prediction of ^{13}C nuclear magnetic resonance (NMR) chemical shift in acyclic alkenes [113]. In Figure 1.2, the topo-stereochemical description of the environment of the α-sp^2 resonating carbon atom considers all sp^3-hybridized carbon neighbors of types A, B, C, and D situated at 1, 2, 3, and 4 bonds away from the resonating atom. The use of an environment with a larger sphere of atoms does not add much information because the influence on the chemical shift of atoms situated at a distance greater than four bonds can be neglected. In a DARC supermolecule some sites collect a group of atoms that have similar influence on the modeled property, such as site ΣC that collects all carbon atoms situated three bonds away from C*, and site ΣD that collects all carbon atoms situated four bonds away from C.

Simon developed the minimal topological difference (MTD) QSAR model by superposing all molecules from the training set into a supermolecule [115]. Special vertices and edges are then created to embed the substituents by maximizing the superposition of their non-hydrogen atoms, and each molecule is embedded in a unique way into the MTD supermolecule. The MTD map has three types of vertices, namely with a positive contribution (increasing the bioactivity), with a negative contribution (decreasing the bioactivity), and neutral (no influence on the bioactivity). The type

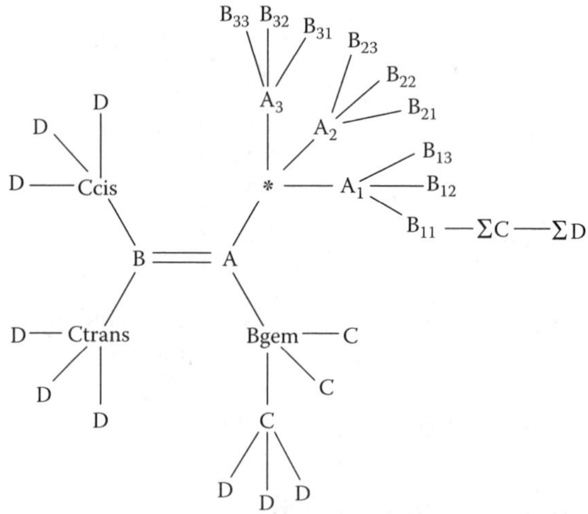

FIGURE 1.2 DARC-type map for the topo-stereochemical environment of α-sp^2 carbon atoms. The ^{13}C NMR chemical shift is predicted for the carbon atom labeled with *.

of each site in the MTD map is determined in an iterative process by embedding the training molecules on the MTD supermolecule and by minimizing the regression error between the experimental and calculated bioactivity. Minailiuc and Diudea extended the MTD supermolecule method by assigning vertex structural descriptors to vertices from the MTD supermolecule that are occupied for a particular molecule [116]. This QSAR model, called topological indices-minimal topological difference (TI-MTD), is very versatile in modeling QSAR properties and can be extended to other atomic properties, such as atomic charge or electronegativity. Recent studies show that the MTD method may be improved by using partial least squares (PLS) instead of multiple regression [117,118].

A similar supermolecule is generated in the molecular field topology analysis (MFTA) model introduced by Palyulin et al. [119]. The atomic descriptors associated with each vertex of the MFTA map are atomic charge, electronegativity, van der Waals radius, and atomic contribution to lipophilicity. The contribution of each site is determined with PLS.

1.3.7 Reaction Graphs

The utilization of reaction databases relies heavily on efficient software for storage and retrieval of reactions and reaction substructure search. Although very useful in suggesting individual reaction steps, reaction databases offer little help in devising strategies for complex reactions. A major accomplishment of chemoinformatics is the development of computer-assisted synthesis design systems and reaction prediction systems (cf. Chapter 11).

The storage and retrieval of reactions in databases, the extraction of reactivity knowledge, computer-assisted synthesis design, and reaction prediction systems are usually based on chemoinformatics tools that represent chemical reactions as a special type of graph [120–122]. As an example we present here the imaginary transition structure (ITS) model proposed by Fujita [120,123,124]. The ITS is a special type of reaction graph that is obtained by superposing reagents and products, and in which the bond rearrangement is indicated with special symbols. The reaction graph of an ITS has three types of bonds: par-bonds, which are bonds that are not modified in the reaction; out-bonds, representing bonds that are present only in reagents; and in-bonds, which are bonds appearing only in products. The diagram of an ITS graph contains distinctive symbols for each bond type: par-bonds are shown as solid lines; out-bonds are depicted as solid lines with a double bar; and in-bonds are depicted as solid lines with a circle. The ITS model is demonstrated here for nucleophilic substitution, with reactants **1.56**, ITS **1.57**, and products **1.58**.

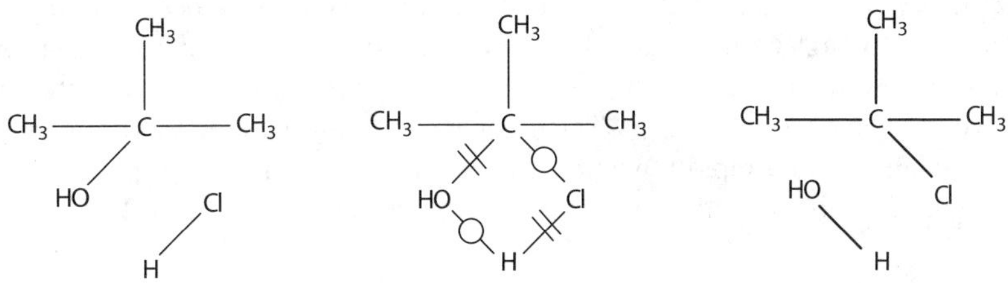

1.56 **1.57** **1.58**

As can be seen from the above reaction, in which *tert*-butyl alcohol reacts with hydrogen chloride to generate *tert*-butyl chloride, reaction mechanism details are not encoded into ITS. The role of ITS is to describe only bond rearrangements that transform reactants into products. The ITSs are not intended to represent reaction mechanisms, but the definition of the ITS may be easily extended to encode them.

The ITS reaction graphs represent a comprehensive framework for the classification and enumeration of organic reactions. The storage and retrieval of chemical reactions are reduced to graph manipulations, and the identification of a reaction type is equivalent to a subgraph search of an ITS database. A unique numerical representation (canonical code) of an ITS can be easily obtained [125,126] with a procedure derived from the Morgan algorithm of canonical coding [127]. The canonical representation of ITS graphs is an effective way of searching and comparing chemical reactions and of identifying reaction types.

1.3.8 Other Chemical Graphs

Many molecular graph models cannot handle systems with delocalized electrons, such as diborane or organometallic complexes, and several special graph models were proposed to encode these systems. Stein extended the bond and electron (BE) matrices introduced by Dugundji and Ugi [128–130] with new bond types for delocalized electrons [131]. Konstantinova and Skorobogatov proposed molecular hypergraphs to depict delocalized systems [132]. Dietz developed a molecular representation for computer-assisted synthesis design systems and for chemical database systems [133].

This molecular representation encodes the constitution, configuration, and conformation of a chemical compound. The constitution is represented as a multigraph describing the unshared valence electrons and the bonding relationships in a molecule, including valence electron sharing and electrostatic interactions. The chemical model suggested by Bauerschmidt and Gasteiger defines a hierarchical organization of molecular systems, starting from the electron system and ending with aggregates and ensembles [134]. Multicenter bonds are described as a list of atoms, type (σ or π), and number of electrons. This molecular representation is implemented in the reaction prediction program elaboration of reactions for organic synthesis (EROS) [135].

Chemical graphs may also be used to model systems in which the interaction between vertices represents hydrogen bonds, especially water, which consists of a large number of locally stable structures with various arrangements of the constituent water molecules. Each water cluster $(H_2O)_n$ is represented by a graph in which vertices are water molecules and bonds represent hydrogen bonds between two water molecules. Although weaker than covalent bonds, hydrogen bonds can form long-lived structures of water clusters for which the thermodynamic properties are determined by the hydrogen bonding patterns. The number of possible configurations of a cluster $(H_2O)_n$ increases very rapidly with n, which makes the identification of all possible local minima on the potential surface of a water cluster difficult [136–139].

1.4 WEIGHTED GRAPHS AND MOLECULAR MATRICES

Simple graphs lack the flexibility to represent complex chemical compounds, which limits their application to alkanes and cycloalkanes, and many widely used topological indices were initially defined for such simple molecular graphs (cf. Chapter 4). The main chemical application of topological indices is that of structural descriptors in QSPR, QSAR, and virtual screening, which requires the computation of these indices for molecular graphs containing heteroatoms and multiple bonds. Such molecular graphs use special sets of parameters to represent heteroatoms as vertex weights, and multiple bonds as edge weights. Early applications of such vertex- and edge-weighted (VEW) molecular graphs were initially developed for the Hückel molecular orbitals theory [140] and were subsequently extended to general chemical compounds [141]. In this section we present selected algorithms for the computation of weighted molecular graphs that are general in scope and can be applied to a large range of structural descriptors. The application of these weighting schemes is demonstrated for a group of molecular matrices that are frequently used in computing topological indices. Other weighting schemes were proposed for more narrow applications, and are valid only for specific topological indices such as Randić–Kier–Hall connectivity indices [24,25], electrotopological indices [26,142], Burden indices [143], and Balaban index J [60].

1.4.1 WEIGHTED MOLECULAR GRAPHS

A VEW molecular graph $G(V, E, Sy, Bo, Vw, Ew, w)$ is defined by a vertex set $V(G)$, an edge set $E(G)$, a set of chemical symbols for vertices $Sy(G)$, a set of topological bond orders for edges $Bo(G)$, a vertex weight set $Vw(w, G)$, and an edge weight set

$Ew(w, G)$, where the elements of the vertex and edge sets are computed with the weighting scheme w. Usually, the weight of a carbon atom is 0, whereas the weight of a carbon–carbon single bond is 1. In the weighting schemes reviewed here, the topological bond order Bo_{ij} of an edge e_{ij} takes the value 1 for single bonds, 2 for double bonds, 3 for triple bonds, and 1.5 for aromatic bonds. As an example of a VEW graph, consider 3,4-dibromo-1-butene **1.59** and its corresponding molecular graph **1.60**.

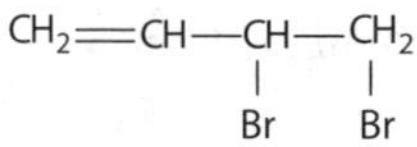

1.59

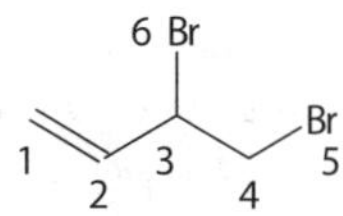

1.60

Graph distances represent the basis for the computation of almost all topological indices, and their computation in VEW graphs is shown here. The length of a path p_{ij} between vertices v_i and v_j, $l(p_{ij}, w, G)$, for a weighting scheme w in a VEW graph G is equal to the sum of the edge parameters $Ew(w)_{ij}$ for all edges along the path. The length of the path $p_1(\mathbf{1.60}) = \{v_1, v_2, v_3, v_6\}$ is $l(p_1) = Ew_{1,2} + Ew_{2,3} + Ew_{3,6}$. The topological length of a path p_{ij}, $t(p_{ij}, G)$, in a VEW graph G is equal to the number of edges along the path, which coincides with the path length in the corresponding unweighted graph. In a VEW graph, the distance $d(w)_{ij}$ between a pair of vertices v_i and v_j is equal to the length of the shortest path connecting the two vertices, $d(w)_{ij} = \min(l(p_{ij}, w))$.

1.4.2 Adjacency Matrix

The adjacency matrix $\mathbf{A}(w, G)$ of a VEW molecular graph G with N vertices is a square $N \times N$ real symmetric matrix with the element $[\mathbf{A}(w, G)]_{ij}$ defined as [34,144]

$$[\mathbf{A}(w, G)]_{ij} = \begin{cases} Vw(w)_i & \text{if } i = j, \\ Ew(w)_{ij} & \text{if } e_{ij} \in E(G), \\ 0 & \text{if } e_{ij} \notin E(G), \end{cases} \tag{1.6}$$

where $Vw(w)_i$ is the weight of vertex v_i, $Ew(w)_{ij}$ is the weight of edge e_{ij}, and w is the weighting scheme used to compute the parameters Vw and Ew. The valency of vertex v_i, $\mathrm{val}(w,G)_i$, is defined as the sum of the weights $Ew(w)_{ij}$ of all edges e_{ij} incident with vertex v_i [49]:

$$\mathrm{val}(w)_i = \sum_{e_{ij} \in E(G)} Ew(w)_{ij}. \tag{1.7}$$

1.4.3 Distance Matrix

The distance matrix $\mathbf{D}(w, G)$ of a VEW molecular graph G with N vertices is a symmetric square $N \times N$ matrix with the element $[\mathbf{D}(w, G)]_{ij}$ defined as [144,145]

$$[\mathbf{D}(w, G)]_{ij} = \begin{cases} d(w)_{ij} & \text{if } i \neq j, \\ Vw(w)_i & \text{if } i = j, \end{cases} \tag{1.8}$$

where $d(w)_{ij}$ is the distance between vertices v_i and v_j, $Vw(w)_i$ is the weight of vertex v_i, and w is the weighting scheme used to compute the parameters Vw and Ew. The distance sum of vertex v_i, $\mathbf{DS}(w, G)_i$, is defined as the sum of the topological distances between vertex v_i and every vertex in the VEW molecular graph G:

$$\mathbf{DS}(w, G)_i = \sum_{j=1}^{N} [\mathbf{D}(w, G)]_{ij} = \sum_{j=1}^{N} [\mathbf{D}(w, G)]_{ji}, \tag{1.9}$$

where w is the weighting scheme. The distance sum is used to compute the Balaban index J [59] and information on distance indices [62].

1.4.4 Atomic Number Weighting Scheme *Z*

Based on the definitions of adjacency and distance matrices introduced above, we demonstrate here the calculation of molecular matrices for weighted graphs. Barysz et al. proposed a general approach for computing parameters for VEW graphs by weighting the contributions of atoms and bonds with parameters based on the atomic number Z and the topological bond order [141]. In the atomic number weighting scheme Z, the parameter $Vw(Z)_i$ of a vertex v_i (representing atom i from a molecule) is defined as

$$Vw(Z)_i = 1 - \frac{Z_C}{Z_i} = 1 - \frac{6}{Z_i}, \tag{1.10}$$

where Z_i is the atomic number Z of atom i and $Z_C = 6$ is the atomic number Z of carbon. The parameter $Ew(Z)_{ij}$ for edge e_{ij} (representing the bond between atoms i and j) is defined as

$$Ew(Z)_{ij} = \frac{Z_C Z_C}{(Bo_{ij} Z_i Z_j)} = \frac{6 \times 6}{(Bo_{ij} Z_i Z_j)}, \tag{1.11}$$

where Bo_{ij} is the topological bond order of the edge between vertices v_i and v_j. The application of the Z parameters is shown for the adjacency matrix of $2H$-pyran **1.61** and for the distance matrix of 4-aminopyridine **1.61** (molecular graph **1.63**).

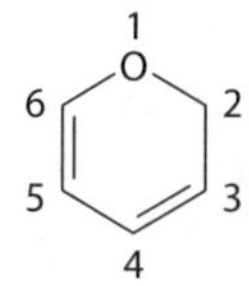

1.61

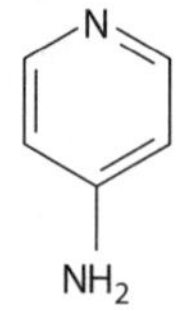

1.62

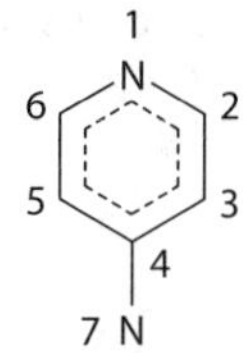

1.63

A(Z, **1.61**) =

	1	2	3	4	5	6
1	0.250	0.750	0	0	0	0.750
2	0.750	0	1	0	0	0
3	0	1	0	0.500	0	0
4	0	0	0.500	0	1	0
5	0	0	0	1	0	0.500
6	0.750	0	0	0	0.500	0

D(Z, **1.63**) =

	1	2	3	4	5	6	7
1	0.143	0.571	1.238	1.905	1.238	0.571	2.762
2	0.571	0	0.667	1.333	1.810	1.143	2.190
3	1.238	0.667	0	0.667	1.333	1.810	1.524
4	1.905	1.333	0.667	0	0.667	1.333	0.857
5	1.238	1.810	1.333	0.667	0	0.667	1.524
6	0.571	1.143	1.810	1.333	0.667	0	2.190
7	2.762	2.190	1.524	0.857	1.524	2.190	0.143

1.4.5 Relative Electronegativity Weighting Scheme *X*

The extension of the Balaban index J to VEW molecular graphs is based on relative electronegativity and covalent radius [60]. First, the Sanderson electronegativities of main group atoms are fitted in a linear regression using as parameters the atomic number Z and the number of the group Ng in the periodic system:

$$S_i = 1.1032 - 0.0204\,Z_i + 0.4121\,\mathrm{Ng}_i. \tag{1.12}$$

Taking as reference the calculated electronegativity for carbon $S_C = 2.629$, the relative electronegativities X are defined as

$$X_i = 0.4196 - 0.0078\,Z_i + 0.1567\,\mathrm{Ng}_i. \tag{1.13}$$

This weight system, developed initially for J, was extended as the relative electronegativity weighting scheme X, in which the vertex parameter $Vw(X)_i$ is defined as [36,146]

$$Vw(X)_i = 1 - \frac{1}{X_i}. \tag{1.14}$$

The edge parameter $Ew(X)_{ij}$ that characterizes the relative electronegativity of a bond is computed with the equation

$$Ew(X)_{ij} = \frac{1}{(Bo_{ij}X_iX_j)}. \tag{1.15}$$

From its definition, the weighting scheme X reflects the periodicity of electronegativity and can generate molecular descriptors that express both the effect of topology and that of electronegativity. A related set of parameters, the relative covalent radius weighting scheme Y, was defined based on the covalent radius [36,146].

1.4.6 Atomic Radius Weighting Scheme *R*

The atomic radius computed from the atomic polarizability is the basis of the atomic radius weighting scheme R, in which the vertex parameter $Vw(R)_i$ is defined as [144,147]

$$Vw(R)_i = 1 - \frac{r_C}{r_i} = 1 - \frac{1.21}{r_i} \tag{1.16}$$

and the parameter $Ew(R)_{ij}$ of the edge e_{ij} representing the bond between atoms i and j is equal to

$$Ew(R)_{ij} = \frac{r_C r_C}{(Bo_{ij}r_ir_j)} = \frac{1.21 \times 1.21}{(Bo_{ij}r_ir_j)}, \tag{1.17}$$

where $r_C = 1.21$ Å is the carbon radius and r_i is the atomic radius of atom i. Similar sets of parameters for VEW graphs were obtained with other atomic parameters, namely the atomic mass weighting scheme A, the atomic polarizability weighting scheme P, and the atomic electronegativity weighting scheme E [144,147].

1.4.7 Burden Matrix

The Burden molecular matrix is a modified adjacency matrix obtained from the hydrogen-excluded molecular graph of an organic compound [143]. This matrix is the source of the Burden, CAS, and University of Texas (BCUT) descriptors, which are computed from the graph spectra of the Burden matrix **B** and are extensively used in combinatorial chemistry, virtual screening, diversity measure, and QSAR [148–150]. An extension of the Burden matrix was obtained by inserting on the main diagonal of **B** a vertex structural descriptor VSD, representing a vector of experimental or computed atomic properties [151]. The rules defining the Burden matrix **B**(VSD, G) of a graph G with N vertices are as follows:

a. The diagonal elements of **B**, $[\mathbf{B}]_{ii}$, are computed with the formula

$$[\mathbf{B}(\mathrm{VSD}, G)]_{ii} = \mathrm{VSD}_i, \tag{1.18}$$

where VSD_i is a vertex structural descriptor of vertex v_i, that reflects the local structure of the corresponding atom i.

b. The nondiagonal element $[\mathbf{B}]_{ij}$, representing an edge e_{ij} connecting vertices v_i and v_j, has the value 0.1 for a single bond, 0.2 for a double bond, 0.3 for a triple bond, and 0.15 for an aromatic delocalized bond.
c. The value of a nondiagonal element $[\mathbf{B}]_{ij}$ representing an edge e_{ij} connecting vertices v_i and v_j is augmented by 0.01 if either vertex v_i or vertex v_j have degree 1.
d. All other nondiagonal elements $[\mathbf{B}]_{ij}$ are set equal to 0.001; these elements are set to 0 in the adjacency matrix **A** and correspond to pairs of nonbonded vertices in a molecular graph.

Examples of the vertex structural descriptor VSD for the diagonal of the Burden matrix are parameters from the weighting schemes A, E, P, R, X, Y, Z, various atomic properties (Pauling electronegativity, covalent radius, atomic polarizability), or various molecular graph indices, such as degree, valency, valence delta atom connectivity δ, intrinsic state I, electrotopological state S, distance sum **DS**, or vertex sum **VS**. An example of the Burden matrix is shown for 4-chloropyridine **1.64** (molecular graph **1.65**) with the Pauling electronegativity EP on the main diagonal.

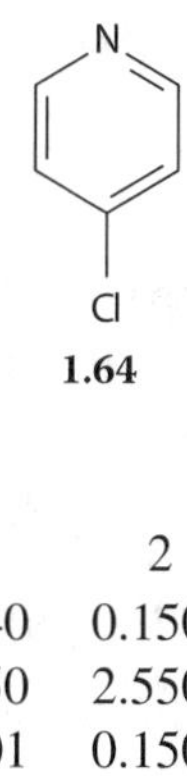

1.64

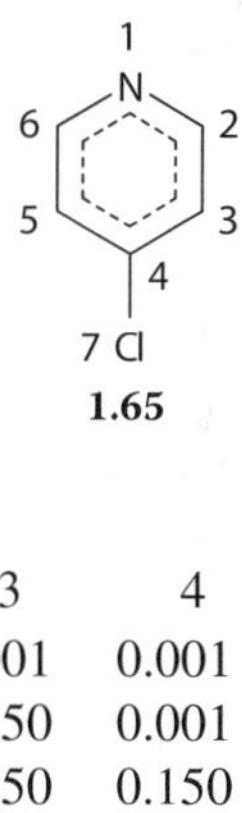

1.65

$\mathbf{B}(EP, \mathbf{1.65}) =$

	1	2	3	4	5	6	7
1	3.040	0.150	0.001	0.001	0.001	0.150	0.001
2	0.150	2.550	0.150	0.001	0.001	0.001	0.001
3	0.001	0.150	2.550	0.150	0.001	0.001	0.001
4	0.001	0.001	0.150	2.550	0.150	0.001	0.110
5	0.001	0.001	0.001	0.150	2.550	0.150	0.001
6	0.150	0.001	0.001	0.001	0.150	2.550	0.001
7	0.001	0.001	0.001	0.110	0.001	0.001	3.160

1.4.8 Reciprocal Distance Matrix

Starting with the Wiener index W, graph distances represented a prevalent source of topological indices. A possible drawback of using graph distances directly is that pairs of atoms that are separated by large distances, and thus have low interaction between them, have large contributions to the numerical value of the index. Because physical interaction between two objects decreases with increasing distance, the reciprocal distance $1/d_{ij}$ was introduced. Using the reciprocal distance, it is possible to define graph descriptors in which the contribution of two vertices decreases with increase of the distance between them [152]. The reciprocal distance matrix of a simple graph G with N vertices $\mathbf{RD}(G)$ is a square $N \times N$ symmetric matrix whose entries $[\mathbf{RD}]_{ij}$ are equal to the reciprocal of the distance between vertices v_i and v_j, that is, $1/d_{ij} = 1/[\mathbf{D}]_{ij}$, for nondiagonal elements, and is equal to zero for the diagonal elements [65,153,154]:

$$[\mathbf{RD}(G)]_{ij} = \begin{cases} \dfrac{1}{[\mathbf{D}(G)]_{ij}} & \text{if } i \neq j, \\ 0 & \text{if } i = j. \end{cases} \tag{1.19}$$

The reciprocal distance matrix of octahydropentalene **1.66** is shown as an example.

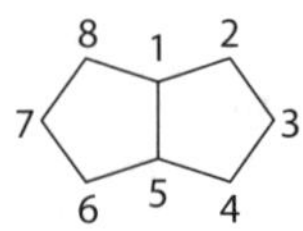

1.66

$\mathbf{RD}(\mathbf{1.66}) =$

	1	2	3	4	5	6	7	8
1	0	1	0.500	0.500	1	0.500	0.500	1
2	1	0	1	0.500	0.500	0.333	0.333	0.500
3	0.500	1	0	1	0.500	0.333	0.250	0.333
4	0.500	0.500	1	0	1	0.500	0.333	0.333
5	1	0.500	0.500	1	0	1	0.500	0.500
6	0.500	0.333	0.333	0.500	1	0	1	0.500
7	0.500	0.333	0.250	0.333	0.500	1	0	1
8	1	0.500	0.333	0.333	0.500	0.500	1	0

Formula 1.19 can be easily extended to weighted molecular graphs. The reciprocal distance matrix $\mathbf{RD}(w, G)$ of a VEW molecular graph G with N vertices is a square $N \times N$ symmetric matrix with real elements [144,145]:

$$[\mathbf{RD}(w, G)]_{ij} = \begin{cases} \dfrac{1}{[\mathbf{D}(w, G)]_{ij}} & \text{if } i \neq j, \\ Vw(w)_i & \text{if } i = j, \end{cases} \tag{1.20}$$

where $[\mathbf{D}(w)]_{ij}$ is the graph distance between vertices v_i and v_j, $[\mathbf{D}(w)]_{ii}$ is the diagonal element corresponding to vertex v_i, and w is the weighting scheme used to compute the parameters Vw and Ew. The reciprocal distance matrix of 2-hydroxypropanoic acid (lactic acid) **1.67** (molecular graph **1.68**) computed with the atomic electronegativity weighting scheme E is presented as an example.

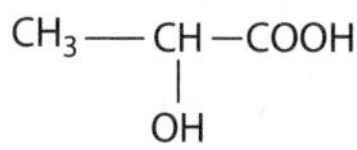

1.67

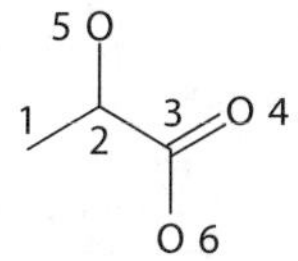

1.68 *Source:* From *Encyclopedia of Chemoinformatics*. With permission.

RD(E, **1.68**) =

	1	2	3	4	5	6
1	0	1	0.500	0.425	0.587	0.370
2	1	0	1	0.740	1.420	0.587
3	0.500	1	0	2.839	0.587	1.420
4	0.425	0.740	2.839	0.296	0.486	0.946
5	0.587	1.420	0.587	0.486	0.296	0.415
6	0.370	0.587	1.420	0.946	0.415	0.296

1.4.9 Other Molecular Matrices

We have presented here a selection of molecular matrices that are used as a source of topological indices and other graph descriptors. Other types of molecular matrices are investigated with the goal of exploring novel procedures for translating graph topology into a matrix [64]. The search for new structural descriptors based on molecular graphs is the catalyst that prompted the development of many molecular matrices, such as the edge Wiener matrix $\mathbf{W_e}$ [155], the path Wiener matrix $\mathbf{W_p}$ [155], the distance-valency matrix **Dval** [34], the quasi-Euclidean matrix $\rho_{q\varepsilon}$ [156,157], the distance complement matrix **DC** [66], the complementary distance matrix **CD** [145,158], the reverse Wiener matrix **RW** [67], the distance-path matrix $\mathbf{D_p}$ [68], the Szeged matrix **Sz** [70], the Cluj matrix **Cj** [70], and the resistance distance matrix Ω [50], which is based on a novel distance function on graphs introduced by Klein and Randić and inspired by the properties of electrical networks.

1.5 CONCLUDING REMARKS

This chapter reviewed the applications of graph theory in chemistry. Many objects manipulated in chemistry, such as atomic orbitals, chemical compounds, and reaction diagrams, can be represented as graphs. Graph operations, such as generating reduced graphs, and the calculation of various matrices derived from the connectivity of the graph can thus be applied to chemicals with applications including virtual screening, topological indices calculations, and activity/property predictions such as spectra predictions. Many algorithms have been and are being developed to solve graph problems, and some of these can be applied to chemistry problems. The goal of the next chapter is to present graph algorithms applied to chemicals.

REFERENCES

1. Harary, F., *Graph Theory*. Adison-Wesley: Reading, MA, 1994.
2. Berge, C., *Graphs and Hypergraphs*. Elsevier: New York, 1973.

3. Behzad, M., Chartrand, G., and Lesniak-Foster, L., *Graphs and Digraphs*. Wadsworth International Group: Belmont, CA, 1979.
4. Buckley, F. and Harary, F., *Distance in Graphs*. Adison-Wesley: Reading, MA, 1990.
5. Foulds, L. R., *Graph Theory Applications*. Springer: New York, 1992.
6. West, D. B., *Introduction to Graph Theory*, 2nd edition. Prentice-Hall: Englewood Cliffs, NJ, 2000.
7. Diestel, R., *Graph Theory*, 3rd edition. Springer: Heidelberg, Germany, 2005.
8. Beineke, L. W. and Wilson, R. J., *Topics in Algebraic Graph Theory*. Cambridge University Press: Cambridge, UK, 2005.
9. Cvetković, D. M., Doob, M., and Sachs, H., *Spectra of Graphs. Theory and Applications*, 3rd edition. Johann Ambrosius Barth Verlag: Heidelberg, Germany, 1995.
10. Graovac, A., Gutman, I., and Trinajstić, N., *Topological Approach to the Chemistry of Conjugated Molecules*. Springer: Berlin, 1977.
11. Cyvin, S. J. and Gutman, I., *Kekulé Structures in Benzenoid Hydrocarbons*, Vol. 46. Springer: Berlin, 1988.
12. Gutman, I. and Cyvin, S. J., *Introduction to the Theory of Benzenoid Hydrocarbons*. Springer: Berlin, 1989.
13. Gutman, I. and Cyvin, S. J., *Advances in the Theory of Benzenoid Hydrocarbons*, Vol. 153. Springer: Berlin, 1990.
14. Cyvin, S. J., Brunvoll, J., and Cyvin, B. N., *Theory of Coronoid Hydrocarbons*, Vol. 54. Springer: Berlin, 1991.
15. Dias, J. R., *Molecular Orbital Calculations Using Chemical Graph Theory*. Springer: Berlin, 1993.
16. Harary, F. and Palmer, E. M., *Graphical Enumeration*. Academic Press: New York, 1973.
17. Pólya, G. and Read, R. C., *Combinatorial Enumeration of Groups, Graphs, and Chemical Compounds*. Springer: New York, 1987.
18. Fujita, S., *Symmetry and Combinatorial Enumeration in Chemistry*. Springer: Berlin, 1991.
19. Biggs, N. L., Lloyd, E. K., and Wilson, R. J., *Graph Theory 1736–1936*. Clarendon Press: Oxford, 1976.
20. Balaban, A. T., *Chemical Applications of Graph Theory*. Academic Press: London, 1976.
21. Trinajstić, N., *Chemical Graph Theory*. CRC Press: Boca Raton, FL, 1992.
22. Gutman, I. and Polansky, O. E., *Mathematical Concepts in Organic Chemistry*. Springer: Berlin, 1986.
23. Gasteiger, J., *Handbook of Chemoinformatics*. Wiley-VCH: Weinheim, 2003.
24. Kier, L. B. and Hall, L. H., *Molecular Connectivity in Chemistry and Drug Research*. Academic Press: New York, 1976.
25. Kier, L. B. and Hall, L. H., *Molecular Connectivity in Structure–Activity Analysis*. Research Studies Press: Letchworth, UK, 1986.
26. Kier, L. B. and Hall, L. H., *Molecular Structure Description. The Electrotopological State*. Academic Press: San Diego, CA, 1999.
27. Bonchev, D., *Information Theoretic Indices for Characterization of Chemical Structure*. Research Studies Press: Chichester, UK, 1983.
28. Devillers, J. and Balaban, A. T., *Topological Indices and Related Descriptors in QSAR and QSPR*. Gordon and Breach Science Publishers: Amsterdam, the Netherlands, 1999.
29. Temkin, O. N., Zeigarnik, A. V., and Bonchev, D., *Chemical Reaction Networks. A Graph-Theoretical Approach*. CRC Press: Boca Raton, FL, 1996.
30. Koča, J., Kratochvíl, M., Kvasnička, V., Matyska, L., and Pospíchal, J., *Synthon Model of Organic Chemistry and Synthesis Design*, Vol. 51. Springer: Berlin, Germany, 1989.

31. Golender, V. E. and Rozenblit, A. B., *Logical and Combinatorial Algorithms for Drug Design*, p. 289. Research Studies Press: Letchworth, UK, 1983.
32. Balaban, A. T., Reaction graphs. In: D. Bonchev and O. Mekenyan (Eds), *Graph Theoretical Approaches to Chemical Reactivity*, pp. 137–180. Kluwer Academic Publishers: Amsterdam, the Netherlands, 1994.
33. Ivanciuc, O., Design of topological indices. Part 11. Distance-valency matrices and derived molecular graph descriptors. *Revue Roumaine de Chimie* 1999, 44(5), 519–528.
34. Ivanciuc, O., Design of topological indices. Part 14. Distance-valency matrices and structural descriptors for vertex- and edge-weighted molecular graphs. *Revue Roumaine de Chimie* 2000, 45(6), 587–596.
35. Dijkstra, E., A note on two problems in connection with graphs. *Numerische Mathematik* 1959, 1, 269–271.
36. Ivanciuc, O., Ivanciuc, T., and Balaban, A. T., Vertex- and edge-weighted molecular graphs and derived structural descriptors. In: J. Devillers and A. T. Balaban (Eds), *Topological Indices and Related Descriptors in QSAR and QSPR*, pp. 169–220. Gordon and Breach Science Publishers: Amsterdam, the Netherlands, 1999.
37. Ivanciuc, O. and Ivanciuc, T., Matrices and structural descriptors computed from molecular graph distances. In: J. Devillers and A.T. Balaban (Eds), *Topological Indices and Related Descriptors in QSAR and QSPR*, pp. 221–277. Gordon and Breach Science Publishers: Amsterdam, the Netherlands, 1999.
38. Higham, D. J., Kalna, G., and Kibble, M., Spectral clustering and its use in bioinformatics. *Journal of Computational and Applied Mathematics* 2007, 204(1), 25–37.
39. Randić, M., Random walks and their diagnostic value for characterization of atomic environment. *Journal of Computational Chemistry* 1980, 1(4), 386–399.
40. Burdett, J. K., Lee, S., and McLarnan, T. J., The coloring problem. *Journal of the American Chemical Society* 1985, 107(11), 3083–3089.
41. Burdett, J. K., Topological control of the structures of molecules and solids. *Accounts of Chemical Research* 1988, 21(5), 189–194.
42. Jiang, Y. and Zhang, H., Aromaticities and reactivities based on energy partitioning. *Pure and Applied Chemistry* 1990, 62(3), 451–456.
43. Wu, Y., Zhao, H. G., Liu, X., Li, J., Yang, K., and He, H. B., Evaluation of molecular moments by three methods. *International Journal of Quantum Chemistry* 2000, 78(4), 237–244.
44. Marković, S., Marković, Z., and McCrindle, R. I., Spectral moments of phenylenes. *Journal of Chemical Information and Computer Sciences* 2001, 41(1), 112–119.
45. Schmalz, T. G., Živković, T., and Klein, D. J., Cluster expansion of the Hückel molecular energy of acyclics: Applications to pi resonance theory. In: R. C. Lacher (Ed.), *MATH/CHEM/COMP 1987. Proceedings of an International Course and Conference on the Interfaces Between Mathematics, Chemistry and Computer Science, Dubrovnik, Yugoslavia, 22–26 June 1987*, Vol. 54, pp. 173–190. Elsevier: Amsterdam, the Netherlands, 1988.
46. Mohar, B., Laplacian matrices of graphs. In: A. Graovac (Ed.), *MATH/CHEM/COMP 1988. Proceedings of an International Course and Conference on the Interfaces Between Mathematics, Chemistry and Computer Sciences, Dubrovnik, Yugoslavia, 20–25 June 1988*, Vol. 63, pp. 1–8. Elsevier: Amsterdam, the Netherlands, 1989.
47. Ivanciuc, O., Chemical graph polynomials. Part 3. The Laplacian polynomial of molecular graphs. *Revue Roumaine de Chimie* 1993, 38(12), 1499–1508.
48. Trinajstić, N., Babić, D., Nikolić, S., Plavšić, D., Amić, D., and Mihalić, Z., The Laplacian matrix in chemistry. *Journal of Chemical Information and Computer Sciences* 1994, 34(2), 368–376.

49. Ivanciuc, O., Design of topological indices. Part 26. Structural descriptors computed from the Laplacian matrix of weighted molecular graphs: Modeling the aqueous solubility of aliphatic alcohols. *Revue Roumaine de Chimie* 2001, 46(12), 1331–1347.
50. Klein, D. J. and Randić, M., Resistance distance. *Journal of Mathematical Chemistry* 1993, 12(1–4), 81–95.
51. Ivanciuc, T., Ivanciuc, O., and Klein, D. J., Posetic quantitative superstructure/activity relationships (QSSARs) for chlorobenzenes. *Journal of Chemical Information and Modeling* 2005, 45(4), 870–879.
52. Ivanciuc, T., Ivanciuc, O., and Klein, D. J., Modeling the bioconcentration factors and bioaccumulation factors of polychlorinated biphenyls with posetic quantitative superstructure/activity relationships (QSSAR). *Molecular Diversity* 2006, 10(2), 133–145.
53. Ivanciuc, T., Ivanciuc, O., and Klein, D. J., Prediction of environmental properties for chlorophenols with posetic quantitative super-structure/property relationships (QSSPR). *International Journal of Molecular Sciences* 2006, 7(9), 358–374.
54. Klein, D. J., Ivanciuc, T., Ryzhov, A., and Ivanciuc, O., Combinatorics of reaction-network posets. *Combinatorial Chemistry & High Throughput Screening* 2008, 11(9), 723–733.
55. Mihalić, Z., Veljan, D., Amić, D., Nikolić, S., Plavšić, D., and Trinajstić, N., The distance matrix in chemistry. *Journal of Mathematical Chemistry* 1992, 11(1–3), 223–258.
56. Floyd, R. W., Algorithm 97: Shortest path. *Communications of the ACM* 1962, 5(6), 345.
57. Warshall, S., A theorem on boolean matrices. *Journal of the ACM* 1962, 9, 11–12.
58. Wiener, H., Structural determination of paraffin boiling points. *Journal of the American Chemical Society* 1947, 69, 17–20.
59. Balaban, A. T., Highly discriminating distance-based topological index. *Chemical Physics Letters* 1982, 89(5), 399–404.
60. Balaban, A. T. and Ivanciuc, O., FORTRAN 77 computer program for calculating the topological index J for molecules containing heteroatoms. In: A. Graovac (Ed.), *MATH/CHEM/COMP 1988. Proceedings of an International Course and Conference on the Interfaces Between Mathematics, Chemistry and Computer Sciences, Dubrovnik, Yugoslavia, 20–25 June 1988*, Vol. 63, pp. 193–211. Elsevier: Amsterdam, the Netherlands, 1989.
61. Hall, L. H., Mohney, B., and Kier, L. B., The electrotopological state: Structure information at the atomic level for molecular graphs. *Journal of Chemical Information and Computer Sciences* 1991, 31(1), 76–82.
62. Balaban, A. T. and Balaban, T.-S., New vertex invariants and topological indices of chemical graphs based on information on distances. *Journal of Mathematical Chemistry* 1991, 8(4), 383–397.
63. Balaban, A. T., Beteringhe, A., Constantinescu, T., Filip, P. A., and Ivanciuc, O., Four new topological indices based on the molecular path code. *Journal of Chemical Information and Modeling* 2007, 47(3), 716–731.
64. Ivanciuc, O., Graph theory in chemistry. In: J. Gasteiger (Ed.), *Handbook of Chemoinformatics*, Vol. 1, pp. 103–138. Wiley-VCH: Weinheim, Germany, 2003.
65. Ivanciuc, O., Balaban, T.-S., and Balaban, A. T., Design of topological indices. Part 4. Reciprocal distance matrix, related local vertex invariants and topological indices. *Journal of Mathematical Chemistry* 1993, 12(1–4), 309–318.
66. Randić, M., Linear combinations of path numbers as molecular descriptors. *New Journal of Chemistry* 1997, 21(9), 945–951.
67. Balaban, A. T., Mills, D., Ivanciuc, O., and Basak, S. C., Reverse Wiener indices. *Croatica Chemica Acta* 2000, 73(4), 923–941.

68. Diudea, M. V., Wiener and hyper-Wiener numbers in a single matrix. *Journal of Chemical Information and Computer Sciences* 1996, 36(4), 833–836.
69. Diudea, M. V., Ivanciuc, O., Nikolić, S., and Trinajstić, N., Matrices of reciprocal distance, polynomials and derived numbers. *MATCH Communications in Mathematical and in Computer Chemistry* 1997, 35, 41–64.
70. Diudea, M. V., Indices of reciprocal properties or Harary indices. *Journal of Chemical Information and Computer Sciences* 1997, 37(2), 292–299.
71. Ivanciuc, O., Design of topological indices. Part 27. Szeged matrix for vertex- and edge-weighted molecular graphs as a source of structural descriptors for QSAR models. *Revue Roumaine de Chimie* 2002, 47(5), 479–492.
72. Cayley, A., On the mathematical theory of isomers. *Philosophical Magazine* 1874, 67, 444–446.
73. Gutman, I., Vidović, D., and Popović, L., Graph representation of organic molecules. Cayley's plerograms vs. his kenograms. *Journal of the Chemical Society, Faraday Transactions* 1998, 94(7), 857–860.
74. Gutman, I. and Vidović, D., Relations between Wiener-type topological indices of plerograms and kenograms. *Journal of the Serbian Chemical Society* 1998, 63(9), 695–702.
75. Toropov, A. A. and Toropova, A. P., QSPR modeling of the formation constants for complexes using atomic orbital graphs. *Russian Journal of Coordination Chemistry* 2000, 26(6), 398–405.
76. Toropov, A. A. and Toropova, A. P., Prediction of heteroaromatic amine mutagenicity by means of correlation weighting of atomic orbital graphs of local invariants. *Journal of Molecular Structure (Theochem)* 2001, 538, 287–293.
77. Toropov, A. A. and Toropova, A. P., QSAR modeling of mutagenicity based on graphs of atomic orbitals. *Internet Electronic Journal of Molecular Design* 2002, 1(3), 108–114.
78. Pogliani, L., From molecular connectivity indices to semiempirical connectivity terms: Recent trends in graph theoretical descriptors. *Chemical Reviews* 2000, 100(10), 3827–3858.
79. Pogliani, L., Algorithmically compressed data and the topological conjecture for the inner-core electrons. *Journal of Chemical Information and Computer Sciences* 2002, 42(5), 1028–1042.
80. Pogliani, L., Complete graph conjecture for inner-core electrons: Homogeneous index case. *Journal of Computational Chemistry* 2003, 24(9), 1097–1109.
81. Pogliani, L., Encoding the core electrons with graph concepts. *Journal of Chemical Information and Computer Sciences* 2004, 44(1), 42–49.
82. Pogliani, L., The evolution of the valence delta in molecular connectivity theory. *Internet Electronic Journal of Molecular Design* 2006, 5(7), 364–375.
83. Barnard, J. M., A comparison of different approaches to Markush structure handling. *Journal of Chemical Information and Computer Sciences* 1991, 31(1), 64–68.
84. Fisanick, W., The chemical abstracts service generic chemical (Markush) structure storage and retrieval capability. 1. Basic concepts. *Journal of Chemical Information and Computer Sciences* 1990, 30(2), 145–154.
85. Ebe, T., Sanderson, K. A., and Wilson, P. S., The chemical abstracts service generic chemical (Markush) structure storage and retrieval capability. 2. The MARPAT file. *Journal of Chemical Information and Computer Sciences* 1991, 31(1), 31–36.
86. Benichou, P., Klimczak, C., and Borne, P., Handling genericity in chemical structures using the Markush Darc software. *Journal of Chemical Information and Computer Sciences* 1997, 37(1), 43–53.

87. Lynch, M. F., Barnard, J. M., and Welford, S. M., Computer storage and retrieval of generic chemical structures in patents. 1. Introduction and general strategy. *Journal of Chemical Information and Computer Sciences* 1981, 21(3), 148–150.
88. Holliday, J. D., Downs, G. M., Gillet, V. J., Lynch, M. F., and Dethlefsen, W., Evaluation of the screening stages of the Sheffield research project on computer storage and retrieval of generic chemical structures in patents. *Journal of Chemical Information and Computer Sciences* 1994, 34(1), 39–46.
89. Barnard, J. M., Lynch, M. F., and Welford, S. M., Computer storage and retrieval of generic chemical structures in patents. 2. GENSAL, a formal language for the description of generic chemical structures. *Journal of Chemical Information and Computer Sciences* 1981, 21(3), 151–161.
90. Welford, S. M., Lynch, M. F., and Barnard, J. M., Computer storage and retrieval of generic chemical structures in patents. 3. Chemical grammars and their role in the manipulation of chemical structures. *Journal of Chemical Information and Computer Sciences* 1981, 21(3), 161–168.
91. Barnard, J. M., Lynch, M. F., and Welford, S. M., Computer storage and retrieval of generic structures in chemical patents. 4. An extended connection table representation for generic structures. *Journal of Chemical Information and Computer Sciences* 1982, 22(3), 160–164.
92. Welford, S. M., Lynch, M. F., and Barnard, J. M., Computer storage and retrieval of generic chemical structures in patents. 5. Algorithmic generation of fragment descriptors for generic structure screening. *Journal of Chemical Information and Computer Sciences* 1984, 24(2), 57–66.
93. Holliday, J. D., Downs, G. M., Gillet, V. J., and Lynch, M. F., Computer storage and retrieval of generic chemical structures in patents. 14. Fragment generation from generic structures. *Journal of Chemical Information and Computer Sciences* 1992, 32(5), 453–462.
94. Holliday, J. D., Downs, G. M., Gillet, V. J., and Lynch, M. F., Computer storage and retrieval of generic chemical structures in patents. 15. Generation of topological fragment descriptors from nontopological representations of generic structure components. *Journal of Chemical Information and Computer Sciences* 1993, 33(3), 369–377.
95. Barnard, J. M., Lynch, M. F., and Welford, S. M., Computer storage and retrieval of generic chemical structures in patents. 6. An interpreter program for the generic structure description language GENSAL. *Journal of Chemical Information and Computer Sciences* 1984, 24(2), 66–71.
96. Dethlefsen, W., Lynch, M. F., Gillet, V. J., Downs, G. M., and Holliday, J. D., Computer storage and retrieval of generic chemical structures in patents. 11. Theoretical aspects of the use of structure languages in a retrieval system. *Journal of Chemical Information and Computer Sciences* 1991, 31(2), 233–253.
97. Gillet, V. J., Welford, S. M., Lynch, M. F., Willett, P., Barnard, J. M., Downs, G. M., Manson, G., and Thompson, J., Computer storage and retrieval of generic chemical structures in patents. 7. Parallel simulation of a relaxation algorithm for chemical substructure search. *Journal of Chemical Information and Computer Sciences* 1986, 26(3), 118–126.
98. Gillet, V. J., Downs, G. M., Ling, A., Lynch, M. F., Venkataram, P., Wood, J. V., and Dethlefsen, W., Computer storage and retrieval of generic chemical structures in patents. 8. Reduced chemical graphs and their applications in generic chemical structure retrieval. *Journal of Chemical Information and Computer Sciences* 1987, 27(3), 126–137.
99. Gillet, V. J., Downs, G. M., Holliday, J. D., Lynch, M. F., and Dethlefsen, W., Computer storage and retrieval of generic chemical structures in patents. 13. Reduced

graph generation. *Journal of Chemical Information and Computer Sciences* 1991, 31(2), 260–270.

100. Downs, G. M., Gillet, V. J., Holliday, J. D., and Lynch, M. F., Computer storage and retrieval of generic chemical structures in patents. 9. An algorithm to find the extended set of smallest rings in structurally explicit generics. *Journal of Chemical Information and Computer Sciences* 1989, 29(3), 207–214.
101. Downs, G. M., Gillet, V. J., Holliday, J. D., and Lynch, M. F., Computer storage and retrieval of generic chemical structures in patents. 10. Assignment and logical bubble-up of ring screens for structurally explicit generics. *Journal of Chemical Information and Computer Sciences* 1989, 29(3), 215–224.
102. Dethlefsen, W., Lynch, M. F., Gillet, V. J., Downs, G. M., Holliday, J. D., and Barnard, J. M., Computer storage and retrieval of generic chemical structures in patents. 12. Principles of search operations involving parameter lists: Matching-relations, user-defined match levels, and transition from the reduced graph search to the refined search. *Journal of Chemical Information and Computer Sciences* 1991, 31(2), 253–260.
103. Holliday, J. D. and Lynch, M. F., Computer storage and retrieval of generic chemical structures in patents. 16. The refined search: An algorithm for matching components of generic chemical structures at the atom-bond level. *Journal of Chemical Information and Computer Sciences* 1995, 35(1), 1–7.
104. Holliday, J. D. and Lynch, M. F., Computer storage and retrieval of generic chemical structures in patents. 17. Evaluation of the refined search. *Journal of Chemical Information and Computer Sciences* 1995, 35(4), 659–662.
105. Gillet, V. J., Willett, P., and Bradshaw, J., Similarity searching using reduced graphs. *Journal of Chemical Information and Computer Sciences* 2003, 43(2), 338–345.
106. Barker, E. J., Gardiner, E. J., Gillet, V. J., Kitts, P., and Morris, J., Further development of reduced graphs for identifying bioactive compounds. *Journal of Chemical Information and Computer Sciences* 2003, 43(2), 346–356.
107. Barker, E. J., Buttar, D., Cosgrove, D. A., Gardiner, E. J., Kitts, P., Willett, P., and Gillet, V. J., Scaffold hopping using clique detection applied to reduced graphs. *Journal of Chemical Information and Modeling* 2006, 46(2), 503–511.
108. Birchall, K., Gillet, V. J., Harper, G., and Pickett, S. D., Training similarity measures for specific activities: Application to reduced graphs. *Journal of Chemical Information and Modeling* 2006, 46(2), 577–586.
109. Dubois, J. E., Laurent, D., and Viellard, H., Système DARC. Principes de recherches des corrélations et équation générale de topoinformation. *Comptes Rendus de l'Académie des Sciences Paris* 1967, 264C, 1019–1022.
110. Dubois, J. E. and Viellard, H., Système DARC. Théorie de génération: Description I. *Bulletin de la Société Chimique de France* 1968, 900–904.
111. Dubois, J. E. and Viellard, H., Système DARC. Théorie de génération: Description II. *Bulletin de la Société Chimique de France* 1968, 905–912.
112. Dubois, J. E. and Viellard, H., Système DARC. Théorie de génération: Description III. *Bulletin de la Société Chimique de France* 1968, 913–919.
113. Ivanciuc, O., Rabine, J.-P., Cabrol-Bass, D., Panaye, A., and Doucet, J. P., 13C NMR chemical shift prediction of the sp^3 carbon atoms in the a position relative to the double bond in acyclic alkenes. *Journal of Chemical Information and Computer Sciences* 1997, 37(3), 587–598.
114. Dubois, J. E., Doucet, J. P., Panaye, A., and Fan, B. T., DARC site topological correlations: Ordered structural descriptors and property evaluation. In: J. Devillers and A. T. Balaban

(Eds), *Topological Indices and Related Descriptors in QSAR and QSPR*, pp. 613–673. Gordon and Breach Science Publishers: Amsterdam, the Netherlands, 1999.
115. Simon, Z. and Szabadai, Z., Minimal steric difference parameter and the importance of steric fit for structure–biological activity correlations. *Studia Biophysica* 1973, 39, 123–132.
116. Minailiuc, O. M. and Diudea, M. V., TI-MTD model. Applications in molecular design. In: M. V. Diudea (Ed.), *QSPR/QSAR Studies by Molecular Descriptors*, pp. 363–388. Nova Science Publishers: Huntington, NY, 2001.
117. Kurunczi, L., Seclaman, E., Oprea, T. I., Crisan, L., and Simon, Z., MTD-PLS: A PLS variant of the minimal topologic difference method. III. Mapping interactions between estradiol derivatives and the alpha estrogenic receptor. *Journal of Chemical Information and Modeling* 2005, 45(5), 1275–1281.
118. Kurunczi, L., Olah, M., Oprea, T. I., Bologa, C., and Simon, Z., MTD-PLS: A PLS-based variant of the MTD method. 2. Mapping ligand–receptor interactions. Enzymatic acetic acid esters hydrolysis. *Journal of Chemical Information and Computer Sciences* 2002, 42(4), 841–846.
119. Palyulin, V. A., Radchenko, E. V., and Zefirov, N. S., Molecular field topology analysis method in QSAR studies of organic compounds. *Journal of Chemical Information and Computer Sciences* 2000, 40(3), 659–667.
120. Fujita, S., "Structure-reaction type" paradigm in the conventional methods of describing organic-reactions and the concept of imaginary transition structures overcoming this paradigm. *Journal of Chemical Information and Computer Sciences* 1987, 27(3), 120–126.
121. Chen, L., Reaction classification and knowledge acquisition. In: J. Gasteiger (Ed.), *Handbook of Chemoinformatics*, Vol. 1, pp. 348–388. Wiley-VCH: Weinheim, 2003.
122. Todd, M. H., Computer-aided organic synthesis. *Chemical Society Reviews* 2005, 34(3), 247–266.
123. Fujita, S., Description of organic reactions based on imaginary transition structures. 1. Introduction of new concepts. *Journal of Chemical Information and Computer Sciences* 1986, 26(4), 205–212.
124. Fujita, S., Formulation of isomeric reaction types and systematic enumeration of six-electron pericyclic reactions. *Journal of Chemical Information and Computer Sciences* 1989, 29(1), 22–30.
125. Fujita, S., Canonical numbering and coding of imaginary transition structures. A novel approach to the linear coding of individual organic reactions. *Journal of Chemical Information and Computer Sciences* 1988, 28(3), 128–137.
126. Fujita, S., Canonical numbering and coding of reaction center graphs and reduced reaction center graphs abstracted from imaginary transition structures. A novel approach to the linear coding of reaction types. *Journal of Chemical Information and Computer Sciences* 1988, 28(3), 137–142.
127. Morgan, H. L., The generation of a unique machine description for chemical structures—a technique developed at chemical abstracts service. *Journal of Chemical Documentation* 1965, 5, 107–113.
128. Dugundji, J. and Ugi, I., An algebraic model of constitutional chemistry as a basis for chemical computer programs. *Topics in Current Chemistry* 1973, 39, 19–64.
129. Ugi, I. and Gillespie, P., Representation of chemical systems and interconversion by BE matrices and their transformation properties. *Angewandte Chemie International Edition in English* 1973, 10, 914–915.

130. Ugi, I., Stein, N., Knauer, M., Gruber, B., Bley, K., and Weidinger, R., New elements in the representation of the logical structure of chemistry by qualitative mathematical models and corresponding data structures. *Topics in Current Chemistry* 1993, 166, 199–233.
131. Stein, N., New perspectives in computer-assisted formal synthesis design—treatment of delocalized electrons. *Journal of Chemical Information and Computer Sciences* 1995, 35(2), 305–309.
132. Konstantinova, E. V. and Skorobogatov, V. A., Molecular hypergraphs: The new representation of nonclassical molecular structures with polycentric delocalized bonds. *Journal of Chemical Information and Computer Sciences* 1995, 35(3), 472–478.
133. Dietz, A., Yet another representation of molecular structure. *Journal of Chemical Information and Computer Sciences* 1995, 35(5), 787–802.
134. Bauerschmidt, S. and Gasteiger, J., Overcoming the limitations of a connection table description: A universal representation of chemical species. *Journal of Chemical Information and Computer Sciences* 1997, 37(4), 705–714.
135. Gasteiger, J., Marsili, M., Hutchings, M. G., Saller, H., Löw, P., Röse, P., and Rafeiner, K., Models for the representation of knowledge about chemical reactions. *Journal of Chemical Information and Computer Sciences* 1990, 30(4), 467–476.
136. Miyake, T. and Aida, M., Enumeration of topology-distinct structures of hydrogen bonded water clusters. *Chemical Physics Letters* 2002, 363(1–2), 106–110.
137. Miyake, T. and Aida, M., Hydrogen bonding patterns in water clusters: Trimer, tetramer and pentamer. *Internet Electronic Journal of Molecular Design* 2003, 2(1), 24–32.
138. Miyake, T. and Aida, M., H-bond patterns and structure distributions of water octamer $(H_2O)_8$ at finite temperatures. *Chemical Physics Letters* 2006, 427(1–3), 215–220.
139. Shi, Q., Kais, S., and Francisco, J. S., Graph theory for fused cubic clusters of water dodecamer. *Journal of Physical Chemistry A* 2005, 109(51), 12036–12045.
140. Mallion, R. B., Schwenk, A. J., and Trinajstić, N., A graphical study of heteroconjugated molecules. *Croatica Chemica Acta* 1974, 46(3), 171–182.
141. Barysz, M., Jashari, G., Lall, R. S., Srivastava, V. K., and Trinajstić, N., On the distance matrix of molecules containing heteroatoms. In: R. B. King (Ed.), *Chemical Applications of Topology and Graph Theory*, pp. 222–227. Elsevier: Amsterdam, the Netherlands, 1983.
142. Ivanciuc, O., Electrotopological state indices. In: R. Mannhold (Ed.). *Molecular Drug Properties. Measurement and Prediction*, pp. 85–109, Wiley-VCH: Weinheim, Germany, 2008.
143. Burden, F. R., Molecular identification number for substructure searches. *Journal of Chemical Information and Computer Sciences* 1989, 29(3), 225–227.
144. Ivanciuc, O., Design of topological indices. Part 12. Parameters for vertex- and edge-weighted molecular graphs. *Revue Roumaine de Chimie* 2000, 45(3), 289–301.
145. Ivanciuc, O., QSAR comparative study of Wiener descriptors for weighted molecular graphs. *Journal of Chemical Information and Computer Sciences* 2000, 40(6), 1412–1422.
146. Ivanciuc, O., Ivanciuc, T., and Balaban, A. T., Design of topological indices. Part 10. Parameters based on electronegativity and covalent radius for the computation of molecular graph descriptors for heteroatom-containing molecules. *Journal of Chemical Information and Computer Sciences* 1998, 38(3), 395–401.
147. Ivanciuc, O. and Klein, D. J., Computing Wiener-type indices for virtual combinatorial libraries generated from heteroatom-containing building blocks. *Journal of Chemical Information and Computer Sciences* 2002, 42(1), 8–22.

148. Pearlman, R. S. and Smith, K. M., Metric validation and the receptor relevant subspace concept. *Journal of Chemical Information and Computer Sciences* 1999, 39(1), 28–35.
149. Stanton, D. T., Evaluation and use of BCUT descriptors in QSAR and QSPR studies. *Journal of Chemical Information and Computer Sciences* 1999, 39(1), 11–20.
150. Pirard, B. and Pickett, S. D., Classification of kinase inhibitors using BCUT descriptors. *Journal of Chemical Information and Computer Sciences* 2000, 40(6), 1431–1440.
151. Ivanciuc, O., Design of topological indices. Part 25. Burden molecular matrices and derived structural descriptors for glycine antagonists QSAR models. *Revue Roumaine de Chimie* 2001, 46(9), 1047–1066.
152. Ivanciuc, O., Design on topological indices. Part 1. Definition of a vertex topological index in the case of 4-trees. *Revue Roumaine de Chimie* 1989, 34(6), 1361–1368.
153. Balaban, T. S., Filip, P. A., and Ivanciuc, O., Computer generation of acyclic graphs based on local vertex invariants and topological indices. Derived canonical labelling and coding of trees and alkanes. *Journal of Mathematical Chemistry* 1992, 11(1–3), 79–105.
154. Plavšić, D., Nikolić, S., Trinajstić, N., and Mihalić, Z., On the Harary index for the characterization of chemical graphs. *Journal of Mathematical Chemistry* 1993, 12(1–4), 235–250.
155. Randić, M., Novel molecular descriptor for structure-property studies. *Chemical Physics Letters* 1993, 211(4–5), 478–483.
156. Zhu, H.-Y. and Klein, D. J., Graph-geometric invariants for molecular structures. *Journal of Chemical Information and Computer Sciences* 1996, 36(6), 1067–1075.
157. Ivanciuc, O., Ivanciuc, T., and Klein, D. J., Intrinsic graph distances compared to Euclidean distances for correspondent graph embedding. *MATCH Communications in Mathematical and in Computer Chemistry* 2001, 44, 251–278.
158. Ivanciuc, O., Ivanciuc, T., and Balaban, A. T., The complementary distance matrix, a new molecular graph metric. *ACH—Models in Chemistry* 2000, 137(1), 57–82.

2 Algorithms to Store and Retrieve Two-Dimensional (2D) Chemical Structures

Milind Misra and Jean-Loup Faulon

CONTENTS

Since molecules are three-dimensional (3D) and lack any intrinsic ordering in their chemical formulas, it becomes necessary to supply a set of linear rules to any computer system designed for storing and retrieving chemical structures. Ideally, such a notation would not only retain important knowledge about a molecule's 3D structure but also contain a mechanism to distinguish between molecules. As we saw in the previous chapter, graphs are dimensionless or zero-dimensional (0D) objects but can also be considered as one-dimensional (1D), two-dimensional (2D), 3D, or higher-dimensional objects in various methods of structural representation. Thus, 0D or constitutional descriptors such as molecular weight and atom counts are defined using local molecular information. 1D notations for structures and reactions include linear representations such as SMILES (Simplified Molecular Input Line Entry Specification) [1,2], WLN (Wiswesser Line Notation) [3,4], SLN (SYBYL Line Notation) [5,6], and InChI (the IUPAC International Chemical Identifier, http://www.iupac.org/inchi). Molecular graphs can be represented in two dimensions as chemical diagrams such that a vertex corresponds to (x, y) coordinates and type of an atom and an edge corresponds to bond type. They can also be extended to three dimensions such that a vertex contains information about (x, y, z) atomic coordinates instead. 3D structures can then be generated by further incorporating knowledge of bond lengths, bond angles, and dihedral angles. In this chapter we illustrate some methods and algorithms for the storage, retrieval, and manipulation of 2D representations of chemical structures, while 3D representation is treated in Chapter 3.

2.1 COMMON REPRESENTATIONS: LINEAR NOTATIONS AND CONNECTION TABLES

The information contained in molecular graphs can be transmitted to and from a computer in several ways for the purpose of manipulating chemical compounds and reactions. It is essential for a particular chemoinformatics application to recognize molecules of interest by recognizing relevant geometric and topological information passed to it. This can be accomplished by representing a molecule using line notation (1D) or as a connection table (2D and 3D). Linear notation is a compact and efficient system that employs alphanumeric characters and conventions for common molecular features such as bond types, ring systems, aromaticity, and chirality. The connection table is a set of lines specifying individual atoms and bonds and can be created as a computer- and human-readable text file.

The significance of standard formats to represent molecules in chemoinformatics systems lies in their numerous and diverse applications such as storage, retrieval, and search in chemical databases; generation of IUPAC names [7]; ring determination [8,9]; generation of compounds [10] and combinatorial libraries [11]; computer-aided design of novel chemicals [12] and organic reactions [13]; and calculation of molecular descriptors [14] for quantitative structure–activity/property relationships (QSAR/QSPR) and virtual screening. In this section we present some important line notations and molecular file formats.

2.1.1 WLN, SMILES, SMARTS, AND SMIRKS

The primary goal of linear notations is to enable easy interpretation by computer programs. They offer several advantages over connection tables such as improved parsability, efficient storage in relational databases, and compression of storage space required per molecule. Chemical line notations can be parsed using string processing algorithms resulting in efficient chemoinformatics applications. These characteristics of linear notations have been exploited for generating large combinatorial libraries using a fast SMILES approach [11]; for designing novel compounds using hydrogen-included SMILES to define operators for an evolutionary algorithm [12]; for virtual screening of chemical libraries using a molecular similarity function based on compressed SMILES strings [15]; for discriminating active and inactive compounds by searching for specific patterns in a SMILES strings database [16]; and for defining patterns useful in QSAR and QSPR models [17].

Linear notation has been explored since almost the beginning of structural chemistry in the 1860s. In his interesting review of line-formula notations [18], William Wiswesser illustrates the efforts of many well-known nineteenth-century chemists like Loschmidt, Erlenmeyer, Kekulé, and Wichelhaus in popularizing the now familiar chemical formulas like CH_3COOH for acetic acid and $C_2H_5COCH_3$ for ethyl methyl ether. Modern development of chemical notations coincided with the advent of computers in the 1940s as many realized the need to carry out automated chemical structure information processing. Wiswesser traces the evolution of linear notations from Dyson in 1947, through Taylor, G-K-D ciphers, Gruber, Silk, Cockburn, Benson, Smith, Bonnett, Gelberg, Hayward, and Lederberg in 1964, among many others. His own system, the WLN [3,4], which he developed starting in the 1940s, remained popular until the introduction of SMILES strings in the 1980s.

The WLN was designed to be used with the information processing systems of the time and with punched cards. To satisfy the requirement of 80 characters per card, the notation was restricted to using uppercase letters, the digits 0–9, and a few characters like "&." In addition, it was designed to be as readily recognizable to chemists as to digital processors. Letters were reserved to indicate functional groups and molecular features like phenyl rings while alphanumeric combinations presented fragment-based descriptions of molecules. Table 2.1 lists some examples of compounds encoded using the WLN. Thus, for acetone, the WLN representation is 1V1, where V is the character used for the central carbonyl group and the digit 1 indicates the presence of saturated single-carbon atom chains on either side. Similarly, for 3-chloro-4-hydroxybenzoic acid, the WLN string is QVR DQ CG. Here Q represents the hydroxyl group, V the carbonyl group, and R the benzene ring. The space character signifies that the following character denotes a specific position on the ring; DQ represents the 4-position hydroxyl group and CG represents the 3-position chloride (the character G denotes the chlorine atom). Note that the WLN does not include an explicit bond specification.

The WLN was the first line notation to succinctly and accurately represent complex molecules. It permitted a degree of standardization leading to the compilation of chemical compounds into databases such as CAOCI (Commercially Available Organic Chemicals Index) [19].

TABLE 2.1
WLN Representations of Chemical Diagrams

Chemical Diagram	Chemical Formula	WLN Representation
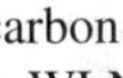	C_2H_6	2H
	C_3H_8	3H
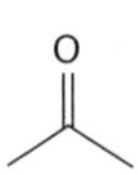	CH_3COCH_3	1V1
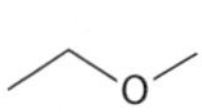	$C_2H_5OCH_3$	2O1
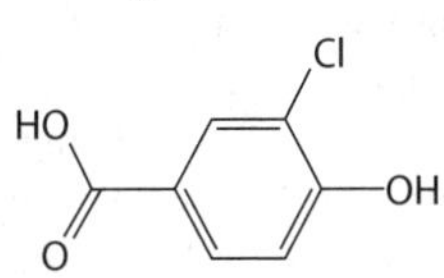	$C_7H_5ClO_3$	QVR DQ CG

As computers became more powerful and capable of handling much larger character sets, newer line notations that can encode chemical concepts, describe reactions, and be stored in relational databases have become prevalent. More than simply shorthand for molecular formulas, these linear systems are linguistic structures that can achieve multiple complex chemoinformatics objectives. SMILES, SMARTS (SMiles ARbitrary Target Specification), and SMIRKS are related chemical languages that have been used in applications such as virtual screening, molecular graph mining, evolutionary design of novel compounds, substructure searching, and reaction transforms.

SMILES is a language with simple vocabulary that includes atom and bond symbols and a few rules of grammar. SMILES strings can be used as words in other languages used for storage and retrieval of chemical information such as HTML, XML, or SQL. A SMILES string for a molecule represents atoms using their elemental symbols, with aliphatic atoms written in uppercase letters and aromatic atoms in lowercase letters. Except in special cases, hydrogen atoms are not included. Square brackets are used to depict elements, such as [Na] for elemental sodium. However, square brackets may be omitted for elements from the organic subset (B, C, N, O, P, S, F, Cl, Br, and I), provided the number of hydrogen atoms can be surmised from the normal valence. Thus, water is represented as O, ammonia as N, and methane as C. Bonds are represented with – (single), = (double), # (triple), and : (aromatic), although single and aromatic bonds are usually left out. Simple examples are CC for ethane, C=C for ethene, C=O for formaldehyde, O=C=C for carbon dioxide, COC for dimethyl ether, C#N for hydrogen cyanide, CCCO for propanol, and [H][H] for molecular hydrogen. Some atomic properties may also be specified using square brackets, for example, charge ([OH^-] for hydroxyl ion) and atomic mass for isotopic specification ([$13CH_4$] for C-13 methane).

A SMILES string is constructed by visiting every atom in a molecule once. A branch is included within parentheses and branches can be nested indefinitely. For example, isobutane is CC(C)C and isobutyric acid is CC(C)C(=O)O. Ring structures are treated by breaking one bond per cycle and labeling the two atoms in the broken bond with a unique integer (cf. Figure 2.1). Thus, C1CCCCC1 is cyclohexane, c1ccccc1 is benzene, n1ccccc1 is pyridine, C1=CCC1 is cyclobutene, and C12C3C4C1C5C4C3C25 is cubane in which two atoms have more than one ring

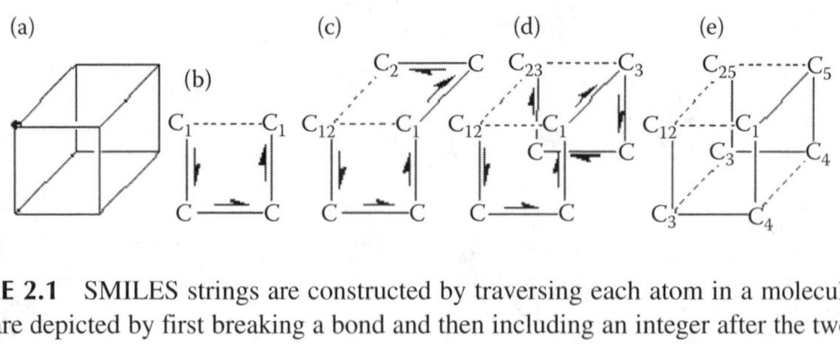

FIGURE 2.1 SMILES strings are constructed by traversing each atom in a molecule once. Rings are depicted by first breaking a bond and then including an integer after the two atoms present in the broken bond. The numbering may change with each addition of a ring. The construction of a SMILES string for cubane is shown. (a) Structure of cubane with the position of the starting atom marked with a dot; (b) C1CCC1; (c) C12CCC1CC2; (d) C12CCC1C3CCC23; and (e) C12C3C4C1C5C4C3C25.

closure. Disconnected compounds may be written as individual structures separated by a ".", such as $[NH_4^+] \cdot [Cl^-]$ for ammonium chloride. Several other rules exist for representing other molecular features such as *cis–trans* isomerism and chirality. Thus *E*-difluoroethene is F/C=C/F while *Z*-difluoroethene is F/C=C\F, and L-alanine is N[C@@H](C)C(=O)O while D-alanine is N[C@H](C)C(=O)O.

SMARTS is a language for describing molecular patterns and is used for substructure searching in chemical databases. Substructure specification is achieved using rules that are extensions of SMILES. In particular, the atom and bond labels are extended to also include logical operators and other special symbols, which allow SMARTS atoms and bonds to be more inclusive. For example, [c,N] represents a SMARTS atom that can be either an aromatic carbon or an aliphatic nitrogen, and "~" denotes a SMARTS bond that will match any bond in a query. Other examples of SMARTS patterns are c:c for aromatic carbons joined by an aromatic bond; c–c for aromatic carbons joined by a single bond (as in biphenyl); [O;H1] for hydroxyl oxygen; [F,Cl,Br,I] for any of these halogens; [N;R] for an aliphatic nitrogen in a ring; and *@;!:* for two ring atoms that are not connected by an aromatic bond. In the last example, "*" denotes any atom, "@" denotes a ring bond, ";" denotes the logical "and" operator with low precedence, and "!" denotes the logical "not" operator. An example of a more complex SMARTS query pattern is that for finding rotatable bonds: [!$(*#*)&!D1]-&!@[!$(*#*)&!D1].

SMIRKS is a hybrid of SMILES and SMARTS and uses the syntax [<SMILES_PART> ; <SMARTS_PART> : <MAP>] to describe chemical reaction transformations of the form "reactants >> products." A few rules ensure interpretation of SMIRKS as a reaction graph, making it a useful linear notation for mapping a general transformation from a set of reactants to a set of products. For example, the SMIRKS representation of amide formation is [C:1](=[O:2])Cl≫[C:1](=[O:2])N. A SMIRKS transformation may be used to represent reaction mechanisms, resonance, and general molecular graph modifications.

2.1.2 InChI and InChIKey

Like SMILES, the IUPAC International Chemical Identifier (InChI) is a 1D linear notation. It has been developed at IUPAC and NIST starting in 2000 (http://www.iupac.org/inchi). Most chemoinformatics databases provide InChI numbers of their chemical substances along with SMILES strings. Compounds can be searched by their InChIs or IUPAC International Chemical Identifier Keys (InChIKeys) (hashed InChIs) via Google, for instance a Google search with the InChIKey BQJCRHHNABKAKU-XKUOQXLYBY returns links to several web pages giving the structure of morphine. Standard versions of InChI and InChIkey were recently developed (http://www.iupac.org/inchi cf. January 2009 release) with the aim of interoperability/compatibility between large databases/web searching and information exchange. We report here the general structures of these standard identifiers.

The standard InChI [which is illustrated in Figure 2.2 for (*S*)-glutamic acid] represents the structure of a covalently bonded compound in four distinct "layers" starting with the string "InChI=1S." The first layer is composed of the molecular formula and the connections between atoms. The connectivity is spliced into three lists, a list

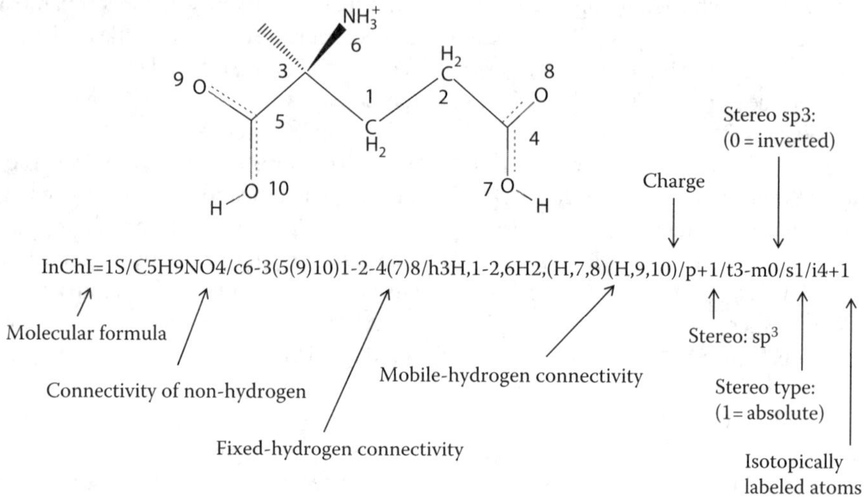

FIGURE 2.2 Standard InChI for (*S*)-glutamic acid. The numbers attached to the atoms in the figure are those used when printing the atom connectivity with InChI. The numbers are obtained after running a canonical labeling algorithm (see Section 2.3 for further details).

of connections between non-hydrogen atoms, a list of connections to fixed hydrogen atoms, and a list of connections to mobile hydrogen atoms. The second layer represents the net charge of the substance. The third layer is related to stereochemistry, and is composed of two sublayers. The first accounts for a double bond, sp^2, and the second for sp^3 tetrahedral stereochemistry and allenes. Other stereo descriptions are given next for relative stereochemistry, followed by a designation of whether the absolute stereochemistry is required. In the fourth and last layer, different isotopically labeled atoms are distinguished from each other.

The standard InChIKey is a fixed-length (27 characters) condensed digital representation of the InChI. The InChIKey consists of 14 characters resulting from a hash of the connectivity information of the standard InChI, followed by a hyphen, followed by eight characters resulting from a hash of the remaining layers of the InChI, followed by a flag character, a character indicating the version of InChI used, a hyphen, and a last character related to protonation. The InChIKey is particularly useful for web searches for chemical compounds. The standard InChIKey of (*S*)-glutamic acid is WHUUTDBJXJRKMK-MYXYCAHRSA-O.

2.1.3 Molecular File Format

A connection table is a widely used representation of the molecular graph. A simple connection table contains a list of atoms and includes the connectivity information for bonding atoms and may also list bond orders. For instance, the drug acetaminophen (Scheme **2.1**), can be represented as follows in a hydrogen-suppressed connection table:

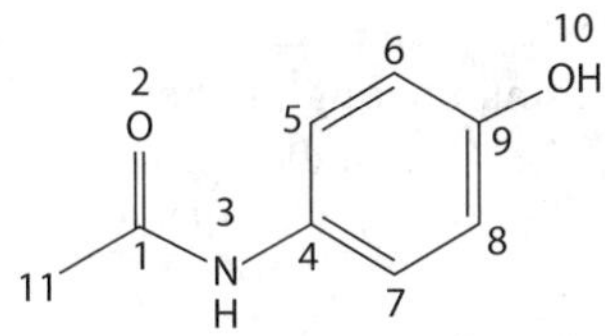

SCHEME 2.1

11 11			
−1.7317	−0.5000	0.0000	C
−1.7317	0.4983	0.0000	O
−0.8650	−1.0000	0.0000	N
0.0017	−0.5000	0.0000	C
0.0017	0.4983	0.0000	C
0.8650	0.9983	0.0000	C
0.8650	−1.0000	0.0000	C
1.7317	−0.5000	0.0000	C
1.7317	0.4983	0.0000	C
2.5967	0.9983	0.0000	O
2.5967	−1.0000	0.0000	C
1	2	2	
1	3	1	
3	4	1	
4	5	1	
5	6	2	
4	7	2	
7	8	1	
8	9	2	
6	9	1	
9	10	1	
1	11	1	

The first line indicates the number of atoms, n, and the number of bonds, m. The next n lines comprise the atoms block and list atomic coordinates and atom types in the molecule. These are followed by the bonds block containing m lines. The first two numbers on a bond specification line indicate atom numbers and the third denotes bond order. While this example is for a 2D chemical diagram, connection tables can easily be extended to represent 3D structures, in which case the z-coordinates are likely to have nonzero values.

Molecular file formats that are based on connection tables can represent a chemical structure in a straightforward way and can make use of various algorithms that are available for reading and writing these formats. Examples of such file formats include the MOL and SDF (structure-data format) formats (from Symyx, http://www.symyx.com) and the MOL2 format (from Tripos, http://www.optive.com). Most molecular modeling packages and databases employ file formats tailored to their specific needs. In such cases, the connection table is usually enhanced by appending additional information such as charge, isotopes, and stereochemistry.

Several hydrogen-suppressed molecular file formats are shown here for acetaminophen, with the atom numbering as shown in Scheme **2.1**. The molecular files were converted using the OpenBabel program (http://openbabel.org) from a simple connection table created in ChemDraw (CambridgeSoft, http://www.cambridgesoft.com/).

The molecular design limited (MDL; now Symyx) chemical table file or CTfile is an example of a detailed connection table in which a set of atoms may represent molecules, substructures, groups, polymers, or unconnected atoms [20]. The CTfile is the basis for both the MOL format and the SDF file that enriches the MOL format with specific data fields for other information. In an SDF file, the molecules are separated by the "$$$$" delimiter. The MOL connection table for acetaminophen is as follows:

```
11 11        0         0        0   0   0   0   0    0999 V2000
−1.7317  −0.5000   0.0000 C   0   0   0   0   0
−1.7317    0.4983   0.0000 O   0   0   0   0   0
−0.8650  −1.0000   0.0000 N   0   0   0   0   0
 0.0017  −0.5000   0.0000 C   0   0   0   0   0
 0.0017    0.4983   0.0000 C   0   0   0   0   0
 0.8650    0.9983   0.0000 C   0   0   0   0   0
 0.8650  −1.0000   0.0000 C   0   0   0   0   0
 1.7317  −0.5000   0.0000 C   0   0   0   0   0
 1.7317    0.4983   0.0000 C   0   0   0   0   0
 2.5967    0.9983   0.0000 O   0   0   0   0   0
−2.5967  −1.0000   0.0000 C   0   0   0   0   0
    1         2         2        0   0   0
    1         3         1        0   0   0
    1        11         1        0   0   0
    3         4         1        0   0   0
    4         5         1        0   0   0
    4         7         2        0   0   0
    5         6         2        0   0   0
    6         9         1        0   0   0
    7         8         1        0   0   0
    8         9         2        0   0   0
    9        10         1        0   0   0
M          END
```

The MOL2 file format from Tripos is used extensively by the SYBYL molecular modeling software. It is divided into several sections using Record Type Indicators (RTIs), each of which has its own data record whose format depends on the section in which it lies. The MOL2 file for acetaminophen is shown here. The data record for the RTI "@<TRIPOS>MOLECULE" contains six lines: the first line has the name of the molecule; the second line contains the number of atoms, bonds, substructures, features, and sets that may be associated with this molecule; the third line is the molecule type; the fourth and fifth lines indicate the type of charge and the energy associated with the molecule; and the last line contains any optional remark about the molecule. The RTI "@<TRIPOS>ATOM" contains data records for each atom in the molecule. As shown, the atom block in a MOL2 file can contain information about atom type, including hybridization state (column 6), substructure ID and name (columns 7 and 8, respectively), and atomic charge (column 9). The data record for the RTI "@<TRIPOS>BOND" comprises the bond block for the MOL2 file. Each line contains the bond ID (column 1), the origin atom in the bond (column 2), the target atom in the bond (column 3), and the bond type (column 4; 1 = single, 2 = double, am = amide, ar = aromatic, etc.). A MOL2 file may have many other RTIs depending on the application that the molecule is used for. For example, the RTI "@<TRIPOS>FF_PBC" can be used to specify periodic boundary conditions and "@<TRIPOS>CENTROID" can be used to specify a dummy atom as the centroid of a molecule or substructure.

```
@<TRIPOS>MOLECULE
acetaminophen
11 11 0 0 0
SMALL
GASTEIGER
Energy=0

@<TRIPOS>ATOM
1 C     −1.7317   −0.5000   0.0000 C.2     1   LIG1    0.2461
2 O     −1.7317    0.4983   0.0000 O.2     1   LIG1   −0.2730
3 N     −0.8650   −1.0000   0.0000 N.am    1   LIG1   −0.1792
4 C      0.0017   −0.5000   0.0000 C.ar    1   LIG1    0.0736
5 C      0.0017    0.4983   0.0000 C.ar    1   LIG1    0.0199
6 C      0.8650    0.9983   0.0000 C.ar    1   LIG1    0.0434
7 C      0.8650   −1.0000   0.0000 C.ar    1   LIG1    0.0199
8 C      1.7317   −0.5000   0.0000 C.ar    1   LIG1    0.0434
9 C      1.7317    0.4983   0.0000 C.ar    1   LIG1    0.1958
10 O     2.5967    0.9983   0.0000 O.3     1   LIG1   −0.2866
11 C    −2.5967   −1.0000   0.0000 C.3     1   LIG1    0.0968

@<TRIPOS>BOND
1     1     2     2
2     1     3     am
3     3     4     1
4     4     5     ar
5     5     6     ar
6     4     7     ar
7     7     8     ar
8     8     9     ar
9     6     9     ar
10    9     10    1
11    1     11    1
```

The XML-based molecular file format CML (Chemical Markup Language, http://cml.sourceforge.net) has been proposed by Murray-Rust and Rzepa [21,22]. CML can be used in many applications requiring representation of molecules, reactions, experimental structures, computational structures, or spectra [23]. CML permits inclusion of chemical information in XML documents that can subsequently be used for chemical data retrieval. The CML connection table of acetaminophen

(Scheme **2.1**) is provided below.

```
<?xml version="1.0"?>
<molecule xmlns="http://www.xml-cml.org/schema"
  id="acetaminophen">
<atomArray>
<atom id="a1" elementType="C" x2="-1.731700" y2="-0.500000"/>
<atom id="a2" elementType="O" x2="-1.731700" y2="0.498300"/>
<atom id="a3" elementType="N" x2="-0.865000" y2="-1.000000"/>
<atom id="a4" elementType="C" x2="0.001700" y2="-0.500000"/>
<atom id="a5" elementType="C" x2="0.001700" y2="0.498300"/>
<atom id="a6" elementType="C" x2="0.865000" y2="0.998300"/>
<atom id="a7" elementType="C" x2="0.865000" y2="-1.000000"/>
<atom id="a8" elementType="C" x2="1.731700" y2="-0.500000"/>
<atom id="a9" elementType="C" x2="1.731700" y2="0.498300"/>
<atom id="a10" elementType="O" x2="2.596700" y2="0.998300"/>
<atom id="a11" elementType="C" x2="-2.596700" y2="-1.000000"/>
</atomArray>
<bondArray>
<bond atomRefs2="a1 a2" order="2"/>
<bond atomRefs2="a1 a3" order="1"/>
<bond atomRefs2="a3 a4" order="1"/>
<bond atomRefs2="a4 a5" order="1"/>
<bond atomRefs2="a5 a6" order="2"/>
<bond atomRefs2="a4 a7" order="2"/>
<bond atomRefs2="a7 a8" order="1"/>
<bond atomRefs2="a8 a9" order="2"/>
<bond atomRefs2="a6 a9" order="1"/>
<bond atomRefs2="a9 a10" order="1"/>
<bond atomRefs2="a1 a11" order="1"/>
</bondArray>
</molecule>
```

In molecular file formats such as the HIN format (HyperCube, http://www.hyper .com), the bond information may be included within the atoms block itself. In the HIN format file, an atom record includes the atomic charge (column 7), the coordination number *c* (the number of covalently bonded atoms; column 11), and *c* pairs denoting the label of the adjacent atom and the corresponding bond type encoded with s, d, t, or a, for single, double, triple, or aromatic bonds, respectively. The HIN file of acetaminophen is shown as an example.

```
mol 1 acetaminophen
atom 1 - C ** -0.24606 -1.73170 -0.50000 0.00000 3 2 d 3 s 11 s
atom 2 - O ** - -0.27297 -1.73170 0.49830 0.00000 1 1 d
atom 3 - N ** - -0.17920 -0.86500 -1.00000 0.00000 2 1 s 4 s
atom 4 - C ** -0.07357 0.00170 -0.50000 0.00000 3 3 s 5 a 7 a
atom 5 - C ** - 0.01988 0.00170 0.49830 0.00000 2 4 a 6 a
atom 6 - C ** - 0.04336 0.86500 0.99830 0.00000 2 5 a 9 a
atom 7 - C ** - 0.01988 0.86500 -1.00000 0.00000 2 4 a 8 a
atom 8 - C ** - 0.04336 1.73170 -0.50000 0.00000 2 7 a 9 a
atom 9 - C ** - 0.19583 1.73170 0.49830 0.00000 3 8 a 6 a 10 s
```

```
atom 10 - O ** - -0.28657 2.59670 0.99830 0.00000 1 9 s
atom 11 - C ** - 0.09680 -2.59670 -1.00000 0.00000 1 1 s
endmol 1
```

The linear notation of acetaminophen is much more compact. Its SMILES string is C(=O)[Nc1ccc(cc1)O]C and its InChI string is InChI=1S/C8H9NO2/c1-6(10)9-7-2-4-8(11)5-3-7/h2-5,11H,1H3,(H,9,10).

2.2 FROM CONNECTION TABLE TO 2D STRUCTURE

Since chemoinformatics systems and databases store chemical structures as linear notations, connection tables, or other digital formats, a scientist who wishes to view the structures as a familiar chemical diagram must be provided with a means to translate the digital data into a viewable image. In the case of connection tables, the atomic coordinates, atom types, and bonding information are sufficient to convert into a chemical diagram using appropriate software. Translation of linear notations requires the conversion software to extract critical information—such as bond lengths, angles, and topology—that is implicit in the notation.

Dittmar et al. designed one of the first systems for drawing a chemical diagram [24] for the Chemical Abstracts Service (CAS) registry system. For this they leveraged the vast experience at CAS of abstracting and extracting chemical information from chemical literature. The method is provided as Algorithm 2.1 and is dependent on a knowledge base of ring systems. The molecular graph is first decomposed into acyclic fragments and ring systems. The ring systems are ranked and processed, during which the acyclic components are systematically reattached to the rings so that an order 1 substituent is directly attached to the ring while an order n substituent is linked to an order $n-1$ substituent.

ALGORITHM 2.1 CAS 2D CHEMICAL DIAGRAM DRAWING*

```
Analyze Structure
Rank Rings
Do Until All Rings Processed
   Select Unprocessed Ring
   Orient Ring
   Draw Ring
   Rank Order 1 Substituents
   Do Until All Order 1 Substituents Processed
      Select Unprocessed Order 1 Substituent
      Orient Link/Chain
      Draw Link/Chain
      Rank Order 2 Substituents
      Do Until All Order n Substituents (n>1) Processed
        Select Unprocessed Order n Substituent
        Orient Link/Chain
```

* Reprinted from Dittmar, P. G., Mockus, J., and Couvreur, K. M., *Journal of Chemical Information and Computer Sciences* 1977, 17(3), 186–192. With permission. Copyright 1977 from American Chemical Society.

```
            Draw Link/Chain
            Rank Order n + 1 Substituents
          End Do
          Do Until All Order n Substituents Positioned
           Select Unpositioned Order n Substituent
           Position Order n Substituent
          End Do
          Position Order 1 Substituent
        End Do
        Position Ring
      End Do
    End
```

Other 2D chemical structure drawing algorithms that are independent of a ring system database have been developed. Some use ring perception algorithms [25] such as the drawing algorithm of Shelley [26], which also employs the Morgan extended connectivity index (cf. Section 2.3 and [27]). Morgan indices are calculated for atoms and ring systems and are used for minimizing atom crowding during assignment of 2D coordinates, for ring system orientation, and for identifying invariant coordinates for ring systems.

The depiction of 2D structures has been reviewed by Helson [28]. Other algorithms have been developed such as those by Weininger for SMILES strings [29], by Fricker et al. [30], by Stierand et al. [31], and by Clark et al. [32] for the Molecular Operating Environment (MOE) software suite (Chemical Computing Group, http://www.chemcomp.com). Fricker et al. also rely on the sequential assembly of chains and rings and proposed an algorithm for drawing structure under directional constraints on bonds [30]. Stierand et al. proposed a method for generating diagrams for protein–ligand complexes that highlights the interaction between amino acids and ligand atoms on a 2D map [31].

The algorithm of Clark et al. [32] partitions a molecular graph into segments, generates local structural options for each segment, assembles local options by random sampling, and selects the minimally congested assembly as the basis for the final output. The algorithm identifies ring systems, atom chains, lone atoms, atom pairs, and stereochemistry for a hydrogen-suppressed molecular graph. The structural units generated are assigned geometric constraints in internal coordinates such that a set of distances and angles between atoms in the structural units is obtained. Rings are identified using an algorithm that finds the smallest set of smallest rings [8,9]. Atom chains are generated using a small set (C, N, O, and S) of neutral, acyclic atoms connected by single bonds. Lone atoms are constrained with standard values for bond lengths and angles. Atom pairs are constrained to preserve local geometry and stereochemistry. A core ring system is identified by removing rings that share only one edge with any other ring and coordinates are assigned to the core ring system based on a database of ring templates. The candidate structures obtained during the assembly of structural units in the sampling step are screened using a congestion function, with the solution structure having the lowest congestion. In the postprocessing step the solution structure is adjusted for aesthetic depiction. Thus, atoms in close contact are permitted to deviate slightly; atoms or bonds may be rotated to relieve overlap;

connected fragments may also be brought to a horizontal position by rotation; and hydrogen atoms are expressed for cases where explicit hydrogen atom depiction is mandated. The MOE 2D chemical drawing algorithm was extensively evaluated and demonstrated high performance using both statistical metrics and human evaluation. Other software systems available include those from OpenEye (Ogham) and Advanced Chemistry Development (ChemSketch).

An important aspect of the use of chemical databases by chemoinformatics systems is their inspection, refinement, and standardization prior to use. This ensures, for example, that a chemical compound with multiple representations (such as Kekulé structures or tautomers) is correctly identified and processed by an application. Special tools such as sdwash (Chemical Computing Group, http://www.chemcomp.com) and Standardizer (ChemAxon, http://www.chemaxon.com) are available to assist with such problems. These tools may work by removing salt and solvent molecules, adding or removing hydrogens, identifying aromaticity, or enumerating protonation states and tautomers. Conjugated systems can be recognized using aromaticity perception [33] or identification of alternating bonds [34].

Perception of tautomers has also received considerable attention in chemoinformatics. Tautomers are isomers of organic compounds that result from migration of a hydrogen atom or a proton accompanied by a switch between adjacent single and double bonds. Tautomerism is an important property because different tautomers of a compound may give different results in virtual screening and for properties based on atom-type parameters. Several approaches have been suggested for the treatment of tautomers in chemical databases and applications [35–38].

Another area of chemoinformatics that has seen active development for a number of years is searching for Markush structures in patent databases [39]. A Markush structure is a generalized notation of a set of related chemical compounds and may be used in patent applications to define a substance that has not yet been synthesized. Searching for Markush structures in chemical databases enables a researcher to rule out priority in patent applications. Chemical drawing programs can represent these structures by, for example, drawing a bond to the center of a phenyl ring indicating substitution at any position in the ring (Figure 2.3).

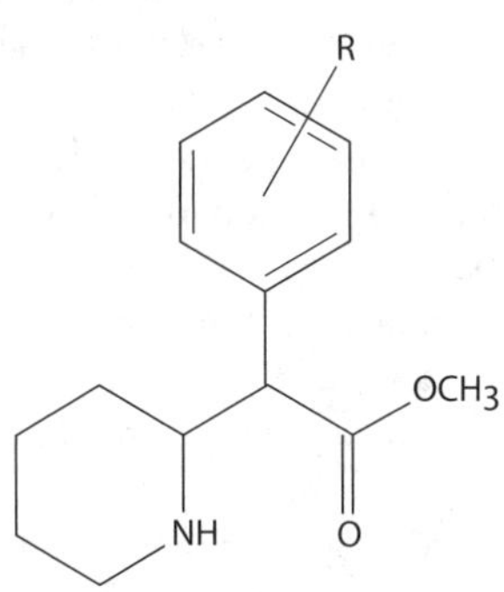

FIGURE 2.3 Markush structure for methylphenidate ester with a generalized R group substitution at any position in the phenyl ring.

2.3 STORING AND RETRIEVING CHEMICAL STRUCTURES THROUGH CANONICAL LABELING

When storing a chemical structure in a database, one is faced with the problem of determining whether the structure is already present in the database. Therefore, one needs to determine whether structures are identical. In graph theory two identical graphs are said to be isomorphic, and an isomorphism is a one-to-one mapping between the atoms of the graph preserving the connectivity of the graphs. When two atoms of the same graph are mapped by an isomorphism, the two atoms are said to be automorph and one uses the term automorphism instead of isomorphism. Atoms that are automorph are symmetrical (or equivalent), and in a given molecular graph, atoms can be partitioned into their automorphism group or equivalent classes. One practical procedure to detect isomorphism or automorphism between molecular graphs is to canonically (i.e., uniquely) label the atoms of the graphs. Two graphs are then isomorphic if they have the same labels.

Automorphism partitioning and canonical labeling are of significant interest in chemistry. Both problems have practical applications in (1) molecular topological symmetry perception for chemical information and quantum mechanics calculations, (2) computer-assisted structure elucidation, (3) NMR spectra simulation, and (4) database storage and retrieval, including determination of maximum common substructure. Since 1965 [27] many canonical labeling methods have been proposed in the context of chemical computation, and this section presents three of them after providing more precise definition and relationships between isomorphism, automorphism, and canonical labeling.

2.3.1 TERMINOLOGY

Two given graphs are isomorphic when there is a one-to-one mapping (a permutation) from the vertices of one graph to the vertices of the second graph, such that the edge connections are respected. An isomorphic mapping of the vertices onto themselves is called an automorphism. The set of all automorphisms of a given graph is the automorphism group of the graph. The automorphism group contains information regarding the topological symmetry of the graph. In particular, the orbits of an automorphism group identify symmetrical (or equivalent) vertices. The canonical labeling problem consists of finding a unique labeling of the vertices of a given graph, such that all isomorphic graphs have the same canonical labels. Examples of canonical representations are graphs that maximize (or minimize) their adjacency matrices. Two graphs with the same canonical representation are isomorphic. This observation is used when querying chemoinformatics databases as the structures stored in chemoinformatics databases are canonized prior storage. The obvious advantage is that isomorphism is reduced to comparing the canonical labels of the input structure with those of the database.

Although most articles related to graph isomorphism have been published in the computer science literature, the computation of the orbits of automorphism groups using partitioning techniques has received most attention in chemistry and the two problems have been shown to be computationally equivalent [40]. The canonical labeling problem has been studied in both chemistry and computer science. It would appear that the canonical labeling problem is closely related to the isomorphism testing problem; the latter can be performed at least as fast as the former, and for many published algorithms, isomorphism tests either include a procedure for canonization or else have an analogue for that problem [41].

Since canonical labeling can be used to detect isomorphism, the remainder of this section focuses on canonization procedures. As already mentioned, several canonical labeling methods have been proposed in the context of chemical computation, and three of them are presented next. The general characteristic of these methods is the use of graph invariants to perform an initial vertex partitioning. Graph invariants can be computed efficiently and have led to the development of fast algorithms. For most published algorithms, the initial vertex partitioning is followed by an exhaustive generation of all labelings. The computational complexity of the exhaustive generation can be reduced using the fact that two vertices with different invariants belong to different equivalent classes; hence, exhaustive labeling generation is performed only for vertices with the same invariant. Nonetheless, because all vertices may have the same invariant, the upper bound of the time complexity for the exhaustive labeling generation scales exponentially with the number of vertices.

Whereas vertices with different invariants belong to different equivalent classes, the reverse is not necessarily true. As a matter of fact, isospectral points are vertices with the same invariant that belong to different classes [42]. Although the invariant approach may not be totally successful in the sense that the proposed methods work in all cases, it has been shown to behave well on average [43,44]. Indeed, for a given random graph there is a high probability that its vertices can be correctly partitioned using graph invariants [45].

To simplify the presentation, the three canonization algorithms given next are illustrated for hydrocarbons and do not take stereochemistry into account. For each algorithm, the reader is referred to relevant literature on how the algorithms can be modified to incorporate heteroelements and stereochemistry.

2.3.2 Morgan's Algorithm

This historically important algorithm was first published in 1965 by H.L. Morgan [27] and was part of the development of a computer-based chemical information system at the CAS. The original algorithm did not take into account stereochemistry, and an extension was proposed in 1974 by Wipke et al. [46]. We present below a scheme based on Morgan's algorithm; while our scheme may not be the exact published algorithm, it captures the essential ideas.

As with many canonization methods, Morgan's algorithm proceeds in two steps. In the first step, graph invariants (cf. the definition in Chapter 1) are calculated for each atom. In the second step, all possible atom labels are computed and printed according to the invariants. Invariants are calculated in a recursive manner. At each recursive iteration, the invariant of any given atom is the sum of the invariants of its neighbors computed in the previous step. The process starts by assigning to all atoms an initial invariant equal to 1 or equal to the number of neighbors of that atom (as in the original algorithm published by Morgan). One notes that when assigning an initial invariant equal to 1, the invariant obtained at the next iteration equals the number of neighbors. The recursive procedure stops when the largest number of distinct invariants is obtained, that is, at an iteration for which the number of invariants remains the same or decreases after that iteration. Labeling of the atoms is performed using the invariants found at the stopping iteration. One starts by choosing the atom with the largest invariant; this atom is labeled 1. One notes that all the other atoms are unlabeled at this point. At the second step, the neighbors of the atom labeled 1 are sorted in decreasing order of invariants; the neighbors are then labeled 2, 3, and so on, in the order that they appear in the sorted list. There may be several ways of sorting the neighbors; as some atoms may have the same invariants, the algorithm computes all the possible sorted lists. At the next step, one searches for atoms with the smallest label having unlabeled neighbors and the procedure described in the second step is repeated, this time for the smallest labeled atom. The procedure halts when all atoms have been labeled. The canonical graph is the one producing the lexicographically smallest list of bonds when printing the graph. In the algorithm given below, the notation || is used to depict the number of distinct elements of a list, and the routine print-graph() prints the edges of a graph for a given list of atom labels.

ALGORITHM 2.2 MORGAN'S ALGORITHM

```
canonical-Morgan(G)
input:    - G: a molecular graph
output:   - printout of graph G with computed labels
     Let inv be the set of invariant initialized to
     {1,1,...,1}
     Let lab be the set of labels initialized to
     {0,0,...,0}
     inv = compute-invariant(G,inv)
     Let L be the set of atoms in G with the largest
     invariant
     For all atom x in L do
       lab(x) = 1
       compute-label(G,inv,lab,1)
       lab(x) = 0
     done

compute-invariant(G,inv)
input:    - G: a molecular graph
          - inv: the initial invariants for all atoms
output:   - INVARIANT: the updated invariants for all
            atoms
       For all atom x of G do
           INVARIANT(x) = Σ[x,y] in G inv[y]
       done
       if |INVARIANT|>|inv|
```

```
        then return(compute-invariant(G,INVARIANT))
        else return(inv)
     fi

compute-label(G,inv,label,n)
input:    - G: a molecular graph
          - inv: the invariants for all atoms
          - label: the labels for all atoms
          - n: the current label
output:   - printout of graph G with computed labels

     Let x be the atom of G with the smallest label
     having a non empty list Lx of unlabeled
     neighbors
     if (x cannot be found) then
        print-graph(G,label)
        return
     fi
     For all lists Sx corresponding to the list Lx where
     neighbors of x are sorted by decreasing invariants do
        For all atoms y in Sx do n=n+1, label(y)= n done
        compute-invariant(G,inv,label,n)
        For all atoms y in Sx do n=n-1, label(y) = 0 done
     done
```

The invariant step of Morgan's algorithm is illustrated in Figure 2.4 for 1,8-dimethyl-decahydronaphthalene; the labeling step is illustrated in Figure 2.5 for the

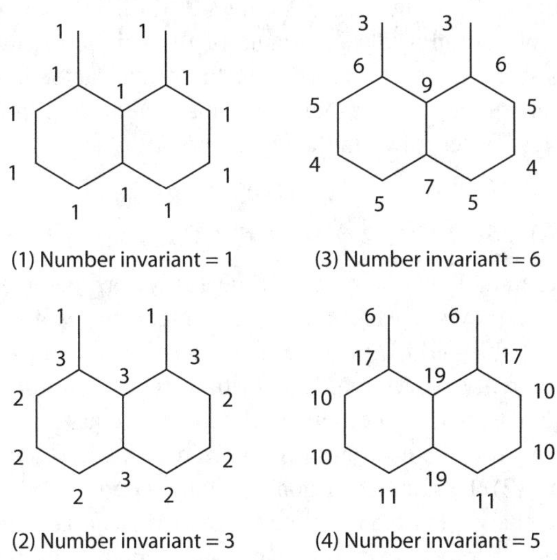

FIGURE 2.4 Invariant calculation with Morgan's algorithm. The final list of invariants is computed the first time the largest numbers of distinct invariant is obtained, the list of graph (3) in the present case.

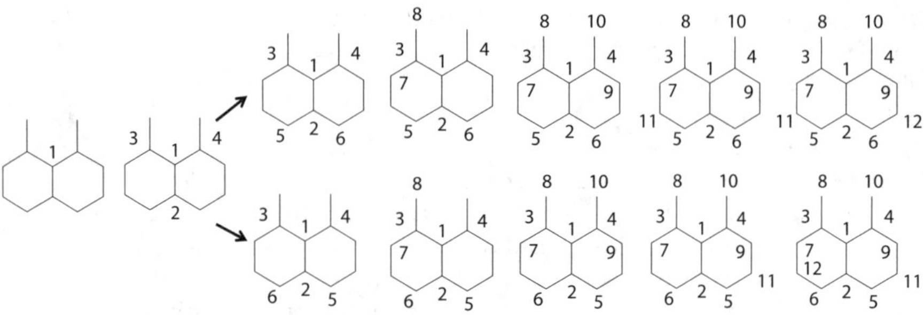

FIGURE 2.5 Atom labeling with Morgan's algorithm. The labels are computed according to the invariants of graph (3) of Figure 2.4. The figure illustrates the fact that there are two ways of sorting the list of neighbors of atom labeled 2 (graph C1 and graph C2). Graph G1 produces the list of edges [1, 2] [1, 3] [1, 4] [2, 5] [2, 6] [3, 7] [3, 8] [4, 9] [4, 10] [5, 11] [6, 12] [7, 11] [9, 12] and graph G2 produces the list [1, 2] [1, 3] [1, 4] [2, 5] [2, 6] [3, 7] [3, 8] [4, 9] [4, 10] [5, 11] [6, 12] [7, 12] [9, 11]. G1 is canonical.

same compound. Graph G1 is the canonical graph, that is, the graph producing the smallest lexicographic list of edges.

The main criticism of Morgan's algorithm is the ambiguity of the summation when computing atom invariants. Indeed, let us suppose that two three-connected atoms have neighbors with respective invariants (1, 1, 3) and (1, 2, 2). After iterating Morgan's algorithm, these two atoms will have the same invariant 5, and the two atoms will be considered to be equivalent whereas they should not. To palliate this problem, Weininger et al. [47] proposed a solution making use of prime numbers. In this implementation the initial labels are substituted by primes, that is, to invariant 1 is associated number 2, to invariant 2 number 3, to invariant 3 number 5, and so on. Next, instead of summing the invariants of the neighbors, the product of the associated primes is computed. According to the prime factorization theorem, the solution of Weininger et al. is unambiguous. This solution was implemented when canonizing SMILES, as described in the algorithm given next.

2.3.3 The Canonical SMILES Algorithm

The algorithm presented below is based on the 1989 work published by Weininger et al. [47]. As in the previous case, the outlined algorithm is not the exact replica of the one that was published but provides the general idea on how SMILES strings are canonically ordered. Like Morgan's algorithm, canonical ordering is performed in two steps. Invariants are first computed and labels are assigned according to the invariants. For each atom, the invariant is initialized taking into account (1) the number of connections, (2) the number of non-hydrogen bonds, (3) the atomic number, (4) the sign of the charge, (5) the absolute value of the charge, and (6) the number of attached hydrogen atoms. Next, invariants are computed using a method similar to the routine compute-invariant given in Algorithm 2.2. However, as mentioned earlier, instead of summing the invariants of the neighbors, Weininger et al. used the products of the primes corresponding to the invariants of the neighbors. At the next step, atoms are labeled according to their invariants, the smallest label being assigned to the atom with the smallest invariant. The atom with the same invariant has the same label. To uniquely label each atom, Weininger et al. used a procedure named "double-and-tie-break." This procedure consists of doubling all labels (label 1 becomes 2, label 2 becomes 4, and so on), then an atom is chosen in the set of atoms having the smallest labels, and the label of that atom is reduced by one. The labels thus obtained form a new set of initial invariants and the whole procedure (invariant computation followed by label assignment) is repeated. The process stops when each atom has a unique label, that is, when the number of labels equals the number of atoms. In the algorithm given next, we assume that an initial set of invariants has been computed, as described above, when calling the routine canonical-SMILES for the first time. The routine compute-invariant is not detailed, but is similar to the one given with Algorithm 2.2 replacing summation by prime factorization.

ALGORITHM 2.3 THE CANONICAL SMILES ALGORITHM

```
canonical-SMILES(G,inv)
input:  - G: a molecular graph
        - inv: a set of initial invariants
output: - printout of a canonical SMILES string of
          graph G
01.  inv = compute-invariant(G,inv)
02.  Let lab be the set of atom labels assigned in
     increasing invariants order
03.    if |lab|=|G| then print-SMILES(G,lab) fi
04.    lab = new set of labels doubling each label value
05.  Let L be the set of atoms with the smallest
     label such that |L| ≥ 2
06.  For all atom x in L do
07.    lab(x) = lab(x)-1 , canonical-SMILES(G,lab),
       lab(x) = lab(x)+1
08.  done
```

In their 1989 paper, Weininger et al. [47] do not recursively apply the canonical-SMILES routine to all atoms in the list of tied atoms with the smallest label (list *L* in Algorithm 2.3 lines 05–08) but to an arbitrary atom selected from that list. Weininger et al. algorithm complexity thus reduces to $N^2\log_2(N)$ for a molecular graph of *N* atoms. While this implementation is efficient, it may not necessarily be correct. Algorithm 2.3 does not scale in polynomial time, as the list *L* may comprise all *N* atoms the first time it is computed, $N - 1$ atoms the second time, $N - 2$ the third time, and so one, leading to an $N!$ complexity. However, Algorithm 2.3 produces all potential canonical SMILES, the final canonical string being the lexicographically smallest one. Printing SMILES (routine print-SMILES) is performed starting with the atom with the smallest label. This atom becomes the root of a tree for a depth-first search. As mentioned in Section 2.1.1, with SMILES notations, branches are indicated with parentheses: "('to open the branch and')" to close the branch. When printing canonical SMILES, the algorithm directs branching toward the lowest label; assuming the atom labels for acetone C−C(=O)–C are 1–4- (=3)-2, one starts with the methyl group labeled 1 and then moves to the central carbon labeled 4. The branch is composed of the smallest atom label attached to the central carbon, that is, the second methyl group labeled 2; the final atom is thus the oxygen labeled 3. The printed canonical SMILES string for acetone is thus CC(C)=O and not CC(=O)C.

The latest developments in the search for a unique representation of chemical structures are the InChI and InChIKey algorithms [48]. These algorithms are based on the McKay general graph canonization algorithm [49]. Because the InChI algorithm has never been formally published, it is difficult to present this complex method just based on the technical manual. Instead we present another algorithm, whose performances have been shown to be comparable to those of the McKay technique [50].

2.3.4 Canonical Signature Algorithm

The method outlined below was first published in 2004 [50]. As with previous algorithms, the method first assigns invariants to atoms and next labels the atoms from the invariants. In the present case, invariants are computed based on atom signatures. The formal definition of an atom signature is given in Chapter 4, but for the purpose of this section we define the signature of an atom x in a molecular graph G as being a tree $T(x)$ spanning all the edges of the graph (cf. Figure 2.6a and b).

When computing an atom signature, the term tree is used in a somewhat loose manner as several vertices in the tree may correspond to the same atom. The root of the tree is atom x itself. The first layer of the tree is composed of the neighbors of x; the second layer is composed of the neighbors of the vertices of the first layer except atom x. The construction proceeds one layer at a time until no more layers can be added, that is, until all the bonds of G have been considered. Assuming the

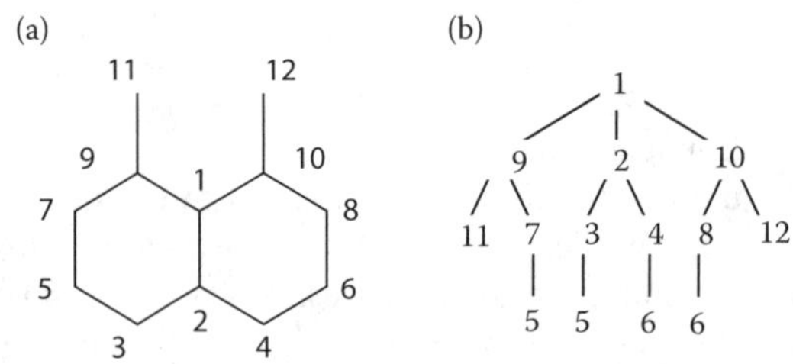

(c)

Atom no.	1	2	3	4	5	6	7	8	9	10	11	12
Atom type and number of parents	C,0	C,1	C,1	C,1	C,2	C,2	C,1	C,1	C,1	C,1	C,1	C,1
Invariant	1	2	2	2	3	3	2	2	2	2	2	2

FIGURE 2.6 Atomic signature. (a) 1,8-Dimethyl-decahydronaphthalene, where carbon atoms have been arbitrarily numbered. (b) Signature tree of atom 1. (c) Atom initial invariants.

tree has been constructed up to layer l, layer $l + 1$ is constructed considering each vertex y of layer l. Let z be a neighbor of y in G. Vertex z and edge $[y,z]$ are added to layer $l + 1$ if the edges $[y,z]$ or $[z,y]$ are not already present in the previous layers of the tree. To each vertex added to the tree, one associates an atom type and the initial label or number of the corresponding atom. Note that a given atom number z may appear several times in the tree (such as atom number 5 in Figure 2.6) since it can be the neighbor of several atoms present in the previous layer. Having defined atom signatures, we next explain how these signatures can be used to compute invariant and ultimately canonized molecular graphs.

The approach taken to canonize atomic signatures is based on the classical Hopcroft and Tarjan's rooted tree canonization algorithm [51]. Let x be an atom of a molecular graph G, and $T(x)$, the corresponding signature tree. To each atom a one associates an atom type and an invariant, inv(a). Invariants are integers no greater than N, the total number of atoms. To each vertex v in $T(x)$ one associates a corresponding atom, atom(v), in graph G and an invariant, inv(v). Additionally, for each vertex of any layer l, one can access its parents in layer $l - 1$ and its children in layer $l + 1$.

Prior to running the algorithm, all the invariants are initialized. The initial invariant of any atom a is computed from the atom type of a and the number of parents a has in $T(x)$. More precisely, a string of characters is compiled from the atom type and the number of parents, and the string is converted into an integer following lexicographic ordering. The integer is not greater than N since there are no more than N different strings. Examples of initial invariants are given in Figure 2.6c.

After initialization, the first step of the algorithm is to compute the invariants of the vertices in $T(x)$ from the atom invariants. The vertex invariants are computed twice, first reading the tree layer by layer from the leaves to the root, and then from the root to the leaves. Unlike the classical Hopcroft–Tarjan algorithm, the tree must also be read from the root to the leaves because, in signature trees, some vertices may have more than one parent; thus the invariants for these vertices may be different depending on the invariants of their parents. We first examine the case where the tree is read from the leaves to the root. Starting at the last layer, to each vertex we associate the invariant of the corresponding atom. Duplicated invariants are removed and all nonidentical invariants are sorted in decreasing order. The vertex invariant becomes the order of the vertex in the sorted list. Going to the layer above, to each vertex one assigns a vector composed of the invariant of the corresponding atom and the invariants of the children of the vertex. Duplicated vectors are removed, the remaining vectors are sorted in decreasing order, and the vertex invariant becomes the order of the vertex in the sorted list. Note that these vertex invariants range from 1 to N since there are no more than Nvertices in a layer. The above procedure is repeated until the root is reached. The algorithm is then run from the root to the leaves but this time for any vertex; the vector invariant is composed of the invariant of the corresponding atom and the invariants of the parents of the vertex. The Algorithm 2.4 is given below and is illustrated in Figure 2.7 for 1,8-dimethyl-decahydronaphthalene.

Once invariants have been computed for all vertices, each atom invariant is compiled from the invariant of all the vertices corresponding to the atom. Precisely, for

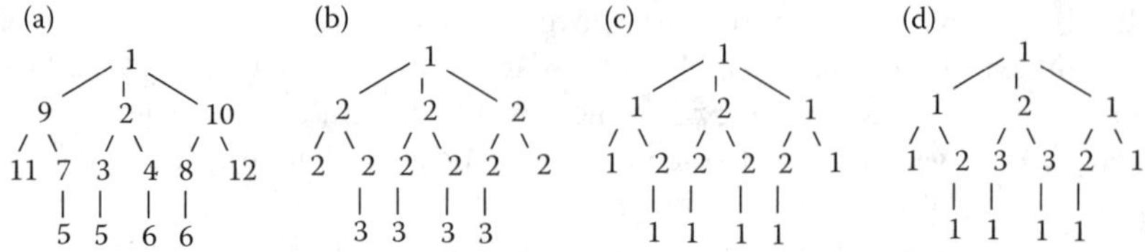

(e)

Atom no.		1	2	3	4	5	6	7	8	9	10	11	12
Layer no.	0	1	0	0	0	0	0	0	0	0	0	0	0
	1	0	2	0	0	0	0	0	0	1	1	0	0
	2	0	0	3	3	0	0	2	2	0	0	1	1
	3	0	0	0	0	1	1	0	0	0	0	0	0
Invariant		7	6	4	4	1	1	3	3	5	5	2	2

FIGURE 2.7 Vertex and atom invariants. (a) Signature tree of atom 1 in Figure 2.6a. (b) The same signature tree with initial invariants from Figure 2.6c. (c) Vertex invariants after running the Hopcroft–Tarjan algorithm from the leaves to the root. (d) Vertex invariants after running the Hopcroft–Tarjan algorithm from the root to the leaves. (e) Corresponding atom vectors and atom invariants. The root is located at layer 0.

each atom, an invariant vector is first initialized to zero; then for each vertex corresponding to the atom, the invariant of the vertex is assigned to the *l*-coordinate of the vector, where *l* is the layer the vertex occurs. Once invariant vectors are computed for all atoms, duplicated vectors are removed, the vectors are sorted in decreasing order, and the atom invariant becomes the order of the atom's vector in the sorted list (cf. Figure 2.7e). The above process is repeated until the number of atom invariants remains constant. Note that at each iteration, the number of invariants increases. Indeed, atom invariants are computed from vertex invariants, which in turn are computed from the atom invariants from the previous iteration. Because the number of invariants is at most the number of atoms, the process cannot repeat itself more than *N* times. The invariant computation algorithm is given next.

ALGORITHM 2.4 SIGNATURE INVARIANT COMPUTATION

```
invariant-vertex(T(x),relative)
Input:  T(x) the signature-tree of atom x
        relative is a parent or child relationship
Output: Updated vertices invariants

List-V_inv = ø
For all layers l of T(x)
    For all vertices v of layer l
        V_inv(v)=inv(atom(v)),{inv(w) s.t. w is a relative of v}
        List-V_inv= List-V_inv + V_inv(v)
    done
    sort List-V_inv in decreasing order
    For all vertices v of layer l
        inv(v)=order of V_inv(v) in List-V_inv
    done
 done
```

```
invariant-atom(T(x),G)
Input:  T(x) the signature-tree of atom x
        G a molecular graph
Output: Updated atoms invariants

Repeat
      invariant-vertex(T(x),child)
      invariant-vertex(T(x),parent)
      For all vertices v of T(x)
         let l be the layer of v
         V_inv(atom(v))(l)= inv(v)
      done
      List-V_inv = ∅
      For all atoms a of G do
         List-V_inv= List-V_inv + V_inv(a)
      done
      sort List-V_inv in decreasing order
      For all atom a of G
         inv(a)= order of V_inv(a) in List-V_inv
      done
until the number of invariant values remain constant
```

The canonization algorithm (Algorithm 2.5) given below and illustrated in Figure 2.8 first computes the invariants for all atoms running the above invariant computation algorithms (step 1). Then in step 2, the atoms are partitioned into orbits such that all the atoms in a given orbit have the same invariant. In the next step (step 3) one searches for an orbit containing atoms with at least two parents in $T(x)$. Note that these atoms have different invariants than atoms with only one parent, since the initial atom invariants embrace the number of parents. When several such orbits exist, those with the maximum number of atoms are selected, and if several orbits have the same number of atoms, one takes the one with the minimum invariant. In the case of Figure 2.7, orbit {5,6} has two atoms, each atom having more than one parent; this orbit is thus selected. If no orbit can be found, or the selected orbit contains only one atom, then the process ends and the signature is printed (steps 4–9). When an

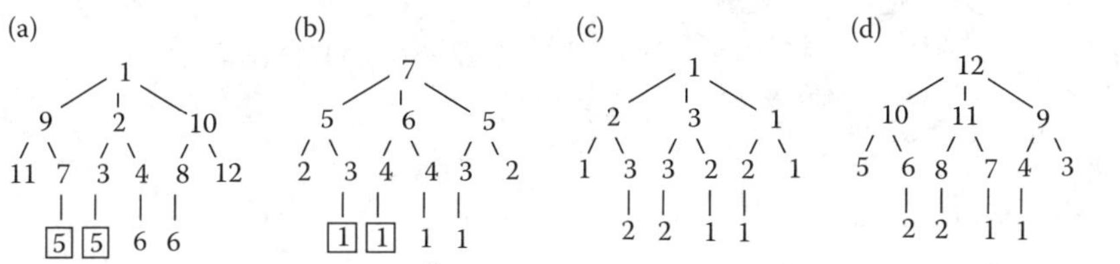

FIGURE 2.8 Signature-canonization algorithm. (a) Signature tree of atom 1 in Figure 2.6a, where atom 5 is labeled. (b) The same signature tree with the invariants computed in Figure 2.7e. (c) Vertex invariants after running the invariant-vertex algorithm where atom 5 is labeled. (d) Vertex invariants after running the invariant-atom algorithm where atom 5 is labeled. Note that every orbit contains only one atom and the canonization algorithm thus stops after labeling atom 5.

orbit is found containing more than one atom, an atom is arbitrarily selected from that orbit and a label is added to its invariant (steps 10–14). The canonization algorithm is run again in a recursive manner and other atoms may be labeled if other orbits with multiple atoms and multiple parents are found. In Figure 2.8, atom 5 is labeled 1. The algorithm stops after the next iteration since each atom becomes singularized in its own orbit. Initially, all atoms are unlabeled (label 0). The first labeled atom is labeled 1 and labels are incremented by 1 each time the algorithm calls itself (step 12). Note that each time an atom is labeled, at the next iteration the atom will be alone in its orbit. Since there are no more than N atoms to be labeled, the algorithm cannot call itself more than N times.

ALGORITHM 2.5 SIGNATURE CANONIZATION

```
canonize-signature(T(x),G,l,Smax)
Input:  T(x) the signature-tree of atom x
        G a molecular graph
        l a label
Output: Smax a canonical string (initialized to
  empty string)

01.  invariant-atom(T(x),G)
02.  partition the atoms of G into orbits according to
     their invariants
03.  let O be the orbit with the maximum number of atoms
     and the minimum invariant value such that all the
     atoms of O have at least two parents.
04.  if |O| ≤ 1 then
05.    label all unlabeled atoms having two parents
       according to their invariant
06.    S = print-signature-string(T(x))
07.    if (S > Smax) Smax = S fi
08.    return Smax
09.  fi
10.  for all atom a in O do
11.    label(a) = l
12.    S= canonize-signature(T(x),G,l+1,Smax)
13.    if (S>Smax) Smax = S fi
14.    label(a)=0
15.  done
16.  return Smax
```

Like SMILES strings, signature strings are printed reading the signature tree in a depth-first order. Prior to printing signature strings, the children of all vertices are sorted according to their invariants taken in decreasing order. In order to avoid printing several times duplicated subtrees, any subtree is printed only the first time it is read. This operation requires maintaining a list of printed edges. The algorithm is detailed in Faulon et al. [50] and depicted in Figure 2.9.

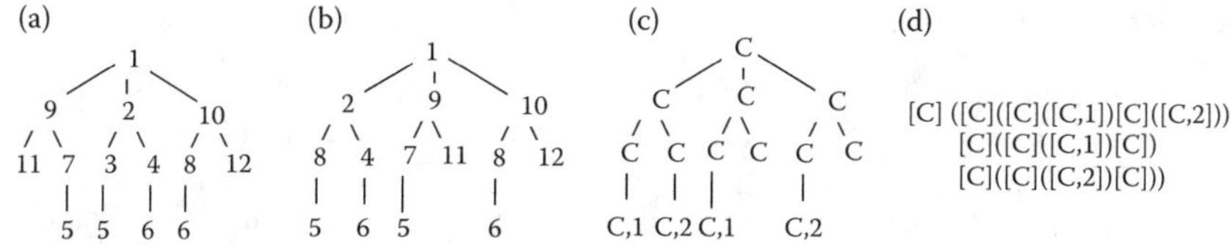

FIGURE 2.9 Printing the signature string. (a) Signature tree of atom 1 in Figure 2.6a, where atom 5 has been labeled. (b) The same signature tree with branches reordered according to atom invariants computed in 2.8d. (c) Signature tree with atom types and labels. Atom 5 is labeled 1; other atoms represented more than one time in the signature tree (atom 6) are labeled in the order they appear reading the tree in a depth-first order (atom 6 is thus labeled 2). (d) The corresponding signature string is printed reading the tree in a depth-first order.

As discussed in Faulon et al. [50], the calculation of invariants based on signature turns out to be powerful, at least for molecular graphs. Indeed for most chemicals there is no need to introduce labels. So far the worst-case scenario for the signature-based algorithm has been found with projective planes for which four labels needed to be introduced; even in that case the algorithm was run no more than $O(N^4)$ times. While the algorithm handles heteroelements and multiple bonds well, it does not yet take stereochemistry into account.

2.4 CONCLUDING REMARKS

We have presented in this chapter the classical formats used to represent 2D chemical structures in chemoinformatics databases. We have also addressed issues that arise when storing chemical structures, such as representations of alternative bonds and perception of tautomers. Because chemicals are usually represented in the classical form of 2D diagrams, we have outlined algorithms that generate these diagrams from linear notations and connection tables. One main issue that has been the source of many algorithms published in the literature is the uniqueness of the representation. Indeed, to store and retrieve chemicals from chemoinformatics databases, one needs a unique (and standard) representation. This issue is generally dealt with in canonical labeling algorithms, which are reviewed in Section 2.3. One outstanding problem that has not been covered in the chapter but is still an active field of research is the development of efficient algorithms to search for substructures; this problem is related to the subgraph isomorphism, which is generally harder to solve than canonical labeling.

ACKNOWLEDGMENTS

The authors would like to thank Ovidiu Ivanciuc for providing us with relevant literature references. The authors also acknowledge the reprinted with permission from Dittmar et al. *Journal of Chemical Information and Computer Sciences* 1977, 17(3), 186–192. American Chemical Society, Copyright (1977).

REFERENCES

1. Weininger, D., SMILES, a chemical language and information system. 1. Introduction to methodology and encoding rules. *Journal of Chemical Information and Computer Sciences* 1988, 28(1), 31–36.
2. Weininger, D., SMILES—a language for molecules and reactions. In: J. Gasteiger (Ed.), *Handbook of Chemoinformatics*, Vol. 1, pp. 80–102. Wiley-VCH: Weinheim, Germany, 2003.
3. Wiswesser, W. J., How the WLN began in 1949 and how it might be in 1999. *Journal of Chemical Information and Computer Sciences* 1982, 22(2), 88–93.
4. Wiswesser, W. J., Historic development of chemical notations. *Journal of Chemical Information and Computer Sciences* 1985, 25(3), 258–263.
5. Ash, S. Cline, M. A., Homer, R. W., Hurst, T., and Smith, G. B., SYBYL line notation (SLN): A versatile language for chemical structure representation. *Journal of Chemical Information and Computer Sciences* 1997, 37(1), 71–79.
6. Homer, R. W., Swanson, J., Jilek, R. J., Hurst, T., and Clark, R. D., SYBYL line notation (SLN): A single notation to represent chemical structures, queries, reactions, and virtual libraries. *Journal of Chemical Information and Modeling* 2008, 48(12), 2294–2307.
7. Wisniewski, J. L., Chemical nomenclature and structure representation: Algorithmic generation and conversion. In: J. Gasteiger (Ed.), *Handbook of Chemoinformatics*, Vol. 1, pp. 51–79. Wiley-VCH: Weinheim, Germany, 2003.
8. Downs, G. M., Gillet, V. J., Holliday, J. D., and Lynch, M. F., Review of ring perception algorithms for chemical graphs. *Journal of Chemical Information and Computer Sciences* 1989, 29(3), 172–187.
9. Downs, G. M., Ring perception. In: J. Gasteiger (Ed.), *Handbook of Chemoinformatics*, Vol. 1, pp. 161–177. Wiley-VCH: Weinheim, Germany, 2003.
10. Bangov, I. P., Topological structure generators. In: J. Gasteiger (Ed.), *Handbook of Chemoinformatics*, Vol. 1, pp. 178–194. Wiley-VCH: Weinheim, Germany, 2003.
11. Weininger, D., Combinatorics of organic molecular structures. In: J. Gasteiger (Ed.), *Handbook of Chemoinformatics*, Vol. 1, pp. 195–205. Wiley-VCH: Weinheim, Germany, 2003.
12. Lameijer, E.-W., Kok, J. N., Bäck, T., and IJzerman, A. P., The molecule evoluator. An interactive evolutionary algorithm for the design of drug-like molecules. *Journal of Chemical Information and Modeling* 2006, 46(2), 545–552.
13. Chen, L., Reaction classification and knowledge acquisition. In: J. Gasteiger (Ed.), *Handbook of Chemoinformatics*, Vol. 1, pp. 348–388. Wiley-VCH: Weinheim, Germany, 2003.
14. Ivanciuc, O., Topological indices. In: J. Gasteiger (Ed.), *Handbook of Chemoinformatics*, Vol. 3, pp. 981–1003. Wiley-VCH: Weinheim, Germany, 2003.
15. Melville, J. L., Riley, J. F., and Hirst, J. D., Similarity by compression. *Journal of Chemical Information and Modeling* 2007, 47(1), 25–33.
16. Karwath, A. and De Raedt, L., SMIREP: Predicting chemical activity from SMILES. *Journal of Chemical Information and Modeling* 2006, 46(6), 2432–2444.
17. Vidal, D., Thormann, M., and Pons, M., LINGO, an efficient holographic text based method to calculate biophysical properties and intermolecular similarities. *Journal of Chemical Information and Modeling* 2005, 45(2), 386–393.
18. Wiswesser, W. J., 107 years of line-formula notations (1861–1968). *Journal of Chemical Documentation* 1968, 8, 146–150.
19. Walker, S. B., Development of CAOCI and its use in ICI plant protection division. *Journal of Chemical Information and Computer Sciences* 1983, 23, 3–5.

20. Dalby, A., Nourse, J. G., Hounshell, W. D., Gushurst, A. K. I., Grier, D. L., Leland, B. A., and Laufer, J., Description of several chemical-structure file formats used by computer-programs developed at molecular design limited. *Journal of Chemical Information and Computer Sciences* 1992, 32(3), 244–255.
21. Murray-Rust, P. and Rzepa, H. S., Chemical markup, XML, and the worldwide web. 1. Basic principles. *Journal of Chemical Information and Computer Sciences* 1999, 39(6), 928–942.
22. Murray-Rust, P. and Rzepa, H. S., Chemical markup, XML, and the world wide web. 4. CML schema. *Journal of Chemical Information and Computer Sciences* 2003, 43(3), 757–772.
23. Murray-Rust, P. and Rzepa, H. S., XML and its applications in chemistry. In: J. Gasteiger (Ed.), *Handbook of Chemoinformatics*, Vol. 1, pp. 466–490. Wiley-VCH: Weinheim, Germany, 2003.
24. Dittmar, P. G., Mockus, J., and Couvreur, K. M., An algorithmic computer graphics program for generating chemical-structure diagrams. *Journal of Chemical Information and Computer Sciences* 1977, 17(3), 186–192.
25. Wipke, W. T. and Dyott, T. M., Use of ring assemblies in a ring perception algorithm. *Journal of Chemical Information and Computer Sciences* 1975, 15(3), 140–147.
26. Shelley, C. A., Heuristic approach for displaying chemical structures. *Journal of Chemical Information and Computer Sciences* 1983, 23(2), 61–65.
27. Morgan, H. L., The generation of a unique machine description for chemical structures—a technique developed at chemical abstracts service. *Journal of Chemical Documentation* 1965, 5, 107–113.
28. Helson, H. E., Structure diagram generation. *Reviews in Computational Chemistry* 1999, 13, 313–398.
29. Weininger, D., SMILES. 3. DEPICT. Graphical depiction of chemical structures. *Journal of Chemical Information and Computer Sciences* 1990, 30(3), 237–243.
30. Fricker, P. C., Gastreich, M., and Rarey, M., Automated drawing of structural molecular formulas under constraints. *Journal of Chemical Information and Computer Sciences* 2004, 44(3), 1065–1078.
31. Stierand, K., Maaß, P. C., and Rarey, M., Molecular complexes at a glance: Automated generation of two-dimensional complex diagrams. *Bioinformatics* 2006, 22(14), 1710–1716.
32. Clark, A. M., Labute, P., and Santavy, M., 2D structure depiction. *Journal of Chemical Information and Modeling* 2006, 46(3), 1107–1123.
33. Roos-Kozel, B. L. and Jorgensen, W. L., Computer-assisted mechanistic evaluation of organic reactions. 2. Perception of rings, aromaticity, and tautomers. *Journal of Chemical Information and Computer Sciences* 1981, 21(2), 101–111.
34. Mockus, J. and Stobaugh, R. E., The chemical abstracts service chemical registry system. 7. Tautomerism and alternating bonds. *Journal of Chemical Information and Computer Sciences* 1980, 20(1), 18–22.
35. Harańczyk, M. and Gutowski, M., Quantum mechanical energy-based screening of combinatorially generated library of tautomers. TauTGen: A tautomer generator program. *Journal of Chemical Information and Modeling* 2007, 47(2), 686–694.
36. Todorov, N. P., Monthoux, P. H., and Alberts, I. L., The influence of variations of ligand protonation and tautomerism on protein-ligand recognition and binding energy landscape. *Journal of Chemical Information and Modeling* 2006, 46(3), 1134–1142.
37. Milletti, F., Storchi, L., Sforna, G., Cross, S., and Cruciani, G., Tautomer enumeration and stability prediction for virtual screening on large chemical databases. *Journal of Chemical Information and Modeling* 2009, 49(1), 68–75.

38. Oellien, F., Cramer, J., Beyer, C., Ihlenfeldt, W.-D., and Selzer, P. M., The impact of tautomer forms on pharmacophore-based virtual screening. *Journal of Chemical Information and Modeling* 2006, 46(6), 2342–2354.
39. Simmons, E. S., Markush structure searching over the years. *World Patent Information* 2003, 25, 195–202.
40. Read, R. C. and Corneil, D. G., The graph isomorphism disease. *Journal of Graph Theory* 1977, 1, 339–363.
41. Miller, G., Graph isomorphism, general remarks. *Journal of Computer and System Sciences* 1979, 18, 128–142.
42. Carhart, R. E., Erroneous claims concerning the perception of topological symmetry. *Journal of Chemical Information and Computer Sciences* 1978, 18, 108–110.
43. Rucker, G. and Rucker, C., Computer perception of constitutional (topological) symmetry: TOPSYM, a fast algorithm for partitioning atoms and pairwaise relations among atoms into equivalent classes. *Journal of Chemical Information and Computer Sciences* 1990, 30, 187–191.
44. Rucker, G. and Rucker, C., On using the adjacency matrix power method for perception of symmetry and for isomorphism testing of highly intricate graphs. *Journal of Chemical Information and Computer Sciences* 1991, 31, 123–126.
45. Babai, L., Erdos, P., and Selkow, S. M., Random graph isomorphism. *SIAM Journal of Computing* 1980, 9, 628–635.
46. Wipke, W. T. and Dyott, T. M., Steorchemically unique naming algorithm. *The Journal of the American Chemical Society* 1974, 96, 4825–4834.
47. Weininger, D., Weininger, A., and Weininger, J. L., SMILES. 2. Algorithm for generation of unique SMILES notation. *Journal of Chemical Information and Computer Sciences* 1989, 29(2), 97–101.
48. Stein, S. E., Heller, S. R., and Tchekhovskoi, D. V., *The IUPAC Chemical Identifier—Technical Manual*. NIST: Gaithersburg, MD, 2006.
49. McKay, B. D., Practical graph isomorphism. *Congressus Numerantium* 1981, 30, 45–87.
50. Faulon, J. L., Collins, M. J., and Carr, R. D., The signature molecular descriptor. 4. Canonizing molecules using extended valence sequences. *Journal of Chemical Information and Computer Sciences* 2004, 44(2), 427–436.
51. Hopcroft, J. E. and Tarjan, R. E., Isomorphism of planar graphs. In: R. E Miller and J. W. Thatcher (Eds), *Complexity of Computer Computations*, pp. 131–150. Plenum Press: New York, 1972.

3 Three-Dimensional (3D) Molecular Representations

Egon L. Willighagen

CONTENTS

3.1 INTRODUCTION

Three-dimensional (3D) molecular representation is at the heart of modern chemistry. The past decades have taught us that pure graph-oriented representations are typically not enough to understand the interactions of molecules with their environments. The 3D molecular geometry has a strong effect on molecular binding, as clearly seen in ligand–protein interactions and packing in crystal structures.

Understanding molecular properties requires us to understand the geometrical features of the molecule. For example, the molecular geometries of a molecule and its surroundings determine the proximity of functional groups and, therefore, why certain

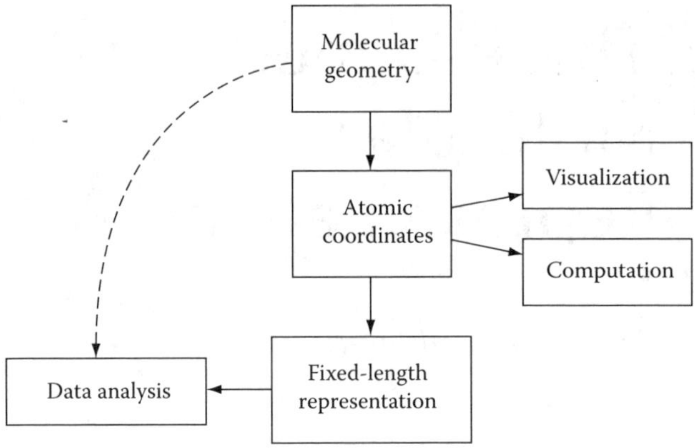

FIGURE 3.1 The raw coordinates of a molecular geometry are suited for visualization and computational studies, such as geometry optimization and energy calculations. However, because they do not provide a uniform-length representation, they do not lend itself for data analysis and pattern recognition.

molecules show strong binding affinity, due to, for example, salt and hydrogen bridges and hydrophobic interactions. Only a brief reminder is needed here that the geometry is not static and that binding affinity often involves induced fit. Visual exploration of geometries is well established in various fields of chemoinformatics, and free tools are abundant. Jmol [1] and PyMol [2] are the best-known open-source applications in this area.

Dealing with 3D geometries in computation, however, is more complex (Figure 3.1). A program does not have the *visual* interpretation of depth or orientation. In order to have an analysis tool to understand these patterns, the patterns need to be expressed numerically. Depth can be represented as a Euclidean distance, which, depending on the application, might be a relative distance or distance ratio. Orientation is even more complex and it involves a coordination reference to which the orientation can be measured, something that can be easily done visually. This brings us to the topic of this chapter: how to represent 3D molecular geometries such that they are useful for analysis and computation.

The molecular structure is, ultimately, governed by the quantum mechanics of the electrons that are organized in atomic and molecular orbitals. This quantum molecular structure defines all molecular properties, including the geometry and chemical reactivity. However, quantum mechanics is for many supramolecular systems too computation intensive, and simpler representations are needed to deal with the fast molecular space we nowadays work with. This simpler 3D representation typically involves an atom-and-bond representation and combines a chemical graph with the geometry information. Instead of the many electrons involved in the molecule, it focuses only on the nuclei and their coordinates. Electronic effects on the geometry are implicitly captured by the coordinates, but can be complemented with atom-type information, which typically includes hybridization information. This is the basic model behind the force field approaches.

Other applications, however, need a different representation. The above-sketched representation still increases in size with the number of atoms and bonds. However, numerical analyses in quantitative structure–activity relationship (QSAR) studies, which correlate geometrical features with binding affinity, often require a fixed-length representation that is independent of the number of atoms and bonds.

This chapter discusses the representation of molecular geometry in various coordinate systems, how to interchange those representations, and how fixed-length, numerical representations may be derived from them.

3.2 COORDINATE SYSTEMS

Three atomic coordinate systems are commonly used: Cartesian coordinates, internal coordinates, and notional coordinates. The last is specific for crystallography data and describes both the molecular geometry as well as the crystal lattice. Internal coordinates rely on and use the chemical graph and therefore aim at single, connected molecules. Cartesian coordinates are the most versatile and are typically used for disconnected 3D structure.

3.2.1 Cartesian Coordinates

Cartesian coordinates describe the atomic coordinates relative to the origin. The X, Y, and Z axes are orthogonal and Euclidean distances can be used to measure distances between atoms. Orientation and placement with respect to the origin is arbitrary.

The Cartesian coordinates for ethanol shown in Figure 3.2 are as follows:

O	1.94459	1.33711	0.00000
C	1.52300	0.00000	0.00000
C	0.00000	0.00000	0.00000
H	1.93156	−0.49598	0.90876
H	1.93156	−0.49598	−0.90876
H	−0.35196	1.05588	0.00000
H	−0.35196	−0.52794	−0.91442
H	−40.35196	−0.52794	0.91442
H	1.18187	1.88994	0.00000

Distances, angles, and torsions are easily calculated from Cartesian coordinates, as well as many other derived properties, such as molecular volume, total polar surface area, and so on (see Section 3.5.1). For example, the molecular center-of-mass may be placed on the origin, so that molecules are located in the same location. Algorithm 3.1 describes the algorithm to calculate a molecule's center-of-mass. Centering the molecule around the origin is then done by subtracting the coordinates of the center-of-mass from the atomic coordinates.

ALGORITHM 3.1 ALGORITHM TO CALCULATE THE MOLECULAR CENTER-OF-MASS

```
sum.x = 0
sum.y = 0
sum.z = 0
```

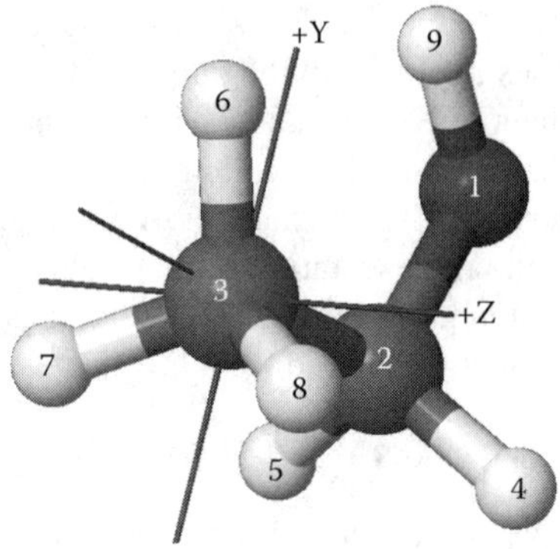

FIGURE 3.2 3D model of ethanol labeled by position in the input file.

```
total.weight=0
iterate over all atoms {
  sum.x = sum.x + atom.x * atom.weight
  sum.y = sum.y + atom.y * atom.weight
  sum.z = sum.z + atom.z * atom.weight
  total.weight = total.weight + atom.weight
}
com.x = sum.x / total.weight
com.y = sum.y / total.weight
com.z = sum.z / total.weight
```

However, although Cartesian coordinates are universal, they are not always the best choice with respect to computation times or algorithm simplicity. For geometry optimization calculations, internal coordinates are more suitable, requiring less computation to reach the same results. Tomczak reports an approximately fourfold speed using internal coordinates over Cartesian coordinates [3].

3.2.2 Internal Coordinates

Internal coordinates describe the atomic coordinates in an internal frame, that is, without an external reference. They describe the molecular geometry in terms of distances between atoms and angles and torsions between bonds. This closely overlaps with force field approaches where the molecular energy is expressed in terms of bond length, angles, and torsions, defining a well-structured search space for geometrical optimization. Many molecular dynamics and quantum mechanics algorithms take advantage of this representation.

The internal coordinates for ethanol shown in Figure 3.2 are given below. The atomic numbering is the same as for the list of Cartesian coordinates and is shown in Figure 3.2 too.

These coordinates are interpreted as follows. The first distance given (1.4020 Å) is between atom 2 (carbon) and atom 1 (oxygen), while the second distance (1.5230 Å) is

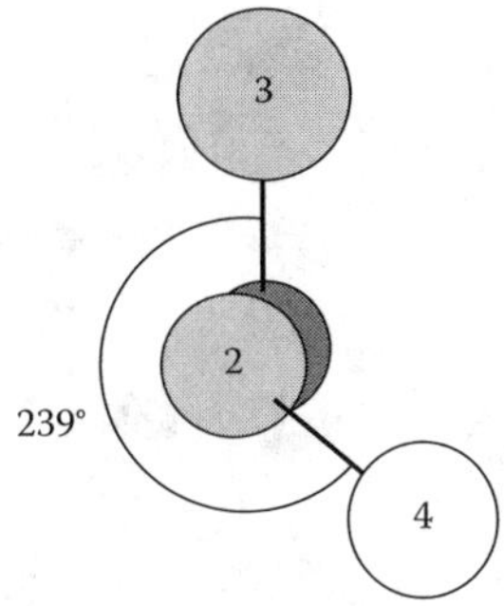

FIGURE 3.3 The torsion angle for atoms 4, 2, 1, and 3 in ethanol (see Figure 3.2) is defined as the angle between a vector through atoms 4 and 2 and the vector through atoms 1 and 3, as measured in a plane perpendicular to the vector through atoms 2 and 1. (Note that atom 1 is depicted behind atom 2.) The lines between atoms are not bonds; in fact, atom 3 is bonded to atom 2 and not to atom 1. However, the vector between atom 4 and atom 2 does coincide with an actual bond.

the bond length of the carbon–carbon bond. The first angle given (107.50°) is between the bonds between the third and second atoms and the second and first atoms. The first torsion angle (239.34°) is the angle between the two lines, one between atom 4 and atom 2 and the other between atom 1 and atom 3, as measured in a plane perpendicular to the bond between atom 1 and atom 2 (Figure 3.3). (Note that atom 1 is located behind atom 2 in this figure.) These lines do not necessarily have to coincide with bonds.

3.2.3 Fractional Coordinates

Fractional coordinates describe the positions of the atoms as fractions of the axes of the crystal's unit cell, which is described by its crystallographic axes *A*, *B*, and *C*. There are two common ways to describe these three axes themselves: as a vector in Cartesian space with nine values, or with six values listing the axes' lengths and the angles between the axes, sometimes referred to as the notional axes. Figure 3.4 shows the unit cell of the cubic unit cell of sodium chloride. The unit cell axes can be described as in notional axes 5.6, 5.6, 5.6 Å and 90°, 90°, 90°, describing the axis lengths and the angles between them, respectively.

Alternatively, the axes can be described as vectors in Euclidean space. This leaves a choice of how to rotate the unit cell in Euclidean space. If we fix the *A* axis on the *x* axis and the *B* axis in the *XY* plane, then rotation in the Euclidean space is fixed. Using this convention, the unit cell axis vectors for the sodium chloride example are $A = 5.6, 0, 0$, $B = 0, 5.6, 0$, and $C = 0, 0, 5.6$. If angles deviate with 90°, then only the *A* axis will be parallel to an Euclidean axis.

The coordinates of atoms in the unit cell are expressed as fractions of the axes *A*, *B*, and *C*. The fractional coordinates of the four sodium atoms in the shown unit cell are 0, 0, 0, 0.5, 0.5, 0, 0, 0.5, 0.5, and 0.5, 0, 0.5. The chloride ions are located at 0.5, 0, 0, 0.0, 0.5, 0, 0, 0.0, 0.5, and 0.5, 0.5, 0.5.

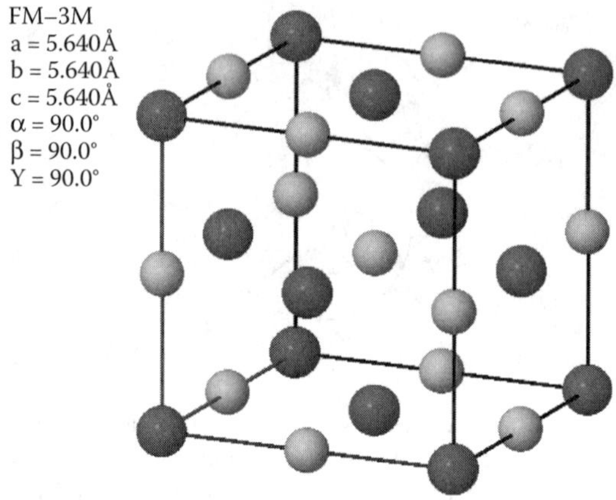

FIGURE 3.4 Unit cell of sodium chloride with the three unit cell axes starting from the origin in the lower left corner of the cube. The notional coordinates of this unit cell are defined by the *A*, *B*, and *C* axis lengths (all 5.6 Å) and the three angles α, β, and γ (all 90) between *B* and *C*, *A* and *C*, and *A* and *B*, respectively.

3.2.4 Two-Dimensional Chemical Diagrams

The fourth coordinate system is very common in chemistry: two-dimensional (2D) chemical diagrams. These diagrams are aimed at graphical visualization of the connection table and typically focus on depiction of atom and bond properties, such as isotope and charge details for atoms, and bond properties like bond order, delocalization, and stereochemistry. This 2D coordinate space is outside the scope of this chapter. It is mentioned here, however, because 2D diagrams are often the input in algorithms that create 3D molecular structures.

These algorithms create 3D Cartesian coordinates from the information presented in 2D molecular representations. Primarily, this information includes the connection table, and atom- and bond-type information. However, to properly reflect stereochemistry features presented in the 2D diagrams, the algorithm has to resolve such information often from wedge bond representations, and 2D coordinates for *cis*/*trans* isomorphism. Additionally, coordination generation for ring systems can use a template library that may or may not contain information on the layout of the attachment points to assemble the geometries of ring and nonring systems. The general concept is given in Algorithm 3.2.

ALGORITHM 3.2 ALGORITHM TO CREATE 3D GEOMETRIES FROM 2D DIAGRAMS

```
extract connection table
derive atom parities from wedge bond and 2D coordinates
   information
derive cis/trans isomorphism from 2D coordinates
isolate ring systems, and look up 3D coordinates from a
   template library
```

```
apply common geometries for non-ring substructures
   taking into account stereochemistry
```

3.3 INTERCONVERTING COORDINATE SYSTEMS

Interconversion between the three coordinate systems is important, because algorithms can perform differently depending on the chosen system, as was discussed earlier. Algorithms to interconvert coordinate systems are abundant, but may differ in detail between implementations. This section discusses two algorithms: conversion of internal coordinates into Cartesian coordinates and conversion of fractional coordinates into Cartesian coordinates.

3.3.1 Internal Coordinates into Cartesian Coordinates

Converting internal coordinates into Cartesian coordinates is fairly straightforward: each next atom is placed into Euclidean space to conform the internal coordinates converted so far. The algorithm has two degrees of freedom: (1) in which Cartesian coordinate the first atom is placed and (2) in which plane the first two bonds are located. The algorithm description given in Algorithm 3.3 puts the first atom at the origin of the coordinate system, the first bond along the x axis, and the second bond in the xy plane.

ALGORITHM 3.3 ALGORITHM TO CONVERT INTERNAL COORDINATES INTO CARTESIAN COORDINATES. ATOM NUMBERING FOLLOWS THOSE FROM TABLE 3.1

```
let the first line define:
   the first atom
then:
   put the first atom at {0,0,0}
let the second line define:
   a new first atom, and a second atom
   a distance to a second atom
then:
   d = distance (first atom, second atom)
   put the second atom on the X axis at {d,0,0}
let the third line define:
   new first and second atoms, and a third atom
   a distance to a second atom,
   an angle between the first, second and third atom on
   this line
then:
   d = distance (first atom, second atom)
   α = angle (first atom, second atom, third atom)
   put the third atom in the XY plane, such that:
      the distance to the second atom is d, and
```

TABLE 3.1
Internal Coordinates for Ethanol Shown in Figure 3.2

Number	Element		Distance		Angle		Torsion
1	O						
2	C	1	1.4020				
3	C	2	1.5230	1	107.50		
4	H	2	1.1130	1	108.34	3	239.34
5	H	2	1.1130	1	108.34	3	120.66
6	H	3	1.1130	2	108.43	1	0.00
7	H	3	1.1130	2	108.43	1	120.00
8	H	3	1.1130	2	108.43	1	240.00
9	H	1	0.9420	2	108.44	3	0.00

```
        the angle between first, second and third
        atom is α
    let the fourth and all later lines define:
        new first, second, and third atoms, and a fourth
        atom,
        a distance between the first and second atom,
        the angle between the first, second and third
        atom, and
        the torsion between the first, second, third, and
        fourth atom
    then:
        d = distance (first atom, second atom)
        α = angle (first atom, second atom, third atom)
        t = torsion (first atom, second atom, third atom,
            fourth atom)
        put the first in euclidean space, such that:
        the distance to the second atom is d,
        the angle between first, second and third
        atom is α
        the torsion is defined by t
```

3.3.2 Fractional Coordinates into Cartesian Coordinates

Converting fractional coordinates into Cartesian coordinates can in the simplest way be performed as a matrix operation:

$$\begin{pmatrix} x' \\ y' \\ z' \end{pmatrix} = \begin{pmatrix} a & b \cdot \cos\gamma & c \cdot \cos\beta \\ 0 & b \cdot \sin\gamma & c(\cos\alpha - \cos\beta \cdot \cos\gamma)/\sin\gamma \\ 0 & 0 & V/(a \cdot b \cdot \sin\gamma) \end{pmatrix} \begin{pmatrix} x \\ y \\ z \end{pmatrix}, \tag{3.1}$$

FIGURE 3.5 Acetylcholinesterase (PDB code: 1ACJ) with tacrine (*InChI=1S/C13H14N2/c14-13-9-5-1-3-7-11(9)15-12-8-4-2-6-10(12)13/h1,3,5,7H,2,4,6,8H2,(H2,14,15)*) the active site. Visualized with Jmol.

with a, b, and c being the length of the crystallographic axes A, B, and C, and α, β, and γ the angles between B and C, A and C, and A and B, respectively, and V the volume of the unit cell, defined as

$$V = abc\sqrt{1 - \cos^2\alpha - \cos^2\beta - \cos^2\gamma + 2\cos\alpha\cos\beta\cos\gamma}. \qquad (3.2)$$

3.4 COMPARING GEOMETRIES

There are many applications where comparing the 3D structure of molecules is interesting; molecular docking is likely the most common one. In such studies the molecule is oriented in the active site of an enzyme or receptor (Figure 3.5). CoMFA studies binding affinities and assumes that molecules are overlayed and oriented in a similar chemical direction, reflecting similar binding modes with protein (see Section 3.5.2).

Comparing two or more molecular 3D geometries is generally not directly possible: the geometries do not share a common reference origin, and they may not be oriented in the same direction. The center-of-masses of the molecules may be far apart, and the structures can be differently aligned. The first can be addressed by putting the center-of-mass of each molecule in the origin of the coordinate system.

It does not, however, orient the molecule in any particular way. While the center-of-mass is in the origin, the molecular conformer can still be oriented in any direction. To address this molecule, one may apply principal component analysis (PCA) and orient the molecule such that the first three latent variables are oriented along the X, Y, and Z axes as described in Algorithm 3.4.

ALGORITHM 3.4 ALGORITHM TO ALIGN CHEMICAL STRUCTURES BASED ON ANISOMORPHISM

```
for each molecule :
   calculate the three PCs from the 3D coordinates
   for each atom:
      the new x coordinate is the score on PC1
```

```
      the new y coordinate is the score on PC2
      the new z coordinate is the score on PC3
overlay the molecules in the new coordinate space
```

The above-sketched algorithm does not take into account structural similarity between molecules, but only looks at the anisomorphism of the structures. That is, the variance in atomic coordinates is used to create new coordinates. Practically, this means that each molecule is reoriented such that the direction in which the molecule is longest, and thus has the highest variance, is aligned with the first principal component (PC1).

Instead, it is often desirable to orient the molecules based on the maximal common substructure (MCSS). For example, alignment of a series of steroids is expected to overlay the sterane skeletons. The alignment must not be disturbed by large side chains that change the overall anisomorphism of the geometry: the variance of the 3D coordinates would change and the alignment too. Using the MCSS, the alignment of the two molecules becomes more, in agreement with what one would expect. Most chemoinformatics toolkits have the means to either find the maximum common substructure or to identify a user-defined substructure using a query language like molecular query language (MQL) [4] or SMARTS.

After having identified the MCSS of the molecular geometries, the full structures can be rotated in the coordinates space to minimize the *root mean square deviation* (RMSD) of the coordinates of the shared substructure (Algorithm 3.5).

ALGORITHM 3.5 ALGORITHM TO ALIGN 3D MOLECULAR STRUCTURES BASED ON THE COMMON SUBSTRUCTURE

```
find the maximal common substructure (MCSS)
find a rotation that minimizes the RMSD of the atomic
   coordinates of the MCSS
```

3.5 FIXED-LENGTH REPRESENTATIONS

One disadvantage of representation in any of the three discussed coordinates systems is that the size depends on the number of atoms. Many chemometrical modeling methods, however, require a numerical and fixed-length vector representation of the molecular structure [5,6]. The above representations do not fulfill this requirement, and hence derived descriptors have been and still are being developed to bridge the gap between those representations and the mathematical modeling methods. These descriptors allow statistical modeling and analysis with, for example, classical methods like PCA, partial least squares (PLS) and neural networks (NNs) and classification methods like linear discriminant analysis (LDA). Only very few methods, such as classification and regression trees (CART), do not require a numerical representation. Distance-based clustering, for example, can work directly with an MCSS-based distance matrix in which two molecules that have a large substructure in common have a smaller distance and are considered more alike.

The *Handbook of Molecular Descriptors* published in 2000 [7] gives a broad overview of known molecular descriptors. Depending on the information content, descriptors are usually classified as 0D, 1D, 2D, and 3D descriptors. The last category, 3D, takes into account the 3D geometry of the molecule. Recently, a fifth category has been proposed: 4D descriptors for which several different but related definitions have been given. Todeschini defines the 4th dimension to describe the interaction field of the molecule [7], while others reserve this dimension to describe its conformations [8]. The latter takes into account the flexibility of molecules, where coordinate systems only treat the molecules as rigid bodies.

3.5.1 Molecular Descriptors

To illustrate how molecular descriptors convert the variable-length 3D molecular geometries into a fixed-length representation, two descriptor algorithms are described in this section. It is important to realize that the representation not only needs to be of fixed length, but also needs to be orientation independent. That is, the descriptor value must not change when the molecular geometry is rotated in coordinate space. Consequently, these descriptors are suitable for comparing molecular geometries without the need for alignment. This requirement is also the reason why these molecular descriptors typically do not describe angular features of the molecule, other than collapsed onto a single value.

3.5.1.1 The Length-over-Breadth Descriptor

The length-over-breadth descriptor describes the anisomorphism of the molecule, but uses their ratio to collapse the length and breadth features into a single number (Algorithm 3.6).

ALGORITHM 3.6 ALGORITHM FOR THE LENGTH-OVER-BREADTH DESCRIPTOR

```
calculate the geometrical dimensions of the molecule
determine the length and breadth
calculate the ratio length over breadth
```

Calculation of the molecular length and breadth is quite similar to the use of PCA alignment (see Section 3.4), which rotates the molecule such that the longest molecular axis is aligned with the PC1. The length is then defined as the difference between the maximum and minimum coordinates on this axis (PC1); the breadth would be the difference on the second axis (PC2). This calculation can include the van der Waals radii of the atoms to reflect the size of the molecule as a function of its molecular surface. However, to simplify calculation, not all possible rotations are taken into account. For example, implementations may only rotate the molecular structure around a single coordinate system axis.

3.5.1.2 Charged Partial Surface Area (CPSA) Descriptors

The molecular surface area and the molecular volume are other methods to reduce the 3D geometry to a fixed-length representation. Neither of the two describe the internal geometry of the molecules, but are aimed at describing the molecular features governing intermolecular interactions. Both surface area descriptors and the molecular formula require the calculation of the molecular surface area (Algorithm 3.7). Depending on the actual surface of interest, different atomic contributions to the total surface can be used. For example, the van der Waals surface will use a smaller sphere around each atom than the solvent accessible surface.

ALGORITHM 3.7 ALGORITHM TO DETERMINE THE 3D MOLECULAR SURFACE

```
For each atom:
  use tessellation to define a sphere of points around
    the 3D coordinate of the atom
  remove all sphere points which are buried inside the
    spheres of neighboring atoms
the molecular surface is defined by the remaining points
```

The CPSA descriptor uses the atomic contributions to this surface, combined with the partial atomic charges, as the starting point to come to 25 descriptor values [9]. A full description of all values is outside the scope of this chapter and is well described in the original paper, but it is illustrative to describe the first six: partial positive surface area (PPSA), total charge weighted PPSA, atomic charge weighted PPSA, and their negative charge equivalents, namely partial negative surface area (PNSA), total charge weighted PNSA, and atomic charge weighted PNSA.

These six descriptors provide a numerical vector representation of the geometrical features of the molecule, but at the same time introduce electronic features that affect intermolecular interactions. The PPSA and PNSA use the aforementioned algorithm to determine the atomic contributions to the molecular surface area. While the PPSA only takes into account atomic contributions of atoms with a positive partial charge $\left(\sum(SA_i^+)\right)$, the PNSA only takes into account contributions from the negatively charged atoms $\left(\sum(SA_i^-)\right)$. This introduces a nice area where implementations of the general algorithm will differ in results, depending on which algorithm has been used to calculate the partial charges. For example, the original paper used an empirical method, whereas the CDK implementation of this descriptor uses Gasteiger charges. The other four descriptors are also derived from the atomic contributions, but are weighted sum of positive (Q_T^+) or negative (Q_T^-) partial charges. An overview of the six descriptor values of the CPSA descriptor is given in Table 3.2.

3.5.2 Comparative Molecular Field Analysis

That an insight into the 3D interaction of a ligand with protein cavities is important in the modeling of biochemical endpoints, such as binding affinity, became apparent and computationally feasible in the last decade. Comparative molecular field analysis (CoMFA) is the primary example of this concept [10]. The CoMFA method studies molecule–environment interaction by putting the molecules in an equidistant grid of points in 3D space. At each point, the interaction energy is calculated using a hypothetical probe, for example, using the Lennard–Jones potential function and the Coulomb potential energy function. It is important to note that because the molecules

TABLE 3.2
First Six of the 25 CPSA Descriptors, with the Formulas to Calculate them

Descriptor	Label	Formula
Partial positive surface area	PPSA	$\sum(SA_i^+)$
Partial negative surface area	PNSA	$\sum(SA_i^-)$
Total charge weighted PPSA	PPSA-2	$PPSA/Q_T^+$
Total charge weighted PNSA	PNSA-2	$PNSA/Q_T^-$
Atomic charge weighted PPSA	PPSA-3	$\sum(SA_i^+ \cot Q_i^+)$
Atomic charge weighted PNSA	PNSA-3	$\sum(SA_i^- \cot Q_i^-)$

Note: SA_i^+ and SA_i^- are the atomic contributions to the surface area for the atoms with positive and negative partial charge, respectively. Q_T^+ and Q_T^- are the sum of positive partial charges Q_i^+ and the sum of negative partial charges Q_i^-, respectively.

are aligned, the interaction similarities of the ligands can be compared by calculating the difference in the interaction energies of the matching grid points for all molecules. Afterwards, PLS is used to correlate the matrix expansion of the grid with the activity, such as ligand–target binding affinities [11,12].

CoMFA requires, however, geometrical alignment of the molecules, as discussed earlier, and only considers one conformation for each molecule, which is only a simplification of reality. Therefore, the focus has moved on to descriptors that are independent of the orientation of the molecules in its reference frame, and possibly even include information of multiple conformations. This was already acknowledged in 1997 by Hopfinger, who made a scheme which incorporated some ideas from CoMFA but which was alignment independent and took into account multiple conformations [13].

3.5.3 Radial Distribution Functions

Another common approach to remove alignment effects is to use the radial distribution function (RDF). This kind of function, as the name says, describes the distribution of certain features as a function of the distance to the central point. RDFs are particularly interesting when distance-related interactions need to be captured. The basic RDF describes the occurrence of a chemical feature at a certain distance, for example, the presence of an atom. For example, Aires-de-Sousa et al. have used five RDFs to describe the environment of protons to predict proton NMR shifts [14] and for the simulation of infrared spectra [15,16].

Figure 3.6 shows a basic spike-like RDF and the effect of smoothing with Gaussian function. This smoothing is particularly useful when a (dis)similarity between two RDFs is calculated: small displacements of the atom positions captured in the RDF will lead to large changes in the similarity between the two functions. However, when a Gaussian smoothing is used, changes in the similarity are less abrupt. Other

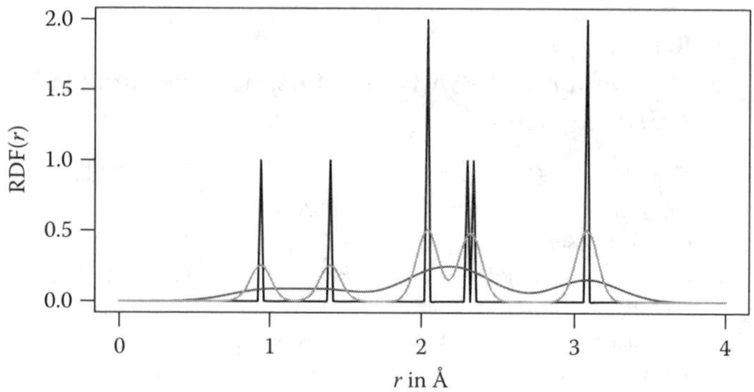

FIGURE 3.6 Three RDFs for the oxygen atom in ethanol shown in Figure 3.2. The highest-intensity, spiked RDF has no Gaussian smoothing applied; each atom contributes equally to the function. The two other RDFs are Gaussian-smoothed functions with different Gauss widths, but equal summed intensities.

approaches can be used too, and one such is used in the application described in the next section.

The algorithm for calculating an RDF for an atom in a molecule is fairly simple and is described in Algorithm 3.8. While the RDF itself is an analogous function, particularly when Gaussian smoothing is used, the function is typically digitized, for example, using binning. Given a central atom, the RDF of atoms around that atom is calculated by iterating over all atoms in the molecule, and determine where it contributes to the RDF. The amount it contributes is defined by a weighing scheme. In its simplest form, the contribution is 1 for each atom present (in black in Figure 3.6). If a Gaussian smoothing is used, then the neighboring bins are increased too, effectively convoluting the spike with a Gaussian function of selectable width (in light gray and dark gray in Figure 3.6).

ALGORITHM 3.8 ALGORITHM FOR CALCULATING AN RDF FOR AN ATOM IN A MOLECULE THAT DESCRIBES THE DISTRIBUTION OF ATOMS AROUND THAT ATOM. THE RDF CONTRIBUTION IN ITS SIMPLEST FORM IS 1, INDICATING THE PRESENCE OF AN ATOM (IN BLACK IN FIGURE 3.6)

```
determine the central atom:
for each other atom in the molecule:
    determine the distance to the central atom
    determine the corresponding RDF bin
    calculate the RDF contribution
    add this contribution to the bin
```

An interesting feature of RDFs is that they can be tuned to particular applications. The aforementioned application in NMR shift prediction uses five such customized RDFs. The contribution an atom gives to the RDF can be weighted in various ways.

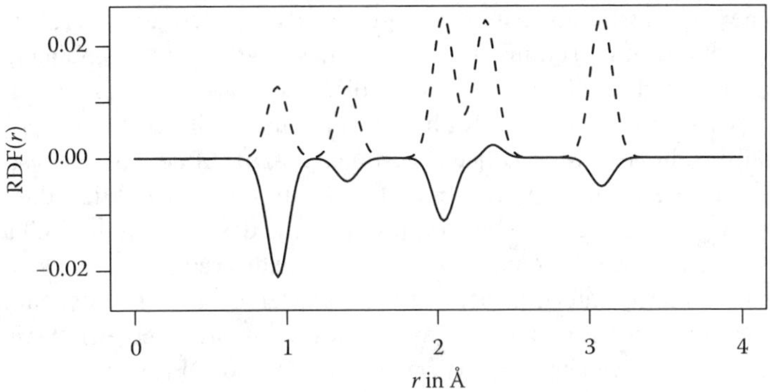

FIGURE 3.7 The coulombic interaction weighted (solid line) and non-weighted RDF (dashed line) for the oxygen atom in ethanol shown in Figure 3.2, showing the effect of the weighing scheme.

Commonly, the contribution is weighted by the distance to the central atom: the farther away from the center, the smaller the contribution. This compensates for the fact that at larger distances, each bin describes an increasing amount of spherical space.

Additionally, the contribution can be weighted by the properties of the atom that affect the contribution. For example, the coulombic interaction can be used, which represents the electronic interaction between the point charges of the atoms (Figure 3.7) and which originates from the desire to describe electronic features of the molecule. The application described in the next section of this chapter applies this approach too, where it uses RDFs to describe complete organic crystal structures.

Importantly, it should be clear that the algorithm allows for any weighting function, offering interesting flexibility in describing molecular geometries.

3.6 APPLICATION: CLUSTERING OF CRYSTAL PACKINGS

Comparing crystal structures is important in both classification and clustering problems. Classification is important for the understanding of the relation between physical properties and the underlying structure of materials. The specific packing of molecules in a crystal directly influences the physical properties of compounds. As an example, in crystal engineering, crystal packings are classified according to intermolecular interactions [17–21]. A second application of the similarity measure is in the clustering stage of *ab initio* crystal structure prediction [22,23]. In this process, hundreds or thousands of different hypothetical crystal packings for the same molecule, called polymorphs, are generated. They need to be clustered to arrive at representative subsets for which analysis and geometry optimization are feasible.

Two things are needed for clustering and classification of crystal structures: a properly defined descriptor and a similarity function applied to this descriptor. A few requirements for both the descriptor of crystal structures and the similarity function are described in the literature [24–26]: the most obvious requirement for a descriptor–similarity combination is that more dissimilar crystal structures result in larger dissimilarity values. Although this seems trivial, several well-known descriptors do not generally satisfy this requirement [24–27]. Many descriptors require a choice of origin, or some other setting. Among such descriptors is the combination of unit cell parameters and fractional coordinates discussed earlier in this chapter. Caused by this choice of origin, a descriptor based on reduced unit cell parameters can vary significantly with only minor lattice distortions [28,29]. Although it is in some cases possible to adapt the similarity function to deal with such instabilities, this issue can better be addressed by using RDFs [30]. Using this descriptor a dissimilarity measure that expresses the differences between two crystal structures can be defined. The resulting dissimilarity value can then be used to cluster or classify the crystal structures by grouping together structures that have a low dissimilarity between them.

Crystal structures can be uniquely represented by an RDF describing the distribution of neighboring atoms around a central atom. Each neighboring atom gives rise to a peak in the function. RDFs are independent of cell choice and can be physically interpreted. In the application presented here, the RDF is adapted to include more specific information about the atoms. To do so, the RDF is weighted by the electrostatic interactions. To indicate the inclusion of electrostatic information in the descriptor, we will refer to this as the electronic RDF, or R_eDF. The reason for including electrostatics is the assumption that these play a major role in crystal packing [18,31,32]. By including partial atomic charges, the R_eDF focuses on atom groups with large partial charges, in particular functional groups, and differentiates between attractive interactions between oppositely charged atoms and repulsive interactions.

An atomic R_eDF describes the distribution of coulombic interactions of one atom with surrounding atoms; the R_eDF for the crystal structure is obtained by summing all atomic R_eDFs of all N atoms in the asymmetric unit:

$$\mathrm{R}_e\mathrm{DF}(r) = \sum_{i=1}^{N}\sum_{j=1}^{M} \frac{q_i q_j}{N \cdot r_{i,j}} \delta(r - r_{i,j}), \tag{3.3}$$

where M is the number of neighboring atoms within a radius r, q_i and q_j are partial atomic charges of the atoms i and j, and δ places the electrostatic interaction at the right distance by its definition $\delta(x) = 1$ if $x = 0$ and $\delta(x) = 0$ if $x \neq 0$. Alternatively, the $\delta(x)$ can reflect Gaussian smoothing. The function is scaled for the number of atoms in the asymmetric unit, N.

Figure 3.8 shows the R_eDF for an artificial crystal with two atoms in the unit cell, a positively and a negatively charged one ($a = 7.97$, $b = 10.26$, $c = 18.77$, and $\alpha = \beta = \gamma = 90°$). The first negative peak is the interaction between the two atoms at exactly the bonding distance. The other negative peaks are also peaks between two oppositely charged atoms. The overall decrease in intensities is caused by the $1/r$ term in the R_eDF equation. The first positive peak is related to the translation along the a axis, that is, $\pm \vec{a}$, and the second peak to the translation along the b axis. The third peak is the translation in the direction $a \pm b$. For this orthogonal structure, there are twice as many contributions to this peak as for the first two positive peaks, resulting in the higher intensity.

The R_eDFs of four experimental cephalosporin crystal structures are shown in Figures 3.9 and 3.10. They show a few distinct high-intensity peaks and many smaller

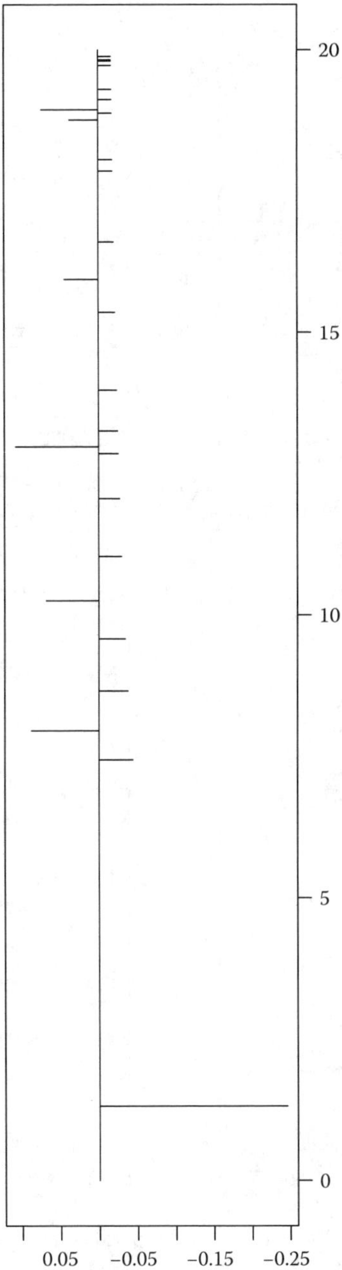

FIGURE 3.8 Example R_eDF of an artificial crystal structure with a positively and a negatively charged atom ($a = 8.0$, $b = 10.3$, $c = 18.8$, and $\alpha = \beta = \gamma = 90°$). Positive peaks are caused by the interaction of atoms with both positive and both negative charges. Consequently, they cause positive peaks at the distances matching the translational symmetry of the crystal. This explains, for example, the positive peaks at 8.0, 10.3, and 18.8 Å.

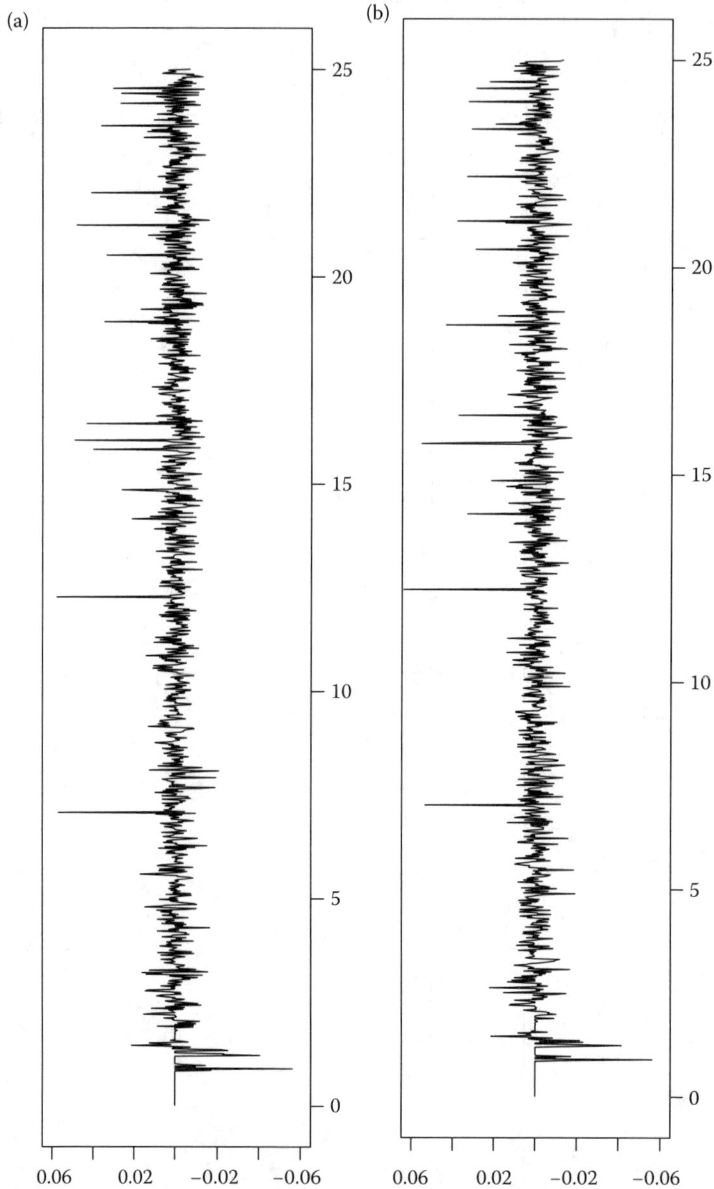

FIGURE 3.9 Example R_eDFs of three cephalosporin compounds: (a) A9, (b) A10 from the same class A.

peaks. The locations of these peaks are specific for the crystal packing: Figure 3.9a and b shows the R_eDFs of two cephalosporin structures from the same class, while Figure 3.9c shows the R_eDF for a different packing. Figure 3.10a shows the function for a simulated estrone crystal structure; a similar pattern can be observed. Figure 3.10b shows the effect of cutting away peaks with intensities lower than some

(c)

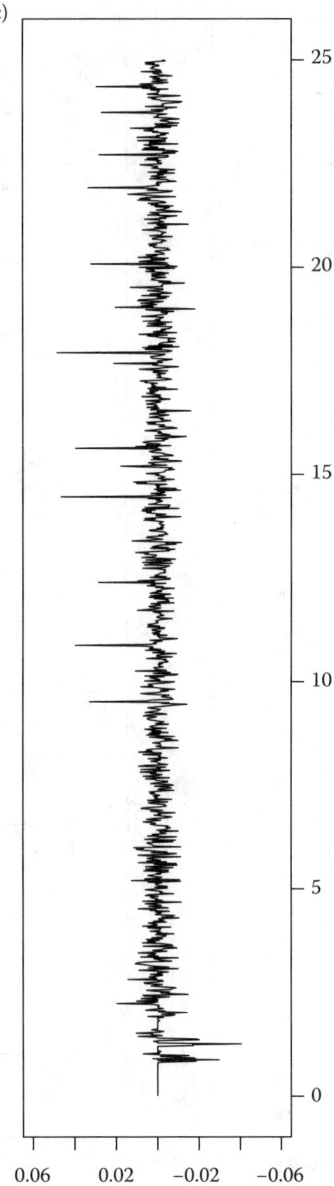

FIGURE 3.9 (Continued) (c) N19 from a different class N.

threshold. It was found that the cutoff value must be around 20% of the highest peak. Cutting away the smaller peaks emphasizes the major features of the R_eDF and leads to better discrimination.

Because of the nature of the R_eDF, one can expect positive contributions at those distances that match the translational symmetry in the crystal. This causes the positive

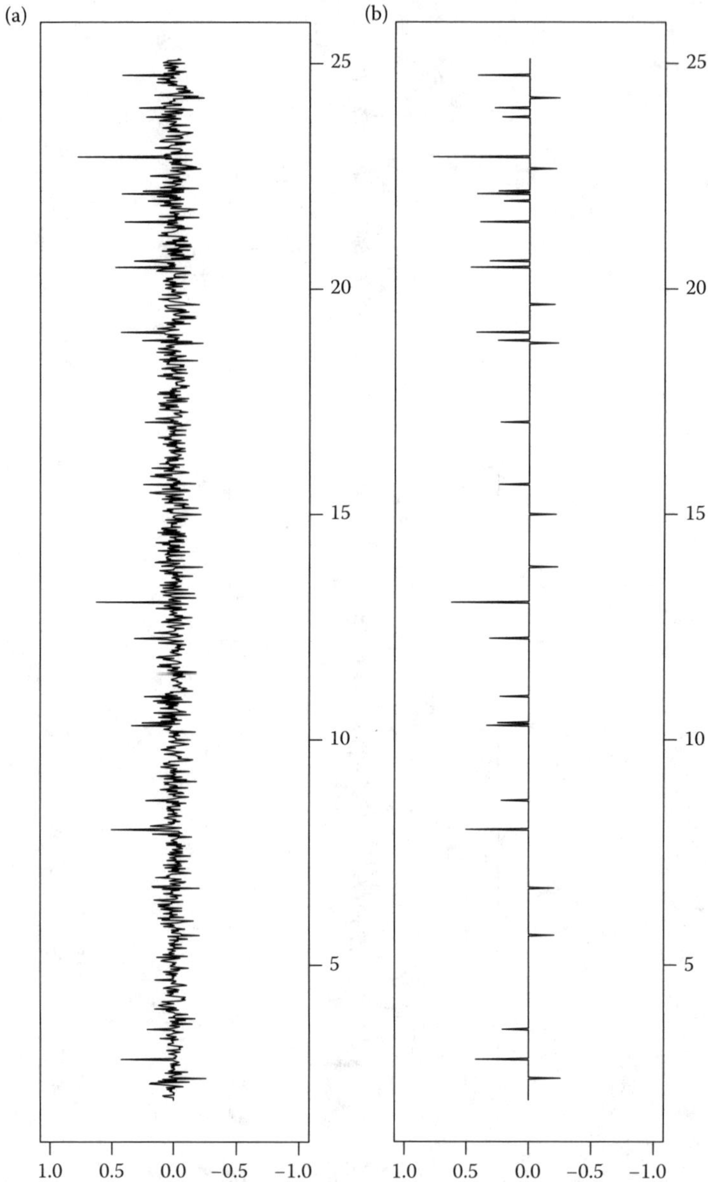

FIGURE 3.10 Example R_eDF of one of the simulated estrone structures shown in (a), and the effect of cutting away of peaks below 20% of the intensity of the highest peak in (b).

peaks at 8.0, 10.3, and 18.8 Å. However, since such contributions can be canceled out by other, negative contributions, they do not always show up in the R_eDF. Moreover, peaks not related to translational symmetry are particularly interesting, because they provide information additional to symmetry in the crystal.

TABLE 3.3
Open-Source Implementations of Algorithms Discussed in the Chapter

Algorithm Number	Algorithm	Libraries	Details
3.1	Calculate the center-of-mass	CDK	
3.2	Create 3D geometries from 2D diagrams	CDK	
3.3	Convert internal to Cartesian coordinates	CDK OpenBabel	
3.4	Align chemical structures based on anisomorphism	R	
3.5	Align 3D molecular structures based on the common substructure	CDK R	MCSS search For algorithm 3.4
3.6	Calculate the length-over-breadth ratio	CDK	
3.7	Calculate the 3D molecular surface	CDK	NumericSurface.class
3.8	Calculate an atomic RDF	CDK	RDFCalculator.class

Using this description, dissimilarities between crystal structures are represented by the difference between the two corresponding R_eDFs. For this, a weighted cross correlation (WCC) is used [19], which is applied to the high-intensity peaks of the R_eDF. Using this approach, both experimental and simulated crystal structures have been clustered and classified successfully [30].

3.7 OPEN-SOURCE IMPLEMENTATIONS

This chapter has presented a variety of basic algorithms involved in the representation of 3D molecular geometries. Because support for these geometries is so fundamental to chemoinformatics, it will not be difficult to find implementations in open-source software for the algorithms described in this chapter. Visualization of 3D geometries can be done in Jmol (http://www.jmol.org/, [1]) and PyMOL (http://www.pymol.org/). Converting different coordinate systems is also supported by various open-source toolkits, including the CDK (http://cdk.sourceforge.net/, [33,34]) and OpenBabel (http://openbabel.org/). Table 3.3 gives a more detailed overview.

REFERENCES

1. Willighagen, E. L. and Howard, M., Fast and scriptable molecular graphics in web browsers without Java3D. *Nat. Precedings*. 2007, http://precedings.nature.com/documents/50/version/1. Doi: 10.1038/npre.2007.50.1.
2. DeLano, W., *The PyMOL Molecular Graphics System*. DeLano Scientific LLC: Palo Alto, CA, http://www.pymol.org, 2008.
3. Tomczak, J., Data types. In: J. Gasteiger (Ed.), *Handbook of Chemoinformatics*, Vol. 1. Wiley-VCH: Weinheim, 2003, pp. 392–409.

4. Proschak, E., Wegner, J. K., Schuller, A., Schneider, G., and Fechner, U., Molecular query language (MQL)—a context-free grammar for substructure matching. *J. Chem. Inf. Model.* 2007, *47*, 295–301.
5. Baumann, K., Uniform-length molecular descriptors for quantitative structure property relationships (QSPR) and quantitative structure–activity relationships (QSAR): Classification studies and similarity searching. *Trends Anal. Chem.* 1999, *18*, 36–46.
6. Willighagen, E., Wehrens, R., and Buydens, L., Molecular chemometrics. *Crit. Rev. Anal. Chem.* 2006, *36*, 189–198.
7. Todeschini, R. and Consonni, V., *Handbook of Molecular Descriptors; Volume 11 of Methods and Principles in Medicinal Chemistry*. Wiley-VCH: New York, 2000.
8. Duca, J. and Hopfinger, A., Estimation of molecular similarity based on 4D-QSAR analysis: Formalism and validation. *J. Chem. Inf. Model.* 2001, *41*, 1367–1387.
9. Stanton, D. T. and Jurs, P. C. Development and use of charged partial surface area structural descriptors in computer-assisted quantitative structure–property relationship studies. *Anal. Chem.* 1990, *62*, 2323–2329.
10. Cramer III, R., Patterson, D., and Bunce, J., Comparitative molecular field analysis (CoMFA). 1. Effect of shape on binding of steroids to carries proteins. *J. Am. Chem. Soc.* 1988, *110*, 5959–5967.
11. Kim, K., List of CoMFA references, 1997. *Perspect. Drug Discov. Des.* 1998, *12–14*, 334–338.
12. Kim, K., Greco, G., and Novellino, E., A critical review of recent CoMFA applications. *Perspect. Drug Discov. Des.* 1998, *12–14*, 257–315.
13. Hopfinger, A., Wang, S., Tokarski, J., Jin, B., Albuquerque, M., Madhav, P., and Duraiswami, C., Construction of 3D-QSAR models using the 4D-QSAR analysis formalism. *J. Am. Chem. Soc.* 1997, *119*, 10509–10524.
14. Aires-De-Sousa, J., Hemmer, M., and Gasteiger, J., Prediction of 1H NMR chemical shifts using neural networks. *Anal. Chem.* 2002, *74*, 80–90.
15. Gasteiger, J., Sadowski, J., Schuur, J., Selzer, P., Steinhauer, L., and Steinhauer, V., Chemical information in 3D space. *J. Chem. Inf. Comput. Sci.* 1996, *36*, 1030–1037.
16. Hemmer, M. C., Steinhauer, V., and Gasteiger, J., Deriving the 3D structure of organic molecules from their infrared spectra. *Vibrat. Spectros.* 1999, *19*, 151–164.
17. Perlstein, J., Steppe, K., Vaday, S., and Ndip, E. M. N., Molecular self-assemblies. 5. Analysis of the vector properties of hydrogen bonding in crystal engineering. *J. Am. Chem. Soc.* 1996, *118*, 8433–8443.
18. Moulton, B. and Zaworotko, M. J., From molecules to crystal engineering: Supramolecular isomerism and polymorphism in network solids. *Chem. Rev.* 2001, *101*, 1629–1658.
19. De Gelder, R., Wehrens, R., and Hageman, J. A., generalized expression for the similarity spectra: Application to powder diffraction pattern classification. *J. Comput. Chem.* 2001, *22*, 273–289.
20. Hollingsworth, M. D., Crystal engineering: From structure to function. *Science* 2002, *295*, 2410–2413.
21. Ilyushin, G., Blatov, N., and Zakutin, Y., Crystal chemistry of orthosilicates and their analogs: The classification by topological types of suprapolyhedral structural units. *Acta Cryst.* 2002, *B58*, 948–964.
22. Lommerse, J. P. M., Motherwell, W. D. S., Ammon, H. L., Dunitz, J. D., Gavezzotti, A., Hofmann, D. W. M., Leusen, F. J. J., et al., A test of crystal structure prediction of small organic molecules. *Acta Cryst.* 2000, *B56*, 697–714.
23. Motherwell, W. D. S., et al., Crystal structure prediction of small organic molecules: A second blind test. *Acta Cryst.* 2002, *B58*, 647–661.

24. Dzyabchenko, A. V., Method of crystal-structure similarity searching. *Acta Cryst.* 1994, *B50*, 414–425.
25. Andrews, L. C. and Bernstein, H. J., Bravais lattice invariants. *Acta Cryst.* 1995, *A51*, 413–416.
26. Kalman, A. and Fabian, L., Volumetric measure of isostructurality. *Acta Cryst.* 1999, *B55*, 1099–1108.
27. Van Eijck, B. P. and Kroon, J., Fast clustering of equivalent structures in crystal structure prediction. *J. Comput. Chem.* 1997, *18*, 1036–1042.
28. Andrews, L. C., Bernstein, H. J., and Pelletier, G. A., A perturbation stable cell comparison technique. *Acta Cryst.* 1980, *A36*, 248–252.
29. Andrews, L. C. and Bernstein, H. J., Lattices and reduced cells as points in 6-space and selection of Bravais lattice type by projections. *Acta Cryst.* 1988, *A51*, 1009.
30. Willighagen, E., Wehrens, R., Verwer, P., de Gelder, R., and Buydens, L., Method for the computational comparison of crystal structures. *Acta Cryst.* 2005, *B61*, 29–36.
31. Pauling, L. and Delbruck, M., The nature of the intermolecular forces operative in biological processes. *Science* 1940, *92*, 77–79.
32. Desiraju, G. R., Supramolecular synthons in crystal engineering—a new organic synthesis. *Angew. Chem. Int. Ed.* 1995, *34*, 2311–2327.
33. Steinbeck, C., Han, Y., Kuhn, S., Horlacher, O., Luttmann, E., and Willighagen, E., The Chemistry Development Kit (CDK): An open-source Java library for chemo- and bioinformatics. *J. Chem. Inf. Comput. Sci.* 2003, *42*, 493–500.
34. Steinbeck, C., Hoppe, C., Kuhn, S., Floris, M., Guha, R., and Willighagen, E., Recent developments of the Chemistry Development Kit (CDK)—An open-source Java library for chemo- and bioinformatics. *Current Pharmaceutical Design* 2006, *12*, 2111–2120.

4 Molecular Descriptors

Nikolas Fechner, Georg Hinselmann, and Jörg Kurt Wegner

CONTENTS

4.1 MOLECULAR DESCRIPTORS: AN INTRODUCTION

The derivation of information and knowledge from real-world data makes it necessary to define properties that differentiate certain objects from others. Therefore, an explicit definition of a formal description of such objects is needed in a way that the natural distinction is preserved. It is obvious that the way an object is described depends on the context of interest. In the case of molecular structures, the chosen description of the same compound would certainly differ if a specific pharmaceutical target affinity or its experimental synthesis should be described. For this reason, literally thousands of molecular descriptors have been proposed covering all properties of interest. Thus, it is not the goal of this chapter to present an exhaustive list of descriptors but to provide a detailed account on the most important types, principles, and algorithms. An encyclopedia that covers most of the important molecular descriptors can be found in Ref. [1].

A molecular descriptor is an abstract, in most cases numerical, property of a molecular structure, derived by some algorithm describing a specific aspect of a compound. There are many ways to define descriptor classes. The most important object is to differentiate between the structural representations used as input. The simplest types are one-dimensional descriptors (0D and 1D) that only depend on the molecular formula, such as molecular mass or the numbers of specific elements. The net charge of a molecule is often regarded as a 1D descriptor. Most descriptors consider the molecular topology (i.e., the structural formula). These are considered as two-dimensional (2D) descriptors like most of the graph theory-based descriptors. Descriptors that also regard the spatial structure are defined as three-dimensional (3D). This class consists, for instance, of molecular interaction field (MIF)-based approaches, but also methods that make use of Euclidean distances. Further descriptor classes that have been introduced consider, for example, different conformations or molecular dynamics. Their dimensionality cannot be expressed in a similar intuitive way; sometimes we can find acronyms like four-dimensional (4D) or five-dimensional (5D) for such methods.

4.2 GRAPH DEFINITIONS

Most of the descriptors we will present in this chapter are at least 2D and therefore make use of the molecular topology. In such approaches, a molecule is often regarded as a graph annotated with complex properties, often using an unrestricted label alphabet. This flexible definition allows us to apply all kinds of structured data algorithms based on graphs [2], which also covers feature-reduced molecular graphs.

Definition 4.1: Given a node label alphabet L_v and an edge label alphabet L_E, we define a *directed attributed graph* g by the four-tuple $g = (V, E, \mu, \nu)$, where

- V defines a finite set of nodes
- $E \subseteq VxV$ denotes a set of edges
- $\mu : V \to L_V$ denotes a node labeling function
- $\nu : E \to L_E$ denotes an edge labeling function

The set V of nodes can be regarded as a set of node attributes of size $|V|$.

The set E of edges defines the structure and (edge) density of the graph. A connection from node $v \in V$ to node $u \in V$ is formed by $e = (u, v)$, if $e \in E$. A labeling function allows integrating information on the nodes or edges by using L_v and L_E. In theory, there is no restriction to the label alphabet. Nevertheless, for practical reasons the label alphabet is restricted to a vector space of a limited dimension $L = R^k$, or a discrete set of symbols, $L = \{s_1, \ldots, s_k\}$. Other definitions of labels might also contain information such as strings, trees, or graphs, as an alphabet reduction may impose constraints on the application domain, allowing a more flexible encoding.

Although there are various labeling functions for molecular graphs possible, there are still ongoing discussions for a standard definition (http://blueobelisk.sourceforge.net, http://opensmiles.org/). Due to differences in chemoinformatics perception algorithms [3] and expert systems, it is not possible to guarantee that two software solutions implement the same labeling function. This becomes important when algorithms are compared; drawn conclusions might rather challenge the labeling function instead of the algorithm of interest. If we regard 3D atom coordinates as an atomic node label triple $L_{V,\mathrm{3D}}(x, y, z)$ this becomes clear, because $L_{V,\mathrm{3D}}(x, y, z)$ labels might differ dramatically between algorithms [4–8]. If algorithms make use of different label functions, it is not sure whether the algorithms or the label function are compared.

Although the representation of a chemical compound as a directed graph is sometimes useful, for example, if asymmetric bond dissociation energies are used as edge labels, it can be regarded as a special case. In most cases, a molecular graph is treated as an undirected graph, where the directed edges $e = (u, v)$ and $e = (v, u)$ are identical, $(u, v) = (v, u)$. This can be written as $e = \{u, v\}$, by replacing the ordered list $(\ldots, \ldots)$ with the unordered set $\{\ldots, \ldots\}$. Another special case is a nonattributed graph with empty node and edge labeling functions $L_V = L_E = \{\}$, which simplifies the graph definition to $g = (V, E)$.

An important task on graphs is to detect a defined graph contained in another graph (i.e., a subgraph).

Definition 4.2: Let $g' = (V', E', \mu', \nu')$ and $g = (V, E, \mu, \nu)$ be graphs. Graph g' is a subgraph of g or g is a supergraph of g', written as $g' \subseteq g$, if

- $V' \subseteq V$
- $E' = E \cap (V' \subseteq V')$
- $\mu'(u) = \mu(u), \quad \forall u \in V'$
- $\nu'(u, v) = \nu(u, v), \quad \forall (u, v) \in E'$

Subgraph matchings and searches are usually applied after using a molecular labeling function. This is crucial, because some labelings depend on the size of a graph. The famous Hückel rule requires a graph size of at least $(2 \cdot |V| + 4)$ to assign aromaticity labels. In such cases, a label function cannot be applied to subgraphs alone and aromatic labels might not be assigned correctly.

The consideration of molecular structures as graph objects with certain properties requires defining the similarity of two structures, which is the base of many chemoinformatics applications by means of graphs. The evaluation of the similarity between two graphs is called *graph matching* [9]. Graph matching methods can be further divided into *exact* and *inexact* or *error-tolerant* matching algorithms. An exact matching algorithm of two graphs g_1 and g_2 decides if both graphs are identical. This is also known as graph *isomorphism*.

A bijective mapping $f : V \rightarrow V'$ denotes a *graph isomorphism* of a graph $g_1 = (V, E)$ and a graph $g_2 = (V', E')$ if

1. $\alpha_i(v) = \alpha'_i[f(v)]$ with $v \in V$, where α_i is a *labeling function*
2. For each edge $e = (v_1, v_2) \in E$ there exists an edge $e' = [f(v_1), f(v_2)] \in E'$ and for each edge $e' = (v'_1, v'_2) \in E'$ there exists an edge $e = [f^{-1}(v'_1), f^{-1}(v'_2)] \in E$

The graph isomorphism problem is hard to solve and possibly NP-complete (i.e., the problem has an exponential complexity with the input) in the case of general graphs. Nonetheless, there are special cases for which polynomial time algorithms are known. An example applicable for molecular graphs is the graph isomorphism approach for graphs with bounded valence by Luks [10]. A variation of this problem is *subgraph isomorphism*, which decides if a graph is completely contained in another one.

4.3 GLOBAL FEATURES AND ATOM ENVIRONMENTS

Global features describe a molecular graph by a real-valued single number. A full enumeration of all global features is beyond the scope of this section and there are well-known textbooks dealing with this topic, a case in point is Ref. [1]. Instead, we introduce some basic principles and implementations of some topological, complexity, eigenvalue-based descriptors, and local atom environments.

4.3.1 Topological Indices

Topological indices are global features that derive information from the adjacency matrix of a molecular graph. A problem of such descriptors is the so-called *degeneracy problem*, which occurs if two molecules are assigned the same descriptor value. This is often the case with stereoisomers on which topology-based algorithms have difficulties in general.

Topological descriptors can be divided in bond-based descriptors and distance-based descriptors. Whereas the first give information on how the atoms in a molecular graph are connected, the latter are based on the topological distance.

The Wiener Index is a convenient measure for the compactness of a molecule and has a low degeneracy [11]. The basic implementation of this topology-based descriptor uses the information contained in the shortest-distance matrix M, see Algorithm 4.1.

ALGORITHM 4.1 WIENER INDEX COMPUTATION

```
method double calculate (Molecule mol) {
       wienerPathNumber = 0.0;
       // get nxn distance matrix from molecular graph
       using Floyd-Warshall or Dijkstra
       DistanceMatrix M = getDistanceMatrix (mol) ;
       for (i = 0; i < M.length; i++) do
           for (j = 0; j < M.length; j++) do
              if (i == j) continue ;
                 wienerPathNumber += M[i] [j] ;
              fi
           od
       od
  return wienerPathNumbers/2 ;
  }
```

Usually, the shortest distances are computed by the Floyd–Warshall algorithm or Johnson's algorithm that is more efficient on sparse graphs:

$$W(G) = \frac{1}{2}\left(\sum_{i=0}^{N}\sum_{j=0,i\neq j}^{N}(M_{ij})\right).$$

4.3.2 Principles of Complexity Descriptors

There are numerous descriptors based on the complexity of molecular graphs. Some popular descriptors are based on this concept. Comprehensive overviews of complexity descriptors were published by Bonchev [12,13].

The Minoli Index [14] is defined as

$$\mathrm{MI} = \left(\frac{|V| \times |E|}{|V| + |E|}\right)\sum_{l} P_l,$$

where P_l is the number of paths of length l.

Information-theoretic indices are derived from the Shannon formula of a system with n elements:

$$I = -\sum_{i=1}^{k}\left(\frac{n_i}{n}\right)\log_2\left(\frac{n_i}{n}\right),$$

where k is the number of different sets of elements and n_i is the number of elements in the ith set. An application is the Bonchev–Trinajstic Index, in which the branching information on the molecule is incorporated into a descriptor.

$$\mathrm{BT} = n\log_2 n - \sum_{l} n_l \log_2 n_l,$$

where n is the total number of distances, n_l is the number of distances of length l, and n equals the sum over all n_l.

A spanning tree is a connected, acyclic subgraph of a graph G that includes all vertices of G. The number of spanning trees is a topological complexity descriptor. It is computed using the Laplacian matrix, which is defined as

$$L(G) = V(G) - A(G),$$

$$t(G) = \left|L_{ij}\right|,$$

where V is the diagonal matrix of G with the vertex degrees and A denoting its adjacency matrix. L_{ij} is the Laplacian matrix with row i and column j deleted, and $t(G)$ returns the number of spanning trees.

The Bonchev Index derives information on the total number of connected subgraphs. The First Bonchev Index is often referred to as the Topological Complexity Index (TC),

$$\mathrm{TC} = \sum_{s}\sum_{i} d_i(s),$$

where $d_i(s)$ is the degree of subgraph s regarding vertex i.

Randić complexity indices are defined using augmented vertex degrees. They are computed by the augmented degree matrix D_{ij}, where d_j is the degree of vertex j and l_{ij} is the distance between vertices i and j:

$$D'_{ij} = \frac{d_j}{2^{l_{ij}}}.$$

The augmented degree is the row sum of the ith row of D'_{ij}.

Zagreb indices are topology-based indices, summing up vertex degrees over vertices and edges. They are defined as follows:

$$M_1 = \sum_{\text{vertices}} (d_i)^2,$$

$$M_2 = \sum_{\text{edges}} (d_i d_j),$$

where d_i is the degree of vertex i. M_1 is the count of all walks of length 2.

Graph complexity can be defined in various ways [12,13], but still there is no standard definition. In Ref. [12] various criteria are compiled from different sources, which describe the requirements for a "good" molecular complexity descriptor. For example, a complexity index should

- Increase with the numbers of vertices and edges
- Reflect the degree of connectedness
- Be independent from the nature of the system
- Differentiate nonisomorphic systems
- Increase with the size of the graph, branching, cyclicity, and number of multiple edges

Still, this is an ongoing discussion, with even conflicting positions. In Ref. [12], it is concluded that common requirements on complexity indices are as follows: principles of homology, reflection of branching, cyclicity, multiple edges, and heteroatoms.

4.3.3 Atom Environments

All atom environments have a common principle, namely that they describe atoms by using the information of the direct neighborhood. The advantage of this procedure is that no functional groups or fragments have to be predefined.

4.3.3.1 HOSE Codes (Hierarchically Ordered Spherical Description of the Environment)

Starting from the "root" (the atom to be described), the symbols of neighboring bonds and atoms are retrieved by a depth-first search and assigned to the so-called spheres. Sphere i includes all direct and non-neighboring atoms with topological distance i. For substructures and rings, priority tables exist such that for each sphere a unique string representation can be assigned. This ensures an efficient comparison and storage because this representation can be mapped to numerical value. The HOSE code was introduced by Bremser [15].

4.3.3.2 Radial Distribution Function

The radial distribution function (RDF) [16,17] is a correlation-based function. It is defined as follows:

$$g(r) = \frac{1}{2} \sum_{n,m,n \neq m(n)}^{|A|} \alpha_i(n)\alpha_i(m)\mathrm{e}^{-\gamma d^2}.$$

The Moreau–Broto autocorrelation is a special case of the RDF:

$$\mathrm{AC}(d) = g(r)_{\lim \to \infty} = \sum_{n,m,n \neq m}^{|A|} \alpha_i(n)\alpha_i(m)\delta_{nm},$$

with

$$\delta_{\mathrm{nm}} = \begin{cases} 1 & \text{if } \mathrm{dist}(a_n, a_m) = d \\ 0 & \text{else} \end{cases}.$$

Parameters $\alpha_i(n)$ and $\alpha_i(m)$ describe the properties of atoms n and m, γ describes the degree of delocalization for the atomic properties and $|A|$ equals the number of atoms in a molecule. The distance $d = r - r_{nm}$ is computed from the sphere radii $r \in \{r_{\min}, \ldots, r_{\min} + kr_{\mathrm{res}} \leq r_{\max}\}$ with $r_{\min}$ and $r_{\max}$ denoting their limits. r_{res} is the chosen step size. With increasing γ, the atomic properties become more localized, and the properties of an atom have no influence on the neighbors of this atom. Therefore, the RDF describes the distribution of an atomic property in the molecule.

For $\gamma \to \infty$, the exponential term turns into the Delta function δ_{nm}. Thus, the autocorrelation is a special case of the general RDF.

4.3.3.3 Local Atom Environment Kernel

The local atom environment kernel is a local atom similarity. It is used by the optimal assignment kernel (OAK) [18,19]:

$$k_{\mathrm{local}}(v, v') = k_{\mathrm{atom}}(v, v') + k_0(v, v') + \sum_{l=1}^{L} \gamma(l) k_l(v, v').$$

The similarity is composed of a local atom similarity $k_0(v, v')$ and spherical neighborhood $k_l(v, v')$ of size l. The maximum spherical (topological) distance is denoted by L, and $\gamma(l)$ is a decay factor.

Note that the optimal neighborhoods $\pi(i)$ are used, such that only meaningful descriptors are regarded. For two atoms v, v', the sum over all kernel similarities match(i) regarding the direct neighbors n_i and $n_{\pi(i)}$ is maximized. The direct neighborhood of an atom in organic molecules is restricted to five; therefore, the optimal assignment of all possible neighborhoods π is computed:

$$k_0(v, v') = \frac{1}{\alpha_{\text{val}}(v')} \max_{\pi} \sum_{i=1}^{\alpha_{\text{val}}(v)} \text{match}_0(i),$$

$$\text{match}_0(i) = k_{\text{atom}}[n_i(v), n_{\pi(i)}(v')] \cdot k_{\text{bond}}[\{v, n_i(v)\}, \{v', n_{\pi(i)}(v')\}].$$

Larger atom environments up to length L can be efficiently computed by the following recursive algorithm, which uses previously computed direct neighborhoods:

$$\text{match}_0(i) = k_{\text{atom}}[n_i(v), n_{\pi(i)}(v')] \cdot k_{\text{bond}}[\{v, n_i(v)\}, \{v', n_{\pi(i)}(v')\}],$$

$$k_l(v, v') = \frac{1}{\alpha_{\text{val}}(v)\alpha_{\text{val}}(v')} \sum_{i}^{\alpha_{\text{val}}(v)} \sum_{j}^{\alpha_{\text{val}}(v)} k_{l-1}[n_i(v), n_j(v')].$$

The local atom environment is designed to distinguish between nominal and numerical atomic and bond properties. Therefore, the local kernels are composed of numerical (L_{num}) and nominal kernel (L_{nom}) functions, which can be weighted by parameters $\gamma_{\text{num}}, \gamma_{\text{nom}}$. s_{Tanimoto} denotes the Tanimoto similarity of two sets of nominal features:

$$k_{\text{atom}}(v, v', \gamma_{V,\text{nom}}, \gamma_{V,\text{num}}) = k_{\text{nom}}(A_{\text{nom}}, A'_{\text{nom}}, \gamma_{V,\text{nom}}) \cdot k_{\text{num}}(A_{\text{num}}, A'_{\text{num}}, \gamma_{V,\text{num}}),$$

$$k_{\text{bond}}(e, e', \gamma_{E,\text{nom}}, \gamma_{E,\text{num}}) = k_{\text{nom}}(B_{\text{nom}}, B'_{\text{nom}}, \gamma_{E,\text{nom}}) \cdot k_{\text{num}}(B_{\text{num}}, B'_{\text{num}}, \gamma_{E,\text{num}}),$$

$$k_{\text{nom}}(L_{\text{nom}}, L'_{\text{nom}}, \gamma_{\text{nom}}) = \exp\left(-\frac{[1 - s_{\text{Tanimoto}}(L_{\text{nom}}, L'_{\text{nom}})]^2}{2\gamma_{\text{nom}}^2}\right),$$

$$k_{\text{num}}(L_{\text{num}}, L'_{\text{num}}, \gamma_{\text{num}}) = \exp\left(-\sum_{i}^{|L_{\text{num}}|} \frac{(L_{\text{num},i}, L'_{\text{num},i})^2}{2\gamma_{\text{num}}^2}\right).$$

A similar approach was published by Bender et al. [20,21], describing an atom environment by a radial fingerprint, which is discussed elsewhere in this chapter.

4.3.4 Eigenvalue Decomposition

4.3.4.1 Characteristic Polynomial

The characteristic polynomial is one of the most important relationships between a graph and the eigenvalues of either the adjacency matrix of a graph or the distance matrix.

Definition 4.3: The eigenvalues $x_1, x_2, \ldots, x_{|V|}$ of a graph with V nodes are also called the characteristic polynomial $P(G,x)$:

$$(x - x_1)(x - x_2), \ldots, (x - x_{|V|}) = P(G,x),$$
$$P(x_i) = 0.$$

If the rings Z of a graph that contain the edge e_{ij} are considered, we can rewrite the equation as the Heilbronner theorem [22].

Definition 4.4: (Heilbronner Theorem) Let v_i and v_j be two nodes of a molecular graph, and e_{ij} be the connecting edge. Then the characteristic polynomial can be computed as graph decomposition:

$$P(G) = P(G - e_{ij}) - P(G - v_i - v_j) - 2 \sum_{\forall Z | e_{ij} \in Z} P(G - Z). \tag{4.1}$$

The last term is a sum over all rings, and different ring systems might lead to the same numerical value. In other words, several graphs can give similar eigenvalues, but a single eigenvalue can map to multiple graphs.

The characteristic polynomial can be extended by atomic properties, which gives different eigenvalues depending on the labeling function used.

4.3.4.2 Burden Matrix and BCUT Descriptors

Closely related molecular descriptors are derived from a modification of the adjacency matrix. The elements in the diagonal are modified by the characteristic properties of a molecule. The descriptors are then computed as the set of eigenvalues.

The Burden Matrix is a symmetric matrix based on the hydrogen-depleted molecular graph with the atomic numbers in the diagonal and $(\pi/10)$ in the off-diagonal between two atoms i, j, where π is the conventional bond order [1]. The ordered sequence of the n smallest eigenvalues of the Burden Matrix is one of the first descriptors.

Burden CAS—University of Texas eigenvalues (BCUT) descriptors are an extension of this approach. In general, the values in the diagonal of the Burden Matrix are replaced by special properties, for example, atomic charge, polarizability, and H-bond capabilities.

Descriptors based on the distance matrix can also be defined, for example, the largest eigenvalue of the distance matrix or the unique negative eigenvalue. Combinations of those descriptors were also proposed, for example, the sum of leading eigenvalues of the distance matrix and the adjacency matrix; for a detailed overview, see Ref. [1].

4.3.4.3 WHIM Descriptors

Weighted Holistic Invariant Molecular (WHIM) descriptors [1] are 3D descriptors that capture information regarding size, shape, symmetry, and atom distribution. A principal component analysis (PCA) is performed on the centered coordinates of a molecule using a weighted covariance matrix. The weighted covariance matrix is obtained by applying a weighting scheme of the general form:

$$s_{jk} = \frac{\sum_{i=1}^{|A|} w_i (q_{ij} - \bar{q}_j)(q_{ik} - \bar{q}_k)}{\sum_{i=1}^{|A|} w_i},$$

where s_{jk} is the weighted covariance between the jth and kth atomic coordinates. $|A|$ is the total number of atoms, w_i is the ith weight, q_{ij} and q_{ik} are the ith and kth coordinates, and $\bar{q}_k$ is the corresponding average value. Again, the weighting schemes can be exchanged, for example, regarding electronegativity, van der Waals volume, and polarizability. The descriptors are computed as statistical indices of the atoms projected onto principal components obtained from the modified covariance matrix.

4.4 MOLECULAR SUBSTRUCTURES

Molecular substructures can be regarded as connected graphs that are completely contained in the molecular graph. Many physicochemical properties can be related to the frequency of certain substructures in a molecule as it is the idea in the Free–Wilson [23] approach to chemometric modeling. Formally, a molecular substructure can be denoted as a subgraph G_{SG} of a molecular graph G_M with

1. $V_S \subseteq V$, with V being the set of all vertices in G_M and V_S being the set of the vertices in G_{SG}
2. $E_S = E \cap (V_S \times V_S)$, with E being the set of all edges in G_M and V_S being the set of the edges in G_{SG}
3. $A_S = \{\alpha_{S,1}, \ldots, \alpha_{|S,A_S|}\}$ denotes the labeling functions for atomic properties restricted to the vertices (atoms) V_S

$$\alpha_{S,i}(v) = \begin{cases} \alpha_i(v) & \text{if } v \in V_S \forall \alpha_i \in A, \\ \text{undefined} & \text{otherwise.} \end{cases}$$

4. Analogously, $B_S = \{\beta_{S,1}, \ldots, \beta_{|S,B_S|}\}$ denotes the labeling functions for bond properties restricted to the edges (bonds) V_S.

$$\beta_S(e) = \begin{cases} \beta(e) & \text{if } e \in E_S \forall \beta_i \in B, \\ \text{undefined} & \text{otherwise.} \end{cases}$$

Substructures are handled differently compared to numerical descriptors in which an algorithm maps the molecule to one or several numerical values. Basically, there are two points of view regarding substructures as descriptors. One popular approach frequently used in molecular fingerprint methods is to consider the substructure as the descriptor and its presence or frequency in a specific molecule as its value. This makes it necessary to define a set of substructures also known as *structural keys*.

An alternative method is to define an algorithm that generates a set of substructures for a molecule. This can be regarded as the generation of a set of descriptors for a specific molecule by an algorithm. This avoids an explicit definition of the substructure set and thus allows us to reveal important but yet unrecognized structural features. In this case, it is necessary to introduce a metric that allows a quantitative comparison of the resulting descriptor sets with variable cardinality for different molecules.

4.4.1 Substructure Types and Generation

4.4.1.1 Atom Types and Reduced Graphs

The fundamental building blocks of molecular graphs are atoms (the smallest substructures that fulfill the upper definition). An atom is usually considered as an instance of a specific chemical element type defining its physical properties (e.g., expected mass, electronegativity, and number of electrons and protons). The chemical properties are expressed only vaguely by the element alone, because most properties related to atomic interactions depend on the hybridization and the neighborhood of an atom. Therefore, it is common to use a finer distinction of atoms of the same element leading to the concept of an atom type.

In this chapter, we will consider an atom type as a structural pattern that denotes which configurations of an atom (including specific properties like charge, hybridization, isotope, etc.) and its intramolecular neighborhood can be considered as equal. This concept is of special importance in the application of empirical force fields in which the potential terms are evaluated using deviations of precalculated *ab initio* or experimental parameters for specific atom types (e.g., the optimal bond length between two sp^3 carbons) that are considered favorable.

The definition of a dictionary of atom types is a crucial step in many applications of chemoinformatics and is a major contribution of chemical expert knowledge in a computational framework.

There are many atom-type dictionaries of different accuracies available. Some popular definitions are SYBYL atom types [24], which differentiate mainly regarding hybridizations and element types, the Meng/Lewis definition [25], or the MacroModel atom types [26], which extend the definition to specific atoms in substructures like ring systems or amino acids using SMARTS patterns [27].

Besides the incorporation of expert knowledge into a chemoinformatics framework, atom types can be powerful features if they are regarded as binary descriptors. This is the base of many structural features that have to deal with the problem whether two atoms can be considered equal. For example, this plays an important role in the computation of the cardinality of the junction of atom-type sets needed by many similarity measures for molecular fingerprints or in the definition of pharmacophoric points.

An extension of the atom-type concept is the definition of substructure types (e.g., using SMARTS expressions) by regarding whole substructures like rings or functional groups as atom types, whose properties reflect the properties of the substructure. This representation is useful in molecular similarity calculations like Feature Trees [28] to ensure that these substructures are only compared in bulk. Another advantage of collapsing rings to pseudoatoms is that the molecular graph is transformed into a tree (i.e., a cycle-free graph), which enables the use of faster and simpler graph algorithms.

4.4.1.2 Atom Pairs

Basic atom types alone do not provide much information about their molecular arrangement, topology or even geometry of the molecule that is not directly contained in the atom type or the collapsed substructure. The easiest way to include the topology of a molecule is to use pairs of atom types together with their intramolecular topological or geometrical distance. The first use of an atom pair encoding [29] known to the authors used an atom type, which denotes the element, the number of attached heavy atoms and the number of π electrons. The interatomic distance was measured as the count of bonds on the shortest path (i.e., the topological distances).

A possible extension to this approach is to use more specific atom types and geometrical distances. It is also possible to define a descriptor that denotes a specific atom-type pair and uses its mean distance to all other atoms in a molecule.

The extraction of all atom pairs that are contained in a molecule can be done by applying Dijkstra's (Algorithm 4.2) shortest path algorithm [30] for each atom in the molecule.

ALGORITHM 4.2 PSEUDOCODE FOR THE DIJKSTRA SHORTEST PATH ALGORITHM (ADAPTED FROM WIKIPEDIA) [30]

```
method getShortestPaths(Graph G, Node s) {
   for all vertices v in G do // Initializations
       dist[v] := infinity // Unknown distance function
                                        from s to v
       previous[v] := undefined // Previous node in
                                           optimal path from s
   od
   dist[s] := 0 // Distance from source to source
   Q := the set of all nodes in Graph
   while Q is not empty do // The main loop
       u := node in Q with smallest dist[]
       remove u from Q
       for all neighbors v of u do // where v has not
       yet been removed from Q.
           alt := dist[u] + dist_between(u, v)
           if alt < dist[v] do // Relax (u,v)
               dist[v] := alt
               previous[v] := u
           fi
       od
    od
    return previous, dist
 }
```

This algorithm is in many applications preferable to the more complex all-pairs-shortest-paths methods (e.g., the Floyd–Warshall algorithm [31]) because all edge weights are non-negative in a molecular graph and the graph is usually weakly connected due to the constrained number of adjacent edges for each node.

4.4.1.3 Sequences of Atom Types: Paths and Walks

Sequences of graph vertices are divided into two classes: a *walk* is according to Borgwardt [32] a nonalternating sequence $(v_1, e_1, v_2, e_2, \ldots, e_{l-1}, v_l)$ of vertices and edges such that $e_i = \text{bond}(v_i, v_{i+1})$. In a molecular graph this corresponds to a sequence of connected atom types. A *path* is a walk in which each vertex is at most contained once. In many cases, paths are used instead of walks to represent chemical structures. For an algorithm for the extraction of all labeled paths up to a length *d* in a molecule, see Algorithm 4.3.

ALGORITHM 4.3 PSEUDOCODE FOR THE EXTRACTION OF ALL LABELED PATHS UP TO A LENGTH D IN A MOLECULE

```
    method list getPaths (molecular graph M , search
    depth d)
{
    list paths_d ;
    list paths_local ;
    for each atom a ∈ M do
        // get all paths of length d via a DFS
        root = getDFSTree (M, a, d) ;
        // get all paths in the depth-first tree
        starting at the root up to length d
        paths_local = enumerateAllPathsInTree
        (root, d)
        for each path ∈ paths_local do
            // check if sequence equals the reverse
            sequence
            if (!(path ∈ path_d) &&!
            (path.reverse( ) ∈ path_d))
path_d. add(path);
            fi
        od
    od
    return paths_d ;
 }
```

Some common molecular fingerprints like the Daylight fingerprint [27] are defined by means of paths up to a specific length that are contained in a molecule. Paths are usually extracted by a depth-first traversal of the molecular graph. Pseudocodes for the depth-first traversal are given in Algorithm 4.4 and for a breadth-first traversal in Algorithm 4.5.

ALGORITHM 4.4 PSEUDOCODE FOR THE EXTRACTION OF DEPTH-FIRST TRAVERSAL TREE UP TO A DEPTH D IN A MOLECULE BEGINNING AT ATOM A (ADAPTED FROM [31])

```
global time = 0
method atom getDFSTree(molecular graph
M, root atom a)
{
  root = a
  for all atoms in M do
    atom.setState(''unvisited'')
    atom.setDepth(∞ ∞)
    atom.setPredecessor(null)
  od
  recursiveVisit(atom root)
  //root atom augmented with its neighbors
  // (tree is implicitly stored as a adjacency list)
  return root
}

method void recursiveVisit(atom u)
{
  time++
  u.setState(''visited'')
  u.setDepth(time)
  for all neighbors of u in M do
     if neighbor has not been visited
        neighbor.setPredecessor(u)
        recursiveVisit(neighbor)
     fi
  od
  u.setState(''finished'')
  u.setDepth(time)
  time++
}
```

ALGORITHM 4.5 PSEUDOCODE FOR THE EXTRACTION OF BREADTH-FIRST TRAVERSAL TREE UP TO A DEPTH D IN A MOLECULE BEGINNING AT ATOM A (ADAPTED FROM [31])

```
method atom getBFSTree(molecular graph M, root
atom a)
{
   root= a
   for all atoms in M except a do
      atom.setState(''unvisited'')
      atom.setDepth(∞)
      atom.setPredecessor(null)
   od
```

```
    root.setState(''visited'')
    root.setDepth(0)
    root.setPredecessor(null)
    queue.add(root) //first-in-first-out queue
    while queu has elements do
        u = queue.getNext()
        for all neighbors of u in M do
        if neighbor has not been visited
            neighbor.setState(''visited'')
        fi
        neighbor.setDepth(u.getDepth()+1)
                        neighbor.setPredecessor(u)
        queue.add(neighbor)
    od
    u.setState(''finished'')
od
// root atom augmented with its neighbors
// (tree is implicitly stored as an adjacency list)
return root
 }
```

4.4.1.4 Trees

A common extension for atom types is to incorporate the atom neighborhood to get a better representation of the topological embedding of an atom. Hence, a property can be assigned to a complete neighborhood. This is to a certain degree included in many of the atom-type definitions (e.g., amide nitrogen) but it requires a predefinition of the neighborhood. A concept to avoid this drawback and to extend the neighborhood to some arbitrary depth is to augment the atom type of each atom A by the paths up to a certain length starting at A. This constructs a star-like graph for each atom which is cycle-free due to the path definition and thus can be regarded as a tree with A as the root. The tree substructure enables the use of highly efficient algorithms defined for trees.

Tree-shaped substructures are also the base for the *signature* molecular descriptor [33–36]. This descriptor is defined using an encoding of atoms and bonds as *signatures*. An atom signature ${}^{d}\sigma(x)$ of an atom x in a molecule G is defined as the depth-first tree starting at x as the root node up to depth d. The depth-first traversal used is slightly different from that given in Algorithm 4.4 because it allows that atoms to occur several times due to rings in the molecule. The tree is represented in a string representation with opening brackets if a new subtree is started and with closing brackets if it is finished. The signature ${}^{d}\sigma(G)$ of the molecule G can then be obtained as the linear combination of the atom signatures.

In this case, an atom signature can be regarded as a molecular descriptor with the number of occurrences of atoms with a specific signature in the molecule as descriptor values.

A similar descriptor generation approach has been proposed by Bender et al. [20,21]. The method starts with the construction of a neighborhood tree. The atomic properties of the root atom are extended by the counts of the different atom types (SYBYL atom types in the original work) in the neighborhood tree up to a specific search depth. This leads to a descriptor vector for each atom containing the frequency of the atom types in its neighborhood. Although it is not an explicit substructure, this atomic neighborhood feature vector can also be regarded as a tree-like substructural pattern and be used for the definition of molecular fingerprints [20].

4.4.1.5 Fragments

Molecular fragments are the most complex and versatile substructure types. There is no general definition for this type, but usually fragments are considered as subgraphs, which are the result of the deletion of an edge (or whole subgraphs). The biggest difference from other substructure types is that fragments in general are lacking a strictly defined structure like *linear sequence* (path, walk), tree, or predefined pattern. The information content of such fragments is in most cases much higher than of the less complex substructure types. This comes at the cost of the higher computational requirements of general graph algorithms to compare the resulting substructure sets (Isomorphism, Matching, etc.).

The generation of a set of molecular fragments applies a decomposition algorithm, which decides which bonds are deleted. Common approaches delete all single bonds, all bonds between a ring and a nonring atom. The RECAP algorithm [37] deletes only bonds that can probably be re-formed by chemical reactions. The idea behind RECAP is to employ a recipe of how a structural complex molecule could be synthesized out of more usual building blocks by reforming the deleted bonds. In general, the problem of enumerating all possible fragments is known to be NP-complete.

4.4.2 Fingerprints

Molecular fingerprints [38,39] are a common method to combine the presence or absence of different substructures in a molecule into one molecular descriptor. They are usually represented as a vector of bits with a fixed length that denotes the presence of a specific structural pattern. There are many different fingerprint implementations that can be classified in hashed and nonhashed fingerprints. The nonhashed fingerprints also known as *structural keys* are mainly based on a predefined dictionary of substructures, such that there is a unique mapping between a bit vector position and a specific substructure. A popular structural key is the MACCS keys [40].

Definition 4.5: Let G be a set of graphs and $P = \{p_1, \ldots, p_n\}$ a set of n structural patterns, such that there exists a function

$$f : G \times P \mapsto \{0, 1\}, f(g \in G, p \in P) = \begin{cases} 1 & g \text{ contains } p \in P \\ 0 & \text{otherwise} \end{cases}.$$

Then the ordered set $\{f(g, p_1), \ldots, f(g, p_n)\}$ is called a *structural key* of g regarding to the set P.

This fingerprint type can also be considered as a pattern-parametrized view of the bit vector, because for each molecule we have to iterate over a set of patterns and to check for every pattern whether it occurs in the molecule.

Such fingerprints have the inherent disadvantage that it is impossible to cover the diversity of the chemical space by a fixed number of patterns. This is avoided with hashed fingerprints. They are based on the idea of defining a method that generates a substructure set for a molecule and converts that into a bit vector of fixed length. This approach will produce different fingerprints for different molecules in most cases.

The patterns that are used depend solely on the generation method (e.g., paths or trees of a certain size, and RECAP fragments) and the molecule that has to be encoded. Therefore, this type can be regarded as pattern-generation-parametrized and has the big advantage that the substructure generation is usually faster than the subgraph isomorphism check of the pattern look-up.

The final mapping of each substructure to a bit position is in most cases done by using the hash code of a pattern as the seed for a pseudorandom number generator and a mapping of this random number to a bit position. This conversion has the drawback that after the hashing the bijective mapping of bit position and pattern is lost because different substructures can be mapped to the same bit positions. This information loss is acceptable regarding similarity searches in databases, but makes a potential interpretation (or feature selection) for knowledge discovery tasks more demanding.

4.4.2.1 Hashed Fingerprints

A popular hashed fingerprint implementation is the Daylight Chemical Information Fingerprint [41], which is calculated by enumerating the set of labeled paths shorter than a specified number of bonds in a molecule.

Each pattern (path) is hashed, which produces a set of bits. The final fingerprint is then obtained by the union (logical OR) of the bit sets according to each pattern of the molecule. This representation is, for example, used by UNITY [42] or JChem [43] in their hashed fingerprint implementations. The pseudocode for a generic algorithm that generates a path-based hashed fingerprint of dimension d is given in Algorithm 4.6

ALGORITHM 4.6 PSEUDOCODE FOR A GENERIC HASHED PATH FINGERPRINT ALGORITHM (ADAPTED FROM BROWN ET AL. [44])

```
method getHashedPathFingerprint(Molecule G, Size
d, Pathlength l)
{
    fingerprint = initializeBitvector(d)
    paths = getPaths(G,l)
    for all atoms in G do
        for all paths starting at atom do
            seed = hash(path) //generate an integer hash
            value
            randomIntSet=randomInt(seed) //generate a set
```

```
            of random integers
            for all rInts in randomIntSet do
               index = rInt % d //map the random int to a
               bit position
               fingerprint[index]=TRUE
            od
         od
      od
      return fingerprint
}
```

4.4.2.2 Comparison of Hashed Fingerprints and Baldi's Correction

One application of hashed fingerprints (and molecular fingerprints in general) is to identify those molecules that are similar to a query structure. This is done by applying bit set-based similarity measures like the well-known Tanimoto/Jaccard coefficient. Ideally, this similarity is computed using the full nonhashed fingerprints providing a measure of the real structural similarity if the encoding is chosen appropriate. However, because of practical considerations the much shorter hashed fingerprints are used in most cases. This implies that there is a strong correlation between the hashed fingerprint similarity $S(A, B)$ and the nonhashed fingerprint similarity $S_*(A, B)$ of two compound fingerprints A and B. This assumption is not necessarily valid; because of the compression rate, the choice of the fixed length of the hashed fingerprint has a significant influence on the number of set bits (the cardinality). Therefore, Swamidass and Baldi [45] propose to use estimates of the nonhashed fingerprints A_* and B_* for the similarity calculation.

The expected cardinality A of a hashed fingerprint of length N given the cardinality A_* of the nonhashed fingerprint with length N_* with identical and independently distributed bits generated by a binomial distribution can be formulated as

$$E(A|A_*) \approx N\left[1 - \left(1 - \frac{A_*}{N_*}\right)^{N_*/N}\right].$$

This expression is based on the assumption that the hash function ensures the statistical independence of the bits and that the compression of the full fingerprint A_* to the fixed size version A is performed using a *modulo* N operation. In this scenario, the probability that a bit is set is given by (A_*/N_*) repeated (N_*/N) times due to the compression. The inverse relationship

$$E(A_*|A) \approx N_*\left[1 - \left(1 - \frac{A}{N}\right)^{N/N_*}\right]$$

is more important because only the cardinality and length of the hashed fingerprint is available in most cases. This expression is a further approximation not obtained by the application of Bayes' theorem but it works well in practice according to Swamidass and Baldi [45]. It is useful to derive the expression of $E(A_*|A)$ independent of N_* by only considering N_* as large because the length N_* of the full fingerprint is often not exactly known:

$$E(A|A_*) \approx \lim_{N_* \to \infty} N \left[1 - \left(1 - \frac{A_*}{N_*} \right)^{N_*/N} \right] = N(1 - \mathrm{e}^{-A_*/N})$$

and thus

$$E(A_*|A) \approx -N \log \left(1 - \frac{A}{N} \right).$$

This estimation can then be applied to derive values for the nonhashed intersection cardinality $A_* \cap B_*$ and the nonhashed union cardinality $A_* \cup B_*$ of two hashed fingerprints A and B and a considerably large N_*:

$$A_* \cup B_* \approx \min \left[-N \left(1 - \frac{A \cup B}{N} \right), -N \log \left(1 - \frac{A}{N} \right) \left(1 - \frac{B}{n} \right) \right],$$

$$A_* \cap B_* \approx \max[0, A_* + B_* - A_* \cup B_*],$$

$$\approx \max \left[0, \underbrace{-N \log \left(1 - \frac{A}{N} \right)}_{E(A_*|A)} \underbrace{-N \log \left(1 - \frac{B}{N} \right)}_{E(B_*|B)} \underbrace{-N \left(1 - \frac{A \cup B}{N} \right)}_{A_* \cup B_*} \right].$$

These two cardinalities combined with the bit string cardinalities are sufficient to compute many similarity measures (e.g., Tanimoto) for hashed fingerprints in terms of the corresponding nonhashed ones.

4.4.2.3 Stigmata

The Stigmata algorithm by Shemetulskis et al. [46] uses hashed fingerprints (daylight in the original work) to generate a kind of consensus fingerprint called *modal fingerprint* which expresses commonalities in a set of molecules active against a specific target. A modal fingerprint has the same length as the molecular fingerprints (e.g., 2048 bits in the original publication by Shemetulskis et al.) of the compounds with a bit set to TRUE if it is set in more than a certain percentage of compound fingerprints. A modal fingerprint can then be used to extract information about frequent pattern types as well as new molecular descriptors. Each atom of each molecule can be labeled by a so-called *ALAB* value, which denotes the percentage of paths the atom is part of that also are contained in the modal fingerprint.

A similar descriptor on the molecular level is MODP that is defined as the percentage of the paths of a molecule that are also part of the modal fingerprint. The Tanimoto similarity between a molecule fingerprint and the modal fingerprint can be used as a molecular descriptor as well. In the publication of Shemetulskis [46], this descriptor is called MSIM.

4.4.2.4 Fingal

The *Fing*erprinting *Al*gorithm (FingAl) published by Brown et al. [44] is an extension of the path-based hashed fingerprints in which geometrical information is incorporated into the paths. This is achieved by an adaption of the atom types that are used in the paths. Each atom type on the path is augmented with its geometrical distance to the previous atoms of the path. The distances are binned into a set of 10 distance classes because it is unlikely that the distances are exactly identical for two paths for a molecular comparison regarding the amount of identical substructures (paths). In the original publication, the boundaries of the bins are {2.35, 2.71, 3.69, 4.47, 5.25, 6.39, 7.59, 9.32, 12.44}. This leads to a better discrimination of the molecules but introduces a bias caused by the conformation. In the original work the –3D structure is computed by CORINA [47].

4.4.3 Hashing

Hashing denotes the mapping of the element of a space χ, which is not necessarily numeric, to an integer (often restricted to a specific interval). The mapping is conducted using a deterministic *hash function* $h : \chi \mapsto Z$. An ideal hash function is injective, which means that the hashes of two inputs are equal only if the inputs are equal. Although this is a preferable property, most hash functions are not injective because the input space is larger than the integer set it is mapped to.

Many hashing algorithms work on bit representations of data objects. Each object is considered as a stream of bits that are subsequently hashed (mostly blockwise) into other bit sets.

4.4.3.1 Cyclic Redundancy Check

The *cyclic redundancy check* (CRC) published by Peterson and Brown [48] is a function that maps a sequence of bit inputs to an integer (its binary representation) in a specific interval. The key idea is to define a *generator polynomial* $p(x) = \sum_{i=0}^{k} c_i x^i$ with coefficients $c \in \{0, 1\}$ and degree k which can be expressed as a bit sequence b of $k + 1$ bits, such that $b_i = c_i$. Analogously, the input bit sequence is subdivided into overlapping blocks of length $k + 1$ and extended with zeros if necessary. The hash value results from a blockwise *modulo* 2 polynomial division of the input sequence by the generator polynomial. Frequently used generator polynomials are $CRC - 32 = x^{32} + x^{26} + x^{23} + x^{22} + x^{16} + x^{12} + x^{11} + x^{10} + x^8 + x^7 + x^5 + x^4 + x^2 + x^1 + x^0$ or $CRC - 16 = x^{16} + x^{12} + x^5 + x^0$. An outline of the algorithm is given in Algorithm 4.7.

ALGORITHM 4.7 PSEUDOCODE FOR THE CRC HASH FUNCTION (ADAPTED FROM WIKIPEDIA)

```
method getCRCvalue(InputBitSet B, GeneratorBitSet G)
{
    ShiftingRegister R := {00000....} //
    Initialization
    while B is not finished do
```

```
        b = nextBit(B)
        R.shiftLeft() //append zero right
        if b != leftmostBit(R) do
            R = R ⊗ G
        od
    od
    return R.asInteger()
}
```

4.4.3.2 InChI Key

The *InChI Key* [49] is a condensed string representation of the InChI encoding of a molecule. The InChI is divided into several blocks that are hashed using a *SHA-256* hash function [50]. The resulting hash values are represented as a string and concatenated, which results in the final InChI key.

$$AAAAAAAAAAAAAA - BBBBBBBBCD$$

The first block A is the encoding of the molecular topology ("skeleton") into a string of 14 uppercase letters. This is achieved by the hashing of the basic InChI layer using a truncated SHA-256 hash function. The second block B represents the remaining layers (i.e., proton positions, stereochemistry, isotopomers, and reconnected layer). Additionally, two single characters are provided. The flag C encodes the InChI version number, the presence of a fixed H-layer, stereochemical and isotopic information. The last position D is a check character defined by a checksum of the first three blocks and verifies the integrity of the key.

4.5 PHARMACOPHORES, FIELDS, AND HIGHER-ORDER FEATURES (3D, 4D, AND SHAPE)

4.5.1 Molecular Shape

One of the basic approaches to take the geometrical structure of a molecule into account for *in silico* comparison and QSAR modeling is to regard the shape of the structure. The shape of two molecules can be compared in several ways, for instance, by the calculation of the overlap volume of two structures and by the comparison of surface descriptors. If not stated otherwise, the following algorithms require that the structures are superimposed in a common coordinate system in a sensible way.

4.5.1.1 Molecular Shape Analysis

The molecular shape analysis (MSA) published by Hopfinger [51] performs a comparison of the overlap volume and provides the base for many other works in shape comparison as well as in structural alignment. In an MSA, the molecules are considered as sets of spheres with different radii (the van der Waals radii of the molecules). The overlap volume of two spheres V_i and V_j with radii r_i and r_j separated by a distance s_{ij} is defined as

$$(V_i \cap V_j) = \frac{\pi}{3}(2r_i^3 + 2r_j^3 + s_{ij}^3)$$
$$- \pi \left[r_i^2 \frac{r_i^2 - r_j^2 + s_{ij}}{2s_{ij}} + \left(s_{ij} - \frac{r_i^2 - r_j^2 + s_{ij}}{2s_{ij}} \right) \left(r_j^2 + s_{ij} \frac{r_i^2 - r_j^2 + s_{ij}}{2s_{ij}} \right) \right].$$

Assuming that the spheres are overlapping (i.e., $s_{ij} < r_i + r_j$) and the smaller sphere is not completely included in the larger one. Thus, the total overlap volume of the molecules A and B can be estimated by $V_{AB} = \sum_{V_i \in A} \sum_{V_j \in B} (V_i \cap V_j)$. This formula does not regard the overlap of more than two atoms and is therefore an overestimation of the real volume. The concept assumes that the molecules are optimally aligned such that one of the structures is translated and rotated maximizing the overlap volume. The maximization of the overlap volume presents also a convenient optimization target function for structural alignment algorithms.

The overlap volume is different from molecular descriptors that have been introduced in this chapter. The most important difference is that it is not a property that only depends on a single molecule. It is only defined in relation to another molecule and thus can be regarded as a measure of the geometrical similarity of molecular structures rather than as a numerical descriptor. It can be used as a descriptor for QSAR modeling by placing all molecules in the same reference frame defined by a fixed reference structure to which the overlap volume is calculated. This overcomes the problem of the relativity of the overlap and provides a common coordinate frame for each molecule. Consequently, it highly depends on the choice of the reference structure, which introduces a bias into the model. Another specific characteristic is that the volume descriptor has an inherent cubic behavior (i.e., cubic influence of the spatial distances on the volume descriptor) introducing a cubic bias into the QSAR formula. To overcome this bias, Hopfinger [51] propose additional features $S_{AB} = V_{AB}^{2/3}$ and $L_{AB} = V_{AB}^{1/3}$ that are calculated based on the overlap volume and which can be used as quadratic and linear terms in the QSAR equation.

4.5.1.2 ROCS—Rapid Overlay of Chemical Structures

ROCS [52] originally was not developed as a molecular descriptor, but a measure for structural similarity to be used in virtual screening. Nevertheless, the same technique of using a common reference molecule as in the MSA can be applied to the ROCS similarity as well. The ROCS approach is similar to MSA in using the overlap volume as a quantitative measure for similarity and can be regarded as an extension of the latter idea.

Instead of using a geometrical definition of the overlap volume of two spheres, ROCS defines the overlap volume as a threefold integral over the products of the *characteristic volumes* expressed by a function $\chi : R^3 \times R^3 \mapsto R$. This function is chosen such that it is zero if a point $\mathbf{r} \in R^3$ lies outside the molecule and is positive if it lies inside. If $\chi(\mathbf{r})$ is set to 1 if $\mathbf{r}$ is inside the molecule, it would yield a hard-sphered volume. For instance, this definition has been used in the work of Masek et al. [53].

In the ROCS approach the characteristic volume is approximated by a 3D Gaussian shape defined as

$$\chi(\mathbf{r}) = 1 - \prod_{i=1}^{N}(1 - p_i e^{-\gamma \mathbf{r}^2}).$$

in which the actual van der Waals radii R of the atoms can be incorporated using the relation $\gamma = \pi(3p/4\pi R^3)^{2/3}$. An advantage is that the intersection of the molecules can be expressed as the product of their characteristic volume functions. Thus, it simplifies the calculation of the overlap volume to an integration (i.e., summing up) over a –3D grid. Hence, the overlap volume of two molecules A and B can be formulated as

$$O_{AB} = \iiint \chi_A(\mathbf{r})\chi_B(\mathbf{r})d\mathbf{r}.$$

This quantity can be regarded as an intersection of two sets of features and thus can be incorporated into other similarity measures like the Tanimoto coefficient. In the case of a discrete characteristic volume function, it can be regarded as a special type of fingerprint in which each bit corresponds to a point in the grid set to 1 if it is inside the molecule and 0 if not. This allows it to define the overlap volume as the intersection of the bit vectors or alternatively as the dot product.

4.5.1.3 Shapelets

MSA [51] and ROCS [52] approaches describe the molecular shape by means of the molecular volume distribution. An alternative representation is chosen by the *Shapelet* [54] concept that encodes the molecular shape using surface patterns. The first step is the calculation of a steric isosurface using a Gaussian modeling of the molecular structure as in the ROCS approach. A molecule is encoded by a 3D function

$$M(\mathbf{x}) = \sum_{i=1}^{N} e^{-2(\mathbf{x}-\mathbf{e}_j)^2/r_i^2},$$

which can be regarded as a superposition of Gaussians, one for each atom i with radius $r \in R$ at the position $\mathbf{c} \in R^3$. As in ROCS, this function assigns every point $\mathbf{x}$ in a spatial grid a value that is positive whenever the points lie inside the molecule. Thus, $M(\mathbf{x})$ corresponds to the characteristic volume function χ used in ROCS although it uses a different mathematical expression. The isosurface of the structure can then be constructed using the marching cubes algorithm [55]. This method generates a set of surface points which is simplified using the welding vertices method [56].

The surface patterns, the *shapelets*, are extracted by a local approximation of the curvature of the surface using hyperbolic paraboloids. The molecular surface is structured into surface patches that are defined by their center surface points p_{r} and their radii r_{s} (default $r_s = 2.5$ Å). A patch is then regarded as the set of surface points P_{i} within the radius r_{s} around the center point P_{r}. The local shape of the surface is then described using the curvature along two perpendicular vectors $\mathbf{e_u}, \mathbf{e_v} \in R^3$

in the tangential plane. These two curvatures together with the normal vector in P_r correspond to a paraboloid that can be regarded as a local surface pattern.

The shape of the surface can be parametrized using the Hessian matrix **H** as $p(u, v, \mathbf{H}) = (uh_{11} + vh_{21})u + (uh_{12} + vh_{22})v$ with $u, v \in R$ being the coordinates in the $\mathbf{e_u}, \mathbf{e_v}$ coordinate system. The coefficients h_{ij} of the Hessian are obtained by minimizing the root mean square error $E(\mathbf{H}) = \sum_i [n_i - p(u_i, v_i, \mathbf{H})]^2$ due to the coordinates n, u, v of a point in the 3D space. The eigenvalues of the resulting Hessian correspond to the two local surface curvatures k_1 and k_2.

These can be used to define the *shape index* $SI(p_r) = \arctan(k_1 + k_2/k_1 - k_2)$ for each surface patch center P_r. This procedure is repeated for each surface point that is not part of a patch, which already has been approximated by another shapelet. An outline of the algorithm is given in Algorithm 4.8.

ALGORITHM 4.8 PSEUDOCODE FOR THE SHAPELET EXTRACTION

```
method getShapeletsForMolecule(Molecule T)
{
   surface = getSurfaceForMolecule(T)
   for all points in surface do
       S = getShapeletAt(point)
       e = getRMSEof(S)
   od
   set = sortedSurfacePointsAccordingToRMSE()
   point = getBest(set)
   shapelets = null
   while set has points do
        shapelets.add(ShapeletOf(point))
        set.remove(point)
        //every surface point is at most part of one
        shapelet
        set.removeAllPointsInPatchOf(point)
        point = getBest(set)
   od
   return shapelets
}
method getShapeletAt(p_r)
{
   e_n = normalVectorFor(p_r)
   p = getPointInSurfacePatchOf(p_r)
   e_u = (e_n × (p − p_r)) / ||e_n × (p − p_r)||
   e_v = e_n × e_u
   points = getLocal2DGridFor(e_u, e_v)
   H = minimizeRMSEonGrid(points)
        k_1 = getFirstEigenvalue(H)
        k_2 = getSecondEigenvalue(H)
```

```
    SI = getShapeIndex(k1,k2,)
    return new Shapelet(Pr,k1,k2,SI, , ,)
}
```

In some respects, shapelets can be regarded as a special type of substructure that does not describe a certain local graph topology but a local surface curvature. Therefore, it suffers from similar drawbacks if used in QSAR modeling or virtual screening. Each shapelet has a complex numerical parametrization and thus two shapelets are unlikely to be parametrized identically. Thus, the intersection of substructure sets of two molecules that forms the common basis of many similarity measures cannot be computed in a straightforward manner. In the original work [54], this problem is solved by defining an algorithm that uses a measure for the similarity of two shapelets in combination with a clique detection approach also used in the detection of pharmacophoric patterns. This results in a least-squares alignment of the rigid structures based on an optimal superposition of the shapelets of the two molecules using the Kabsch algorithm [57]. A similarity score for two molecules A and B can be defined by the molecular volume function $M(\mathbf{x})$ and by summing up the values for the atom centers of molecule A at the atom centers of molecule B.

4.5.2 MIF-Based Features

Field-based features describe a ligand by modeling possible receptor interaction sites around the ligand. This is addressed by placing the molecule in a rectangular grid that defines the spatial points where the interaction potential is calculated. The approach is related to the concept of molecular force fields but takes only nonbonded terms into account. It is therefore independent of the molecular topology (apart from its influence on the atomic charges).

The key idea is to define a probe particle with certain interaction-related properties like charge, size, or hydrophobic potential. For this probe, each interaction potential is calculated at discrete spatial points around the molecule.

In the calculation of the MIF, the probe is placed at each grid point. Next, the interaction potential of the *target* molecule and the probe is calculated. The actual composition of the potential depends on the definition of the field. Often, the potential is restricted to electrostatic, steric, and hydrogen-bonded contributions. However, entropic terms can also be incorporated [58].

4.5.2.1 GRID

A fundamental work in this field was the definition of the GRID force field [59]. In GRID, the probes are not only single atom-like particles but also atom types that include information about the atomic neighborhood (e.g., carbonyl oxygen) or small groups of atoms. The interaction potential is composed of a steric part using the Lennard-Jones potential

$$E_{\mathrm{lj}}(d) := \frac{A}{d^{12}} - \frac{B}{d^{6}}, \quad A, B \in R,$$

an electrostatic contribution that is based on the Coulomb potential E_{el} and the term E_{hb} that describes the hydrogen-bonded interaction. To address the heterogeneous medium, which consists of the solute and the target molecule between the probe and the target atom, the method of images is used [60,61]:

$$E_{\text{el}}(d,p,q) = \frac{pq}{\text{Const.} \cdot \xi}\left(d^{-1} + \frac{(\xi - \varepsilon)}{(\xi + \varepsilon)\sqrt{d^2 + 4s_{\text{p}}s_{\text{q}}}}\right).$$

The dielectric constants of the solute ε and the molecule ξ are considered as constant and separated by a planar boundary. The depth of the target atoms and the probe in the target molecule are addressed by s_{p} and s_{q}. Both are expressed by the number of target atoms in a 4 Å neighborhood. The other parameters and contributions are obtained from the standard Coulomb equation expressing the point charges p and q and their Euclidean distance d. The hydrogen bond contribution is calculated by a modified 6,4-Lennard-Jones potential

$$E_{\text{hb}}(d,\theta) = \left(\frac{C}{d^6} - \frac{D}{d^4}\right)\cos^m\theta$$

to incorporate the angle θ at which the target donor atom prefers the H-bond acceptor. If the probe is a donor, the angle is set to zero because the probe is expected to be rotated optimally. The cosine term is often raised to the power of three but other values are also possible for m. Algorithm 4.9 outlines the calculation of the interaction potential of a single grid point and, thus, one MIF-based descriptor.

ALGORITHM 4.9 PSEUDOCODE FOR A SINGLE GRID INTERACTION POINT

```
method getEmpiricalEnergy(Molecule T, Probe P)
{
    Esteric  = 0
    for all atoms in T do
        if distance(atom, P) < s then //
        s := steric_ distance_threshold e.g. 8~Å
          Esteric + E lj (distance(atom,P))
        fi
    od
    Esteric  = 0
    for all atoms in T do
        Eelectro += Eel
 (distance(atom,P),charge(P),charge(atom))
    od
    Eh-bond  = 0
    for all atoms in T do
        theta = getAngle(P,atom)
```

```
        if theta < 90
            continue
        fi
        if P is a H-bond donor then
            theta = 0
        fi
        Eh-bond += Ehb (distance(atom,P),theta)
    od
    return Esteric + Eelectro + Eh-bond
}
```

The GRID method was developed to examine the surroundings of a receptor-binding pocket and not for small molecules (like most ligands). Therefore, some properties are drawbacks regarding the use of the potentials as molecular descriptors. In a QSAR analysis, the use of the Lennard-Jones potential for steric interactions leads to problems at grid points that are close to target atoms. The reason is the steep increase of the potential if the distance approaches zero [62]. Thus, small differences in the position of the atoms of two molecules can lead to large differences in descriptor values. The respective descriptors have large variances and therefore are regarded as especially meaningful by many QSAR techniques, leading to probably highly overfitted models.

4.5.2.2 Alignment-Based Methods

MIFs [59] yield a large number of molecular features that describe molecular properties that are important for the recognition by a receptor protein. Therefore, they provide a promising starting point for structure–activity models. The consideration of the 3D interaction grid points has some major drawback, though. The most important one is that many machine learning techniques used in QSAR modeling depend on the relation of the values that one descriptor has on different molecules. This makes it necessary to define which grid points of two different molecules are compared to each other. A convenient approach to solve this problem is to align two molecules (i.e., their interaction fields) geometrically. This step is nearly as important as the definition of the interaction field.

4.5.2.3 CoMFA—Comparative Molecular Field Analysis

The CoMFA method [63] is one of the first approaches that used grid-based potentials as molecular descriptors to generate a QSAR model. The field definition is similar to the GRID force field [59] but does not contain a hydrogen-bond contribution. As a default probe serves a sp^3 carbon with a point charge of $+1$. The steric potential is calculated using a 6–12 Lennard-Jones potential similar to GRID with a "steric repulsion" cutoff of 30 kcal/mol. The electrostatic contribution is calculated using the Coulomb potential with a dielectric constant set to the reciprocal distance. The charges of the target atoms are calculated using Gasteiger–Marsili partial charges [64].

To be able to use the resulting descriptors in combination with machine learning methods, the grid points have to be mapped into some reference space. This is done by a 3D structural alignment on an expert selected reference molecule. The method *Field Fit* [65] has been used in the original work of Cramer et al., but other alignment approaches are also possible.

After the corresponding descriptors have been identified by the structural alignment, the QSAR model is inferred using partial least squares. This approach has the advantage that the dimensionality of the problem is reduced but a direct interpretation of the model is not possible. The QSAR equation is transformed back into the original feature space to overcome this drawback. This gives the possibility of a 3D visualization of the grid points that are considered as meaningful.

The relatively simple interaction potential definition used in CoMFA leads to several problems. The Lennard-Jones and the Coulomb potential use the distance between a probe and a target atom in the denominator of the equation. Therefore, a cutoff value is necessary to avoid large or even infinite values at grid points close to target atoms with singularities at the atom positions. The differential behavior of the two potentials inhibits a simultaneous treatment of the cutoff distance, leading to different parametrizations, which worsens the comparability of the interaction fields. Furthermore, it inhibits the calculation of potentials at grid points that lie inside the molecule.

4.5.2.4 CoMSIA—Comparative Molecular Similarity Indices Analysis

The CoMSIA approach [62] avoids some of the drawbacks in the CoMFA method by working without an explicit potential and using the similarity of the interaction potentials instead. The idea is to calculate a similarity $A^q_{F,k}$, analogous to SEAL [66], to the probe instead of an interaction potential. The similarity regarding a physicochemical property k and a probe at the grid point q is calculated as

$$A^q_{F,k} = -\sum_{i=1}^{n} w_{\text{probe},k} w_{ik} e^{-\alpha r^2_{i,q}},$$

where w_{ik} denotes the value of the physicochemical property k of atom i and $w_{\text{probe},k}$ the value of the property at the probe. The attenuation factor α and the distance $r_{i,q}$ are identical to SEAL [66]. Three physicochemical properties have been regarded in the original approach. The electrostatic contribution is realized by using Gasteiger–Marsili partial charges [64] for the target atoms and a +1 charge of the probe. The steric term uses the cubic power of the van der Waals radius of the atoms and 1 Å for the probe. The entropic contribution is represented by the hydrophobicity parametrization published by Viswanadhan et al. [67] (+1 for the probe).

The resulting grid points of the similarity indices can be used as molecular descriptors for the inference of a QSAR model. Partial least squares are—as in CoMFA—the method of choice because of the large number of descriptors. The exchange of the term that incorporates the distance between the probe and target atoms to a bell-curve prevents the numerical problems that occur in the Lennard-Jones and Coulomb potentials in CoMFA. Furthermore, it simplifies the implementation.

This advantage comes at the cost of an interpretable interaction field that can be used for a guided drug design because of a lack of an explicit physicochemical interpretation of the grid points.

As in CoMFA, this algorithm does need a structural alignment to compare the descriptor values of the grid points of different molecules. In the original publication, the alignment was computed by applying the SEAL method.

4.5.2.5 Structural Alignment

4.5.2.5.1 Kabsch Algorithm

The Kabsch algorithm [57] is a popular algorithm for a rigid superposition of two points set (i.e., molecular structures) by minimizing the root mean squared distance of pairwise assigned atoms. Recently, Coutsias et al. reformulated the algorithm in terms of quaternion algebra to overcome some pitfalls of the original version without giving different results [68]. Nonetheless, it is still necessary to define a pairwise assignment of the atoms of two molecules onto each other.

The target function to be minimized is given by the rotation matrix

$$E = N^{-1} \sum_{k=1}^{N} w_k \, |Q\mathbf{x}_k + \mathbf{t} - \mathbf{y}_k|^2 \quad \text{with } Q \in R^3 \times R^3,$$

where $\mathbf{t} \in R^3$ represents the translation vector and $\mathbf{x}$ and $\mathbf{y}$ the ordered sets, such that $\mathbf{x}_k$ and $\mathbf{y}_k$ are assigned onto each other. The optional weight factor w_k allows setting individual penalties for distances of certain atomic pairs. The reference molecule is denoted by $\mathbf{y}$. The rotation matrix Q is expressed in terms of quaternions:

$$Q = \begin{pmatrix} q_0^2 + q_1^2 - q_2^2 - q_3^2 & 2(q_1q_2 - q_0q_3) & 2(q_1q_3 + q_0q_2) \\ 2(q_1q_2 + q_0q_3) & q_0^2 - q_1^2 + q_2^2 - q_3^2 & 2(q_2q_3 - q_0q_1) \\ 2(q_1q_3 - q_0q_2) & 2(q_2q_3 + q_0q_1) & q_0^2 - q_1^2 - q_2^2 + q_3^2 \end{pmatrix}, \tag{4.2}$$

which allows formulating the rotation as a single quaternion $\mathbf{q} = (q_1, q_2, q_3, q_4)$. The explicit calculation of the optimal translation $\mathbf{t}$ is avoided by shifting the barycenters of both structures into the origin of the coordinate system. For the calculation of the rotation matrix, a 3×3 matrix $R_{ij} = \sum_{k=1}^{n} x_{ik}y_{jk}, i,\ j = 1, 2, 3$, is defined. The quaternion representing the optimal rotation can then be found as the eigenvector of the matrix

$$F = \begin{pmatrix} R_{11} + R_{22} + R_{33} & R_{23} - R_{32} & R_{31} - R_{13} & R_{12} - R_{21} \\ R_{23} - R_{32} & R_{11} - R_{22} - R_{33} & R_{12} + R_{21} & R_{13} + R_{31} \\ R_{31} - R_{13} & R_{12} - R_{21} & -R_{11} + R_{22} - R_{33} & R_{23} + R_{32} \\ R_{12} - R_{21} & R_{13} + R_{31} & R_{23} + R_{32} & -R_{11} - R_{22} + R_{33} \end{pmatrix}$$

that correspond to the largest eigenvalue. A summarizing pseudocode for the algorithm is given in Algorithm 4.10.

ALGORITHM 4.10 PSEUDOCODE FOR THE KABSCH ALGORITHM

```
method getKabschRotation(Molecule mol, Molecule ref)
{
      //apply weighting and shift barycenters
      for all atoms in mol do
           coord(atom) *= w_k
           shift atom
      od
      for all atoms in ref do
           coord(atom) *= w_k
           shift atom
      od
      F = calculate_F_Matrix(mol,ref)
      lambda_max = getLargestEigenvalue(F)
      q = getEigenvectorForEigenvalue(lambda_max)
      return RotationMatrixForQuaternion(q)
}
```

4.5.2.6 SEAL—Steric and Electrostatic Alignment

The SEAL algorithm [66] performs a 3D structural alignment by rotating and translating a rigid structure such that the weighted distances of their atoms to a reference molecule is minimized. The target function A_{F} is a double sum over all atoms of both molecules:

$$A_{\mathrm{F}} = -\sum_{i=1}^{m}\sum_{j=1}^{n} w_{ij}\mathrm{e}^{-\alpha r_{ij}^2}.$$

The weight coefficients $w_{ij} = w_{\mathrm{E}}q_i q_j + w_{\mathrm{S}}v_i v_j$ allow regarding the physicochemical similarity of two atoms. Charges q_i and q_j with the same sign lead to a larger weight and, therefore, to a larger penalty of the distance as do large van der Waals radii v_i and v_j. The attenuation parameter α and the physicochemical (electrostatic and steric) weights w_{E} and W_{S} should be optimized by the user, depending on the data.

The optimal alignment (i.e., that one that minimizes A_{F}) is calculated by expressing the distance (radius) variable r_{ij} in terms of a rotation and translation of the structure to be aligned. This can be achieved—similar to the Kabsch formulation—in terms of quaternion algebra using a rotation matrix Q and a translation vector $\mathbf{t}$ leading to $r_{ij}^2 = (\mathbf{x}_i - Q\mathbf{x}_j - \mathbf{t})^2$.

The alignment procedure starts with a Kabsch superposition of predefined atomic pairs. It results in a barycentered starting point for the *rational function optimization* (RFO) [69], an optimization method that is guaranteed to find the global minimum of A_{F} and converges quadratically. SEAL is only designed to perform rigid structure alignment. Therefore, the authors propose to incorporate it into a framework that also performs a conformational sampling and a diversity subset selection.

Other drawbacks that have been reported are the strong sensitivity to changes in the parametrization and the lack of terms that regard entropic contributions [70].

4.5.2.7 Alignment-Free Methods

The 3D descriptor algorithms presented so far share the need for a structural alignment of the molecules. This is a time-consuming step likely to introduce a bias depending on the alignment procedure used. To overcome these drawbacks several methods has been proposed that avoid the use of a shared external coordinate system.

4.5.2.8 GRIND—GRid-INdependent Descriptors

The GRIND algorithm [71] extends the idea of an autocorrelation of structures to the surrounding MIF. The potentials at the grid points can be calculated by any force field suitable for the definition of a ligand–receptor interaction potential. Unlike the CoMFA approach, these potentials are not directly used as molecular descriptors but further processed in an autocorrelation algorithm. Relevant interaction regions are identified regarding significant negative energy potentials of neighboring grid points. This filtering step is performed by defining an optimization problem. The resulting ensemble of regions with strong interaction potentials is then regarded as the *virtual receptor site* (VRS) that represents the starting point of the autocorrelation.

To overcome the need of a structural alignment the spatial positions of the points of the VRS are not represented using explicit coordinates but their distances to other points of the VRS either of the same (autocorrelation) or of a different (cross-correlation) potential type. This idea is similar to an atom pair or a pharmacophore approach that also relates property positions using distances to other positions. Although spatial points in the molecular neighborhood are different to atoms or groups, which are part of a specific molecule, molecular descriptors are obtained by this concept. These are correlogramms that relate products of potentials with the spatial distances between them.

4.5.2.9 VolSurf

In contrast to the approaches that have been discussed, VolSurf [72,73] does not attempt to describe relationships between the structure of a small molecule and its activity towards some protein target but its relationship to a complex property. It has been successfully applied in modeling physicochemical properties like the ΔG of hydration and in predicting pharmacokinetic behavior like skin permeability. VolSurf is also different in some way to the other interaction field methods because the descriptors that are calculated are not attributes of grid points but combinations of grid point features and their spatial distribution. For the calculation of the interaction potential the force field definition of GRID [59] is used. However, the resulting potentials are not regarded as grid point descriptors. Instead, the volumes and the surface areas of interaction contours (i.e., spatial clusters of grid points with interaction potential above a certain threshold) are calculated. The properties of the interaction regions are further combined to give a set of molecular features that are divided into several parts depending on which types of potentials are used:

4.5.2.9.1 Size and Shape Descriptors

Size and shape descriptors are properties of a molecule that depend solely on its topology and geometry only taking steric features into account. In the original work, this set consists of four descriptors for the *solvent excluded molecular volume*, which is calculated using the grid points that do have a steric interaction potential above 0.2 kcal/mol, and the *solvent accessible surface*, also based on the same steric potential. Additionally, the ratio between volume and surface and the ratio between the surface of the molecule and the surface of a sphere of the same volume is considered.

4.5.2.9.2 Hydrophilic Region Descriptors

The threshold for grid points to be regarded as parts of a hydrophilic region is an interaction potential for a water probe between −1.0 and −6.0 kcal/mol. Ratios are also taken into account, as it is the case for steric regions, giving a *capacity factor* of the molecule defined as the relative size of the hydrophilic regions compared to the complete molecular surface (the ratio between the hydrophilic and the complete surface).

4.5.2.9.3 Hydrophobic Region Descriptors

Hydrophobic descriptors are defined analogous to the hydrophilic using a specific range of interaction potentials towards the hydrophobic (in GRID: DRY) probe.

4.5.2.9.4 INTEGY Moments

The interaction energy (INTEGY) moments are a measure for describing the distribution of the interaction regions around the molecule. An INTEGY moment is expressed as a vector that points from the barycenter of the molecule to the center of the regions of a specific interaction type (e.g., hydrophilic or hydrophobic interactions).

4.5.2.9.5 Mixed Descriptors

Several descriptors which are not of the former types are also included. The best interacting (of the different potential types towards a water probe) grid points and their spatial distance from each other are taken into account as well as combinations of the hydrophilic and hydrophobic regions. The latter consist of the ratio between their sizes, the vector that points from the hydrophobic center towards the hydrophilic center (the amphiphilic moment), and a further ratio-based feature that relates the volume of the hydrophobic regions to the product of the hydrophilic surface area and the lipophilic surface area. In addition to these combinations of previously calculated features, two descriptors are defined to cover the hydrogen bonding and the polarizability of the molecule. The method used for the calculation of the polarizability considers the topology of a molecule but is independent of the MIF [74].

4.5.3 Pharmacophores

The assumption that there is a direct connection between the structure of a compound and its biological properties is a fundamental paradigm of medicinal chemistry. It is also a basic necessity of any structure–activity relationship model, independent of whether it is quantitative or qualitative. The development of increasingly elaborated molecular descriptors can therefore also be regarded as a trial to encode the underlying structural causes which cannot be elucidated in an explicit manner. This is one drawback of all models that are based on descriptor encodings of molecules. Even if a biological property can be quantitatively described in a precise and general way, this does not necessarily give a recipe for structural modification that would enhance a molecule. One of the advantages of the MIF approaches is that the possibility of a graphical representation can serve as guidance on which interactions at which structural parts are important for the biological functionality. This is especially important regarding the modeling of the binding affinity of a small molecule to a protein.

The underlying concept that the ability of a molecule to bind to a protein (i.e., to fit into the binding pocket and establish interactions strong enough to induce or inhibit an effect) depends on a certain geometrical arrangement of interaction (pharmacophoric) points. Several methods that extract such geometrical patterns describe them as *pharmacophores*. The IUPAC defines a pharmacophore as "the ensemble of steric and electronic features that is necessary to ensure the optimal supramolecular interactions with a specific biological target structure and to trigger (or to block) its biological response" [75].

The methods that are used for the recognition of pharmacophoric patterns can be divided into ligand- and receptor-based approaches. The basic idea of ligand-based pharmacophore extraction is to detect spatial patterns of pharmacophoric points that are conserved in many active structures (ensemble methods), whereas a receptor-based approach defines a spatial arrangement of areas in the binding pocket at which specific interactions (e.g., H-bonds) can be established. The latter is sometimes also referred to as *inverse* pharmacophore or interaction *hot spots*.

The first step in most ligand-based pharmacophore recognition algorithms is the definition of a set of structural features that are regarded as pharmacophoric points. Usually, this is done using a categorization of the ligands atoms (or substructures) into atom types like *hydrogen-bond donor* or *aliphatic*. The definition of the atom types is an important step and has a major influence on the quality of the pharmacophores that are found using the ligand-based methods. If the typing is chosen too fine, the algorithm will likely not be able to find shared patterns, whereas a too general definition would decrease the information content and induce many false positives. Most pharmacophore extraction procedures assume that the typing of the ligands has already been accomplished and are designed for the recognition of shared spatial arrangements of the pharmacophoric points.

4.5.3.1 Ensemble Methods

Ensemble methods are designed to detect a spatial arrangement of pharmacophoric points that is preserved in a set of active ligands. There are several ways to approach this task, for instance: the *distance geometry* methods [76,77], the *clique detection/DISCO* method [78], and the configuration-based approaches [79].

4.5.3.1.1 Distance Geometric Methods

The idea behind a distance geometric approach to pharmacophore recognition is to derive spatial bounds for the relative positions of some pharmacophoric points. Dammkoehler et al. [77] used this concept to find conserved spatial arrangements of predefined pharmacophoric points in a set of structures. The *constrained search of conformational hyperspace* starts with a specified set of k pharmacophoric points. A parametrization of the geometric arrangement of these points can be expressed by the $(1/2)k(k-1)$ pairwise distances and is referred to as a *regular model*, which can be restricted by defining specific distances as fixed. Therefore, each geometry of a pharmacophore corresponds to a point in the $(1/2)k(k-1)$ dimensional hyperspace H defined by the model. The algorithm then subsequently constrains the subspace of H which contains geometric arrangements of the pharmacophoric points that can occur in the molecules due to their conformational flexibility. This is achieved by iterating over the actives beginning at the most rigid structure and determining which geometries can be produced by variations of the torsional angles subject to steric constraints. Each spatial arrangement that cannot be produced by a new molecule is removed from the subspace of H. This subspace is further constrained until it only contains geometric configurations that can be adopted by every structure. These allowed pharmacophoric point arrangements are considered as pharmacophores of the examined set of actives.

4.5.3.1.2 DISCO—DIStance COmparison

The DISCO algorithm by Martin et al. [78] solves the pharmacophore detection problem by the generation of the *association graph* H for every pair of ligand conformations.

H is defined such that a preserved geometrical arrangement of pharmacophore points corresponds to a *clique* in H. A clique of H is every subgraph of H that is fully connected. The problem of the identification of cliques in a graph is known to be NP-complete. In spite of that, it is computationally feasible due to the sparseness of H in this case.

The algorithm begins with a set of active compounds with assigned pharmacophoric points and a sample of diverse conformations. The molecule with the smallest number of conformations is used as the reference structure R. In the next step, the association graph $H(R, M_j)$ is constructed for each conformation M_j of each other molecule M. $H(R, M_j)$ is defined as the vertex set $V := \{(a, b)|a \in \text{atoms}(R), b \in \text{atoms}(M_j)\}$ and the edge set $E := \{[(a, b), (c, d)]||\text{dist}(a, c) - \text{dist}(b, d)| < \theta\}$. This corresponds to a graph that has one vertex for each pair of atoms with identical atom types (i.e., pharmacophoric points) with one in R and one in M_j. Vertices are connected by edges if the intramolecular distances of the atoms in R and the atoms in M_i are equal up to a certain tolerance threshold θ.

Each clique in the association graph corresponds to a set of similar atoms (identical pharmacophoric points) in the two molecules that adopt a similar spatial arrangement and can thus be regarded as a pharmacophore of two molecules (Figure 4.1). The largest clique and therefore the most expressive pharmacophore can be identified using the Bron–Kerbosch algorithm [80] outlined in Algorithm 4.11.

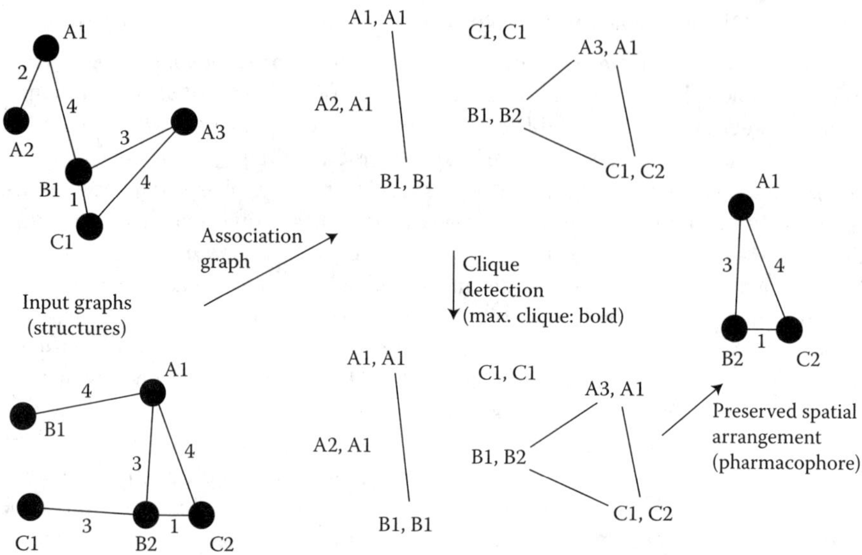

FIGURE 4.1 Schematic example of the DISCO pharmacophore detection algorithm [78].

ALGORITHM 4.11 PSEUDOCODE FOR THE BRON–KERBOSCH ALGORITHM [80] FOLLOWING SAMUDRALA/MOULT [81]

```
method clique(, M_D, C_D, N_D)
{
    if a node in N_D is connected to all nodes in
    C_D then
         // branch and bound step
         no more cliques can be found;
    else
         // backtracking steps
         for all nodes in C_D do
             move candidate from C_D to M_{D+1};
             create C_{D+1} by removing all nodes from C_D,
             which are not connected with the candidate
             node;
             create N_{D+1} by removing all nodes from N_D,
             which are not connected with the candidate
             node;
          if C_{D+1} = Ø and N_{D+1} = Ø then
             store M_{D+1} as maximum clique;
          else
             clique(M_{D+1},C_{D+1},N_{D+1}),;
          fi
          move nodes from M_D to N_D;
     od
   fi
}
```

DISCO can be extended to regard the chirality of a molecule. This can be important if pharmacophores consisting of more than three points are identified. To achieve this, a clique is only accepted if the torsional angles in the corresponding pharmacophoric points in both molecules are similar, according to another tolerance threshold.

4.5.3.1.3 Common Configurations

A major drawback of the clique-based pharmacophore approaches is that they either have to be repeated for each pair of actives (and their conformations) or a set of *reference* compounds has to be selected. The first is computationally too demanding (quadratic complexity) in most cases and the latter introduces a bias to the reference selection.

Therefore, Barnum et al. [79] proposed a new approach that simultaneously considers each (precalculated) conformation of each active structure as a reference while preserving a linear runtime complexity in the number of molecules.

The algorithm starts with the identification of *configurations* of the pharmacophoric points that are shared among the molecules. A configuration is similar to the distance geometry. A specific spatial arrangement is defined by the distances between the considered pharmacophoric points. Two configurations are considered as equal if the difference in their geometries is below a threshold. Additional to this relaxation it is allowed that the configurations do not share all pharmacophoric patterns. The configurations are therefore not regarded as incompatible if they are different in one aspect and share the remaining features. This concept can be described more formally by the definition of a *partition* P as a set of specific pharmacophoric patterns and a *subpartition* of P as each subset of P that contains all but one pattern of P.

The algorithm proceeds by iterating through the existing partitions in ascending cardinality. It checks, for each partition, which reference molecules have configurations (and conformations) that are associated with the partition and additionally have subconfigurations related to the subpartitions. After this step, a list of reference (sub)configurations of the reference compounds is obtained that is associated with a specific (sub)partition. Then, all configurations of both reference and nonreference compounds are examined if they can be considered as *common*. This is the case if they have a compatible geometric arrangement of the partition elements to one of the previously identified reference configurations. The compatibility depends on the spatial arrangement as well on the similarity of the pharmacophoric points. The procedure is repeated for the next partition until there are no more common configurations possible.

The result of the described algorithm is a list of configurations of pharmacophoric patterns that are considered as common. Each of these common configurations fulfills the demands of a pharmacophore. An additional ranking can be used to the select the most descriptive representations. For this purpose Barnum et al. [79] propose a scoring scheme related to the Kullback–Leibler distance of probability distributions. The score of a configuration C of K pharmacophoric features is defined as

$$s(C) = \#\text{actives} \sum_{x=0}^{K+1} q(x) \log_2 \frac{q(x)}{p(x)},$$

with x being the *class* of a match between a molecule M and the configuration C. A match has class $K + 1$ if all the configurations features are matched by M and class 0 if no feature and no subconfiguration are matched. The classes $1 \ldots K$ correspond to matches between M and one of the K subconfigurations of C (every $K - 1$ sized subset of C). The two quantities $q(x)$ and $p(x)$ denote the fraction of the structures that have a class x match to C ($q(x)$ regards only active structures and $p(x)$ regards all compounds).

4.5.3.2 Receptor Surface Models

The definition of the surface of the binding pocket of a protein using only ligand structures is usually not considered as a pharmacophore approach but can be regarded as an intermediate between a field-based and a pharmacophore approach. The basic idea presented by Hahn [82,83] is to generate a kind of consensus shape out of a set of structures that are active against a specific target. Assuming that the knowledge of a bioactive conformation is given, a structural alignment of the ligands can be used to calculate the spatial areas that are not occupied by any ligand. The boundary of this space can then be considered as the hypothetical surface of the receptor binding pocket. The algorithm starts with a set of aligned conformations of ligands that are embedded in a spatial grid similar like to CoMFA [63]. A steric potential is calculated for each grid point. Hahn proposes two different potential functions: the *van der Waals* function and the *Wyvill* function. Each function calculates a steric potential regarding one atom a for each grid point $\mathbf{x}$:

$$P_{vdW}(\mathbf{x}) = \text{dist}(a, \mathbf{x}) - \text{radius}(a)$$

$$P_{Wy}(\mathbf{x}) = -\frac{4}{9}\left(\frac{\text{dist}(a,\mathbf{x})}{R}\right)^6 + \frac{17}{9}\left(\frac{\text{dist}(a,\mathbf{x})}{R}\right)^4 - \frac{22}{9}\left(\frac{\text{dist}(a,\mathbf{x})}{R}\right)^2 + 1,$$

$$\text{where dist(a,}\mathbf{x}) < R.$$

The van der Waals potential defines a grid that is zero exactly on the van der Waals surface of the atom, negative inside and positive outside. The Wywill function is bounded by a parameter R that defines the distance at which the potential vanishes. This function is evaluated at each grid point for each atom. The resulting potential for the grid point is the minimum potential of all atoms. The atom nearest to that specific grid point is further used to define the physicochemical properties of the grid point.

The receptor surface is modeled by averaging the potentials over the ligand ensemble and calculating the isosurface similar to in the Shapelet method [54] using the *marching cubes* algorithm. If this is set to zero, the resulting isosurface resembles a kind of joint van der Waals surface of the ligand ensemble. This hypothetical receptor surface can then be annotated with physicochemical properties by interpolating the property values of the eight grid points that surround each grid cell that corresponds to a surface point.

The physicochemical properties that can be incorporated in the receptor surface model are related to those in the MIF algorithms but express values of receptor atoms. Therefore, it is calculated which property values of the receptor surface would be preferable at each specific surface point. The properties considered are: the partial charge formulated as the inverse of the mean partial charge of the neighboring ligand atoms, the complementary electrostatic coulomb-like potential to the ligand grid property, a hydrogen bond property that denotes if in average a donor (-1) or acceptor ($+1$) would be preferable, and a binary flag for the hydrophobicity of that surface part.

The resulting annotated isosurface of the hypothetical binding pocket can be used in several ways. It can be viewed as a kind of inverse pharmacophore denoting which ligand groups would be preferable at certain spatial points. Furthermore, it can be used for the calculation of the potential energy of unknown molecules towards the hypothetical surface. In all cases, the model should be first relaxed by the user by cutting out those parts of the surface which cover assumably the opening of the pocket and therefore do not restrict spatial positions of ligands. The potential energy can then be separated into different potential types (e.g., steric and electrostatic) which can be used as molecular descriptors to infer a QSAR model for this protein target. Other proposed descriptors are the interaction energy for the receptor surface model, the conformational energy of the "bound" conformation, the conformational energy of the "relaxed" conformation (minimized outside the binding pocket model), and the difference between the bound and the relaxed conformational energy.

4.5.4 Higher Dimensional Features

The incorporation of geometrical information in the field- and shape-based molecular representations introduces a strong bias to the structural conformation on which the calculation is performed. Several approaches to avoid this problem have been proposed. The general idea is to regard several geometrical conformations during the feature generation. For instance, this concept has been outlined in the *4D QSAR paradigm* published Santos-Filho and Hopfinger in 2002 [84]. The first step is analogous to a field-based 3D QSAR and consists of the definition of coordinate system (grid) where initial conformers of structures are placed. In contrast to the field approach, no interaction potentials are calculated. The atoms of the molecules are categorized into several pharmacophoric classes (*negative polar, positive polar, nonpolar, hydrogen-bond donor, hydrogen-bond acceptor, aromatic*) and a wildcard type (*any*) resulting in a set of interaction pharmacophoric elements (IPEs). The fourth dimension is introduced by a conformational sampling using a molecular dynamics simulation. This leads to a set of conformers for each molecule. The comparison of the positions of the IPEs requires a structural alignment of the different conformations.

The aligned structures are further processed to return a set of 4D features which are based on the *occupancies* of the cells of the reference grid the conformers are placed in. These occupancies represent the features which can be regarded as 4D molecular descriptors. For each IPE type, three occupancy types of the grid cells were proposed in the original work of Santos-Filho and Hopfinger [84]:

Absolute-occupancy A_0*:* The absolute occupancy is a measure for the number of all IPEs of all conformers of a molecule that are placed inside a specific grid cell.

Joint-occupancy J_0*:* The joint-occupancy counts the number of IPEs in a specific cell that occur in this cell as well for the actual compound and for the reference molecule. Therefore, it is a kind of similarity measure to a reference molecule that regards the conformational space of both compounds.

Self-occupancy S_0*:* The self-occupancy is calculated as the difference between the absolute-occupancy and the joint-occupancy.

The resulting 4D features can be used as descriptors in QSAR modeling. Similar to CoMFA [63], the large number of feature favors machine learning methods that are capable of dealing with large feature spaces as it is the case for partial least squares.

All approaches that have been presented so far describe the ligand compound in increasing complexity reaching its limit in the consideration of conformational ensembles in the 4D QSAR paradigm. The 5D QSAR idea [85,86] goes beyond that by incorporating information about the receptor structure and even its flexibility regarding induced fit effects. This receptor-dependent QSAR (RD-QSAR) concept does not necessarily need real information about the ligands target. The construction of *receptor envelopes* as it is proposed in the work on 5D QSAR of Vedani and Dobler [85,86] uses only the set of conformational ensembles of the ligands to infer a model of the hypothetical receptor binding side. This is done using the concept of a hypothetical receptor surface model originally published by Hahn [82,83]. The receptor surface model is extended in this approach to incorporate *induced fit* effects. A ligand-specific induced fit surface, called the "inner envelope," is calculated for each molecule by mapping the receptor surface that has been computed using all ligands onto the van der Waals surface of the single molecule. The magnitude of the deformation measured as the RMSD of corresponding surface points can be used to calculate a hypothetical "induced fit" energy of this molecule. This energy is combined with other force field energy terms into an equation that describes the binding energy of this molecule to the hypothetical receptor.

Therefore, the inferred surface can be regarded as QSAR equation. The equation is trained by a genetic algorithm that varies the surface properties, which have been randomly assigned, in order to optimize the fit of the models energy equation to the target values. Thus, the 5D QSAR approach is different from most of the previously represented ideas because of its different understanding of descriptors. The surface properties are varied in order to learn the model. Therefore, they can be regarded as coefficients rather than features. If considered as descriptors, the interaction potentials towards the ligand atoms are regarded as the values of the surface points. Thus, the approach is to some extent the learning of a receptor binding pocket.

4.6 IMPLICIT AND PAIRWISE GRAPH ENCODING: MCS MINING AND GRAPH KERNELS

4.6.1 MCS Mining

4.6.1.1 Maximum Common Subgraph

A maximum common subgraph (MCS) is the result of a search for maximum isomorphic pairs (S, S'), such that S is subgraph of G and S' a subgraph of G'. From a formal point of view, a *graph isomorphism* is a bijection between the vertices G and G', such that the structure of the assignments is conserved by the assignment function.

4.6.1.2 Exact Maximum Common Substructure

Sheridan and Miller [87] suggest a method for detecting meaningful common substructures in active compounds based on clique-detection in pairs of molecules. The Highest Scoring Common Substructures (HSCS) are subgraphs that have a significantly higher score then common substructures by chance for randomly selected compounds.

Sheridan and Miller [87] define common substructures as corresponding atoms with the same atom type, where the atom type was set to cation, anion, neutral HBD, neutral HBA, polar atom (acceptors or donors), hydrophobe, and other. The second requirement for a common substructure is that the respective pairs of atoms must have the same topological distance. This is determined by the shortest path between two atoms. The score for each substructure is defined by

$$\text{Score} = \text{Size} - p(N_{\text{frag}} - 1),$$

where the size equals the number of atoms, p is a "discontinuity penalty" (between 1.0 and 2.0), N_{frag} is the number of discontinuous fragments.

$$\text{Meanscore HSCS}(n_A, n_B) = M_{\text{mean}} \cdot \min(n_a, n_b) + B_{\text{mean}},$$

$$\text{Stdv HSCS}(n_A, n_B) = M_{\text{stdv}} \cdot \min(n_A, n_B) + B_{\text{stdv}},$$

$$Z - \text{Score} = \frac{[\text{Score} - \text{Mean}(n_A, n_B)]}{\text{Stdv}(n_A, n_A)},$$

where n_{A}, n_{B} are the number of atoms in molecules A and B, respectively. M_{mean} and B_{mean} are constants depending on p, the database, and the atom types. $\text{MeanscoreHSCS}(n_{\text{A}}, n_{\text{B}})$ describes a linear function of the expected score. The Z-Score for an HSCS between two molecules is a score for the unlikeliness of an occurrence for an HSCS. HSCSs are regarded as significant with a specific score (Sheridan and Miller use a threshold of ≤ 4.0).

Pharmacophores are detected via a modified clique-detection algorithm (see Algorithm 4.12). C denotes a clique that is defined as set of paired atoms from A, B. $C_A(i) = j$ is the ith atom in a clique in A, and the corresponding atom number is j. $V_A(i) = 1$ records if atom i in A is available for matching. $V(*) = 0$ resets the complete mask.

ALGORITHM 4.12 HSCS CLIQUE DETECTION, PSEUDOCODE ADAPTED FROM SHERIDAN AND MILLER [87]

```
// enumerate each possible pairing of atoms in
both structures for i, j do
// first match of a possible clique
if (A.getAtom(i)=B.getAtom(j))
```

```
Initialize
        VA(*) = 1,    VB (*) = 1,    VA (i) = 0,    VB(i) = 0
npair = 1;    CA(i) = i;    CA(j) = j;
for all pairs do
    boolean kmcheck=false;
    for all pairs of k ∈ A and m ∈ B do
    if  (M(k,m) = 1)
       if (VA(k) = 1 and VB(m) = 1)
       for k, l do
       if  (dist_topo (k,CA (*)) == dist_topo (m,CB(*)))
       S=sum of distances
       if (S < Smin)
       Smin = S
       k' = k
       m' = m
       kmcheck=true
    fi
    fi
    od
    fi
    fi
    od
    // if no cliques of smaller size exist exit here
    if(!kmcheck)
    exit;
    else
    n pair++;
          VA(k') = 0, VB (m') = 0,
          CA (npair) = k';
          CB (npair) = m';
   fi
   if npair ≥ min(nA,nB)
   exit;
   fi
od
od
```

4.6.1.3 Inexact Maximum Common Substructure

Birchall et al. [88] suggest an approach based on edit operations, such that the similarity is determined by the cost of insertions, deletions, and mutation to transform a graph into another. The approach published by Birchall et al. [88] works on the set of paths p, p' extracted from a reduced molecular graph using the optimal cost of transforming p into p'. The set of paths is computed as the set of all linear shortest paths of all vertices of degree one. These are referred to as maximum paths. For all maximum paths in two molecules A, B the minimum weighted distance is determined by a minimum backtracking using a cost matrix. The final edit score is the maximum cost path considering two graphs. The penalties for the cost matrix were optimized using a genetic algorithm.

4.6.2 Kernel Functions

A kernel is a real-valued, symmetric and positive semidefinite function $k : \chi \times \chi \rightarrow R$, defined on the space χ. Many sophisticated kernels are built up of a set of basic kernel functions subject to the closure properties of kernel functions.

Numerical kernels can deal with arbitrary vectors of numerical data of the same dimension; nominal kernels compare nominal features, such as symbols or discrete values.

Graph kernels are able to consider various features of a molecular graph like paths, cycles, and pharmacophores [19,89–96]. Kernels have the advantage that they are not restricted to a fixed-sized vectorial representation like a binary fingerprint, a vector of molecular attributes, or structural keys of a defined size.

In the following section, we introduce the kernel closure properties and basic kernel functions on numerical and nominal attributes. Then, we will introduce kernel functions defined on molecular graphs beginning with simple topological kernel functions up to 3D kernel functions.

4.6.2.1 Kernel Closure Properties

A suitable kernel can be designed systematically to reflect a sensible similarity. Kernel machines, like SVMs or Gaussian Processes, use the kernel property to solve a convex learning problem (optimization problem) optimally. The kernel properties are closed under a number of operations (adapted from [97]):

$$c \cdot k(x_i, x_j)$$

$$f(x_i) \cdot k(x_i, x_j) \,.\, f(x_j)$$

$$q[k(x_i, x_j)]$$

$$\exp[k_1(x_i, x_j)]$$

$$k_1(x_i, x_j) + k_2(x_i, x_j)$$

$$k_1(x_i, x_j) \cdot k_2(x_i, x_j)$$

$$k_3[\phi(x_i), \phi(x_j)]$$

$$x_i^T A x_j$$

$$k_a(x_a, \overline{x_a}) + k_b(x_b, \overline{x_b})$$

$$k_a(x_a, \overline{x_a}) \cdot k_b(x_b, \overline{x_b})$$

where c is a scalar, $\varphi(x)$ defines a mapping $x \rightarrow R^M$, k_3 is a valid kernel in R^M. A is positive semidefinite matrix, x_a and x_b are variables with $x = (x_a, x_b)$, and k_a and k_b

are kernel functions in their respective spaces. f is any function and q is a polynomial function with non-negative weights.

4.6.3 Basic Kernel Functions

With the help of the closure properties it is possible to build powerful kernels from a set of basic kernel functions. It is not only useful to describe the similarity of global descriptors, but also for local numerical descriptors like atomic attributes. The following basic numeric kernel functions can be directly applied to numerical descriptors.

4.6.3.1 Numerical Kernel Functions

The Gaussian radial basis function (RBF) kernel is defined as

$$k_{\text{RBF}}(x_i, x_j) = \exp\left(-\frac{\|x_i - x_j\|^2}{2\sigma^2}\right).$$

The RBF kernel is used for the pairwise comparison of two samples $\mathbf{x_i}, \mathbf{x_j} \in R$. The σ parameter adjusts the diameter and height of the resulting Gaussian peaks at the support vectors (the noise of the data), if a support vector machines is applied.

The polynomial kernel is defined as

$$k_{\text{poly}}(x_i, x_j) = (\langle x_i, x_j \rangle + \theta)^d.$$

The linear kernel is a special case of the polynomial kernel:

$$k_{\text{linear}}(x_i, x_j) = \langle x_i, x_j \rangle.$$

The hyperbolic tangent kernel, which is not always positive semidefinite and therefore should be regarded as a pseudokernel, is defined as

$$k_{\text{tanh}}(x_i, x_j) = \tanh(\langle x_i, x_j \rangle + \theta).$$

The Laplacian kernel is defined as

$$k_{\text{Laplacian}}(x_i, x_j) = \exp(-\sigma\|x_i - x_j\|).$$

The Brownian Bridge kernel is defined as

$$k_{\text{Brownian}}(x_i, x_j) = \max(0, c - k\|x_i - x_j\|),$$

where $d \in N$ and $\{c, k, \sigma\} \in R$ and $x_i, x_j \in R^n$.

The RBF-related kernels (here: RBF, Laplacian) normalize the similarity in $x \in [0, 1] \in R$. This is a useful property for support vector machines, where the complexity of the training algorithm depends on the kernel values. Smaller kernel values decrease the training computation time.

4.6.3.2 Nominal Kernel Functions

Some kernel functions are defined on arbitrary sets of nominal features.

The Delta Function (or Dirac kernel) is defined as

$$k_{\text{Delta}}(x_i, x_j) = \begin{cases} 1 & \text{if } x_i = x_j \\ 0 & \text{else} \end{cases}.$$

The p-spectrum kernel is defined on strings of length p. Basically, it can be used to compare feature sets of an infinite length:

$$k_{\text{spectrum}}(a, b) = \sum_{s \in S} \varphi_s^p(a)\varphi_s^p(b),$$

where $\varphi_s^p(x)$ counts the number of equal strings $S = S_a^p \cup S_b^p$ with S_x^p denoting the set (spectrum) of strings of length p of instance x. With a closer look at this formula, it is easy to recognize that this is equivalent with the definition of the dot product for vectors of variables sizes.

In the original publication [98,99] it is applied to strings obtained from proteins. Mahé et al. use Spectrum kernels for a fast approximation of two-point and three-point pharmacophore kernels, which closely relates this kernel to fingerprints approaches.

4.6.4 2D Kernels on Nominal Features

The Tanimoto and the MinMax kernel can compare arbitrary feature sets $F_i = \{f_{i1}, f_{i2}, \ldots, f_{im}\}$ and $F_j = \{f_{j1}, f_{j2}, \ldots, f_{jn}\}$. Ralaivola et al. [92] introduced both kernels for the prediction of chemical properties using fingerprinting schemes. A further study was published by Azencott et al. [89].

Note that fingerprints are a general concept and not a fixed scheme. The following flavors exist: List representation without fixed length (This avoids the possibility of a collision), Structural keys (a defined look-up table of patterns), and different pattern generation algorithms (e.g., fragmentation and DFS). An advantage of a kernel function is that in can handle fingerprints of an undefined size and can weight the patterns with, for example, the molecular weight of the substructure or a branching factor.

The MinMax kernel K_{MM} is capable of including information about the individual counts of each feature. It is defined as

$$k_{\text{MM}}(F_i, F_j) = \frac{\sum_p \min[\phi_p(F_i, F_j)]}{\sum_p \max[\phi_p(F_i, F_j)]},$$

where a feature of the joint feature space $p \in P$ is counted by φ_p.

$$k_{\text{TM}}(F_i, F_j) = \frac{|F_i \cap F_j|}{|F_i \cup F_j|}.$$

The Tanimoto kernel K_{TM} is the cardinality of the intersection divided by the cardinality of the union of F_i, F_j. A useful property is that new features lead to an increased

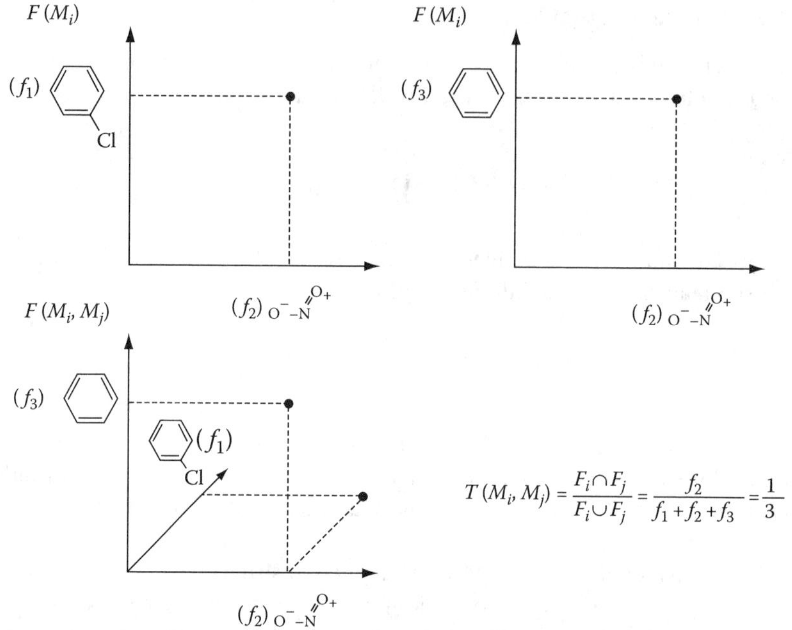

FIGURE 4.2 The Tanimoto kernel and MinMax kernel compare molecular graphs in a joint space of their features. A simple example is shown using the Tanimoto kernel.

dissimilarity and that features that are not contained in the two compared structures are simply omitted. Intuitively, this representation corresponds to a molecular fingerprint of unlimited size.

The Tanimoto coefficient and the MinMax kernel are valid kernels [92,100]. The Tanimoto kernel is a special case of the MinMax kernel, if the count of all features equals one. The Tanimoto kernel is used in chemoinformatics to compare sets of molecular features like fragments, paths, and bit sets.

An efficient way to compare molecules with the Tanimoto and MinMax kernel is to represent the structures as trees [92]. The MinMax and Tanimoto kernel are self-normalizing real-valued kernel functions, which yield values between zero and one. A recursive computation of the Tanimoto kernel on nominal features using a prefix search tree (trie) is illustrated in Algorithm 4.13 with a visual representation of the kernel function shown in Figure 4.2. The patterns are inserted as tuples of nominal features (e.g., the sequence of atom types and bond labels of a path). The last element is labeled as leave and may have the count of the corresponding feature as additional property.

ALGORITHM 4.13 RECURSIVE TANIMOTO KERNEL COMPUTATION ON TWO TRIES

```
dirac ← 0
double computeSimilarityTanimoto(Trie trie_a,
      Trie trie_b) {
```

```
computeSimDirac(trie_a · root, trie_b · root)
leaves_a ← trie_a · leaves
leaves_b ← trie_b · leaves
                          dirac
tanimoto ← ─────────────────────────────────
            leaves_a + leaves_b − dirac
      return tanimoto;
}
   void  computeSimDirac(TreeNode
   node_a, TreeNode node_b) {
   if  (node_a · isLeave  AND node_b · isLeave) then
   dirac ← dirac + 1
   fi
   children_a ← node_a · children ←
   children_b ← node_b · children
   for i ← 0 to children_a · size do
   for j ← 0 to children_b · size do
   if
   children_a [i] · label = children_b [j]· label
   then

   computeSimDirac(
   children_a [i], children_b [j],)
  fi
  od
  od
 }
```

4.6.4.1 Marginalized Graph Kernel

Another important class of kernels is the class of expectation kernels. This approach may be useful, if the feature space is too large to be computed directly. The marginalized graph kernel is based on random walks and is defined as the expectation of a kernel of all pairs of walks from two graphs, see Kashima et al. [91].

4.6.5 2D Kernels Nominal and Numerical Features

4.6.5.1 Optimal Assignment Kernels

The idea of the OAK is to compute an optimal weighted assignment on two sets of objects and to use the resulting similarity as a kernel function. The underlying problem is to solve $\max w(M) = \max \sum_{e \in M} w(e)$, where $w(M)$ is the sum of the weights of the matching edges $e(i,j)$, between two objects i, j of two disjoint sets. Each feature of the smaller set has to be assigned to exactly one feature of the other set. The OAK was introduced by Fröhlich et al. [18,19,101] and successfully applied to attributed molecular graphs.

The OAK regards atoms as vertices attributed with chemical properties and neighborhood information. After the atomic labels have been assigned, a pairwise

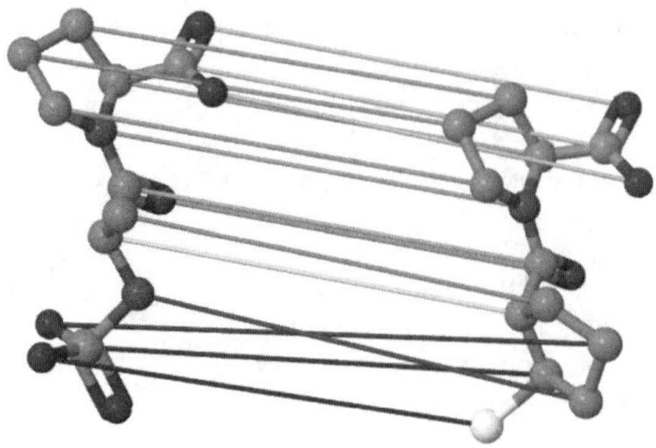

FIGURE 4.3 Assignment of two arbitrary structures $\mathbf{x}, \mathbf{x}'$ by the OAK. Each assignment edge has a similarity score and contributes to the final kernel value $k_{\text{OA}}(\mathbf{x}, \mathbf{x}')$.

similarity matrix of the attributes of each atom of two molecular graphs is computed. If the number of atoms of both structures is not equal, dummy nodes are added to the smaller molecular graph. Finally, the Hungarian Method is used to find an optimal weighted assignment of the atoms on the resulting quadratic similarity matrix.

The kernel function of the optimal weighted assignment is defined as follows:

$$k_{\text{OA}}(\mathbf{x}, \mathbf{x}') := \begin{cases} \max\limits_{\pi \in \Pi(\mathbf{x}')} \sum\limits_{i=1}^{|\mathbf{x}|} \kappa(x_i, x'_{\pi(i)}) & \text{if } |\mathbf{x}'| > |\mathbf{x}| \\ \max\limits_{\pi \in \Pi(\mathbf{x})} \sum\limits_{j=1}^{|\mathbf{x}'|} \kappa(x_{\pi(j)}, x'_j) & \text{otherwise} \end{cases}$$

In the context of molecular graphs $\mathbf{x} := (x_1, x_2, \ldots, x_{|x|})$ and $\mathbf{x}' := (x'_1, x'_2, \ldots, x'_{|x'|})$ are the sets of atoms or atom environments that compose the corresponding molecular graph. $\Pi(\mathbf{x})$ denotes all possible permutations of $\mathbf{x}$ and $\Pi(\mathbf{x}')$ of $\mathbf{x}'$ respectively. The atomwise similarity is determined by κ which can be any suitable kernel function on atomic attributes. Note that the OAK uses a local atom environment, described in 4.2. $k_{\text{OA}}(\mathbf{x}, \mathbf{x}')$ computes the maximum score of all possible permutations which are visualized in Figure 4.3 for two sample structures. A related kernel, the Iterative Similarity OAK (ISOAK), has been published by Rupp et al. [93]. Fechner et al. [102] published an extension of the local kernels which is able to encode the flexibility of a molecule by local flexibility patterns. This further improved the modeling performance of the OAK on molecules with both rigid and flexible substructures.

OAKs are pseudokernels [103]. Therefore, each kernel matrix has to be fixed by subtracting the smallest negative eigenvalue from the diagonal of the kernel matrix. Java implementations of the OAK and ISOAK are available for downloading free of charge. Both assignment kernels have a good prediction performance [90,93].

4.6.6 3D Kernels on Nominal and Numeric Features

4.6.6.1 A General Framework for Pharmacophore Kernels

From an abstract point-of-view, a pharmacophore is 3D relationship between pharmacophoric interaction features in a molecule responsible for an interaction with a pharmaceutical target. Pharmacophore kernels between two molecules a, b are kernel functions of the form

$$k(a,b) = \sum_{i=0}^{m}\sum_{j=0}^{n} \kappa(p_i, p_j).$$

The total similarity is determined by summing up all pairwise similarities between all pharmacophores in two molecules. The information of a pharmacophore lies in the distances and the pharmacophoric features. Mahe et al. [104] propose a general kernel function κ of the form

$$\kappa(p_i, p_j) = k_I(p_i, p_j) \times k_S(p_i, p_j),$$

where the intrinsic kernel k_I provides a similarity measure two pharmacophoric features and the spatial kernel k_S is a measure for the distance. For other kernels then the simple Dirac kernel, a trace matrix is used to compute efficiently the distance between three-point pharmacophores, see Algorithm 4.14 for the matrix computation. The kernel value can then be traced by Algorithm 4.15.

ALGORITHM 4.14 TRACE MATRIX COMPUTATION

```
TraceMatrix(FeatureAtomFingerprint[]fs1, FeatureAtom
Fingerprint[] fs2) {
int n1=fs1.length; int n2=fs2.length; int n=n1*n2;
int i=0; int j=0;
M=new double[n][n];
BitSet nonezeroRows=new BitSet(n);
BitSet nonezeroCols=new BitSet(n);
for (int i1=0; i1<n1; i1++) {
for (int i2=0; i2<n2; i2++) {
for (int j1=0; j1<n1; j1++) {
if (i1 == j1)
continue;
for (int j2=0; j2<n2; j2++) {
if (i2 == j2)
continue;
i=i1+i2*n1;
j=j1+j2*n1;
M[i][j]=fs1[i1].similarity(fs2[i2]);
if (M[i][j]>0) {
double e1=fs1[i1].distance(fs1[j1]);
double e2=fs2[i2].distance(fs2[j2]);
```

```
M[i][j] *= K_Dist(e1, e2);
nonezeroRows.set(i); nonezeroCols.set(j);
}
}
}
}
}
reduceTraceMatrix(); // delete columns and rows
  containing zeros only (set
  in nonezeroCols)
}
```

ALGORITHM 4.15 TRACE MATRIX SEARCH

```
public double trace(){
int n=M.length;
double r=0;
for (int i=0; i<n; i++) {
for (int j=0; j<n; j++) {
if (M[i][j] == 0.0)
continue;
for (int k=0; k<n; k++) {
r += M[i][j]*M[j][k]*M[k][i]; //=kappa(p,p')
}
return r;
}
```

4.6.6.2 Fast Approximation of the Pharmacophore Kernel by Spectrum Kernels

For a fast computation Mahé at al. proposed a p-spectrum kernel [98,99] like approach using a search tree of all three-point pharmacophores of a structure. To enable a rapid calculation, the distances of the pharmacophoric points are labeled by a distance grid and the pharmacophoric points by their general atom type. Both, the distance and the atom type can now be compared by the Dirac kernel. Consequently, a recursive kernel calculation is possible, as outlined in Algorithm 4.13. In the original work of Mahé et al., a simple Spectrum kernel was used.

The product of the counts of a specific pharmacophoric pattern equals $k(a, b)$. The final kernel value is obtained via normalizing the kernel $k(a, b)$ with the self-similarities of a and b:

$$k(a, b) \leftarrow \frac{k(a, b)}{\sqrt{k(a, a)} \cdot \sqrt{k(b, b)}}.$$

The self-normalization is frequently used with kernel functions.

REFERENCES

1. Todeschini, R. and Consonni, V., *Handbook of Molecular Descriptors*. Wiley-VCH, Weinheim, Germany, 2000.
2. Cook, D. J. and Holder, L. B. (eds), *Mining Graph Data*. Wiley, Hoboken, NJ, 2007.
3. Bush, B. L. and Sheridan, R. P., PATTY: A programmable atom typer and language for automatic classification of atoms in molecular databases. *J. Chem. Inf. Comput. Sci.*, 1993, 33, 756–762.
4. Leite, T. B., Gomes, D., Miteva, M. A., Chomilier, J., Villoutreix, B. O., and Tuffery, P., Frog: A. FRee online druG 3D conformation generator. *Nucleic Acids Res.*, 2007, 35, 568–572.
5. Izrailev, S., Zhu, F., and Agrafiotis, D. K., A distance geometry heuristic for expanding the range of geometries sampled during conformational search. *J. Comput. Chem.*, 2006, 27, 1962–1969.
6. Gasteiger, J., Sadowski, J., Schuur, J., Selzer, P., Steinhauer, L., and Steinhauer, V., Chemical information in 3D space. *J. Chem. Inf. Comput. Sci.*, 1996, 36, 1030–1037.
7. Schwab, C., Konformative Flexibilitaet von Liganden im Wirkstoffdesign. PhD thesis, Erlangen, Germany, 2001.
8. Good, A. C. and Cheney, D. L., Analysis and optimization of structure-based virtual screening protocols (1): Exploration of ligand conformational sampling techniques. *J. Mol. Graphics Model*, 2003, 22, 23–30.
9. Bunke, H. and Neuhaus, M., Graph matching—exact and error-tolerant methods and the automatic learning of edit costs. In: *Mining Graph Data*, pp. 17–34, Wiley, Hoboken, NJ, 2007.
10. Luks, E. M., Isomorphism of graphs of bounded valence can be tested in polynomial time. *J. Comput. Syst. Sci.*, 1982, 25, 42–65.
11. Faulon, J.-L. Stochastic generator of chemical structure, 2. Using simulated annealing to search the space of constitutional isomers. *J. Chem. Inf. Comput. Sci.*, 1996, 36, 731–740.
12. Bonchev, D. and Rouvray, D., *Complexity in Chemistry: Introduction and Fundamentals*. Taylor & Francis, London and New York, 2003.
13. Bonchev, D., Overall connectivities/topological complexities: A new powerful tool for QSPR/QSAR. *J. Chem. Inf. Comput. Sci.*, 2000, 40, 934–941.
14. Minoli, D. Combinatorial graph complexity. *Atti. Accad. Naz. Lincei, VIII. Ser., Rend., Cl. Sci. Fis. Mat. Nat.*, 1975, 59, 651–661.
15. Bremser, W. HOSE—a novel substructure code. *Anal. Chim. Acta*, 1978, 103, 355–365.
16. Hemmer, M. C., Steinhauer, V., and Gasteiger, J., Deriving the 3D structure of organic molecules from their infrared spectra. *Vib. Spectrosc.*, 1999, 19, 151–164.
17. Hemmer, M. C. and Gasteiger, J., Prediction of three-dimensional molecular structures using information from infrared spectra. *Anal. Chim. Acta*, 2000, 420, 145–154.
18. Fröhlich, H., Wegner, J. K., Sieker, F., and Zell, A., Kernel functions for attributed molecular graphs—a new similarity based approach to ADME prediction in classification and regression. *QSAR Comb. Sci.*, 2006, 25, 317–326.
19. Fröhlich, H., Wegner, J. K., and Zell, A., Assignment kernels for chemical compounds. *International Joint Conference on Neural Networks 2005 (IJCNN'05)*, 2005, pp. 913–918.
20. Bender, A., Mussa, H. Y., Glen, R. C., and Reiling, S., Similarity searching of chemical databases using atom environment descriptors (MOLPRINT 2D): Evaluation of performance. *J. Chem. Inf. Comput. Sci.*, 2004, 44, 1708–1718.

21. Bender, A., Mussa, H. Y., Glen, R. C., and Reiling, S., Molecular similarity searching using atom environments, information-based feature selection, and a nave bayesian classifier. *J. Chem. Inf. Comput. Sci.*, 2004, 44, 170–178.
22. Gutman, I., *Chemical Graph Theory. Introduction and Fundamentals*. Abacus Press—Gordon and Breach, New York, 1991.
23. Free, S. M. and Wilson, J. W., A mathematical contribution to structure–activity studies. *J. Med. Chem.*, 1964, 7, 395–399.
24. Clark, M., Cramer, R., and van Opdenbosch, N., Validation of the general purpose tripos 5.2 force field. *J. Comput. Chem.*, 1989, 10, 982–1012.
25. Meng, E. C. and Lewis, R. A., Determination of molecular topology and atomic hybridization states from heavy atom coordinates. *J. Comput. Chem.*, 1991, 12, 891–898.
26. Schroedinger, Macromodel 9.5 Reference Manual, 2007.
27. Daylight Chemical Information Systems, Inc., 2008, *Daylight Theory Manual*.
28. Rarey, M. and Dixon, J. S. Feature trees: A new molecular similarity measure based on tree matching. *J. Comput. Aided Mol. Des.*, 1998, 12, 471–490.
29. Carhart, R. E., Smith, D. H., and Venkataraghavan, R., Atom pairs as features in structure–activity studies: Definition and applications. *J. Chem. Inf. Comput. Sci.*, 1985, 25, 64–73.
30. Dijkstra, E. W., A note on two problems in connexion with graphs. *Num. Math.*, 1959, 1, 269–271.
31. Cormen, T. H., Leiserson, C. E., Rivest, R., and Stein, C., *Introduction to Algorithms*. MIT Press, McGraw-Hill Book Company, Cambridge, MA, 2001.
32. Borgwardt, K. M., Graph kernels. PhD thesis, 2007, Ludwig-Maximilians-Universitaet Muenchen.
33. Faulon, J.-L., Visco, D. P., and Pophale, R. S., The signature molecular descriptor. 1. Using extended valence sequences in QSAR and QSPR studies. *J. Chem. Inf. Comput. Sci.*, 2003, 43, 707–720.
34. Faulon, J.-L., Churchwell, C. J., and Visco, D. P. The signature molecular descriptor. 2. enumerating molecules from their extended valence sequences. *J. Chem. Inf. Comput. Sci.*, 2003, 43, 721–734.
35. Churchwell, C. J., Rintoul, M. D., Martin, S., Visco, D. P., Kotu, A., Larson, R. S., Sillerud, L. O., Brown, D. C., and Faulon, J.-L. The signature molecular descriptor. 3. inverse-quantitative structure–activity relationship of ICAM-1 inhibitory peptides. *J. Mol. Graphics Model*, 2004, 22, 263–273.
36. Faulon, J.-L., Collins, M. J., and Carr, R. D. The signature molecular descriptor. 4. Canonizing molecules using extended valence sequences. *J. Chem. Inf. Comput. Sci.*, 2004, 44, 427–436.
37. Lewell, X. Q., Judd, D. B., Watson, S. P., and Hann, M. M., RECAP—retrosynthetic combinatorial analysis procedure: A powerful new technique for identifying privileged molecular fragments with useful applications in combinatorial chemistry. *J. Chem. Inf. Comput. Sci.*, 1998, 38, 511–522.
38. Flower, D. R., On the properties of bit string-based measures of chemical similarity. *J. Chem. Inf. Comput. Sci.*, 1998, 38, 379–386.
39. Dean, P. M., and Lewis, R. A., (eds), *Molecular Diversity in Drug Design*. Kluwer Academic Publishing, Dordrecht, Boston, London, 1999.
40. Durant, J. L., Leland, B. A., Henry, D. R., and Nourse, J. G., Reoptimization of MDL keys for use in drug discovery. *J. Chem. Inf. Comput. Sci.*, 2002, 42, 1273–1280.
41. James, C. A. and Weininger, D., *Daylight Theory Manual*. Daylight Chemical Information Systems, Inc., 1995.
42. Tripos, *UNITY Reference Manual*. Tripos Inc., St. Louis, MO, 1995.
43. JChem 3.2, 2006, ChemAxon.

44. Brown, N., McKay, B., and Gasteiger, J., Fingal: A novel approach to geometric fingerprinting and a comparative study of its application to 3D-QSAR modelling. *QSAR Comb. Sci.*, 2005, 24, 480–484.
45. Swamidass, S. J. and Baldi, P., Mathematical correction for fingerprint similarity measures to improve chemical retrieval. *J. Chem. Inf. Model.*, 2007, 47, 952–964.
46. Shemetulskis, N. E., Weininger, D., Blankley, C. J., Yang, J. J., and Humblet, C., Stigmata: An algorithm to determine structural commonalities in diverse datasets. *J. Chem. Inf. Comput. Sci.*, 1996, 36, 862–871.
47. Gasteiger, J., Rudolph, C., and Sadowski, J., Automatic generation of 3D-atomic coordinates for organic molecules. *Tetrahedron Comput. Methodol.*, 1990, 3, 537–547.
48. Peterson, W. W. and Brown, D. T., Cyclic codes for error detection. *Proc. IRE*, 1961, 49, 228–235.
49. Introducing InChI Key. *Chem. Intern.*, 2007, 29.
50. Secure hash standard (shs). *FIPS 180-182*, 2002, US Government.
51. Hopfinger, A. J., A QSAR investigation of dihydrofolate reductase inhibition by baker triazines based upon molecular shape analysis. *J. Am. Chem. Soc.*, 1980, 102, 7196–7206.
52. Rush, T. S., Grant, J. A., Mosyak, L., and Nicholls, A. A shape- based 3-D scaffold hopping method and its application to a bacterial protein–protein interaction. *J. Med. Chem.*, 2005, 48, 1489–1495.
53. Masek, B. B., Merchant, A., and Matthew, J. B., Molecular shape comparison of angiotensin ii receptor antagonists. *J. Med. Chem.*, 1993, 36, 1230–1238.
54. Proschak, E., Rupp, M., Derksen, S., and Schneider, G., Shapelets: Possibilities and limitations of shape-based virtual screening. *J. Comput. Chem.*, 2007, 29, 108–114.
55. Lorensen, W. E. and Cline, H. E., Marching cubes: A high resolution 3D surface construction algorithm. *Comput. Graph*, 1987, 21, 163–169.
56. Dunn, F. and Parberry, I., *3D Math primer for Graphics and Game Development*. Wordware Publishing, Inc., 2002.
57. Kabsch, W., A solution for the best rotation to relate two sets of vectors. *Acta Crystallogr.*, 1976, 32, 922.
58. Kellogg, G. E. and Abraham, D. J., Key, lock, and locksmith: Complementary hydropathic map predictions of drug structure from a known receptor–receptor structure from known drugs. *J. Mol. Graph. Model*, 1992, 10, 212–217.
59. Goodford, P. J., A computational procedure for determining energetically favorable binding sites on biologically important macromolecules. *J. Med. Chem.*, 1985, 28, 849–857.
60. Warwicker, J. and Watson, H. C., Calculation of the electric potential in the active site cleft due to alpha-helix dipoles. *J. Mol. Biol.*, 1982, 157, 671–679.
61. Rogers, N. K. and Sternberg, M. J. Electrostatic interactions in globular proteins. Different dielectric models applied to the packing of alpha helices. *J. Mol. Biol.*, 1984, 174, 527–542.
62. Klebe, G., Abraham, U., and Mietzner, T., Molecular similarity indices in a comparative analysis (CoMSIA) of drug molecules to correlate and predict their biological activity. *J. Med. Chem.*, 1994, 37, 4130–4146.
63. Cramer, R. D., Patterson, D. E., and Bunce, J. D., Comparative molecular field analysis (CoMFA). 1. Effect of shape on binding of steroids to carrier proteins. *J. Am. Chem. Soc.*, 1988, 110, 5959–5967.
64. Gasteiger, J. and Marsili, M., A new model for calculating atomic charges in molecules. *Tetrahedron Lett.*, 1978, 34, 3181–3184.

65. Clark, M., Cramer, R., Jones, D., Patterson, D., and Simeroth, P., Comparative molecular field analysis (CoMFA). 2. Toward its use with 3D structural databases. *Tetrahedron Comput. Methodol.*, 1990, 3, 47–59.
66. Kearsley, S. K. and Smith, G. M., An alternative method for the alignment of molecular structures: Maximizing electrostatic and steric overlap. *Tetrahedron Comput. Methodol.*, 1990, 3, 615–633.
67. Viswanadhan, V. N., Ghose, A. K., Revankar, G. R., and Robins, R. K., Atomic physicochemical parameters for three dimensional structure directed quantitative structure–activity relationships. 4. Additional parameters for hydrophobic and dispersive interactions and their application for an automated superposition of certain naturally occurring nucleoside antibiotics. *J. Chem. Inf. Comput. Sci.*, 1989, 29, 163–172.
68. Coutsias, E. A., Seok, C., and Dill, K. A., Using quaternions to calculate RMSD. *J. Comput. Chem.*, 2004, 25, 1849–1857.
69. Kearsley, S. K., An algorithm for the simultaneous superposition of a structural series. *J. Comput. Chem.*, 1990, 11, 1187–1192.
70. Klebe, G., Mietzner, T., and Weber, F., Different approaches toward an automatic structural alignment of drug molecules: Applications to sterol mimics, thrombin and thermolysin inhibitors. *J. Comput. Aided Mol. Des.*, 1994, 8, 751–778.
71. Pastor, M., Cruciani, G., McLay, I., Pickett, S., and Clementi, S., Grid-independent descriptors (GRIND): A novel class of alignment independent three-dimensional molecular descriptors. *J. Med. Chem.*, 2000, 43, 3233–3243.
72. Crivori, P., Cruciani, G., Carrupt, P.-A., and Testa, B., Predicting blood–brain barrier permeation from three-dimensional molecular structure. *J. Med. Chem.*, 2000, 43, 2204–2216.
73. Cruciani, G., Pastor, M., and Guba, W., VolSurf: A new tool for the pharmacokinetic optimization of lead compounds. *Eur. J. Pharm. Sci.*, 2000, 11(Suppl 2), S29–S39.
74. Miller, K. J., Additivity methods in molecular polarizability. *J. Am. Chem. Soc.*, 1990, 112, 8533–8542.
75. Wermuth, C. G., Ganellin, C. R., Lindberg, P., and Mitscher, L. A., Glossary of terms used in medicinal chemistry. *Pure Appl. Chem.*, 1998, 70, 1129–1143.
76. Sheridan, R. P., Nilakantan, R., Dixon, J. S., and Venkataraghavan, R., The ensemble approach to distance geometry: Application to the nicotinic pharmacophore. *J. Med. Chem.*, 1986, 29, 899–906.
77. Dammkoehler, R. A., Karasek, S. F., Shands, E. F., and Marshall, G. R., Constrained search of conformational hyperspace. *J. Comput. Aided Mol. Des.*, 1989, 3, 3–21.
78. Martin, Y. C., Bures, M. G., Danaher, E. A., DeLazzer, J., Lico, I., and Pavlik, P. A., A fast new approach to pharmacophore mapping and its application to dopaminergic and benzodiazepine agonists. *J. Comput. Aided Mol. Des.*, 1993, 7, 83–102.
79. Barnum, D., Greene, J., Smellie, A., and Sprague, P., Identification of common functional configurations among molecules. *J. Chem. Inf. Comput. Sci.*, 1996, 36, 563–571.
80. Bron, C. and Kerbosch, J., Finding all cliques of an undirected graph (algorithm 457). *Comm. ACM*, 1973, 16, 575–576.
81. Samudrala, R. and Moult, J., A graph-theoretic algorithm for comparative modeling of protein structure. *J. Mol. Biol.*, 1998, 279, 287–302.
82. Hahn, M. and Rogers, D., Receptor surface models. 2. Application to quantitative structure–activity relationships studies. *J. Med. Chem.*, 1995, 38, 2091–2102.
83. Hahn, M., Receptor surface models. 1. Definition and construction. *J. Med. Chem.*, 1995, 38, 2080–2090.

84. Santos-Filho, O. A. and Hopfinger, A. J., The 4D-QSAR Paradigm: Application to a novel set of non-peptidic HIV protease inhibitors. *Quant. Struct.–Act. Relat.*, 2002, 21, 369–381.
85. Vedani, A. and Dobler, M., 5D-QSAR: The key for simulating induced fit? *J. Med. Chem.*, 2002, 45, 2139–2149.
86. Vedani, A. and Dobler, M., Multidimensional QSAR: Moving from three- to five-dimensional concepts. *Quant. Struct.-Act. Relat.*, 2002, 21, 382–390.
87. Sheridan, R. P. and Miller, M. D., A Method for visualizing recurrent topological substructures in sets of active molecules. *J. Chem. Inf. Comput. Sci.*, 1998, 38, 915–924.
88. Birchall, K., Gillet, V. J., Harper, G., and Pickett, S. D., Training similarity measures for specific activities: Application to reduced graphs. *J. Chem. Inf. Model.*, 2006, 46, 577–586.
89. Azencott, C.-A., Ksikes, A., Swamidass, S., Chen, J., Ralaivola, L., and Baldi, P., One- to four-dimensional kernels for virtual screening and the prediction of physical, chemical, and biological properties. *J. Chem. Inf. Model.*, 2007, 47, 965–974.
90. Fröhlich, H., Kernel methods in chemo- and bioinformatics. PhD thesis, University of Tuebingen, 2006.
91. Kashima, H., Tsuda, K., and Inokuchi, A., Marginalized kernels between labeled graphs. *Proceedings of the Twentieth International Conference on Machine Learning* (ICML-2003), Washington DC, 2003.
92. Ralaivola, L., Swamidass, S. J., Saigo, H., and Baldi, P., Graph kernels for chemical informatics. *Neur. Netw.*, 2005, 18, 1093–1110.
93. Rupp, M., Proschak, E., and Schneider, G., Kernel approach to molecular similarity based on iterative graph similarity. *J. Chem. Inf. Model.*, 2007, 47, 2280–2286.
94. Swamidass, S. J., Chen, J., Bruand, J., Phung, P., Ralaivola, L., and Baldi, P., Kernels for small molecules and the prediction of mutagenicity, toxicity and anti-cancer activity. *Bioinformatics*, 2005, 21, 359–368.
95. Borgwardt, K. M., Ong, C. S., Schoenauer, S., Vishwanathan, S. V. N., Smola, A. J., and Kriegel, H.-P., Protein function prediction via graph kernels. *Bioinformatics*, 2005, 21, 47–56.
96. Gaertner, T., A survey of kernels for structured data. *ACM SIGKDD Explor. Newslett.*, 2003, 5, 49–58.
97. Bishop, C. M., *Pattern Recognition and Machine Learning*. Springer, New York, 2006.
98. Leslie, C., Eskin, E., and Noble, W. S., The spectrum kernel: A string kernel for protein classification. *Pacific Symposium on Biocomputing*, 2002.
99. Leslie, C. S., Eskin, E., Cohen, A., Weston, J., and Noble, W. S., Mismatch string kernels for discriminative protein classification. *Bioinformatics*, 2004, 20, 467–476.
100. Gower, J. C., A general coefficient of similarity and some of its properties. *Biometrics*, 1971, 27, 857–871.
101. Fröhlich, H., Wegner, J. K., and Zell, A., Optimal assignment kernels for attributed molecular graphs. *The 22nd International Conference on Machine Learning (ICML 2005)*, pp. 225–232, Omnipress, Madison, WI, 2005.
102. Fechner, N., Jahn, A., Hinselmann, G., and Zell, A., Atomic local neighborhood exibility incorporation into a structured similarity measure for QSAR. *J. Chem. Inf. Model*, 2009, 49, 549–560.
103. Vert, J.-P., The optimal assignment kernel is not positive definite. Tech. Rep., 2008.
104. Mahe, P., Ralaivola, L., Stoven, V., and Vert, J.-P., The pharmacophore kernel for virtual screening with support vector machines. *J. Chem. Inf. Model*, 2006, 46, 2003–2014.

5 Ligand- and Structure-Based Virtual Screening

Robert D. Clark and Diana C. Roe

CONTENTS

A prominent tool in the computational drug discovery toolbox are the various methods and algorithms developed for virtual screening, that is, the selection of molecules likely to show a desired bioactivity from a large database. While the preceding Chapter 4 focused on methods to *describe* molecules (molecular descriptors), the present chapter will firstly deal with methods of how molecules of the desired type can actually be *selected* from the database. Hence, Section 5.1 will describe similarity searching methods, among which similarity and distance coefficients are prominent examples.

Subsequently, and this is also often the order followed in drug discovery projects in industry, we will continue with computational approaches to compound selection *when the receptor (or generally target) structure is known*, such as from a crystal structure. Hence, in Section 5.2 we will discuss some of the most prominent approaches for algorithms employed in ligand–protein docking.

5.1 SIMILARITY SEARCHING FOR VIRTUAL SCREENING

Robert D. Clark

One way to run a virtual screen is to take molecular structures of ligands that are known to bind to the target protein of interest and look for other compounds that are similar to them in structure. In the simplest case, this is simply a matter of looking for particular explicit substructures (Chapters 1, 2, and 4). That approach is limited to identifying close structural analogs, however, and only rarely produces leads novel enough to establish new patent estates. When finding such leads is the goal, researchers rely on more generalized molecular descriptors (Chapter 4) to identify novel chemistries that are "close" to the known actives in some sense. Not surprisingly, the appropriate way to assess "close" depends on the descriptors used and on the data set of interest.

5.1.1 Distance Measures

Maximizing the similarity between two objects is equivalent to minimizing the differences between them, which—for real-valued descriptors—can be accomplished by minimizing their distance according to a given metric. Three different distance metrics—Euclidean, Manhattan, and Mahalanobis—account for most chemoinformatic distance applications, and each reflects a somewhat different notion of the nature of "space."

5.1.1.1 Euclidean Distance

"Space" is an abstract mathematical construct as well as a name for something that each of us creates as a way to put our direct visual and tactile perceptions of the location and orientation of objects in the world around us into a personal context. Our experience of how separation in space "works" is generally most consistent with the *Euclidean distance* d_{L2},* which is defined for three-dimensional (3D) Cartesian space by

$$d_{L2}^2 = \sum_{q \in \{x,y,z\}} (q_1 - q_2)^2,$$

* The subscripts reflect the fact that Euclidean and Manhattan distances are particular instances of Minkowski distances, $d_{Lq}^q = \sum_{j=1}^{K} |x_{1j} - x_{2j}|^q$, with $q = 1$ and $q = 2$, respectively.

where q_i is the value of the corresponding spatial coordinate for the ith object. This measure of distance works extremely well for most of the geometries and processes that we encounter in our day-to-day life, and generalizing it to spaces having more dimensions than the three that we are accustomed to is straightforward. More generally, in a space of K dimensions:

$$d_{L2}^2 = \sum_{j=1}^{K} (x_{1j} - x_{2j})^2, \tag{5.1}$$

where x_{ij} is the jth coordinate (variable or descriptor value) of the ith object.

5.1.1.2 Manhattan Distance

Euclidean distance is not always the most appropriate measure of the distance between objects, however, such as commonly experienced when getting from one place to another within a city or town. For streets laid out on a more or less rectilinear grid, the *Manhattan distance* d_{L1} between locations is more useful since it describes the distance that needs to be traveled to go from one point to another in this situation. This, too, generalizes to K-dimensional spaces:

$$d_{L1} = \sum_{j=1}^{K} |x_{1j} - x_{2j}|, \tag{5.2a}$$

where the vertical bars denote absolute value. d_{L1} is appropriate when differences in one variable cannot be meaningfully offset by differences in another, which is typically the case for categorical variables or others for which ratios are not meaningful.

The Manhattan distance is often used to compare binary vectors such as those encoding substructural fingerprints, where it indicates the number of bit mismatches, that is, the number of bits set in one fingerprint but not in the other. Such vectors of binary variables can also be thought of as bit sets, where a 1 at a particular position indicates that the structure in question belongs to the set of structures that contain the corresponding substructure. Then

$$d_{L1} = |\mathbf{x}_1 - \mathbf{x}_2| + |\mathbf{x}_2 - \mathbf{x}_1|. \tag{5.2b}$$

Here, the vertical bars indicate cardinality, not absolute value, and the "subtractions" represent set differences.

5.1.1.3 Scaled Distances

There is a problem with applying such measures to higher-dimensional spaces in which the individual dimensions differ in relevance (e.g., cost) or scale. For the Manhattan distance, this comes up when "north–south" blocks are shorter than "east–west" blocks—as is true in Manhattan itself. In that particular case, rescaling from "blocks" to "meters" solves the problem, but consider the case where hills are involved or there are stairs at the end of one path but not the other; then "calories" may be a more relevant unit of "distance."

In cases where the underlying descriptors are not naturally commensurate, each can be rescaled [1] based on the range of values encountered or by the respective root mean squares (RMS) [2]. Most often, however, the individual variances (σ_j^2) are used: directly, if they are known, but more often as estimated by the sample root mean square deviations (RMSDs) from the sample mean for descriptor j:

$$s_j = \sqrt{\frac{1}{n-1} \sum (x_{ij} - \bar{x}_j)^2}.$$

The *Mahalanobis distance*, d_{M}, is a more general—and more powerful—way to address this problem. It is defined by

$$d_{\mathrm{M}}^2 = (\mathbf{x}_1 - \mathbf{x}_2)^T \Sigma^{-1} (\mathbf{x}_1 - \mathbf{x}_2), \tag{5.3}$$

where $\mathbf{x}_i = [x_{i1}, x_{i2}, \ldots, x_{ik}]$, the superscript "$T$" indicates transposition, Σ is a symmetrical matrix of pairwise scaling factors, and the superscript "-1" indicates matrix inversion. In the 3D case, for example, this becomes

$$d_{\mathrm{M3}}^2 = \begin{bmatrix} \delta_1 \\ \delta_2 \\ \delta_3 \end{bmatrix} \times \begin{bmatrix} \sigma_1^2 & \sigma_{12} & \sigma_{13} \\ \sigma_{12} & \sigma_2^2 & \sigma_{23} \\ \sigma_{13} & \sigma_{23} & \sigma_3^2 \end{bmatrix}^{-1} \times \begin{bmatrix} \delta_1 & \delta_2 & \delta_3 \end{bmatrix},$$

where $\delta_j = x_{1j} - x_{2j}$. When the coordinates are mutually orthogonal, the off-diagonal scaling factors σ_{jk} are all equal to zero. In that case, the x_{ij} elements can simply be rescaled (normalized) by the corresponding diagonal scaling factors σ_j^2, and the Mahanalobis distance is the same as the Euclidean distance for the normalized vectors: $d_{\mathrm{M}} = d_{\mathrm{L2}}$.

5.1.2 Population Dissimilarity

To this point, we have considered distances as measures of physical separation between pairs of objects. In many applications involving high-dimensional spaces, it is more appropriate to think of distances as measures of how different two samples that have been drawn independently from a population are from each other, that is, of *dissimilarity*.

This is, in fact, the main place where the Mahalanobis distance is used. In such cases, Σ in Equation 5.1 represents the *covariance matrix*, which specifies the average degree of pairwise *correlation* between descriptors, for example, the ratio of block lengths along avenues to those along streets. The (co)variances $\sigma_{jk} = 0.5\langle(x_{1j} - x_{2j}) \times (x_{1k} - x_{2k})\rangle$, with the angle brackets indicating expectation across "all possible" pairs $\mathbf{x}_1$ and $\mathbf{x}_2$ that might possibly be drawn from the population. In the case where the descriptor variables (coordinates) are independent (rectilinear), the covariances are 0 and the variances σ_j^2 indicate how far the population values for x_j extend away from the average for that descriptor.

One difference between this usage of the Mahalanobis distance and what we usually think of as distance is that it is much more localized. In principle, the deviation between x_{ij} and the population mean $\langle x_{ij} \rangle$ is still unbounded, but in fact will usually lie within a few standard deviations of the mean—roughly 95% within $2\sigma_j$ of the respective mean for a normally distributed population.

The second major difference is that the Mahalanobis distance is context dependent; if you change the population under consideration—for example, move to a different city, start commuting by autogyro or shift to a structurally very different application domain—the meaning of the distances obtained by applying Equation 5.4 may change.

5.1.3 Similarity Coefficients

When one thinks of distances, larger differences seem naturally more significant than small ones. In many applications, however, proximity is more relevant than distance. This is particularly true when the variables being used to describe the space of interest are only weakly commensurate, as when time is considered as a "fourth dimension": two people meet when they find themselves in more or less the same place at the same time, but if they are there at very different times it does not much matter how different those times are. A measure of *similarity*—for which higher values connote greater proximity—is more useful than a measure of dissimilarity in such cases. The conceptual difference is a subtle one, but it has substantial mathematical and practical implications. Distance measures are generally unbounded, for example, whereas similarity measures are bounded above by 1 (identical) and are bounded below either by -1 (antithetical in every respect) or by 0 (having nothing at all in common).

5.1.3.1 Similarity between Real-Valued Vectors

The pairwise similarity metrics most often encountered in chemoinformatics applications all start from the dot product between two vectors ($\mathbf{x}_1$ and $\mathbf{x}_2$) of descriptor values, one for each of the structures being compared. The most basic is the *cosine coefficient*, for which the dot product is scaled by the geometric mean of the individual vector magnitudes:

$$S_{\cos} = \frac{\sum (x_{1j} x_{2j})}{\sqrt{\sum x_{1j}^2 \times \sum x_{2j}^2}}. \tag{5.4a}$$

The cosine coefficient takes on values between -1 and 1, or between 0 and 1 if all allowed descriptor values are non-negative. Its name derives from the fact that its value is equal to the cosine of angle formed by the pair of rays running from the origin out to the points defined by the two vectors.

Alternatively, the dot product can be scaled by the arithmetic mean to yield the *Dice similarity*:

$$S_{\mathrm{Dice}} = \frac{\sum (x_{1j} x_{2j})}{(1/2)\left(\sum x_{1j}^2 + \sum x_{2j}^2\right)}. \tag{5.5a}$$

The most popular similarity measure used in chemoinformatics, however, is the *Tanimoto coefficient*. It is a close cousin of the Dice coefficient but differs in that the rescaling includes a correction for the size of the dot product rather than simply taking the average. The main effect of the change is to expand the resolution between similarities at the high end of the range.

$$S_{\text{Tan}} = \frac{\sum x_{1j}x_{2j}}{\sum x_{1j}^2 + \sum x_{2j}^2 - \sum x_{1j}x_{2j}}. \tag{5.6a}$$

Note that the "Tan" subscript does not indicate any connection to the tangent function familiar from trigonometry.

5.1.3.2 Similarity between Bit Sets

As noted above in connection with the Manhattan distance, the elements of $\mathbf{x}_1$ and $\mathbf{x}_2$ are binary in many chemoinformatics applications, that is, they only take on values of 0 or 1. Such vectors can be thought of as *bit sets*, with $x_{ij} = 1$ indicating that $\mathbf{x}_i$ is a member of set j and $x_{ij} = 0$ indicating that $\mathbf{x}_i$ is not a member of set j. In fact, this is the chemoinformatics area in which similarity searching sees its greatest use. For substructural fingerprints, the K sets are defined as being structures that contain fragment f_j as a substructure. In that situation the cosine coefficient can be recast as

$$S_{\text{cos}} = \frac{|\mathbf{x}_1 \cap \mathbf{x}_2|}{\sqrt{|\mathbf{x}_1|^2 \times |\mathbf{x}_2|^2}}. \tag{5.4b}$$

Cardinalities are always non-negative, so S_{cos} is bounded below by zero when applied to bit sets.

Similarly, the binary equivalent of the Dice coefficient is given by

$$S_{\text{Dice}} = 2 \times \frac{|\mathbf{x}_1 \cap \mathbf{x}_2|}{|\mathbf{x}_1| + |\mathbf{x}_2|}. \tag{5.5b}$$

Finally, the binary equivalent of the Tanimoto coefficient, which is more precisely referred to as the *Jaccard index* [3], is given by

$$S_J = \frac{|\mathbf{x}_1 \cap \mathbf{x}_2|}{|\mathbf{x}_1| + |\mathbf{x}_2| - |\mathbf{x}_1 \cap \mathbf{x}_2|} = \frac{|\mathbf{x}_1 \cap \mathbf{x}_2|}{|\mathbf{x}_1 \cup \mathbf{x}_2|}. \tag{5.6b}$$

Tversky [4] noted that the Dice and Jaccard (binary Tanimoto) similarities could be cast as special cases of a more generalized similarity measure.

$$S_{\text{Tvesrky}} = \frac{|\mathbf{x}_1 \cap \mathbf{x}_2|}{|\mathbf{x}_1 \cap \mathbf{x}_2| + \alpha \times |\mathbf{x}_1 - \mathbf{x}_2| + \beta \times |\mathbf{x}_2 - \mathbf{x}_1|}. \tag{5.7}$$

Note that the subtraction $\mathbf{x}_1 - \mathbf{x}_2$ in Equation 5.8a represents a *set* difference, that is, the set of bits that are set to 1 in $\mathbf{x}_1$ but not in $\mathbf{x}_2$; it does not represent the difference in cardinalities between the two sets. Setting $\alpha = \beta = 1$ yields the Jaccard similarity,* whereas setting $\alpha = \beta = 0.5$ yields the Dice coefficient. In the chemoinformatics arena, Tversky's generalization has mostly been used to assess *asymmetric* or *modal similarity*, where $\alpha = 1$ and $\beta = 0$ (or vice versa). This is useful for doing a partial match variation of substructure or partial shape similarity searching [5–7].

5.1.3.3 Similarity of Populations

The cosine coefficient is actually most commonly encountered as the *Pearson correlation coefficient* r, a special case in which the elements x_{ij} of the vectors $\mathbf{x}_1$ and $\mathbf{x}_2$ are themselves observed deviations from the means ($\langle x_{1j}\rangle$ and $\langle x_{2j}\rangle$, respectively) for two different variables. The definition in Equation 5.5a then becomes

$$r = \frac{\sum x_j y_j}{\sqrt{\sum x_j^2}\sqrt{\sum y_j^2}}, \tag{5.8}$$

where $\mathbf{x} = \mathbf{x}_1$ and $\mathbf{y} = \mathbf{x}_2$. The indexing formalism is different, but the underlying measure has the same properties. For models, it can be shown that the absolute value of r is the same as the correlation between $\mathbf{y}$ and $\widehat{\mathbf{y}}$, the vector of y values predicted by ordinary least squares (OLS) regression on the vector of x values. This generalizes to predictions for multiple linear regression, where $\mathbf{y}$ is a set of predicted response values based on a matrix $\mathbf{X}$ that encompasses several descriptors. The multiple correlation coefficient R is given by

$$R = \frac{\sum y_j \widehat{y}_j}{\sqrt{\sum y_j^2}\sqrt{\sum \widehat{y}_j^2}},$$

where deviation $\widehat{y}_j$ is the predicted value of deviation y_j based on the vector $[x_{1j}, x_{2j}, \ldots, x_{kj}]$.

Population sampling's effects can be important for similarity searches involving fully flexible pharmacophore multiplets [5,8]. These bitmaps (compressed bit sets) represent a union of bitmaps derived from a random sample of accessible conformations. Even when that sample is large, there is considerable variation in the union bitmaps obtained for a flexible molecule, so the similarities calculated using the determinate similarity coefficients discussed above may be deceptively low. Worse, the expected similarity of a structure to itself is less than 1, often substantially so. Moreover, the expected value for that similarity is dependent on the number of conformations being considered. These problems can be addressed by using the *stochastic cosine* to compare bitmaps:

$$s_{\text{SCos}} = \frac{\langle|\mathbf{x}_1 \cap \mathbf{x}_2|\rangle}{\sqrt{\langle|\mathbf{x}_1 \cap \mathbf{x}_1'|^2\rangle \times \langle|\mathbf{x}_2 \cap \mathbf{x}_2'|^2\rangle}}, \tag{5.9}$$

* Note that $(\mathbf{x}_1 - \mathbf{x}_2) + (\mathbf{x}_1 \cap \mathbf{x}_2) = \mathbf{x}_1$.

where the angle brackets ($\langle\ \rangle$) denote expectation and the primed vectors are based on distinct, independently drawn conformational samples. In practice, the population expectations are estimated by creating two bitmaps for each structure. The cardinality of the self-intersections is obtained for each pair, and the cross-intersections are averaged across the four possible combinations.

The stochastic similarity between two very similar structures calculated according to Equation 5.8b may be greater than 1, as can the similarity between a structure and itself. The excesses are usually small in practice, however, and the *expectation* for the stochastic similarity is bounded above by 1, which is also the similarity expected when comparing a molecule to itself.

Stochastic analogs of other non-deterministic similarity coefficients can be defined similarly.

5.1.4 Applications

An exhaustive review of published similarity searching applications is beyond the scope of this work. The specific papers cited below are intended to serve as illustrative examples of how the various distance and similarity measures can be used productively.

5.1.4.1 Distance Applications

Distances between vectors of real-valued descriptors, particularly those based on properties calculated from molecular structure—size, polarity, polarizability, lipophilicity, and so on—are typically expressed in terms of *Euclidean distance* [9]. Historically such analyses have more typically involved cluster analysis than similarity searching [10], but virtual screening based on BCUT descriptors [11] constitutes a significant exception to this generalization.

Pre- and postfiltering operations can be thought of as similarity search applications of *Manhattan distances*, where a candidate structure is allowed to "pass" so long as the bit set representing the presence (1) or absence (0) of certain critical properties (or substructures) is "close enough" to a set of reference properties. Usually candidates are discarded if the *Manhattan distance* is greater than or equal to 1, that is, if any discordances are found. Higher distances are sometimes allowed, however, as in Lipinski's Rule of Five [12]. There one violation is permitted, corresponding to a critical value $d_{L1} \geq 2$.

Matches in flexible 3D searching are usually evaluated as simple filters, that is, a set of features must be identified in the target that satisfy all of the relationships specified in the query. Partial match constraints, however, can be cast as similarity searching against a set of query vectors, one for each partial match constraint. The elements in the query and target bit sets in this case represent the satisfaction (1) or failure to satisfy (0) the particular constraints (involving spatial positions, interfeature distance or angles, exclusion volumes, etc.) that make up the corresponding partial match constraint. The minimum and maximum "match" counts specified for each constraint, then, define the allowed *Manhattan distances* between the query and target vectors.

A more straightforward application of the Manhattan distance was presented for Eigen Vector Analysis (EVA) descriptors, which are calculated from normal mode analyses of query and target structures [13].

5.1.4.2 Similarity Applications

Substructural and pharmacophoric fingerprint similarity searching is usually based on the *Jaccard index*, although the more general term "*Tanimoto similarity*" is often used in this connection [9,14,15]. The cosine coefficient has also been used, however, especially in connection with pharmacophore multiplet bitmaps [7]. The *Tanimoto coefficient* itself—that is, the real-valued version—has been used to assess shape similarity [16–18]. Others have applied the *cosine coefficient* (as such or in the guise of the squared *Pearson's correlation* coefficient R^2) to shape and molecular fields [13].

Similarities between atom-pair descriptors and topological torsions have been assessed in terms of their *Dice similarity* [19,20], as have count vectors based on pharmacophore triplets [21]. EVA descriptor similarities have been evaluated in terms of cosine and Dice similarities as well as in terms of their Manhattan distances [13].

Carbó et al. [22] and Hodgkin and Richards [23] evaluated similarities between molecular fields using continuous versions of the cosine and Dice coefficients, respectively, wherein the summations are replaced by integrals

$$R^2_{\text{Carbo}} = \frac{(\int \rho_1 \rho_2 \, \mathrm{d}v)^2}{\int \rho_1^2 \, \mathrm{d}v \times \int \rho_2^2 \, \mathrm{d}v},$$

$$R_{\text{Hodgkin}} = \frac{2 \times \int \rho_1 \rho_2 \, \mathrm{d}v}{\int \rho_1^2 \, \mathrm{d}v + \int \rho_2^2 \, \mathrm{d}v}.$$

A more efficient and accurate way to carry out the required numerical integrations was subsequently described by Good et al. [24].

Occasionally the Jaccard index has been recast to use count vectors rather than bit sets. Rather than use the dot product formulation of the Tanimoto, Grant et al. [25] substituted the minimum count for each element for cardinality of the bit set intersection when evaluating the similarity of Lingo character substrings:

$$S_{\text{Lingo}} = \frac{\sum \min(x_{1j}, x_{2j})}{\sum x_{1j} + \sum x_{2j} - \sum \min(x_{1j}, x_{2j})}.$$

5.1.5 Behavior of Similarity and Distance Coefficients

Many distances and similarity measures not discussed here have been formulated over the years [26], but those described above are the ones that dominate virtual screening work. Willett et al. [27] have carried out numerous studies involving a wide range of similarity measures and conclude that although other measures may perform somewhat better on some targets, the Tanimoto coefficient generally works best overall, at least for drug-like molecules. Their work has centered on substructural fingerprints of various types; careful surveys have yet to be carried out for other descriptor classes.

Measures such as the Euclidean distance may be more appropriate for molecules that are large or complex enough to set a majority of bits in a hashed fingerprint [28,29]. This effect is best understood by noting that bits not set in either fingerprint reduce the Euclidean distance without affecting either the Tanimoto or cosine coefficient. If the probability of any one bit being set is p, then the probability that a bit will not be set in one fingerprint is $q = 1 - p$ and the probability that it will not be set at random in either fingerprint is q^2. If the fingerprints being considered are relatively sparse, p is small and q^2 is close to 1. Hence, finding a bit set in one or both fingerprints is rare and informative, whereas finding that it is set in neither is common and uninformative. Such saturation effects are probably better addressed by modifying the descriptor, however, so as to keep p below 0.1 than by trying to adjust the similarity measure used.

Such considerations underscore the fact that the exact value calculated for any given similarity measure means different things in different contexts, as does the value of any distance measure. This context includes the descriptors used as well as the scope of chemistries to which it is being applied. A Jaccard similarity threshold of "0.85" is useful when using substructural fingerprints of drug-like molecules likely to exhibit similar biological activity [30]. This cannot be taken to imply that "0.85" would be a useful cutoff for similarity searching of peptide pharmacophore multiplets using the cosine coefficient. In most cases, only the order of similarities—that is, the similarity rank—is really meaningful, and fresh benchmarks need to be determined for any new application. Fortunately, many such problematic differences in scale fall away when a simple rank transformation is applied to the raw similarities [31].

5.1.6 Combining Similarities

Although the Tanimoto coefficient works reasonably well in most applications, combining it with complementary measures often improves performance. *Consensus scoring* is now widely used to improve scoring in structure-based (docking) screens, and the analogous approach—termed data fusion [13,32,33]—has shown considerable potential for improving ligand-based similarity searching. Because the different similarity measures are not directly commensurate, however, it is usually the ranks that are combined, typically using the *minimum rank* or *sum of ranks* for each target. The *median rank* has shown promise in consensus scoring [34] and is probably worth exploring as an alternative fusion technique when three or more similarity measures are involved.

Related work has also been carried out on the best way to combine "hit lists"—that is, to optimize the definition of similarity between a single target and multiple query structures [35,36]. Logically one might expect that a "hit" that is particularly similar to two or more queries is more likely to be active itself, so taking the average of the similarities or of the ranks would improve performance. For substructural fingerprints, however, this was found not to be the case [33].

Nonetheless, a rather extreme extension of the data fusion model to multiple "actives" does sometimes work. "Turbo search," which involves retrieving compounds similar to compounds that are similar to queries (i.e., near neighbors of actives that are not themselves known to be active) improves performance, at least in some cases [37].

5.2 STRUCTURE-BASED VIRTUAL SCREENING

Diana C. Roe

5.2.1 INTRODUCTION

The first step in drug discovery is to identify lead compounds with novel chemical structures that bind to a target receptor. Originally this occurred primarily through chance discovery, requiring large efforts to find and screen natural products. Virtual screening approaches provide a rational alternative for lead identification, by performing large screens of compounds *in silico*, existing ones or those easily synthesizable in a combinatorial library, and reducing the number of compounds that need to be screened experimentally. Structure-based virtual screening, made possible by rapid advances in protein crystallography and computational power in the last two decades, has proven to be a useful tool speeding the discovery process and has become an industry standard [38]. Structure-based screening tries to rank a database of small molecules by their predicted binding affinities to a target receptor. The starting point is the 3D (usually crystal) structure of a protein and a database of small molecule ligands with modeled 3D structures. Each ligand is "docked" into the binding site of the target receptor and a score representing binding affinity is calculated. This calculation is commonly referred to as docking.

The docking problem can be broken down into three components: (1) orientational search, or the search for the 3D orientation of a molecule with respect to another; (2) conformational search, or the search through rotatable torsions; and (3) scoring, or evaluating "pose" or orientation/conformation combination by some measure of predicted binding. The original docking program was UCSF DOCK [39], which addressed only the orientational search and scoring for ligand/receptor systems. In the original implementation, spheres were used to represent the ligand and the "negative image" of the receptor, by generating spheres along the *inside* of the surface of the ligand, and the *outside* of the surface of the receptor. For small molecules, ligand atoms were used instead of spheres. This provided an identical description of the ligand and receptor site used to optimize the geometric fit between the two. The orientational problem was thus reduced to the problem of matching ligand spheres to receptor spheres. Matching was performed using a graph theoretical algorithm that looked at ligand sphere–sphere distances and matched them to receptor sphere–sphere distances (Figure 5.1). A set of interconnected distances between ligand spheres matching (within a tolerance) the same size set of interconnected distances within the receptor spheres is a clique. A clique of size four is sufficient to define a unique orientation for a ligand. The ligand was then transformed to superimpose its spheres onto the receptor's spheres for final placement.

The next development in docking programs was to include ligand flexibility (i.e., a conformational search) into the process. The first approach to address flexibility broke a ligand into two pieces, docked each of them separately, and identified for fragments that could be rejoined [40]. AutoDock [41] developed a completely different strategy that combined the conformational and orientational searches together into one step by employing a simulated annealing approach. Later versions of AutoDock included an evolutionary algorithm [42], popular also with several other programs

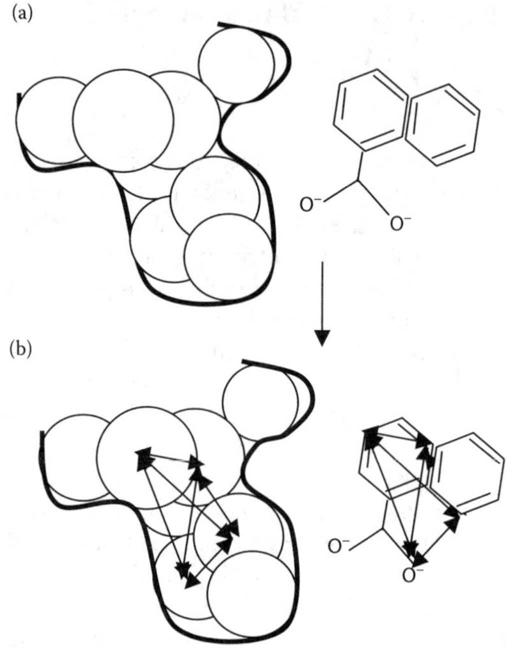

FIGURE 5.1 Clique detection algorithm. (a) Set of ligand atoms and receptor spheres; (b) clique of size 4 found matching.

[43–47]. Monte Carlo approaches have also been successfully applied [46–50]. Other programs perform a conformational search in sequence with an orientational search, by docking an "anchor" or base fragment and incrementally building up flexible ligand conformers [51,52]. This approach works well at reproducing docked structures in cases where the base fragment has a strong interaction with the target receptor, and where each flexible unit has a piecewise interaction with the protein. In other cases, such as when there is an interaction gap along a flexible unit, the incremental buildup will not place the flexible unit in the gap position but rather in a position to maximize its interaction with the receptor. Finally, some algorithms completely separate the orientational and conformational search by precalculating low-energy conformations of the small molecules and docking a rigid database of conformers [53–55]. This has the trade-off of ensuring a better conformational search of low-energy ligand conformations versus the efficiencies of on-the-fly conformational search within the receptor site, which may balance higher intramolecular energies to optimize receptor interactions.

Recently, receptor flexibility has also been added to docking programs. As with ligand flexibility, conformations can be precalculated or generated on-the-fly during docking. The first approach precalculates a series of protein conformations, such as snapshots from a molecular dynamics simulation, or from a normal mode analysis, and subsequently docks the ligand to an ensemble of proteins [56–59]. The advantage of this ensemble approach is the ability to search a wider conformational space that includes backbone and sidechain variation. The on-the-fly approaches may include protein sidechain rotamers [45,60], sidechains and user-defined loops [49,61], or induced fit using protein structure prediction [62]. The latter approach, while accurate, is not currently fast enough for virtual screening.

After calculating a ligand pose, the last step in docking is to evaluate it with some sort of scoring function. The original scoring function from UCSF DOCK was a shape-based contact score. Later, force-field-based functions were introduced. These functions took the Lennard–Jones and electrostatic parameters from force fields such as AMBER [63] or CHARMM [64,65]. To save computational time by turning an $O(N^2)$ calculation to $O(N)$, a grid was precalculated for the sum of the receptor potential at each point in space. To generate this grid, a geometric mean approximation ($A_{ij} = \sqrt{A_{ii}}\sqrt{A_{jj}}$) was made to the van der Waals portion of the force field [66]. Eventually, solvation and entropy terms were added to many force-field-based scoring functions, typically using DELPHI [67] or ZAP [68] for solvation [69,70]. As many factors known to be important in the free energy of ligand binding are missing in force-field scores, many programs chose instead to derive an empirical scoring using several intuitive parameters such as hydrophobicity, solvation, metal-binding, or the number of number rotatable bonds, along with van der Waals and electrostatic energy terms. Empirical functions were derived from a least-squares fit of the parameters to ligand–protein systems with known crystal structures and known binding energies [42,45,52,71–73]. Again these scoring functions are usually calculated on a grid for computational speed. The advantage of starting with a force-field-based method is that it is applicable to a wide range of ligands. The empirical scoring schemes work well when the ligands and receptors resemble the training set. Another approach was to use a knowledge-based function, derived from a statistical analysis of ligand atom/protein atom contact frequencies and distances in a database of crystal structures [74–76]. As each of these scoring approaches were shown to work well in different cases, many programs started to create "consensus" functions combining several different scoring schemes, which were shown to be more predictive than any single scoring scheme [77]. After primary scoring, several approaches "rescore" top hits. For example, the PostDOCK filter [78] was derived from a supervised machine learning study on protein/ligand structures in the Protein Data Bank [79], and it was shown to improve enrichment by as much as 19-fold. Short molecular dynamics runs using implicit waters, implemented as MM-PBSA (for Poisson–Boltzmann solvation) or MM-GBSA (Generalized Born), were also shown to improve enrichment rates [80,81].

Many other factors to improve the quality of structure-based affinity predictions have been addressed including waters, metals, and protonation states of the receptor protein (see Refs. [82,83] for a detailed review). Additional screens have been developed to identify lead compounds that not only show strong binding affinity to the target receptor, but also have good pharmacological properties. Lipinski's Rule of Five [84], which uses a set of property heuristics such as molecular weight, hydrogen bonds, and so on that match the range in the majority of known orally absorbed drugs, has become a standard for prescreening ligands prior to docking for "drug-like" properties. Filters have been developed to remove compounds known to be promiscuous binders (i.e., false positives) [85,86], or that interact with hERG channel [87]. Structure-based screening has been combined successfully with 3D pharmacophore searching and 3D similarity searches, to add complementary information to the searching process. Structure-based virtual screening remains a useful tool in the arsenal for lead-drug development.

5.2.2 Docking Algorithms

5.2.2.1 Orientational Search: The Clique Detection Algorithm

Many docking algorithms separate between the orientational search and the conformational search. The clique detection algorithm for orientational search that finds interconnected sets between ligand sphere–sphere distances with receptor sphere–sphere distances (Figure 5.1) is still one of the most commonly used docking algorithms. Later versions of the algorithm use ligand atoms themselves rather than ligand spheres for docking small molecules. Although to find the largest clique in a graph is considered an NP complete problem, the search for all cliques of a limited size is tractable. The original algorithm used a bipartite graph, where ligand nodes and receptor nodes were assigned to separate graphs. However, in DOCK 4.0 [51,88] the algorithm was changed to a single "docking" graph where each node represented a ligand/receptor–sphere pairing. This docking graph and the exhaustive search method were first discussed by Bron and Kerbosch [89].

The algorithm begins (see Figure 5.2 and Algorithm 5.1) with a set of T total nodes, each representing a ligand/atom pair, where T equals the number of ligand atoms times the number receptor spheres. It then precalculates all "edges" between the nodes. An edge exists between two nodes if the distance between ligand atoms in the two nodes is within a residual (distanceTol) of the distance between the receptor spheres in the two nodes. All edges are stored in an EdgeMatrix of size $T \times T$. If no edge exists, null is put in that spot.

Two arrays then store the growing clique search, each of size N, the length of the current clique. The first is Clique [N], containing all nodes of the current growing clique. The second is NodeSearchGraph [N][T], containing the set of all new nodes consistent with the current clique of length M, meaning that edges exist from clique nodes $1, \ldots, N$ to these nodes. (Up to T nodes can be stored at each N Index, representing all nodes being allowed.)

The search begins by adding a new branch node j Node from the NodeSearchGraph [N] list of allowable nodes, onto the current clique at Clique [$N + 1$]. The NodeSearch Graph [$N + 1$] is then calculated as the intersection of all remaining nodes kNode from NodeSearchGraph [N] and all nodes with an edge to j Node, by testing all k Nodes for an edge to j Node in the precalculated EdgeMatrix. If an edge exists, k Node is added to NodeSearchGraph [$N + 1$]. When a clique is complete, either because no more nodes can be found or it reaches NODESMAX size, pop_node() removes the Nth node in the clique and the next node is tried in that position. If none is found, pop_node() removes the $N-1$ node. Backtracking can continue back to position 0, until all nodes have exhaustively been searched. In practice MAXCLIQUES is used to limit the total number of cliques found.

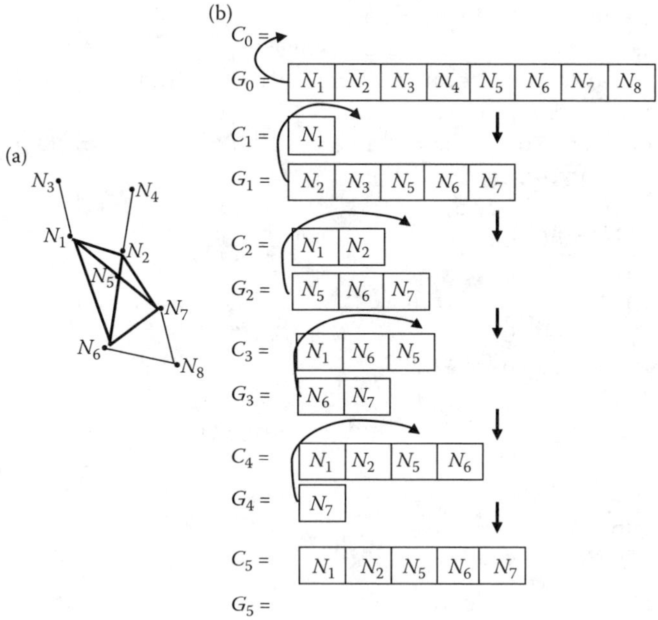

FIGURE 5.2 Clique detection algorithm. (a) Set of nodes (receptor/ligand pairs) and edges connecting nodes. (b) Exhaustive search for cliques. Initial growing clique (C_0) is NULL and node search graph (G_0) is the set of all nodes. The next node N_x from G_n is added to C_{n+1}. G_{n+1} is recalculated as all nodes in G_n with an edge to N_x. The procedure is repeated until $\geq$ MAXNODES or no further expansion is possible. Additional cliques are searched by backtracking and testing next node N_y in G_n.

ALGORITHM 5.1 PSEUDOCODE FOR EXHAUSTIVE SEARCH FOR CLIQUES OF GIVEN SIZE RANGE (NONRECURSIVE)

```
C=Clique
N=number of nodes in current clique C ==
size(clique)
NODESMIN,NODESMAX: minimum/maximum allowed
nodes in a clique
S=NodeSearchGraph[NODESMAX][T]: list of
allowed nodes
PreCompute EdgeMatrix[T x T} for all Nodes, such that
  EdgeMatrix[iNode][jNode]=edge if jedge exists within
    distanceTol,
    or 0 if it does not exist;

getNextClique(clique){
    #remove last node to start new clique
    IF (size(clique)>0)THEN
       pop_node(clique)
    END IF
```

```
    Boolean validClique=false;
    WHILE (not ValidClique) THEN
      #Expand clique until no further expansion
       possible
      #(i.e.clique size is >= NODESMAX or no more
       expansion nodes
      Expand (clique)
      IF (expand(clique) == TRUE) THEN
      CONTINUE;
      END IF
      IF (size (clique) >= NODESMIN)
       validClique=TRUE;
    ELSE
      #remove last node and try again
      pop_node(clique)
       END ELSE
    END WHILE
      RETURN clique.

}

  Expand (clique){
    N=size(clique)
    #termination 1: test if clique already fully
     expanded
    IF (N>NODESMAX) THEN
     return FALSE;
    ENDIF
     #termination 2: test if have explored all possible
      nodes at #position N+1
     IF ((jNode=next node in NodeSearchGraph[N] array )
      ==0) THEN
      return FALSE;
     END IF
    #jNode=new node to add to clique at position N+1
    #kNode=remaining node consistent with current
     clique
    #jEdge=edge between jNode and kNode
     Add jNode to clique
    #Calculate set of edges consistent with jNode and
     rest of clique.
    FOREACH kNode (loop through NodeSearchGraph[N]) DO
      jEdge=EdgeMatrix[jNode][kNode];
     IF (jEdge != 0) THEN
     Add kNode to NodeSearchGraphIndex[N+1] array
     END IF
```

```
    END DO
    RETURN TRUE
}
```

This algorithm was first implemented in UCSF DOCK4 [51], and is included in the latest version of UCSF DOCK6 [70,90]. UCSF DOCK at http://dock.compbio.ucsf.edu/DOCK_6/index.htm is freely available to academics, but a license fee is charged for commercial users. A similar algorithm is implemented in FLOG [55], using a minimal-residual heuristic to limit the search, where only the node with the minimal residual is expanded at each branch point, rather than all remaining exceed the number of nodes. FLOG is an in-house software package at Merck, in current use. The primary advantage of this clique detection approach is a more efficient sampling of relevant orientational space compared to random rotation/translation. In fact, despite the overhead of identifying cliques and then transforming the ligand coordinates, it has been shown to have a speed-up between 10- and 100-fold compared to uniform random translation [91]. The memory cost for the docking-graph algorithm is not large and limited to the precomputed EdgeMatrix, which grows as the square of (ligand atoms × receptor spheres). The search time for this algorithm grows as a function of the number of distance-constrained solutions rather than all possible unconstrained solutions, because the search does not explore invalid branches.

5.2.2.2 Conformational Search: Incremental Buildup

Programs that use a rigid orientational search algorithm can then either start with a database of precomputed conformations of ligands, using a program such as Omega [92], or perform orientational search on a rigid unit of a small molecule, and follow it with a buildup procedure. Algorithm 5.2 outlined below is a common buildup procedure, used in UCSF DOCK [51]. A similar procedure is used in FlexX [52], commercially available at http://www.biosolveit.de/flexx/. As the scoring and optimization steps are the primary time-consuming steps in the algorithm, the time demand for this algorithm can be approximated by the number of function evaluations, which becomes [51]

$$\text{Time} = C_0 \times N_0 + C_1 N_\text{c} \times N_\text{b} \times N_\text{t},$$

where C_0 and C_1 are constants, N_0 is the number of anchor orientations searched, N_c is the average number of pruned configurations saved each round, N_b is the number of rotatable bonds, and N_t is the average number of torsions per bond.

ALGORITHM 5.2 PSEUDOCODE FOR CONFORMATIONAL SEARCH USING INCREMENTAL BUILDUP

```
Identify flexible units in ligand
Start with largest rigid unit
Order flexible groups in layers starting from rigid unit
Orient rigid unit using orientational search
```

```
Loop over flexible layers
  Add next layer
  Search torsions for each flexible group in layer
  Prune by score
End loop
Minimize and score final structure
```

5.2.2.3 Combined Orientational and Conformational Search: Lamarckian Genetic Algorithm

Simulated annealing, Monte Carlo, and evolutionary algorithms have all been applied to docking programs to combine orientational and conformational search steps. Although initially these stochastic approaches were prohibitively slow and were used primarily to study a known ligand/receptor interaction in detail, owing to increases in computer power it is now possible to use these approaches for virtual screening as well. One algorithm in common use is the Lamarckian Genetic Algorithm (LGA) [42,93,94].

The LGA combines the global search properties of a genetic algorithm (GA) with a local search. The LGA is structured in the same way as the normal GA algorithm: a chromosome is created representing the orientation/conformation of the ligand with respect to the target receptor. The chromosome is shown in Figure 5.3 and consists of a number of variables for ligand translation/rotation that is the same for all ligands and a number of variables for ligand flexibility that is specific to each ligand. The sum of the chromosome represents the *genotype* of a ligand. The *phenotype* of a ligand is its 3D coordinates after applying all the transformations in the genotype. Ligand fitness can be calculated from its phenotype using any of the standard docking scoring functions.

The standard GA starts with a random population of size N, and runs through a set number of generations. Each generation runs through the following steps: (1) *mapping* genotype to phenotype for each member of the population; (2) *fitness* of each member; (3) *selection* of members for use in breeding the next population; (4) breeding: consisting of *mutation* and *crossover*; and (5) *elitist selection* (optional). The new addition for the LGA (described in Algorithm 5.3) is to add a local search function, performed at the end of each generation, on a set percentage of the (new) population. The local search algorithm can be any local optimization technique, such as Pattern search [95,96], or Solis–Wets [97]. The local search may be performed on the phenotype and transformed back onto the genotype, or performed directly on the genotype. As long as the local search results are passed onto the final genotypes it is

C_1	C_2	C_3	Q_1	Q_2	Q_2	Q_2	T_1	T_2	T_3	...	T_N

FIGURE 5.3 Chromosome for docking genetic algorithm. The first three genes (C_1–C_3) represent coordinates for ligand translation. The next four genes (Q_1–Q_4) are the quaternian for ligand rotation. The last N genes (T_1–T_N) is the ligand torsional values, the number of which vary per ligand. Receptor torsions can also be added to this portion.

considered Lamarckian [42]. In a standard GA, the crossover function provides large search moves, while mutations generate small, refining genotypic changes. In LGA, the local search provides the refining moves, while mutations are employed for more exploratory moves and for its role in replacing alleles that have been lost during the selection process. The global GA continues until one of several termination steps is reached, either the number of the current generation exceeds the maximum number of generations, the number of total fitness evaluations exceeds the maximum allowed evaluations, the fitness of the worst individual is the same as average fitness in the population (i.e., population convergence), and so on.

ALGORITHM 5.3 PSEUDOCODE FOR LGA

```
LGA {
Generate G=population size N of random genotypes
generation=1;

Mapping genotype G to phenotype P of population
Fitness
valuation Fp (scoring) on P

#Loop over generations
WHILE (not TERMINATION) do
 Process generation {
   selection of G' from G for reproduction using
     weighted function based on
     Fp of each individual
   Gc=child population=crossover & mutation of G'
   Pc=Mapping genotype to phenotype of Gc
   Fpc=Fitness evaluation on Pc
   elitist selection (optional)
   Gc=local_search on (Gc'=percentage of Gc;
     calculate Fpc')
   G2 (next generation)=compose(G, Gc)
   Generation ++

  }

END WHILE

}

local_search(G) {
 (optional) mapping genotype G to phenotype P
local optimization (e.g., Pattern-search or Solis-Wets)
  on P to calculated
 optimized P2 Fp2
```

```
(optional) mapping of optimized P2 to G2 return G2,
Fp2

}

termination conditions:
  generation>maximum generations
  number of fitness evaluations>maximum number
    evaluations
  worst fitness is average fitness in population
```

The Lamarckian GA has been implemented in AutoDock [42] and is available with a GNU General Public License at http://autodock.scripps.edu/. GOLD [45] uses a GA without the local search and is commercially available at http://www.ccdc.cam.ac.uk/products/life_sciences/gold/. DARWIN [98], an in-house software package at the Wistar Institute, employs a similar GA, where every phenotype is minimized using CHARMM as part of the fitness evaluation procedure.

As the GA is a nondeterministic search strategy, the time required to reach a solution is dependent on user parameters to stop the calculation. In general, if a set number of scoring evaluations (the slowest step in the algorithm) is taken as a stopping condition, then the search speed will go roughly as the square of the size of the ligand, since the intramolecular energy calculation for the ligand scales as $O(N^2)$, while the intermolecular grid score scales as $O(N)$. The LGA was shown to reliably provide faster and more accurate solutions than simulated annealing or GA alone [42] within the context of the AutoDock program and scoring function. In comparing GAs to incremental construction, in practice the GAs take more time to run than the incremental-construction algorithms for the range of ligands commonly used in virtual screening, but as the number of flexible ligand bonds increases the computational time for incremental construction goes up linearly, while the GA solution times stay constant. In terms of pose accuracy in real examples, no comprehensive comparison has been performed between GAs and incremental construction alone, and studies that compare various programs are confounded by other differences in the protocols, in particular the scoring functions, making the outcomes difficult to compare. In practice both approaches reproduce known crystal structures well.

In summary, docking has become a ubiquitous tool in virtual screening. A number of algorithms have successfully been applied to discover novel small molecule ligands (Table 5.1). However, there are still many active areas for development in docking algorithms. In particular, all docking algorithms have the general limitation that they are highly sensitive to small changes in the 3D structure of the receptor and ligand. This is mitigated in algorithms that incorporate a measure of ligand flexibility, and further reduced with receptor flexibility, but further improvement is needed. In addition, scoring functions, while useful for enriching databases, are not accurate enough to predict individual binding affinities reliably. Even so, the current algorithms generate enough enrichment of screening databases to be an important tool in the drug discovery process.

TABLE 5.1
Commonly Used Docking Programs and Their Algorithms

Program	Orientation Algorithm	Conformation Algorithm	Scoring Function	Examples of Novel Leads Identified
DOCK [39,51,70]	Sphere matching	Incremental construction, flexible DB/multiple receptors	Force field, MM-PBSA, MM-GBSA	20a-hydroxysteroid dehydrogenase [99]
AutoDock [42,60]	LGA	Flexible ligand and receptor sidechains	Semi-empirical force field	Cdc25 phosphatase [100]
FlexX [52]	Descriptor match	Incremental construction/ multiple receptors	Empirical	Histamine H4 receptor [101]
FRED [54]	Shape matching (Gaussian)	Precompute ligand conformations (OMEGA)	Consensus	Cdc25 phosphatase [102]
GLIDE [46,47,62]	Descriptor match/ Monte Carlo	Induced fit w/protein structure prediction	Semiempirical force field	Liver X receptor modulators [103]
GOLD [45]	GA	Ligand and receptor sidechains	Empirical	SARS-3CL(pro) [104]
ICM [48,105]	Monte Carlo	Flexible sidechains and loops/multiple receptors	Consensus	Serotonin *N*-acetyltransferase [106]
QXP [49]	Monte Carlo/ systematic search	Flexible sidechains and loops	Empirical and force field	β-Catenin [107]
SLIDE [108]	Descriptor matching	Induced-fit ligand and receptor	Empirical	Brugia malayi asparaginyl-tRNA synthetase [109]

REFERENCES

1. Bath, P. A., Morris, C. A., and Willett, P., Effect of standardization on fragment-based measures of structural similarity. *J. Chemometr.* 1993, *7*, 543–550.
2. Lin, T.-H., Yu, Y.-S., and Chen, H.-J., Classification of some active compounds and their inactive analogues using two three-dimensional molecular descriptors derived from computation of three-dimensional convex hulls for structures theoretically generated for them. *J. Chem. Inf. Comput. Sci.* 2000, *40*, 1210–1211.
3. Jaccard, P., Étude comparative de la distribution florale dans une portion des Alpes et des Jura. *Bull. Soc. Vaud. Sci. Nat.* 1901, *37*, 547–579.
4. Tversky, A., Features of similarity. *Psych. Rev.* 1977, *84*, 327–352.
5. Shemetulskis, N. E., Weininger, D., Blankley, C. J., Yang, J. J., and Humblet, C., Stigmata: An algorithm to determine structural commonalities in diverse datasets. *J. Chem. Inf. Comput. Sci.* 1996, *36*, 862–871.
6. Clark, R. D., Fox, P. C., and Abrahamian, E., Using pharmacophore multiplet fingerprints for virtual HTS. In: J. Alvarez and B. Shoichet (Eds), *Virtual Screening in Drug Discovery*, pp. 207–225. CRC Press, Taylor & Francis: Boca Raton, FL, 2005.
7. Nicholls, A., MacCuish, N. E., and MacCuish, J. D., Variable selection and model validation of 2D and 2D molecular descriptors. *J. Comput. Aided Mol. Des.* 2004, *18*, 451–474.
8. Abrahamian, E., Fox, P. C., Nærum, L., Christensen, I. T., Thøgersen, H., and Clark, R. D., Efficient generation, storage and manipulation of fully flexible pharmacophore multiplets and their use in 3-D similarity searching. *J. Chem. Inf. Comput. Sci.* 2003, *43*, 458–468.
9. Cheng, C., Maggiora, G., Lajiness, M., and Johnson, M., Four association coefficients for relating molecular similarity measures. *J. Chem. Inf. Comput. Sci.* 1996, *36*, 909–915.
10. Shemetulskis, N. E., Dunbar, J. B., Jr., Dunbar, B. W., Moreland, D.W., and Humblet, C., Enhancing the diversity of a corporate database using chemical database clustering and analysis. *J. Comput. Aided Mol. Des.* 1995, *9*, 407–416.
11. Pearlman, R. S. and Smith, K. M., Metric validation and the receptor-relevant subspace concept. *J. Chem. Inf. Comput. Sci.* 1999, *39*, 28–35.
12. Lipinski, C. A., Lombardo, F., Dominy, B. W., and Feeney, P. J., Experimental and computational approaches to estimate solubility and permeability in drug discovery and development settings. *Adv. Drug Deliv. Rev.* 2001, *46*, 3–26.
13. Ginn, C. M., Turner, D. B., and Willett, P., Similarity searching in files of three-dimensional chemical structures: Evaluation of the EVA descriptor and combination of rankings using data fusion. *J. Chem. Inf. Comput. Sci.* 1997, *37*, 23–37.
14. Willett, P. and Winterman, V., A Comparison of some measures of inter-molecular structural similarity. *Quant. Struct. Act. Relat.* 1986, *5*, 18–25.
15. Nettles, J. H., Jemkins, J. L., Bender, A., Deng, Z., Davies, J. W., and Glick, M., Bridging chemical and biological space: "Target Fishing" using 2D and 3D molecular descriptors. *J. Med. Chem.* 2006, *49*, 6802–6810.
16. Hahn, M., Three-dimensional shape-based searching of conformationally flexible molecules. *J. Chem. Inf. Comput. Sci.* 1997, *37*, 80–86.
17. Putta, S., Eksterowicz, J., Lemmen, C., and Stanton, R., A novel subshape molecular descriptor. *J. Chem. Inf. Comput. Sci.* 2003, *43*, 1623–1635.
18. Haigh, J. A., Pickup, B. T., Grant, J. A., and Nicholls, A., Small molecule shape-fingerprints. *J. Chem. Inf. Model* 2005, *45*, 673–684.

19. Hull, R. D., Fluder, E. M., Singh, S. B., Nachbar, R. B., Kearsley, S. K., and Sheridan, R. B., Chemical similarity searches using latent variable semantic structural indexing (LaSSI) and comparison to TOPOSIM. *J. Med. Chem.* 2001, *44*, 1185–1191.
20. McGaughey, G. B., Sheridan, R. P., Bayly, C. I., Culberson, J. C., Kreatsoulas, C., Lindsley, S., Maiorov, V., Truchon, J.-F., and Cornell, W. D., Comparison of topological, shape, and docking methods in virtual screening. *J. Chem. Inf. Model* 2007, *47*, 1504–1519.
21. Ewing, T., Baber, J. C., and Feher, M., Novel 2D fingerprints for ligand-based virtual screening. *J. Chem. Inf. Model* 2006, *46*, 2423–2431.
22. Carbó, R., Leyda, L., and Arnau, M., How similar is a molecule to another? An electron density measure of similarity between two molecular structures. *Int. J. Quantum Chem.* 1980, *17*, 1185–1189.
23. Hodgkin, E. E. and Richards, W. G., Molecular similarity based on electrostatic potential and electric field. *Int. J. Quantum Chem.* 1987, *14*, 105–110.
24. Good, A. C., Hodgkin, E. E., and Richards, W. G., Utilization of Gaussian functions for the rapid evaluation of molecular similarity. *J. Chem. Inf. Comput. Sci.* 1992, *32*, 188–191.
25. Grant, J. A., Haigh, J. A., Pickup, B. T., Nicholls, A., and Sayle, R. A., Lingos, finite state machines, and fast similarity searching. *J. Chem. Inf. Model.* 2006, *46*, 1912–1918.
26. Gower, J. C., Measures of similarity, dissimilarity and distance. In: S. Kotz and N. L. Johnson (Eds), *Encyclopedia of Statistical Sciences*, Vol. 5, pp. 397–405. Wiley, New York, 1985.
27. Willett, P., Barnard, J. M., and Downs, G. M., Chemical similarity searching. *J. Chem. Inf. Comput. Sci.* 1998, *38*, 983–996.
28. Flower, D. R., On the properties of bit string-based measures of chemical similarity. *J. Chem. Inf. Comput. Sci.* 1998, *38*, 379–386.
29. Dixon, S. L. and Koehler, R. T., The hidden component of size in two-dimensional fragment descriptors: Side effects on sampling in bioactive libraries. *J. Med. Chem.* 1999, *42*, 2887–2900.
30. Patterson, D. E., Cramer, R. D., Ferguson, A. M., Clark, R. D., and Weinberger, L. E., Neighborhood behavior: A useful concept for validation of molecular diversity descriptors. *J. Med. Chem.* 1996, *39*, 3049–3059.
31. Clark, R. D., Brusati, M., Jilek, R., Heritage, T., and Cramer, R. D., Validating novel QSAR descriptors for use in diversity analysis. In: K. Gundertofte and F. S. Jørgensen (Eds), *Molecular Modeling and Prediction of Bioactivity*, pp. 95–100. Kluwer Academic/Plenum Publishers, New York, 2000.
32. Salim, N., Holliday, J., and Willett, P., Combination of fingerprint similarity coefficients using data fusion. *J. Chem. Inf. Comput. Sci.* 2003, *43*, 435–442.
33. Willett, P., Enhancing the effectiveness of ligand-based virtual screening using data fusion. *QSAR Comb. Sci.* 2006, *25*, 1143–1152.
34. Klon, A. E., Glick, M., and Davies, J. W., Combination of Naïve Bayesian classifier with consensus scoring improves enrichment of high-throughput docking results. *J. Med. Chem.* 2004, *47*, 4356–4359.
35. Hert, J., Willett, P., and Wilton, D. J., comparison of fingerprint-based methods for virtual screening using multiple bioactive reference structures. *J. Chem. Inf. Comput. Sci.* 2004, *44*, 1177–1185.
36. Hert, J., Willett, P., Wilton, D. J., Acklin, P., Azzaoui, K., Jacoby, E., and Schuffenhauer, A., Comparison of topological descriptors for similarity-based virtual screening using multiple bioactive reference structures. *Org. Biomol. Chem.* 2004, 2, 3256–3266.

37. Hert, J., Willett, P., Wilton, J. D. J., Acklin, P., Azzaoui, K., Jacoby, E., and Schuffenhauer, A., New methods for ligand-based virtual screening: Use of data fusion and machine learning to enhance the effectiveness of similarity searching. *J. Chem. Inf. Model.* 2006, *46*, 462–470.
38. Reddy, A. S., et al., Virtual screening in drug discovery—a computational perspective. *Curr. Protein Pept. Sci.* 2007, *8*, 329–351.
39. Kuntz, I. D., et al., A geometric approach to macromolecule–ligand interactions. *J. Mol. Biol.* 1982, *161*, 269–288.
40. DesJarlais, R. L., et al., Docking flexible ligands to macromolecular receptors by molecular shape. *J. Med. Chem.* 1986, *29*, 2149–2153.
41. Goodsell, D. S. and Olson, A. J., Automated docking of substrates to proteins by simulated annealing. *Proteins* 1990, *8*, 195–202.
42. Morris, G. M., et al., Automated docking using a Lamarckian genetic algorithm and an empirical binding free energy function. *J. Comput. Chem.* 1998, *19*, 1639–1662.
43. Judson, R. S., Jaeger, E. P., and Treasurywala, A. M., A genetic algorithm based method for docking flexible molecules. *J. Mol. Struct.—THEOCHEM*, 1994, *308*, 191–206.
44. Oshiro, C. M., Kuntz, I. D., and Dixon, J. S., Flexible ligand docking using a genetic algorithm. *J. Comput. Aided Mol. Des.* 1995, *9*, 113–130.
45. Jones, G., et al., Development and validation of a genetic algorithm for flexible docking. *J. Mol. Biol.* 1997, *267*, 727–748.
46. Friesner, R. A., et al., Glide: A new approach for rapid, accurate docking and scoring. 1. Method and assessment of docking accuracy. *J. Med. Chem.* 2004, 47,1739–1749.
47. Halgren, T.A., et al., Glide: A new approach for rapid, accurate docking and scoring. 2. Enrichment factors in database screening. *J. Med. Chem.* 2004, *47*, 1750–1759.
48. Ruben, A., Maxim, T., and Dmitry, K., ICM: A new method for protein modeling and design: Applications to docking and structure prediction from the distorted native conformation. *J. Comput. Chem.* 1994, *15*, 488–506.
49. McMartin, C. and Bohacek, R. S., QXP: Powerful, rapid computer algorithms for structure-based drug design. *J. Comput. Aided Mol. Des.* 1997, *11*, 333–344.
50. Liu, M. and Wang, S., MCDOCK: A Monte Carlo simulation approach to the molecular docking problem. *J. Comput. Aided Mol. Des.* 1999, *13*, 435–451.
51. Ewing, T. J., et al., DOCK 4.0: Search strategies for automated molecular docking of flexible molecule databases. *J. Comput. Aided Mol. Des.* 2001, *15*, 411–428.
52. Rarey, M., et al., A fast flexible docking method using an incremental construction algorithm. *J. Mol. Biol.* 1996, *261*, 470–489.
53. Lorber, D. M. and Shoichet, B. K. Flexible ligand docking using conformational ensembles. *Protein Sci.* 1998, *7*, 938–950.
54. McGann, M. R., et al., Gaussian docking functions. *Biopolymers* 2003, *68*, 76–90.
55. Miller, M. D., et al., FLOG: A system to select 'quasi-flexible' ligands complementary to a receptor of known three-dimensional structure. *J. Comput. Aided Mol. Des.* 1994, *8*, 153–174.
56. Polgar, T. and Keseru, G. M., Ensemble docking into flexible active sites. Critical evaluation of FlexE against JNK-3 and beta-secretase. *J. Chem. Inf. Model* 2006, *46*, 1795–1805.
57. Claussen, H., et al., FlexE: Efficient molecular docking considering protein structure variations. *J. Mol. Biol.* 2001, *308*, 377–395.
58. Wei, B. Q., et al., Testing a flexible-receptor docking algorithm in a model binding site. *J. Mol. Biol.* 2004, *337*, 1161–1182.
59. Cavasotto, C. N., Kovacs, J. A., and Abagyan, R. A., Representing receptor flexibility in ligand docking through relevant normal modes. *J. Am. Chem. Soc.* 2005, *127*, 9632–9640.

60. AutoDock4.0 [computer software], Scripps Research Institute: La Jolla, CA. http://autodock.scripps.edu/
61. Totrov, M. and Abagyan, R.A. Flexible protein-ligand docking by global energy optimization in internal coordinates. *Proteins* 1997, Suppl. 1, 215–220.
62. Sherman, W., et al., Novel procedure for modeling ligand/receptor induced fit effects. *J. Med. Chem.* 2006, *49*, 534–553.
63. Case, D. A., et al., *AMBER 10*. University of California, San Francisco, 2008.
64. Brooks, B. R., et al., CHARMM: A program for macromolecular energy, minimization, and dynamics calculations. *J. Comput. Chem.* 1983, *4*, 187–217.
65. Mackerel, A. D., Brooks, C. L., Nilsson, L., Roux, B., Won, Y., and Karplus, M. CHARMM: The energy function and its parameterization with an overview of the program. In: Schleyer (Ed.), *The Encyclopedia of Computational Chemistry*, Vol. 1, pp. 271–277. Wiley, Chichester, 1998.
66. Meng E. C., Shoichet, B. K., and Kuntz I. D., Automated docking with grid-based energy evaluation. *J. Comput. Chem.* 1992, *13*, 505–524.
67. Nicholls, A. and Honig, B., A rapid finite difference algorithm, utilizing successive over-relaxation to solve the Poisson–Boltzmann equation. *J. Comput. Chem.* 1991, *12*, 435–445.
68. Grant, J. A., Pickup, B. T., and Nicholls, A., A smooth permittivity function for Poisson–Boltzmann solvation methods. *J. Comput. Chem.* 2001, *22*, 608–640.
69. Shoichet, B. K., Leach, A. R., and Kuntz, I. D., Ligand solvation in molecular docking. *Proteins* 1999, *34*, 4–16.
70. Lang, P. T., et al., *DOCK 6.2*. University of California, San Francisco, 2006.
71. Bohm, H. J., The development of a simple empirical scoring function to estimate the binding constant for a protein–ligand complex of known three-dimensional structure. *J. Comput. Aided Mol. Des.*, 1994, *8*, 243–256.
72. Eldridge, M. D., et al., Empirical scoring functions: I. The development of a fast empirical scoring function to estimate the binding affinity of ligands in receptor complexes. *J. Comput. Aided Mol. Des.*, 1997, *11*, 425–445.
73. Huey, R., et al., A semiempirical free energy force field with charge-based desolvation. *J. Comput. Chem.* 2007, *28*, 1145–1152.
74. DeWitte, R. S. and Shakhnovich, E. I., SMoG: *De novo* design method based on simple, fast, and accurate free energy estimates. 1. Methodology and supporting evidence. *J. Am. Chem. Soc.* 1996, *118*, 11733–11744.
75. Gohlke, H., Hendlich, M., and Klebe, G., Knowledge-based scoring function to predict protein–ligand interactions. *J. Mol. Biol.* 2000, *295*, 337–356.
76. Muegge, I. and Martin, Y. C., A general and fast scoring function for protein–ligand interactions: A simplified potential approach. *J. Med. Chem.* 1999, *42*, 791–804.
77. Charifson, P. S., et al., Consensus scoring: A method for obtaining improved hit rates from docking databases of three-dimensional structures into proteins. *J. Med. Chem.* 1999, *42*, 5100–5109.
78. Springer, C., et al., PostDOCK: A structural, empirical approach to scoring protein–ligand complexes. *J. Med. Chem.* 2005, *48*, 6821–6831.
79. Berman, H. M., et al., The protein data bank. *Acta Crystallogr. D Biol. Crystallogr.* 2002, *58*, 899–907.
80. Graves, A. P., et al., Rescoring docking hit lists for model cavity sites: Predictions and experimental testing. *J. Mol. Biol.* 2008, *377*, 914–934.
81. Kuhn, B., et al., Validation and use of the MM-PBSA approach for drug discovery. *J. Med. Chem.* 2005, *48*, 4040–4048.

82. Kroemer, R. T., Structure-based drug design: Docking and scoring. *Curr. Protein Pept. Sci.* 2007, *8*, 312–328.
83. Kontoyianni, M., et al., Theoretical and practical considerations in virtual screening: A beaten field? *Curr. Med. Chem.* 2008, *15*, 107–116.
84. Lipinski, C. A., et al., Experimental and computational approaches to estimate solubility and permeability in drug discovery and development settings. *Adv. Drug Deliv. Rev.* 2001, *46*, 3–26.
85. McGovern, S. L., et al., A common mechanism underlying promiscuous inhibitors from virtual and high-throughput screening. *J. Med. Chem.* 2002, *45*, 1712–1722.
86. Roche, O., et al., Development of a virtual screening method for identification of "frequent hitters" in compound libraries. *J. Med. Chem.* 2002, *45*, 137–142.
87. Aronov, A. M., Predictive in silico modeling for hERG channel blockers. *Drug Discov. Today* 2005, *10*, 149–155.
88. Ewing, T. J. A. and Kuntz, I. D., Critical evaluation of search algorithms for automated molecular docking and database screening. *J. Comput. Chem.* 1997, *18*, 1175–1189.
89. Bron, C. and Kerbosch, J., Finding all cliques of an undirected graph. *Commun. ACM* 1973, *16*, 575–577.
90. Moustakas, D. T., et al., Development and validation of a modular, extensible docking program: DOCK 5. *J. Comput. Aided Mol. Des.* 2006, *20*, 601–619.
91. Ewing, T. J. A. and Kuntz, I. D., Critical evaluation of search algorithms for automated molecular docking and database screening. *J. Comput. Chem.* 1997, *18*, 1175–1189.
92. Hart, W. E., *Adaptive Global Optimization with Local Search.* University of California, San Diego, 1994.
93. Hart, W. E., Kammeyer, T. E., Belew, R. K., The role of development in genetic algorithms. In: W. D. and V. M. (Eds), *Foundations of Genetic Algorithms III.* Morgan Kauffman, San Francisco, CA, 1994.
94. Dennis, J. E. and Torczon, V., Derivative-free pattern search methods for multidisciplinary design problems. In: *Proceedings 5th AiAA/NASA/ISSMO Symposium on Multidisciplinary Analysis and Optimization*, 1994. Panama city, FL.
95. Torczon, V. and Trosset, M. W., From evolutionary operation to parallel direct search: Pattern search algorithms for numerical optimization. In: *Computing Science and Statistics. Proceedings 29th Symposium on the Interface*, 1997. Houston, TX.
96. Solis, F. J. and Wets, J.-B., Minimization by random search techniques. *Math. Oper. Res.* 1981, *6*, 19–30.
97. Taylor, J. S. and Burnett, R. M., DARWIN: A program for docking flexible molecules. *Proteins* 2000, *41*, 173–191.
98. Dhagat, U., et al., A salicylic acid-based analogue discovered from virtual screening as a potent inhibitor of human 20-hydroxysteroid dehydrogenase. *Med. Chem.* 2007, *3*, 546–550.
99. Park, H., et al., Discovery of novel Cdc25 phosphatase inhibitors with micromolar activity based on the structure-based virtual screening. *J. Med. Chem.* 2008, *51*, 5533–5541.
100. Kiss, R., et al., Discovery of novel human histamine H4 receptor ligands by large-scale structure-based virtual screening. *J. Med. Chem.* 2008, *51*, 3145–3153.
101. Montes, M., et al., Receptor-based virtual ligand screening for the identification of novel CDC25 phosphatase inhibitors. *J. Chem. Inf. Model* 2008, *48*, 157–165.
102. Cheng, J.-F., et al., Combination of virtual screening and high throughput gene profiling for identification of novel liver X receptor modulators. *J. Med. Chem.* 2008, *51*, 2057–2061.

103. Mukherjee, P., et al., Structure-based virtual screening against SARS-3CLpro to identify novel non-peptidic hits. *Bioorg. Med. Chem.* 2008, *16*, 4138–4149.

104. Cavasotto, C. N. and Abagyan, R. A., Protein flexibility in ligand docking and virtual screening to protein kinases. *J. Mol. Biol.* 2004, *337*, 209–225.

105. Szewczuk, L. M., et al., *De novo* discovery of serotonin *N*-acetyltransferase inhibitors. *J. Med. Chem.* 2007, *50*, 5330–5338.

106. Trosset, J. Y., Dalvit, C., Knapp, S., Fasolini, M., Veronesi, M., Mantegani, S., Gianellini, L. M., Catana, C., Sundström, M., Stouten, P. F., and Moll, J. K., Inhibition of protein–protein interactions: The discovery of druglike beta-catenin inhibitors by combining virtual and biophysical screening. *Proteins* 2006, *64*, 60–67.

107. Schnecke, V. and Kuhn, L. A., Virtual screening with solvation and ligand-induced complementarity. *Perspect. Drug Discov. Des.* 2000, *20*, 171–190.

108. Sukuru, S., et al., Discovering new classes of Brugia malayi asparaginyl-tRNA synthetase inhibitors and relating specificity to conformational change. *J. Comput. Aided Mol. Des.* 2006, *20*, 159–178.

109. OMEGA 2.0 [computer software], OpenEye Scientific Software: Sante Fe, NM. http://www.eyesopen.com

6 Predictive Quantitative Structure–Activity Relationships Modeling

Data Preparation and the General Modeling Workflow

Alexander Tropsha and Alexander Golbraikh

CONTENTS

In this and the next chapter, we shall consider modern approaches for developing statistically robust and externally predictive quantitative structure–activity relationships (QSAR) models. We shall discuss the general QSAR model development and validation workflow that should be followed irrespective of specifics of any particular QSAR modeling routine. We will refrain on purpose from discussing any specific model optimization algorithms because such details could be found in many original publications. This chapter focuses on the initial steps in QSAR modeling, that is, input data preparation and curation, as well as introduces the general workflow for developing validated and predictive models. Conversely, the next chapter addresses general data modeling and model validation procedures that constitute the important elements of the workflow.

This chapter starts with the discussion of the general workflow for developing predictive QSAR models. Then, we concentrate on the requirements to QSAR datasets and procedures that should be employed for the initial data treatment and preparation for model development. We consider briefly major types of descriptors (i.e., quantitative characteristics of chemical structures) and discuss algorithms for preprocessing descriptor files prior to QSAR studies. We emphasize that rigorous validation of QSAR models is impossible without using both test and additional external model evaluation sets and discuss several approaches for the division of a dataset into training, test, and external evaluation sets. We address critical aspects of preliminary data analysis such as the detection of possible structural and activity outliers and dealing with the imbalanced datasets.

To complete the discussion of major modern QSAR modeling principles, the next chapter covers some special topics of QSAR analysis such as different target functions and measures of prediction accuracy, approaches to model validation, model applicability domains, consensus prediction and the use of QSAR models in virtual screening. We emphasize that the true utility of QSAR models is in their ability to make accurate predictions for external datasets. In this regard, we ascertain that the integration of all components of the QSAR modeling workflow discussed in this and the subsequent chapter is absolutely necessary for building rigorously validated and externally predictive QSAR models.

6.1 INTRODUCTION: PREDICTIVE QSAR MODELING

The rapid development of information and communication technologies during the last few decades has dramatically changed our capabilities of collecting, analyzing, storing, and disseminating all types of data. This process has had a profound influence on the scientific research in many disciplines, including the development of new generations of effective and selective medicines. Large databases containing millions of chemical compounds tested in various biological assays such as PubChem[1] are increasingly available as online collections (recently reviewed by Oprea and Tropsha[2]). In order to find new drug leads, there is a need for efficient and robust procedures that can be used to screen chemical databases and virtual libraries against molecules with known activities or properties. To this end, QSAR modeling provides an effective means for both exploring and exploiting the relationship between chemical structure and its biological action toward the development of novel drug candidates.

The QSAR approach can be generally described as an application of data analysis methods and statistics to developing models that could accurately predict biological activities or properties of compounds based on their structures. Our experience in QSAR model development and validation has led us to establish a complex strategy that is summarized in Figure 6.1. It describes the predictive QSAR modeling workflow focused on delivering validated models and ultimately computational hits confirmed for the experimental validation. We start by randomly selecting a fraction of compounds (typically, 10–20%) as an external evaluation set. The sphere exclusion protocol implemented in our laboratory[3,4] is then used to rationally divide the

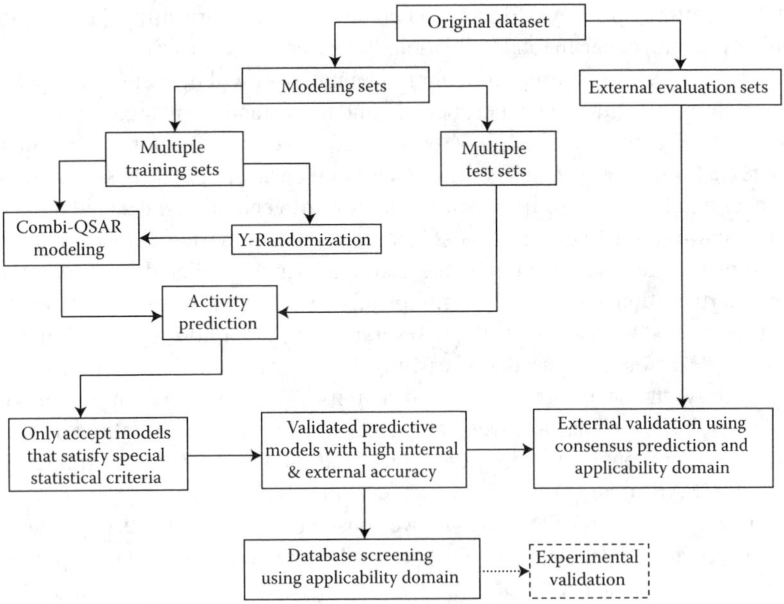

FIGURE 6.1 Predictive QSAR modeling workflow.

remaining subset of compounds (the modeling set) into multiple training and test sets that are used for model development and validation, respectively. We employ multiple QSAR techniques based on the combinatorial exploration of all possible pairs of descriptor sets and various supervised data analysis techniques (combi-QSAR) and select models characterized by high accuracy in predicting both training and test sets data. Validated models are finally tested using the external evaluation set. The critical step of the external validation is the use of applicability domains (AD). If external validation demonstrates the significant predictive power of the models, we employ them for virtual screening of available chemical databases (e.g., ZINC[5]) to identify putative active compounds and work with collaborators who could validate such hits experimentally. The entire approach is described in detail in several recent papers and reviews (see, e.g., Refs. 6–9).

The development of truly validated and predictive QSAR models affords their growing application in chemical data mining and combinatorial library design.[10,11] For example, three-dimensional (3D) stereoelectronic pharmacophore based on QSAR modeling was used recently to search the National Cancer Institute Repository of Small Molecules to find new leads for inhibiting human immunodeficiency virus (HIV) type 1 reverse transcriptase at the non-nucleoside binding site.[12]

It is increasingly critical to provide experimental validation as the ultimate assertion of the model-based prediction. In our recent studies we were fortunate to recruit experimental collaborators who have validated computational hits identified through our modeling of several datasets including anticonvulsants,[13] HIV-1 reverse transcriptase inhibitors,[14] D1 antagonists,[15] antitumor compounds,[16] β-lactamase inhibitors,[17] and histone deacetylase (HDAC) inhibitors.[18] Thus, models resulting from the predictive

QSAR modeling workflow (Figure 6.1) could be used to prioritize the selection of chemicals for the experimental validation. However, since we still cannot guarantee that every prediction resulting from our modeling effort will be validated experimentally, we do not include the experimental validation step as a mandatory part of the workflow in Figure 6.1, which is why we used the dotted line for this component. We note that our approach shifts the emphasis on ensuring good (best) statistics for the model that fits known experimental data toward generating a testable hypothesis about purported bioactive compounds. Thus, the output of the modeling has exactly the same format as the input, that is, chemical structures and (predicted) activities making model interpretation and utilization completely seamless for medicinal chemists.

Thus, studies in our as well as several other laboratories have shown that QSAR models could be used successfully as virtual screening tools to discover compounds with the desired biological activity in chemical databases or virtual libraries.[6,13,15–17,19] The discovery of novel bioactive chemical entities is the primary goal of computational drug discovery, and the development of validated and predictive QSAR models is critical to achieve this goal.

In the remaining part of this chapter, we consider the requirements to primary data used for QSAR analysis, approaches used in the preparation of data, preprocessing of descriptors, and detection of outliers. We emphasize that rigorous validation of QSAR models is impossible without using test and additional external evaluation sets and discuss several approaches for division of data into training, test, and external evaluation sets.

6.2 REQUIREMENTS TO A DATASET

The number of compounds in the dataset for QSAR studies should not be too small, or, for practical reasons, too large. The upper limit is defined by the computer and time resources available for building QSAR models using the selected methodologies. For example, for the k-nearest neighbors (kNN) QSAR approach frequently practiced in our laboratory,[20,21] the maximum number of compounds in the *training set* (i.e., compounds used to build QSAR models) may not exceed about ca. 2000 due to the inefficiency of the approach when processing large datasets. When a dataset includes more compounds, several approaches can be implemented: (i) select a diverse subset of compounds; (ii) cluster a dataset and build models separately for each cluster; (iii) sometimes, in the case of classification or category QSAR, when compounds belong to a small number of activity classes or categories (e.g., active and inactive), it is possible to exclude many compounds from model development. (The difference between classes and categories is that that in contrast to classes, categories can be ordered. An example of classes: ligands of different receptors. An example of categories: compounds that are very active, active, moderately active, and inactive.)

The lower limit of the number of compounds in the dataset is also defined by several factors. For example, in most cases, as part of model validation schemes, we divide a dataset into three subsets: training, test, and external evaluation sets. Training sets are used in model development, and if they are too small, chance correlation and overfitting become major problems not allowing one to build truly predictive models. While it is impossible to give an exact minimum number of compounds in a dataset for which building reliable QSAR models is feasible, some simple ideas described here may help. In the case of continuous response variable (activity), the number of compounds in the training set should be at least 20, and about 10 compounds should be in each of the test and external evaluation sets, so the total minimum number of compounds should be no less than 40. In the case of classification or category response variable, the training set should contain at least about 10 compounds of each class, and test and external evaluation sets should contain no less than five compounds for each class. So, there should be at least 20 compounds of each class. The best situation is when the number of compounds in the dataset is between these two extremes: about 150–300 compounds in total, and in the case of classification or category QSAR, approximately equal number of compounds of each class or category.

There are also requirements for activity values. In the case of continuous response variable, the total range of activities should be at least 5 times higher than the experimental error. No large gaps (that exceed 10–15% of the entire range of activities) are allowed between two consecutive values of activities ordered by value. In the case of classification or category QSAR, there should be at least 20 compounds of each class or category; preferably, the number of compounds in all classes or categories should be approximately the same. However, many existing datasets are imbalanced or biased (i.e., sizes of different classes or categories are different). In these cases, special QSAR algorithms are used to equalize the number of compounds in different classes or categories. There are also approaches (such as cost-sensitive learning[22,23]) that account for these differences by including additional parameters in target functions (see Section 7.2) and criteria of prediction accuracy.

The main QSAR hypothesis underlying all QSAR studies is as follows: similar compounds should have similar biological activities or properties. If this condition for compounds in the dataset is not satisfied, building truly predictive QSAR models is impossible. In fact, one can define two compounds as similar if their chemical structures are similar. In computer representation, compounds are characterized by a set of quantitative parameters called descriptors. *Similarity* between two compounds is a quantitative measure that is defined based on compounds' descriptor values. Different definitions of compound similarity exist. These measures reflect the similarity in molecular structure of these compounds. Obviously, quantitative values of similarity measures between two compounds also depend on which descriptors are used. So there is no unique similarity measure. Below, we will address several definitions of similarity.

6.3 DATASET CURATION

Any modeling study requires a dataset of compounds where all chemical structures are correct, there are no duplicates, and activity values are accurate. It is highly recommended that before the modeling studies begin, the datasets be examined to establish that the above listed quality control criteria are satisfied. A recent study provides a great illustration as to how having even a few incorrect structures could significantly impart the accuracy of QSAR models.[24] In addition, the calculation of molecular descriptors should be possible for every compound in a dataset. In this regard, it should be kept in mind that most of the molecular descriptors cannot be calculated for compounds consisting of two or more molecules that are not covalently connected (e.g., salts); many molecular descriptors cannot be calculated for inorganic compounds or compounds that include heavy metal atoms due to lack of the corresponding parameters; many types of descriptors cannot take chirality and some other types of isomerism into account, and so on. Depending on the descriptors used and the dataset, all or some of these compounds should be excluded from the dataset.

Consider, for example, a dataset that includes many molecules containing chiral atoms, including some pairs of enantiomers and diastereomers. If atomic chiralities for all these compounds are always available along with compounds' activities, descriptors taking chirality into account should be used, and all isomers should be retained in the dataset. If, however, chirality information is unavailable, only one compound, usually with the highest (or mean) activity should be retained, and chirality descriptors should not be used. There are different tools available for dataset curation. For example, Molecular Operating Environment (MOE)[25] includes Database Wash tool. It allows changing molecules' names, adding or removing hydrogen atoms, removing salts and heavy atoms, even if they are covalently connected to the rest of the molecule, and changing or generating the tautomers and protomers (cf. the MOE manual for more details). Various database curation tools are included in ChemAxon[26] as well. If commercial software tools such as MOE are unavailable (notably, ChemAxon software is free to academic investigators), one can use standard UNIX/LINUX tools to perform some of the dataset cleaning tasks. It is important to have some freely available molecular format converters such as OpenBabel[27] or MolConverter from ChemAxon.[26]

We shall discuss the use of some of the standard data cleaning operations using freely available tools. Suppose that a file called *mydata* contains a dataset in the SMILES format. Each line of this file contains SMILES string for one compound and ID for this compound. Some of the compounds contain metal atoms such as Na, K, Ca, Fe, Co, Ni, Mn, and Mg, and one wants to exclude all of them from the dataset. It can be easily done in the UNIX/LINUX operating system by giving the command:

egrep –v "\[Na|\[K|\[Ca|\[Fe|\[Co|\[Ni|\[Mn|\[Mg" mydata > mynometaldata.

Suppose that a file *mydata* also contains some compounds that are not fully covalently connected. In SMILES, disconnected parts of the compound are separated by a dot. So all compounds containing dots can be removed:

grep –v "\." mynometaldata > mynometalnosaltdata.

Alternatively, one may want to retain the largest fragment of a compound. In this case, the following awk code can be used:

```
{
    if(index($1,".")==0) printf("%s ",$1);
    else
    {
        n=split($1,a,".");
        p=0;
        m=0;
        for(i=1;i<=n;i++)
        {
```

```
            r=length(a[i]);
            if(r>p) {m=i; p=r;}
      }
         printf("%s ",a[m]);
      }
         if(NF==1) printf("No_ID\n");
         if(NF==2) printf("%s\n",$2);
         else
         {
            for(i=2;i<NF;i++) printf("%s_",$i);
            printf("%s\n",$i);
      }
}
```

Create a file *removesalts.awk* and copy the above code in this file. The following command will remove smaller fragments of compounds:

awk –f removesalts.awk mynometaldata > mynometalnosaltdata.

If a user is not interested in small molecules, the user may decide to remove compounds described by short SMILES strings with lengths up to 8. This can be done by using a short awk script:

awk '{if(length($1) > var1) print $0;}' mynometalnosaltdata > mycleaneddata.

The *mycleaneddata* still may contain duplicates. Duplicates of SMILES strings can be removed using the following awk script:

```
BEGIN {i=0;}
{ if(var==$1) i++; if(i<=1 || i>1 && var!=$1 ) print;}
END {if(i>1) print var,i>file;}
```

where *var* and *file* are external variables; *var* is one SMILES string, and *file* is a file name containing SMILES strings included more than once in the input file. Create file *removeduplicates.awk* and copy the above script into it. *var* variable runs through all compounds, and each time, duplicates of this compound are removed. The following Cshell script used with the *removeduplicates.awk* will do the job.

```
#!/bin/csh
cp data2 temp
foreach i ('cut -d" " -f1 data2')
awk -v var="$i" -v file="duplicates" -f removeduplicates.
    awk temp > temp2†
cp temp2 temp
cat duplicates >> duplicates.txt
end
cp temp2 mycleaneddatanodup
rm temp*
```

Name it *removeduplicates.csh* and run it using the following command:

csh removeduplicates.csh.

---

† It is one command which should be entered on one line of the UNIX/LINUX terminal.

6.4 CALCULATION OF DESCRIPTORS

Descriptors are quantitative characteristics describing molecular structures that are used in QSAR and other chemoinformatics studies. They can be experimental or calculated physicochemical properties of molecules such as molecular weight, molar refraction, energies of HOMO and LUMO, normal boiling point, octanol/water partition coefficients, topological indices or invariants of molecular graphs (structural formulas), molecular surface, molecular volume, etc. The first abstract molecular topological indices introduced in molecular property prediction studies were the Wiener index[28] and the Platt index.[29]

Herein, we will not discuss different types of descriptors in detail but mention briefly major descriptor classes. There is an excellent monograph titled *Handbook of Molecular Descriptors* by Roberto Todeschini and Vivian Consonni[30] that provides reference materials on more than 2000 different descriptors. Most of descriptors included in this book can be calculated by the Dragon software.[31] Dragon calculates many different groups of descriptors such as constitutional descriptors (sometimes referred to as zero-dimensional [0D] descriptors), counts of different molecular groups, physicochemical properties of compounds, and so on. (one-dimensional [1D] descriptors), connectivity indices, information indices, counts of paths and walks, and so on (two-dimensional [2D] descriptors), geometrical properties, GETAWAY, WHIM, 3DMoRSE descriptors, and so on (3D descriptors), and some other descriptors. MolconnZ[32] is another widely used descriptor calculation software. In total, it calculates more than 800 descriptors including valence path, cluster, path/cluster and chain molecular connectivity indices, kappa molecular shape indices, topological and electrotopological state indices, differential connectivity indices, the graph's radius and diameter, Wiener and Platt indices, Shannon and Bonchev-Trinajstić information indices, counts of different vertices, and counts of paths and edges between different kinds of vertices. MOE[25] descriptors include both 2D and 3D molecular descriptors. 2D descriptors include physical properties, subdivided surface areas, atom counts and bond counts, Kier and Hall connectivity and kappa shape indices, adjacency and distance matrix descriptors, pharmacophore feature descriptors, and partial charge descriptors. 3D molecular descriptors include potential energy descriptors, surface area, volume and shape descriptors, and conformation-dependent charge descriptors. Chirality molecular topological descriptors (CMTD) developed in our laboratory include chirality and ZE-isomerism molecular connectivity indices, overall Zagreb indices, extended indices, and overall connectivity indices.[33–35] They are calculated as conventional descriptors with modified vertex degrees. Another group of descriptors frequently used in our laboratory is atom-pair (AP) descriptors.[36] Each descriptor is defined as a count of pairs of atoms of the same type being away from each other on a certain topological distance (2D AP descriptors) or a Euclidean distance within certain intervals (3D AP descriptors). A new version of the program includes chirality descriptors, which are counts of APs with one or both atoms in the pair chiral.[37] Comparative molecular field analysis (CoMFA) descriptors represent values of Lennard-Jones and Coulomb energies of interactions between a molecule and a probe atom at certain grid points built around a set of spatially aligned molecules.[38] The molecules are aligned according to a pharmacophore model, a spatially arranged set of features that are believed to be responsible for the biological activity or property of compounds in question. There are several types of other 3D CoMFA-like descriptors, such as comparative molecular similarity indices analysis (CoMSIA),[39] comparative similarity indices analysis (QSiAR),[40] self-organizing molecular field analysis (SOMFA),[41] and so on. These descriptors can be effectively used for sets of rigid compounds, or compounds that have a large common fragment, but for flexible compounds with different scaffolds they require extensive conformational analysis along with rigid templates to superimpose molecules onto each other. There are different conformational analysis and pharmacophore modeling tools included in molecular modeling packages such as MOE,[25] Sybyl,[42] Discovery Studio,[43] LigandScout,[44] and so on.

It has been demonstrated that in many cases QSAR models based on 2D descriptors have comparable (or even superior) predictivity than models based on 3D descriptors.[20,21,33,45] At the same time, 3D methods are much more time and resource consuming. Moreover, even for rigid compounds, generally it is not known whether the alignment corresponds to real positions of molecules in the receptor binding site.[46] So, when 3D QSAR studies are necessary, if possible, 3D alignment of molecules should be preferably obtained by docking studies. VolSurf[47,48] and GRIND[49] descriptors are examples of alignment-free 3D descriptors. GRIND descriptors are obtained from 3D interaction energy grid maps. Calculation of VolSurf descriptors includes the following steps: (i) building a grid around a molecule; (ii) calculation of an interaction field (with water, dry, amide and carbonyl probes representing solvent, hydrophobic, and hydrogen bond acceptor and donor effects) in each grid point; (iii) eight or more energy values are assigned and for each energy value, the number of grid points inside the surface corresponding to this energy (volume descriptors) or belonging to this surface (surface descriptors) is calculated. VolSurf descriptors include size and shape descriptors, hydrophilic and hydrophobic region descriptors, interaction energy moments, and other descriptors. Both VolSurf and GRIND descriptors are available in Sybyl (VolSurf and Almond modules).[40] Virtually, any molecular modeling software package contains sets of its own descriptors and there are many other descriptors not mentioned here that can be found in the specialized literature.

There are sets of descriptors that take values of zero or one depending on the presence or absence of certain predefined molecular features (or fragments) such as oxygen atoms, aromatic rings, rings, double bonds, triple bonds, halogens, and so on. These sets of descriptors are called molecular fingerprints or structural keys. Such descriptors can be represented by bit strings and many are found in popular software packages. For instance, several different sets of such descriptors are included in MOE,[23] Sybyl,[40] and others, and examples of their use can be found in the published literature.[50,51] Molecular holograms are similar to fingerprints; however, they use counts of features rather than their presence or absence. For example, holograms are included in the Sybyl HQSAR[40] module. There are also more recent approaches when molecular features are not predefined *a priori* (as fingerprints discussed above) but are identified for each specific dataset. For example, frequent subgraph mining approaches developed independently at the University of North Carolina[52] and at the Louis Pasteur University in Strasbourg[53] can find all frequent closed subgraphs (i.e., subgraph descriptors) for given datasets of compounds described as chemical graphs.

As may be obvious from the above discussion, most chemical descriptors are only available from commercial software, which in our opinion is a strong impediment to accelerating the development of chemoinformatics approaches and applications. Fortunately, Dr. W. Tong from FDA recently announced the availability of the first non-commercial descriptor generating software from his laboratory, which is an important step in the right direction of making core chemoinformatics tools available free of charge.[54]

After descriptors are calculated, a dataset can be represented in the form of a QSAR table (Table 6.1). At this point, we shall introduce the concept of *multidimensional descriptor space.* Suppose that we have just two descriptors for a compound dataset. In this case, we can define a 2D space with orthogonal (perpendicular to each other) coordinates, abscissa and ordinate, and represent each compound i by a point in this 2D descriptor space with coordinates (X_{i1}, X_{i2}), where X_{i1} and X_{i2} are descriptor values for a compound i. In case we have three descriptors, we can introduce a 3D descriptor space, in which each descriptor will be represented by an axis, and all three axes are orthogonal to each other, and so on. We can represent each compound i in this 3D descriptor space by a point (X_{i1}, X_{i2}, X_{i3}), where X_{i1}, X_{i2}, and X_{i3} are descriptor values for compound i. Obviously, the same consideration can be extended to any number of descriptors and so we can introduce higher-dimensional descriptor spaces. If the total number of descriptors is N, we can introduce an N-dimensional descriptor space. We can endow this space with some metric by introducing distances between representative points of compounds in this space. For example, if distance D_{ij} between points i and j is defined as

$$D_{ij} = \sqrt{\sum_{n=1}^{N} (X_{in} - X_{jn})^2}, \qquad (6.1)$$

then this descriptor space is the N-dimensional Euclidean space. In fact, the distance can be axiomatically defined in many ways. For example, we can define the Minkowsky distance with parameter p as

$$D_{ij}^{\text{Mink}} = \left[\sum_{n=1}^{N} (X_{in} - X_{jn})^p\right]^{1/p}. \qquad (6.2)$$

TABLE 6.1
QSAR Table

Compound	Descriptor 1	Descriptor 2	...	Descriptor N	Activity
1	X_{11}	X_{12}	...	X_{1N}	Y_1
2	X_{21}	X_{22}	...	X_{2N}	Y_2
...	...	...	...	...	...
M	X_{M1}	X_{M2}	...	X_{MN}	Y_M

If $p = 2$, the Minkowsky distance becomes Euclidean distance. If $p = 1$, the Minkowsky distance becomes the Manhattan distance, and so on. *Mahalanobis distances* are frequently used instead of Euclidean distances since they account for correlations between descriptors. They are also used in definition of AD and in finding outliers.[55] The Mahalanobis distance between compounds *i* and *j* is defined as

$$M_{ij} = \sqrt{(\mathbf{X_i} - \mathbf{X_j})^{\mathrm{T}}\mathbf{C}^{-1}(\mathbf{X_i} - \mathbf{X_j})}, \qquad (6.3)$$

where $\mathbf{X_i}$ and $\mathbf{X_j}$ are vectors of representative points of compounds *i* and *j*, and $\mathbf{C}^{-1}$ is the inverse covariance matrix for descriptors.

$$\mathbf{X_i} - \mathbf{X_j} = \begin{pmatrix} X_{i1} - X_{ji} \\ X_{i2} - X_{j2} \\ \ldots \\ X_{iN} - X_{jN} \end{pmatrix}, \qquad (6.4)$$

$$\mathbf{C} = \begin{pmatrix} E[(X_1-\mu_1)(X_1-\mu_1)] & E[(X_1-\mu_1)(X_2-\mu_2)] & \ldots & E[(X_1-\mu_1)(X_N-\mu_N)] \\ E[(X_2-\mu_2)(X_1-\mu_1)] & E[(X_2-\mu_2)(X_2-\mu_2)] & \ldots & E[(X_2-\mu_2)(X_N-\mu_N)] \\ \ldots & \ldots & \ldots & \ldots \\ E[(X_N-\mu_N)(X_1-\mu_1)] & E[(X_N-\mu_N)(X_2-\mu_2)] & \ldots & E[(X_N-\mu_N)(X_N-\mu_N)] \end{pmatrix}, \qquad (6.5)$$

Here, $E[(X_p - \mu_p)(X_q - \mu_q)]$ is the covariance between descriptors p and q, and μ_p and μ_q are their mean values.

The Euclidean distance between two compounds can be used as a measure of dissimilarity between them, that is, the larger the distance, the more dissimilar the compounds are, and the smaller the distance, the more similar the compounds are. If the distance is zero, compounds have identical descriptors, and from this descriptor set point of view, they are identical. However, these compounds can still be nonidentical; for example, if two compounds are enantiomers and the descriptors do not take chirality into account, then they will have identical descriptors. Another widely used similarity measure is the Tanimoto coefficient. Usually, it is defined for fingerprint bit strings. The Tanimoto coefficient is defined as follows:

$$T = \frac{c}{a + b - c}, \qquad (6.6)$$

where a, b, and c are the number of descriptors, for which the corresponding bits are the ones for the first compound, the second compound, and both compounds. In fact, the Tanimoto coefficient can be defined for compounds *i* and *j* with any set of descriptors as well:

$$T_{ij} = \frac{\sum_{a=1}^{N} X_{ia}X_{ja}}{\sum_{a=1}^{N} X_{ia}^2 + \sum_{a=1}^{N} X_{ja}^2 - \sum_{a=1}^{N} X_{ia}X_{ja}}. \qquad (6.7)$$

Here, X_{ia} and X_{ja} are values of descriptor a for compounds i and j, respectively. In general, $-1/3 \leq T \leq 1$. For bit strings and range-scaled descriptors, $0 \leq T \leq 1$, and this can be used instead of the Euclidean distance. Many additional details about distance and similarity measures used in chemoinformatics studies can be found elsewhere.[56]

Any QSAR method can be generally defined as an application of data analysis methods and statistics to the problem of finding empirical relationships (QSAR models) of the form $\hat{Y} = f(X_1, X_2, \ldots, X_N)$, where $\hat{Y}$ is the approximated (predicted) biological activity (or other property of interest) of molecules, $X_1, X_2, \ldots, X_N$ are calculated (or, in some cases, experimentally measured) structural properties (molecular descriptors) of compounds, and f is some empirically established mathematical transformation that should be applied to descriptors to calculate the property values for all molecules. A QSAR model optimization has a goal of minimizing the error of prediction, for example, the sum of squares of differences or the sum of absolute values of differences between predicted $\hat{Y}_i$ and observed Y_i activities for the training set of compounds (compounds used for building a QSAR model), and so on. However, prior to model building, data need to be preprocessed and prepared properly to enable not only building but validating models as well.

6.5 PREPROCESSING OF DESCRIPTORS

Many descriptors such as those calculated with Dragon[29] or MolconnZ[30] software include a large variety of "real-value" descriptors as well as descriptors indicating the presence or absence or counts of certain molecular fragments and groups. Real-value descriptors can have very substantial differences in their values and ranges, sometimes by orders of magnitude. Such sets of descriptors should be normalized. There are also descriptors, which may have the same value for all compounds of the modeling set or have a very low variance. These descriptors should be excluded from consideration prior to building QSAR models because they cannot explain the variability of the target property. However, such (nearly) constant value descriptors are important for defining the model AD and should be retained for this purpose (see Section 7.4). Descriptor normalization should be performed separately for modeling and external evaluation sets. Normalization of descriptors of external compounds for prediction, such as those included in a chemical database or virtual library, should be performed in the same way as for the external evaluation set. On the contrary, molecular fingerprints or holograms do not need to be normalized since they have the same format and similar ranges. Nevertheless, as in the case of real-value descriptors discussed above, those fingerprint or hologram descriptors that take the same value for all compounds of the modeling set should also be excluded.

Normalization of descriptors for the modeling set: There are two widely used methods of descriptor normalization: range scaling and standard normalization (or autoscaling). In the case of range-scaling, descriptors are normalized according to the following formula:

$$X_{ik}^{n} = \frac{X_{ik} - \min X_k}{\max X_k - \min X_k}, \tag{6.8}$$

where X_{ik} and X_{ik}^n are the non-normalized and normalized values of descriptor k ($k = 1, \ldots, N$) for compound i ($i = 1, \ldots, M$), and $\min X_k = \min_i X_{ik}$ and $\max X_k = \max_i X_{ik}$ are the minimum and maximum values of the kth descriptor. Descriptors normalized using Formula 6.8 have a minimum value of 0 and a maximum value of 1. Standard normalization (also called autoscaling) is performed according to the following formula:

$$X_{ik}^n = \frac{X_{ik} - \mu_k}{\sigma_k}, \tag{6.9}$$

where

$$\mu_k = \frac{1}{M} \sum_{i=1}^{M} X_{ik} \tag{6.10}$$

and

$$\sigma_k = \sqrt{\frac{\sum_{i=1}^{M} (X_{ik} - \mu_k)^2}{M - 1}} \tag{6.11}$$

are the estimations of the average and the standard deviation of the kth descriptor over all compounds $1, \ldots, M$. After normalization, descriptors with very low variance (e.g., lower than 0.001) or standard deviation, which is the square root of variance (see Formula 6.11), are excluded.

Normalization of descriptors for an external compound: For an external compound, the same Formulas 6.8 through 6.11 are used with X_k^{ext} (the kth descriptor for the external compound) instead of X_{ik} in Formula 6.8 or 6.9. Other values in Formulas 6.8 and 6.9 are parameters calculated for the modeling set.

For establishing the model AD it is important to take into account all descriptors. However, for descriptors having the same value for all compounds of the modeling set, the procedures described above will not work. So, if some kth descriptor has the same value for all compounds of the modeling set, but different values for the prediction set, then Formula 6.8 or 6.9 with min X_k and max X_k values for the prediction set should be used. The descriptor value for the modeling set should be normalized using min X_k and max X_k for the prediction set. However, if there is only one external compound for prediction with this descriptor value different from that of the modeling set, this problem has no solution, and this descriptor should not be taken into account, or if the difference is too large (depending on the nature of the descriptor), the external compound can be considered as an outlier.

Pairwise correlation analysis: After the removal of constant descriptors (those having the same value for all compounds of the modeling set) and descriptors with low variance, the number of descriptors could still be too high. Of course, theoretically, it is possible to build QSAR models using all these descriptors. However, it is not necessarily true in practice. For example, if a QSAR procedure includes variable selection, it will take too much time to develop a model with too high number of descriptors, but the prediction power of models built with all descriptors will not be much better than that for models built using a smaller number of descriptors selected in some special way. Sometimes, if models are built using a limited number of iterations (steps of calculations), then there is a possibility that an optimal subset of descriptors will not be selected from the pool of all descriptors. For the *k*NN QSAR procedure employed in our laboratory, the number of descriptors should not exceed 200–400, whereas after the removal of constant and low variance descriptors from the entire set of Dragon descriptors, the number of remaining descriptors could be as high as about 800–1000. The initial number of some collections of fragment-based descriptors or fingerprints can amount to thousands and even dozens of thousands. Most of these descriptors should be excluded.

The commonly used approach to reducing the number of descriptors is *pairwise correlation analysis*. As a result, one of the descriptors from a pair of highly correlated descriptors found by this analysis is excluded. However, the outcome of this procedure can be nonunique and will depend on the order of descriptors, if applied incorrectly. Suppose that there are three descriptors X_a, X_b, and X_c and three correlation coefficients $|R(X_a, X_b)| > t$, $|R(X_b, X_c)| > t$, and $|R(X_a, X_c)| < t$, where t is a predefined threshold. Suppose that descriptor X_a is the first in the list, descriptor X_b is the second, and descriptor X_c is the third. If the descriptor with the higher number is deleted, then descriptors X_a and X_c will be retained and descriptor X_b will be deleted. However, if descriptor X_b is the first in the list, descriptor X_a is the second, and descriptor X_c is the third and the descriptor with higher number is deleted, then descriptors X_a and X_c will be deleted and descriptor X_b will be retained. We suggest that the following procedure should be used:

i. Select the descriptor with the highest variance.
ii. Calculate the correlation coefficients between this descriptor and all other descriptors. The correlation coefficient between two descriptors X_a and X_b is calculated as follows:

$$R(X_a, X_b) = \frac{\sum_{i=1}^{M}(X_{ia} - \mu_a)(X_{ib} - \mu_b)}{\sqrt{\sum_{i=1}^{M}(X_{ia} - \mu_a)^2 \sum_{i=1}^{M}(X_{ib} - \mu_b)^2}}, \quad (6.12)$$

where μ_a and μ_b are the mean values of descriptors X_a and X_b, and all summations are over all compounds of the modeling set.
iii. Remove all descriptors for which the absolute value of the correlation coefficient with this descriptor is higher than the predefined threshold value.
iv. Among the remaining descriptors, if any, select one with the highest variance
v. Go to step (ii).

If there are no two descriptors with equal variance, this procedure gives a unique result, irrespective of the order of descriptors.

Note 1: The threshold value depends on the dataset and the number of descriptors one wants to retain. For *k*NN QSAR, we still need a relatively large number of descriptors (see above), so with Dragon or Molconn-Z descriptors, the typical threshold could be about 0.90–0.95. For multiple linear regression (MLR), a smaller number of descriptors can be retained, so smaller threshold values can be used.

Note 2: The correlation coefficient between two descriptors (Formula 6.12) is not the only choice in descriptor selection. For example, if fingerprints are used, a reasonable measure of similarity between two descriptors could be the Tanimoto coefficient, which can be used instead of the correlation coefficient. The Tanimoto coefficient is calculated as follows:

$$T = \frac{c}{a + b - c}, \tag{6.13}$$

where a, b, and c are the number of compounds, for which descriptors X_a, X_b, and both X_a and X_b have the value of 1. In fact, the Tanimoto coefficient can be defined for descriptors taking continuous values as well:

$$T = \frac{\sum_{i=1}^{M} X_{ia}X_{ib}}{\sum_{i=1}^{M} X_{ia}^2 + \sum_{i=1}^{M} X_{ib}^2 - \sum_{i=1}^{M} X_{ia}X_{ib}}. \tag{6.14}$$

For range-scaled descriptors, $T \geq 0$, and it can be used instead of the absolute value of the correlation coefficient calculated by Formula 6.12.

Note 3: Sometimes (e.g., if MLR is used), after performing pairwise correlation analysis, the number of descriptors is still too high. In this case, other methods of descriptor selection can be used. One of the popular methods consists of building simple regressions of each descriptor with the response variable, and selection of those descriptors that have a regression coefficient (slope of regression) significantly (according to the Student's t-test) different from zero. Alternatively, a certain number of descriptors with the highest t-values are retained. Let

$$y = b_1X + b_0 \tag{6.15}$$

be the regression of descriptor X against activity y. b_1 and b_0 are calculated according to the following formulas:

$$b_1 = \frac{\sum_{i=1}^{M}(X_i - \mu_x)(y_i - \mu_y)}{\sum_{i=1}^{M}(X_i - \mu_x)^2}, \qquad b_0 = \mu_y - b_1\mu_x, \tag{6.16}$$

where μ_x and μ_y are the mean values of descriptor X and activity y, respectively. Let the null hypothesis be H_0: $b_1 = 0$ and the alternative hypothesis be H_1: $b_1 \neq 0$. To test the hypotheses, it is necessary to calculate the t-value:

$$t = \frac{b_1\sqrt{\sum_{i=1}^{M} X_i^2 - \left(\sum_{i=1}^{M} X_i\right)^2 \Big/ M}}{\sqrt{\sum_{i=1}^{M}(y_i - b_1X_i - b_0)^2 \Big/ M - 2}}, \tag{6.17}$$

which has the t-distribution with $M-2$ degrees of freedom. If $t \geq t_{\alpha/2,M-2}$ or $t \leq -t_{\alpha/2,M-2}$, where α is the significance level, reject H_0. Usually, $\alpha = 0.05$. For one-sided tests, H_1': $b_1 > 0$ and H_1'': $b_1 < 0$, the following tests are used for $t \geq t_{\alpha,M-2}$ or $t \leq -t_{\alpha,M-2}$, respectively.[57]

Principal Component Analysis (PCA): Principal component regression is one of the popular methods of QSAR analysis.[58] Principal components (PCs) are orthogonal linear combinations of descriptors. Above, we have introduced a multidimensional descriptor space and introduced the Euclidean metric in it. In this space, we represented a compound by a point with coordinates equal to the values of descriptors for this compound. In this space, like in the more familiar 2D and 3D spaces, we can also introduce other objects. For example, we can introduce lines and vectors. We can also imagine a distribution of points on the plane or in the 3D space. The points can be distributed evenly in some square or cubic area of the plane or space, or the cloud of points can be stretched more in some direction (even not coinciding with the directions of coordinate axes) than in others. The same is true for the high-dimensional descriptor space. We can imagine a direction (or axis), in which the distribution of points is stretched maximally. This direction defines the first PC. We can project all points onto this direction. Suppose that we introduced a zero point on this axis. Then the projection of point i onto this direction can be represented by a number V_{i1}, which is the distance from the projection point to the zero point on this axis. Then we can calculate the variance Var_1 for these projection points as

$$\text{Var}_1 = \frac{\sum_{i=1}^{M}(V_{i1} - \mu_{v1})^2}{M - 1}, \tag{6.18}$$

where μ_{v1} is the mean of all V_{i1}. It turns out that for the first PC, the variance of the points projected onto it has the largest value among all directions in the multidimensional descriptor space. Then it is possible to define the second PC, as the direction, orthogonal to the first PC, for which the variance of projections of points is the largest among all directions, orthogonal to the first PC. Then it is possible to define the third PC, as the direction, orthogonal to the first two PCs, for which the variance of projections of points is the largest among all directions, orthogonal to the first two PCs. This process can be continued. The maximum number of PCs cannot exceed both the number of points minus one, $M - 1$, and the number of descriptors N. So, if $M - 1 \geq N$, the maximum number of PCs is N (or less); otherwise it is $M - 1$ (or less).

PCs can be used instead of original descriptors in QSAR studies. In practice, the number of PCs used in QSAR studies is smaller than these limitations, since only the most important PCs are used, which, taken together, account for a large portion (90–95%) of the total variance of representative points. So, in many cases, the number of PCs is no more than 10–20, and in some cases just 2 or 3, while the number of descriptors can be several dozens or even hundreds. The total variance can be calculated as follows:

$$\text{Var} = \frac{\sum_{j=1}^{N}\sum_{i=1}^{M}(X_{ij} - \mu_j)^2}{M - 1} = \sum_{j=1}^{N'}\text{Var}_j = \frac{\sum_{j=1}^{N'}\sum_{i=1}^{N'}(V_{ij} - \mu_{vj})^2}{M - 1}, \tag{6.19}$$

where μ_j is the mean of projections of all points onto axis X_j, and N' is the number of PCs. Usually, the descriptor and PC axes are centered, so that in Formula 6.19 all

μ_j and μ_{vj} are 0: if we define

$$X'_{ij} = X_{ij} - \mu_j \quad \text{and} \quad V'_{ij} = V_{ij} - \mu_{vj}, \tag{6.20}$$

then

$$\text{Var} = \frac{\sum_{j=1}^{N}\sum_{i=1}^{M} X'^2_{ij}}{M-1} = \sum_{j=1}^{N'} \text{Var}_j = \frac{\sum_{j=1}^{N'}\sum_{i=1}^{M} V'^2_{ij}}{M-1}. \tag{6.21}$$

In fact, PCs are linear combinations of descriptors, that is, they can be represented in the form

$$V_j = \sum_{i=1}^{N} \alpha_{ij} X_i, \tag{6.22}$$

where α_{ij} are coefficients of transformation from descriptors to PCs. Due to this feature of PCs, in many cases PCR models are very difficult to interpret. Sometimes, if a small number of coefficients α_{ij} have values significantly different from zero, it is possible to interpret the model.[58]

Unsupervised Forward Selection (UFS) or complete correlation analysis: UFS selects a set of linearly independent descriptors.[59] The method can be used to select descriptors that most fully describe the descriptor space. The only parameter that is necessary to assign is the threshold correlation coefficient, which is the maximum correlation coefficient between the next descriptor to be selected and all linear combinations of descriptors already selected. Usually, this threshold value is 0.99. The maximum number of descriptors selected by UFS is the same as the maximum number of PCs: it cannot exceed either the number of points minus one, $M - 1$, or the number of descriptors N. So, if $M - 1 \geq N$, the maximum number of descriptors selected by the procedure is N (or less); otherwise it is $M - 1$ (or less). Usually, the number of descriptors selected is significantly less than these limits. The advantage of UFS over PCA is that it selects individual descriptors rather than constructing linear combinations of all descriptors. The disadvantage of UFS is that descriptors selected are not orthogonal. The UFS algorithm is as follows:

i. Select two descriptors with the lowest absolute value of the correlation coefficient.
ii. Using the Gram–Schmidt procedure, construct an orthonormal basis in the descriptor space defined by these descriptors.
iii. Select the next descriptor.
iv. Add a new basis vector to the existing basis using the Gram–Schmidt procedure.
v. Calculate the cosine of the angle between this descriptor and the hyperplane defined by all descriptors selected. This is the maximum correlation coefficient between this descriptor and all linear combinations of descriptors selected.
vi. If it was not the first descriptor to find the next descriptor to add, compare this correlation coefficient with the previous one. If it is closer to zero than the

previous one, retain it and discard the previous one with the corresponding basis vector. Otherwise, discard this correlation coefficient and the new basis vector.

vii. Repeat steps (iii) through (vi) for all remaining descriptors. If all correlation coefficients are above the predefined threshold, stop. Otherwise retain the final descriptor and the basis vector.

viii. If there are more descriptors, start selection of the next one: go to step (iii). Otherwise stop.

Earlier in this chapter, we introduced the multidimensional descriptor space, in which each compound was represented as a point with coordinates equal to its descriptor values. We can also consider descriptors as vectors in the "compounds" space. So descriptor values in this case are defined by the first of the Formulas 6.20, but for brevity we will omit prime signs. For example, vector $\mathbf{X}_i = \{X_{1i}, X_{2i}, \ldots, X_{Mi}\}$. Without losing the generality, we subtract the average value of each descriptor from all corresponding descriptor values. For each vector in the "compounds" space, we can define a unity vector. For example, for vector $\mathbf{D}$ in this space, unity vector will be $\mathbf{e} = \mathbf{D}/|\mathbf{D}|$. We can also define a dot product of two vectors. For example, $\mathbf{X_i X_j} = X_{1i}X_{1j} + X_{2i}X_{2j} + \cdots + X_{Mi}X_{Mj}$. The correlation coefficient between two descriptors will then be defined as cosine between them:

$$R_{ij} = \cos(\alpha_{ij}) = \frac{\mathbf{X_i X_j}}{|\mathbf{X_i}||\mathbf{X_j}|}. \tag{6.23}$$

Suppose that in step (i) we selected descriptors $\mathbf{X}_1$ and $\mathbf{X}_2$. In step (ii) we define the orthonormal basis according to the following formulas:

$$\mathbf{e}_1 = \frac{\mathbf{X}_1}{|\mathbf{X}_1|}, \tag{6.24}$$

$$\mathbf{e}_2 = \frac{\mathbf{X}_2 - (R_{12}|\mathbf{X}_2|)\mathbf{e}_1}{|\mathbf{X}_2 - (R_{12}|\mathbf{X}_2|)\mathbf{e}_1|} = \frac{\mathbf{X}_2 - (\mathbf{e}_1\mathbf{X}_2)\mathbf{e}_1}{|\mathbf{X}_2 - (\mathbf{e}_1\mathbf{X}_2)\mathbf{e}_1|}. \tag{6.25}$$

In Formula 6.25 we replaced R_{12} by the right part of 6.23. In step (iii), we select the next descriptor among the remaining descriptors. In step (iv), using the Gram–Schmidt orthogonalization procedure, we add the new basis vector $\mathbf{e}_3$:

$$\mathbf{e}_3 = \frac{\mathbf{X}_3 - (\mathbf{e}_1\mathbf{X}_3)\mathbf{e}_1 - (\mathbf{e}_2\mathbf{X}_3)\mathbf{e}_2}{|\mathbf{X}_3 - (\mathbf{e}_1\mathbf{X}_3)\mathbf{e}_1 - (\mathbf{e}_2\mathbf{X}_3)\mathbf{e}_2|}. \tag{6.26}$$

$\mathbf{e}_3$ is orthogonal to the plane defined by vectors $\mathbf{e}_1$ and $\mathbf{e}_2$, so the cosine of angle between $\mathbf{X}_3$ and $\mathbf{e}_3$ is equal to the sine of angle α between $\mathbf{X}_3$ and the plane defined by vectors $\mathbf{e}_1$ and $\mathbf{e}_2$. Thus,

$$\sin\alpha_3 = \frac{\mathbf{e}_3\mathbf{X}_3}{|\mathbf{X}_3|}, \tag{6.27}$$

and the maximum correlation coefficient between vector $\mathbf{X_3}$ and all linear combinations of vectors X_1 and X_2 will be (step (v))

$$R_3 = \sqrt{1 - \frac{(\mathbf{e_3 X_3})^2}{|\mathbf{X_3}|^2}}. \tag{6.28}$$

So in step (iii), a descriptor is selected for which R_3 in Formula 6.28 is closest to zero. Now, go again to step (iii) and select the next descriptor. Calculate $\mathbf{e}_3$ and R_3 for this descriptor and compare R_3 (new) with R_3 (old). If $|R_3(\text{new})| < |R_3(\text{old})|$, retain the new descriptor and discard the old one. Otherwise, retain the old one and discard the new one. Go through all remaining descriptors and find that one with the lowest $|R_3|$. Retain this descriptor and the corresponding $\mathbf{e_3}$. Continue the procedure. In step k, descriptor X_k will be selected such that

$$\mathbf{e}_k = \frac{\mathbf{X}_k - \sum_{i=1}^{k} (\mathbf{e}_i \mathbf{X}_k)\mathbf{e}_i}{\left|\mathbf{X}_k - \sum_{i=1}^{k} (\mathbf{e}_i \mathbf{X}_k)\mathbf{e}_i\right|}, \tag{6.29}$$

$$\sin \alpha_k = \frac{\mathbf{e}_k \mathbf{X}_k}{|\mathbf{X}_k|}, \tag{6.30}$$

and

$$R_k = \sqrt{1 - \frac{(\mathbf{e}_k \mathbf{X}_k)^2}{|\mathbf{X}_k|^2}}. \tag{6.31}$$

The procedure ends when no more descriptors left or when no descriptors were found for which the corresponding correlation coefficient is below the threshold.

6.6 STOCHASTIC CLUSTER ANALYSIS

In the next two sections we will discuss several algorithms based on the calculation of the distance matrix. For a large dataset, calculation of the distance matrix may take too much time. For example, the kNN QSAR method[43,44] requires the calculation of the distance matrix at each step of the algorithm, which makes it very inefficient for large datasets. Besides, for large datasets, better models could be obtained using local approaches (i.e., when models are built separately for different subsets of the entire dataset) as opposed to global approaches (in which models are built for the entire dataset of compounds).[60] So, we consider here one method that can be used to select a diverse subset of compounds without calculating the entire distance matrix. This method, called Stochastic Data Analysis (SCA), was developed by Reynolds and colleagues.[61,62] The diverse subset of compounds can be selected in one run through the dataset. The input to this algorithm is a threshold similarity value between compounds. Compounds more similar to those already selected will not be added to the list of compounds selected. The algorithm is as follows:

i. Select a compound randomly or select the first compound.
ii. Include it in the list of diverse subset of compounds.

iii. Select the next compound randomly or in the order it is included in the dataset.
iv. Calculate similarity of this compound with compounds already selected.
v. If all similarity values are below the similarity threshold, include this compound into the list of diverse subset of compounds. If at least one similarity value is above this threshold, do not include this compound in the list.
vi. If no more compounds left, stop. Otherwise, go to step (iii).

If m compounds are selected out of the entire dataset of Mcompounds, the total number of distances calculated will be less than $s = mM/2$. If $m \ll M$, then $s \ll M(M-1)/2$, that is, the number of distances calculated is much smaller than the total number of distances.

Note 1: In steps (i) and (iii), compounds can be selected randomly or in the order they are included in the dataset. Actually, compounds can be selected randomly if before each run the dataset is randomized, and then each compound is selected in the order it is included in the randomized dataset.

Note 2: Different similarity measures can be used in this algorithm. For example, the Tanimoto coefficient can be used. If some distance measure like Euclidean distance is used, the higher value means higher dissimilarity, not similarity, so in step (iv) a compound is added to the list if all distances to compounds already in the list are above the threshold.

Note 3: Since it is unknown *a priori* how diverse the compounds included in the dataset are, the procedure should be performed with different threshold values. For example, a user wants to select a subset of m compounds. Then the user can perform the procedure with two significantly different threshold values and see how many compounds are selected. If m_1 and m_2 are the sizes of subsets selected, and threshold values were $T_1 < T_2$, then it would be logical to select the new value T_3 as

$$T_3 = T_1 + \frac{m - m_1}{m_2 - m_1}(T_2 - T_1). \tag{6.32}$$

The process can be repeated until a number of compounds selected will be sufficiently close to m. Formula 6.32 may not work if m_1 and m_2 are close to each other. In practice, in the beginning, calculations with the range of m_1 and m_2 values and a relatively large range of similarity values can be performed. Then, based on the results, more precise calculations can be performed for narrower ranges of parameters.

Note 4: After selection of the diverse subset, which is many times smaller than the entire dataset, additional runs can be performed to select compounds similar to the selected compounds. The threshold for these runs could be the same as that for the selection of the diverse subset. If the number of compounds in the entire dataset is M, and the number of compounds selected is m, and $m \ll M$, then the maximum number of distances calculated (the number of elements of the distance matrix between selected and not selected compounds) will be $m(M - m)$, which will be much smaller than $M(M-1)/2$. Still, if the procedure was run a small number of k times, the total number of distances calculated is $kmM/2 + m(M - m) \ll M(M-1)/2$. In this way, the dataset can be divided into m initial clusters. If some of the $M - m$ compounds are close to more than one compound of the diverse subset, they can be assigned to a cluster that is defined by the closest point selected. Local QSAR models could be built separately for each cluster containing at least a certain number of compounds. Additional procedures for cluster processing could be necessary. Currently, this approach is under development in our laboratory. Alternatively, the initial clusters defined by these points can be merged. Using SCA, it is also easy to find outliers: these are compounds not included in any clusters (or singletons) that have no other compounds close to them. Small clusters can also be found and discarded, if necessary. If the similarity threshold for these runs is larger than for the selection of the diverse subset, additional outliers can be found among the remaining $M - m$ compounds.

Note 5: The method described in this section is also useful for the comparison of large molecular databases.[62]

6.7 DETECTION AND REMOVAL OF OUTLIERS PRIOR TO QSAR STUDIES

The success of QSAR modeling depends on the appropriate selection of a dataset for QSAR studies. In a recent editorial of the *Journal of Chemical Information and Modeling*, Maggiora[63] noticed that one of the main deficiencies of many chemical datasets is that they do not fully satisfy the main hypothesis underlying all QSAR studies: similar compounds have similar biological activities or properties. Maggiora defines the "cliffs" in the descriptor space where the properties change so rapidly that in fact adding or deleting one small chemical group can lead to a dramatic change in the compound's property. In other words, small changes of descriptor values can lead to large changes in molecular properties. Generally, in this case there could be not just one outlier, but a subset of compounds whose properties are different from those on the other "side" of the cliff. In other words, cliffs are areas where the main QSAR hypothesis (similar compounds have similar properties) does not hold. So cliff detection is a major QSAR problem. In the QSAR area, many people were aware of these and other problems related to outlier detection, but have not yet paid sufficient attention to addressing them in automated QSAR procedures. There are two types of outliers we must be aware of: *leverage* (or structural) outliers and *activity* outliers. In the case of activity outliers the problem of "cliffs" should be addressed as well. Algorithms considered in this section are applied to modeling sets (see Section 6.1).

Leverage (structural) outliers: Similarity between compounds included in the datasets must be considered in the context of the entire descriptor space. Singletons included in the datasets are the first candidates to be outliers. In many QSAR studies, these compounds are not excluded from datasets; if they are assigned to training sets they could significantly worsen the model statistics and if they are assigned to test or validation sets they will worsen the general model's predictivity. This type of outliers will be referred to as leverage outliers. Actually, detection of leverage outliers is relatively simple. However, the standard procedure of detecting leverage outliers by using the diagonal elements of a so-called hat matrix (or the Mahalanobis distance to the data centroid) which are called "leverage" might not detect all outliers. For example, a distribution of a dataset in the descriptor space can have areas of very low density even near the geometrical center of the distribution (Figure 6.2). So outliers should be detected in these areas. In these cases, standard procedures may not work. Thus,

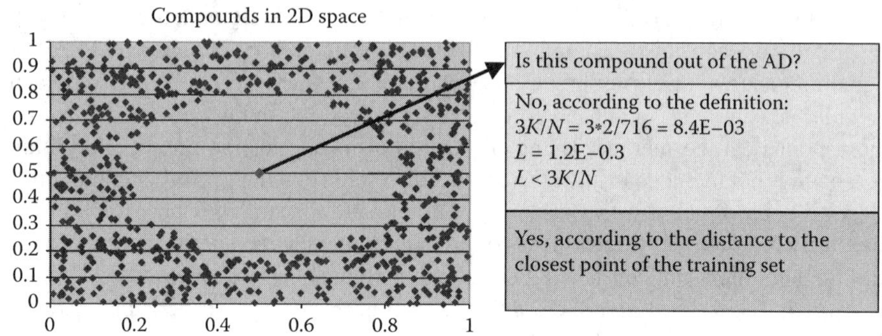

FIGURE 6.2 Leverage outliers. Black points represent compounds of the training set. Gray point represents a query compound. In a standard procedure, an outlier is defined by the condition $3N/M \geq L$, where N is the number of descriptors, M is the number of compounds, and L is leverage of a compound (which is a corresponding diagonal element of the hat matrix). According to this procedure, a compound represented by a red point is not an outlier. On the other hand, according to the distance to the closest point of the training set, it is an outlier.

we have suggested the following procedure based on the sphere-exclusion algorithm (Figure 6.3).[17,64]

Input to the procedure is the distance cutoff value:

i. Calculate the distance or similarity matrix.
ii. For all compounds in a dataset, find their nearest neighbors.
iii. If for some compound, there are no nearest neighbors within a certain distance cutoff value, then this compound is an outlier with this cutoff value.

Since we do not know *a priori* the properties of the dataset in the given descriptor space, the distance cutoff value can be defined as follows:

i. Calculate average $\langle D \rangle$ and standard deviation s of all distances between nearest neighbors within the dataset.
ii. For a set of Z-cutoff values, defined by a user, calculate different distance cutoff values as

$$D_{\text{cutoff}} = \langle D \rangle + Zs. \tag{6.33}$$

Typical Z-cutoff values are from 0 to 5 with step 0.1.
iii. Repeat the leverage outlier finding procedure for each D_{cutoff}.

Of course, the higher the Z-cutoff is, the lower is the number of outliers. The more compounds are included in the training set, the larger is the model AD. On the other hand, we expect that after excluding more outliers, models with better statistics can be built. So we recommend building models with different Z-cutoff values (and different counts of leverage outliers). Thus, we expect to build QSAR models with better statistics and determine the natural Z-cutoff and D_{cutoff} values.[65] The entire

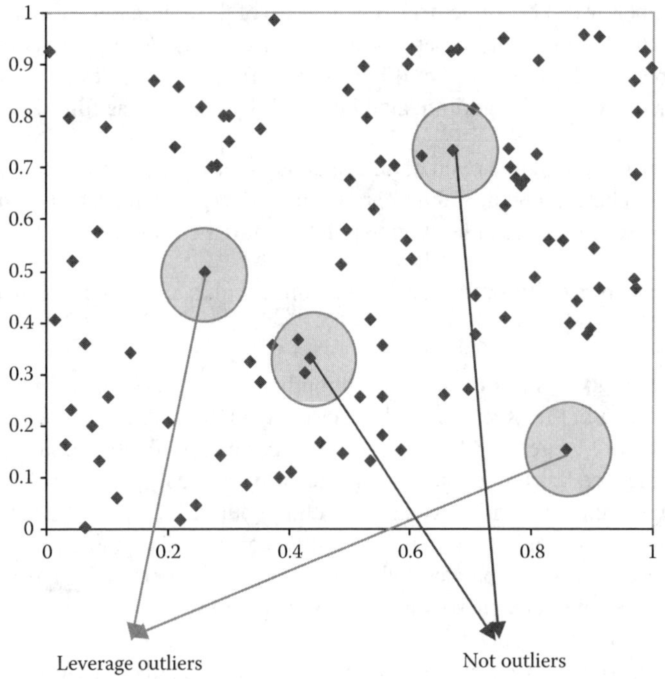

FIGURE 6.3 Leverage (structural) outliers in the 2D descriptor space. Build a sphere of certain radius D_{cutoff} (see the main text) with the center at each compound. If there are no other compounds within the sphere, this compound is a leverage outlier for this D_{cutoff}.

procedure can be fully automated. An alternative approach consists of the following steps.

i. Find minimum distance $D_{\min}$ and maximum distance $D_{\max}$ between all nearest neighbors.
ii. Define a set of D_{cutoff} distances evenly dividing the interval $[D_{\min}, D_{\max}]$.
iii. For each D_{cutoff}, find outliers.
iv. If necessary, calculate Z-cutoff values corresponding to D_{cutoff} values: $Z = (D_{\text{cutoff}} - \langle D \rangle)/s$.

In these calculations, we can use different distance and similarity measures (see Section 6.5). The procedure for finding leverage outliers can be also applied to detect small clusters of compounds that are far from all other compounds in the descriptor space. If, for example, compound *a* has only one nearest neighbor, compound *b*, and the only nearest neighbor of compound *b* is compound *a*, then compounds *a* and *b* make a cluster. Optionally, this cluster can also be removed from a dataset, since if one of these compounds is assigned to the training set, it will be an outlier in the training set (unpublished observations).

Activity outliers. Separating compounds on different sides of activity "cliffs": The algorithm below can be implemented in the case of the continuous response variable. In this section, we describe an approach on how to detect *local* activity outliers. We also address the Maggiorra's "cliff" problem. After removing leverage outliers, we use a sphere-exclusion algorithm along with the Dixon's test as follows:

i. Take a representative point of a compound in the descriptor space and build a probe sphere of certain radius R around it. Radius R is defined by the condition that there should be at least three points within the sphere.

If there are no activity outliers, all compounds within the sphere must have close activity values.

ii. Use the Dixon's test to find activity outliers within the sphere as follows.[66] Rank data in the ascending order. For compounds with the highest and lowest activity values, calculate τ statistic according to Table 6.2. For a chosen significance level α compare τ statistic with a critical value. If the τ statistic is higher than the critical value, the null hypothesis is rejected, and the compound with the highest or lowest activity is considered an outlier. If necessary, the Dixon's test can be repeated for the remaining compounds. The precision of this test will decrease with each new repetition.
iii. Repeat the procedure for all compounds (step (i)). If there is a "cliff" in the descriptor space, we might be able to separate the whole dataset into sufficiently large groups of nonoutliers and outliers and build separate QSAR models for them.

To consider a compound an outlier, we would recommend using an additional criterion: the difference between the activity of an outlier candidate and the activity of the compound with the activity closest to it should be not less than 10% or 20% of the entire range of activities of the dataset. The algorithm allows finding isolated activity outliers (i.e., when there is only one outlier within the probe sphere) as well as groups of outliers. Our experiments show that in the latter case, a small number

TABLE 6.2
Calculation of τ Statistic for the Dixon's Test[66]

Number of Compounds n	Test for the Highest Value: Calculate $\tau_h = \frac{x_n - x_p}{x_n - x_k}$		Test for the Lowest Value: Calculate $\tau_l = \frac{x_p - x_1}{x_k - x_1}$	
	k	p	k	p
3 to 7	1	$n-1$	n	2
8 to 10	2	$n-1$	$n-1$	2
11 to 13	2	$n-2$	$n-1$	3
14 to 25–30	3	$n-2$	$n-2$	3

of other compounds close to the group can also be detected as possible outliers. We would recommend supplementing this procedure with an additional similarity search for detecting additional candidate outliers.

For logarithms of activity data, the Grubb's test can be used. This test is recommended by the Environmental Protection Agency.[69] According to this test, all activities are ranked in the ascending order, and the mean $\bar{x}$ and standard deviation s of the data are calculated. Then the τ statistics are calculated for compounds with the highest and lowest activities as follows:

$$\tau_{\text{low}} = \frac{\bar{x} - x_1}{\sigma} \quad \text{and} \quad \tau_{\text{high}} = \frac{x_n - \bar{x}}{\sigma}. \tag{6.34}$$

The τ statistics are compared with the corresponding critical τ_{crit} value for the sample size and selected alpha.[70,71] If $\tau_{\text{low}} > \tau_{\text{crit}}$ or $\tau_{\text{high}} > \tau_{\text{crit}}$, the corresponding compound is considered an outlier.

6.8 CLASSIFICATION AND CATEGORY QSAR: DATA PREPARATION FOR IMBALANCED DATASETS

In many datasets, the counts of compounds that belong to different classes or categories are significantly different (there could be several times and even orders of difference). Usually, active compounds constitute a smaller class and inactive compounds constitute a larger class. Active compounds (typically binding to a certain biological target) belong to a relatively small number of structural classes. On the other hand, compounds included in the larger class (i.e., inactive compounds) can be very diverse: some of them can belong to the same structural classes as active compounds, while other compounds (often, the majority of them) have very different structures highly dissimilar from those included in the smaller class. So they cover a large area in the descriptor space relative to the active compounds, which are much more similar to each other. In these cases, direct development of predictive QSAR models using entire datasets is difficult, if not impossible. Indeed, training and test sets reflect the composition of the entire dataset, in which almost all compounds are inactive, so the modeling and validation will be biased toward correct prediction of the larger class. Thus, reducing the number of compounds included in the larger class is necessary. In the scenario just described, the following approach can be applied:

i. Divide a dataset into separate classes.
ii. Calculate the distance or similarity matrix between the compounds belonging to different classes.
iii. Exclude compounds of the larger class dissimilar from those of the smaller class.

If appropriate, in step (ii) the same distance or similarity measure that will be used in the optimization procedure should be employed. For example, the current implementation of *k*NN QSAR is based on Euclidean distances. So, in step (ii), the Euclidean distance matrix should be used preferably. In step (iii), different distance cutoff values (see Section 6.7) should be used. Ideally, after excluding dissimilar compounds of the larger class, the number of remaining compounds of this class should be more or less equal to the number of compounds of the smaller class. QSAR models are developed only for compounds of the smaller class, and those compounds of the larger class which were not excluded by the procedure. In other words, the modeling subset will not include compounds excluded by the procedure. Now, the entire area occupied by the modeling set is divided into two parts: occupied by similar compounds of the larger and smaller classes and occupied by compounds of the larger class dissimilar from those of the smaller class. Prediction of a query compound is performed by finding part of the area in the descriptor space to which the compound belongs. If it belongs to the area occupied by similar compounds belonging to both classes, the QSAR model is used to predict a class of this compound. If this compound belongs to the part occupied by points of the larger class, this compound is predicted as belonging to this class. If a compound belongs to neither of these parts, it is outside of the AD. Sometimes, to equalize the number of compounds in both classes, a distance cutoff value smaller than that used in defining the AD is used. In this case, some of the compounds of the larger class excluded from the model building can still be predicted by the model.[17]

Suppose that there is a slightly different situation: there are two classes of the same or different size, and there are many compounds of each class dissimilar from those of another class. In this case, the entire area occupied by representative points of the modeling set can be divided into three subareas: occupied by points of the first class only, occupied by points of the second class only, and occupied by points of both classes. Then a model should be built for compounds included in the latter subarea. Again, prediction of a query compound is performed by finding part of the area in the descriptor space to which the compound belongs. If it belongs to the area occupied by similar compounds belonging to both classes, the QSAR model is used to predict a class of this compound. If this compound belongs to the part occupied by points of the first class, this compound is predicted as belonging to the first class. If this compound belongs to the part occupied by points of the second class, the compound is predicted as belonging to the second class parts. If a compound belongs to neither of these parts, it is outside of the AD.

A similar approach can be used if there are more than two classes. For each part of the area occupied by representative points of more than one class, a QSAR model should be built.

Now, suppose that in the part of the area occupied by representative points of two classes, one of the classes is still significantly overrepresented. Then it is possible to reduce the number of compounds of this class by choosing only a fraction of these compounds for QSAR modeling. This approach is called undersampling and is described elsewhere.[72,73] The opposite approach called oversampling consists of including the same compounds of the smaller class several times into the modeling set and is also described elsewhere.[74] The extended discussions of oversampling and undersampling are beyond the scope of this chapter.

6.9 MODEL VALIDATION: MODELING, TRAINING, TEST, AND EXTERNAL EVALUATION SETS

The main goal of QSAR studies is the development of predictive QSAR models that can be used in computer-aided drug discovery and design as reliable tools for the prediction of activities or properties of new compounds, for example, those included in chemical databases or combinatorial libraries. Prior to using the model for external predictions, its predictive power should be established and validated. Thus, model validation has become a standard (and in some laboratories such as ours, mandatory) part of QSAR modeling. As we[75] and other authors[38] demonstrated, high prediction accuracy for the training set in leave-one-out or leave-group-out cross-validation is the necessary, but not sufficient condition for a QSAR model to be predictive. This statement is of particular importance. Recently, the European Organization for Economic Co-operation and Development (OECD) elaborated a set of principles for the development and validation of QSAR models, which in particular requires "appropriate measures of goodness-of-fit, robustness, and predictivity."[76] QSAR models should be rigorously validated using external sets of compounds that were not used in the model development. Nevertheless, there are still publications that do not include any external validation of QSAR models (i.e., by prediction of activities of compounds that were not used in model building); see, for example, Harju et al.[77] and Sharma et al.[78] In the next chapter, we will consider different aspects of model validation such as internal cross-validation and validation using test set compounds (i.e., compounds that were not used in model building), Y-randomization, and AD.

QSAR modeling can be viewed as a machine-learning procedure, during which the model is "trained" (i.e., model parameters are tuned to provide the highest predictivity in terms of some statistical criterion used as a target function which is optimized during the procedure). It is important to emphasize that the true predictive power of a QSAR model can be established only through the model validation procedure, which consists of prediction of activities of compounds that were not included in model building, that is, compounds in the *test set*. In contrast to the test set, compounds used for model building constitute the *training set*. In many QSAR studies, multiple models are built and from them "best" models are selected, which are defined as those based on the prediction statistics for the test set. Thus, the test set is actually used to select models. This use of the test set for model selection practically negates the consideration of such a routine as an adequate *external* model validation. In fact, it does not guarantee at all that models selected in this way will make accurate predictions if used for chemical *database mining* (i.e., predicting activities of compounds in a truly external database). In our workflow, to simulate the use of QSAR models for database mining, the so-called *external evaluation set* is employed. It should consist of compounds with known activities that are not included in either training or test sets. An external evaluation set can be selected randomly from the entire initial dataset. In general, the size of the external evaluation set should be about 15–20% of the entire dataset. The remaining part of the dataset is called a *modeling set* that can be divided into training and test sets. Algorithms for dividing a modeling set into training and test sets developed in our group previously[4] are discussed in the next section.

6.10 DIVISION OF A MODELING SET INTO TRAINING AND TEST SETS. EXTERNAL EVALUATION SETS

In most QSAR publications, the authors do not describe a procedure for dividing a dataset into training and test sets (see, e.g., Zvinavashe et al.[78] and Padmanabhan et al.[80]). In many studies, the test set is selected randomly from the modeling set.[81,82] To some extent, a method of selecting training and test sets depends on the QSAR method used and the size of the dataset. In many cases, multiple QSAR models are built by using different parameters of the QSAR algorithm or a stochastic QSAR procedure is repeated many times. In our opinion, multiple QSAR models should always be built; even if there are no parameters to change, different pairs of training and test sets can be generated, and for each pair a QSAR model can be developed and validated. In the combinatorial QSAR approach,[83,84] QSAR models are developed for multiple combinations of descriptor collections and optimization algorithms, so multiple models are generated. In this case, test sets are used to select models from those that have acceptable statistics for the training sets (see Section 7.1).

After models with acceptable statistics for training and test sets are selected, they should be validated using an external evaluation set, which is used to find a true (external) prediction accuracy of selected QSAR models. We can say that an external evaluation set plays the role of a small test database for virtual screening, and in the case when multiple models are selected, the *consensus prediction* (see Section 7.4) of the external evaluation set is employed. In many practical cases, the external evaluation sets are generated naturally in ongoing experimental projects that take place while the models are being developed. If an independent external evaluation set of compounds with known activities is not available, it should be selected randomly from the entire dataset. The remaining modeling set should be divided into training and test sets. These test sets should satisfy the following criteria: (i) The distribution of activities in training and test sets should be similar. (ii) The training set should be distributed within the entire area of the dataset distribution. (iii) All points of the test set should be within the AD defined by the training set at least in the entire descriptor space. (iv) Ideally, each point of the training set should be close to at least one point of the test set. Requirement (i) is particularly important for the continuous response variable. It can be satisfied by dividing a dataset into a small number of bins and selecting one compound from each bin as well as the most active and the most inactive compound into the training set.

In some QSAR studies, the division of a modeling set into training and test sets is based solely on activity values.[85] Sometimes, the subgroups of compounds with certain scaffolds are entirely included in the training or the test set.[86] In these cases, conditions (ii) through (iv) are not satisfied. The D-optimal design approach[87,88] is based on maximization of the determinant of the covariance matrix. It has been shown that this algorithm selects representative points predominantly located close to the borders of the area in the descriptor space in which they are distributed.[89] So, the training set selected with the D-optimal design algorithm does not satisfy conditions (i) through (iv). Another frequently used algorithm is the Kennard–Stone algorithm[90] in which compounds with the highest distances from all other compounds are selected. This algorithm is similar to one of the versions of the sphere-exclusion algorithms described below; however, it does not take into account the density of representative point distribution, and activities are also not taken into account. So, condition (i) is not satisfied.

We have at least partially satisfied condition (i) by selecting the most active and the most inactive compounds as well as several compounds with different activities into the training set. To satisfy conditions (ii) through (iv), we recommend applying the approach based on the sphere-exclusion algorithm[3,4] described below. In the case of the classification or category QSAR, it is important that at least five compounds of each class would be included in the test set. To achieve this goal, the sphere-exclusion algorithm is used separately for each class or category. At the end of the procedure, training sets for all classes are merged to form one training set, and the corresponding test sets are also merged to form one test set. The procedure is as follows (see also Figure 6.4):

i. Calculate the distance matrix D for the modeling set. Different distance or similarity measures can be used.
ii. Define probe sphere radii. Let D_{min} and D_{max} be the minimum and maximum elements of D, respectively. P probe sphere radii are defined by the following

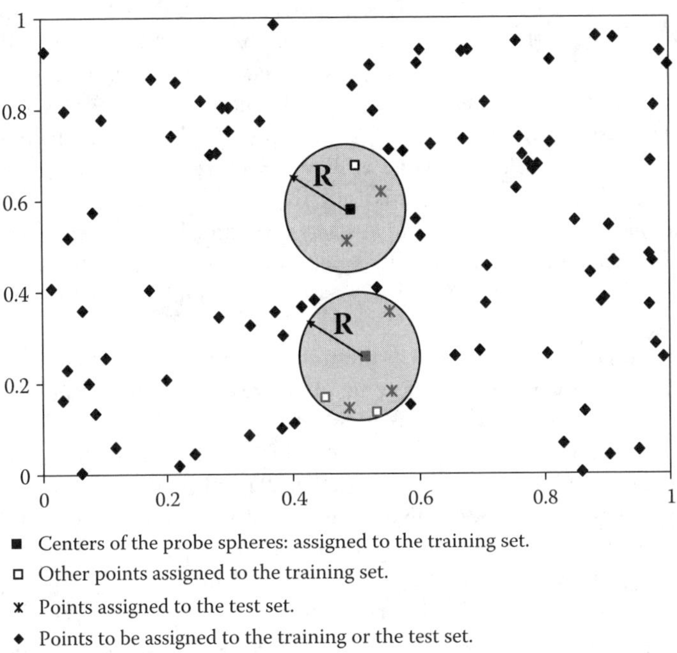

FIGURE 6.4 Division of a dataset into training and test sets. Suppose we have just two descriptors. Then the descriptor space will be 2D. Axes on the figure represent these descriptors, and points represent compounds. Probe spheres are built with the centers at some points as described in the main text. Two such spheres are shown. Centers of spheres are included in the training set, and other points within the spheres are included in the test and training sets as described in the main text.

formulas. $R_{min} = R_1 = D_{min} + (R_N - R_1)/P$, $R_{max} = R_N = D_{max}/4, Ri = R_1 + (i-1)^*(R_N - R_1)/P$, where $i = 2, \ldots, P$. Each probe sphere radius corresponds to one division into the training and the test set. We recommend $P = 50$.

iii. Select one or several compounds for the training set, as explained above.
iv. Construct probe spheres with centers at each of these compounds.
v. Select compounds from this sphere and include them alternately into the test and training sets.
vi. Exclude all compounds from within these spheres from further consideration.
vii If no more compounds are left, stop. Otherwise, let m be the number of probe spheres constructed and n be the number of remaining compounds. Let d_{ij} ($i = 1, \ldots, m; j = 1, \ldots, n$) be the distances between the remaining compounds and the probe sphere centers. Select a compound corresponding to the lowest d_{ij} value, include it into the training set, and perform steps (iv) through (vii) for it.

Note 1: There are several slightly different versions of this algorithm.[3,4] Probe sphere radii can be defined not by the minimum and maximum elements of the distance matrix, but by using different values of a parameter called dissimilarity level. Let V be the volume of the hyperparallelepiped in the descriptor space occupied by the representative points of compounds of the modeling set. If descriptors are range-scaled, $V = 1$. If descriptors are varied within different intervals, V can become very large or very small. If there are M compounds and N descriptors in the dataset, the average volume of the space for one point is $V' = V/M$, and if to consider this volume as a hypercube, its edge will be $l = (1/M)^{1/N}$. Probe sphere radii are defined as $R_c = cl$, where c is the dissimilarity level. For each c, one probe sphere radius and one split into training and test sets is obtained. We recommend using this method only when there is a relatively small number of descriptors or if descriptors are range-scaled. In our sphere-exclusion software, default values for c are 0.2, 0.3, 0.4, ..., 5.2.

Note 2: Previously, we noted that the training and the test set should include at least a certain minimum number of compounds. However, some training or test sets do not satisfy these conditions. This problem can be solved in two different ways. (i) Splits in which the training or test set contains very few compounds are removed. So, the final number of splits could be less than P from step (ii), or less than the total number of different c values (see Note 1). (ii) The probe sphere radii from step (ii) are recalculated with the larger R_{min} value, or a different set of c values is given with the larger c_{min} value.

Note 3: In step (v), there are different schemes of how to select compounds into test and training sets from the probe sphere. For example, it is possible to select two compounds into the test set, then one compound into the training set, and so on. It is also possible to have compounds within the sphere arranged by the distance to the center or randomized. In the latter case, different splits will be obtained by different runs of the procedure. This option is important if too few suitable splits (with acceptable sizes of training and test sets) were generated using the initial set of parameters (see also Note 2). Selecting points other than the center from probe spheres into the training set is important; in this way the training set accounts for the density of the distribution of points in the descriptor space.

Note 4: In step (vii), there are different ways of selecting the next compound. It can be selected randomly. It can be selected as the compound that has the largest distance from one of the spheres already created or the compound that has the smallest (or largest) average distance to the spheres already created or the compound that has the smallest largest distance among all distances to the spheres already created and so on.

The sphere-exclusion algorithm guarantees that at least in the entire descriptor space, the representative points of the test set are close to representative points of the training set (test set compounds are within the AD defined by the training set); the training set represents the entire modeling set (i.e., there is no subset in the modeling set which is not represented by a similar compound in the training set); given the size of the test set, as many of the representative points of the training set as possible are close to representative points of the test set. In other words, all the requirements (ii) through (iv) above are also satisfied. Besides, this algorithm takes into account the density of distribution of points in the descriptor space.

Using the SCA algorithm to divide a modeling set into training and test sets: The SCA algorithm described above can be modified so that it could be used for the selection of training and test sets. If appropriate, the similarity or distance measure used in the SCA algorithm should coincide with that used in the QSAR studies. For example, *k*NN QSAR employs Euclidean distances between compounds, so Euclidean distances should be used in the SCA. In the case of the continuous response variable, in step (i) of the SCA algorithm several compounds instead of just one should be included into the training set, that is, the most active, the most inactive as well as one compound from each bin of the activity range. Other compounds of the diverse subset are selected as in the SCA algorithm described above. All compounds of this subset are included into the training set. Then a procedure similar to that described in Note 4 following the description of the SCA algorithm (cf. Section 6.6) is implemented. For each compound of the selected subset, compounds similar to it among $M - m$ remaining compounds are selected and distributed between the test and training sets in a way similar to that described in step (v) and Note 3 following the description of the sphere-exclusion algorithm in this section (see above). As soon as these compounds are included in the training or the test sets, they are excluded from further consideration, so if a compound is close to several compounds of the diverse subset, it is accounted for only once, as it should be. In the case of the classification or category QSAR, a dataset is divided into classes, and the SCA algorithm is applied to each class. Then training sets for all classes are merged, and test sets are also merged.

More on external evaluation sets: Ideally, to exclude chance correlation, QSAR study should be performed several times with different external evaluation sets. For example, external validation could be made as an external leave-group-out cross-validation procedure where each external evaluation set would include 10–20% of the entire dataset. If, for example, there are 100 compounds in a dataset, after the random division of the dataset into five equal parts, there would be five external evaluation sets containing 20 compounds each. The first 20 compounds could be randomly selected from the entire dataset, the second 20 compounds could be randomly selected from the remaining 80 compounds, and so on. In each case, the modeling sets would consist of 80 compounds. Each modeling set is divided into multiple training and test sets as described above, and QSAR models are built using training sets and validated using test sets. Activities of compounds of the external evaluation sets are predicted by consensus prediction using selected models from QSAR studies performed for the corresponding modeling set. Finally, statistics of prediction for each external evaluation set as well as for the entire set are calculated. High prediction accuracy for each external evaluation set would corroborate the robustness of selected QSAR models and their usefulness for chemical database mining in the process of drug discovery.

6.11 CONCLUSIONS

This chapter addressed the most important aspects of data analysis prior to initiating a QSAR modeling procedure. We have considered the general QSAR workflow as it is implemented and practiced in our laboratory and presented a brief overview of the main steps of QSAR modeling including data preparation, model generation, and model validation, as well as establishing the AD of QSAR models. In Section 6.2, we have discussed the requirements to datasets for QSAR analysis concerning their size and activity range. We have established that in the case of the continuous response variable, the size (i.e., the number of compounds) of a QSAR dataset should be no less than 40, and in the case of the classification or category response variable, the dataset should include at least 20 compounds in each class or category. We have also pointed out that in the case of the continuous response variable, the range of activities should be at least 5 times larger than the experimental error and that there should be no large gaps in activity values.

In Section 6.3, we have focused on the curation of datasets. We have pointed out that many available or user-compiled datasets used for QSAR analysis could contain errors that should be detected and corrected; one of the duplicates of compounds (they occur in datasets frequently) should be removed; compounds containing heavy atoms or consisting of more than one fully covalently connected part (such as organic salts) should be excluded or in some cases the salt component can be removed. We have also discussed what to do when a dataset contains isomers (e.g., enantiomers) that may have all descriptor values equal to each other. We gave examples of Unix scripts that can be used for data curation and mentioned some commercial and freely available software that could help with the task of data cleaning.

In Section 6.4, we have briefly considered major types of descriptors. We have discussed a notion of multidimensional descriptor space and considered several possible definitions of distances between points representing compounds in the descriptor space. Then, in Section 6.5, we have considered important algorithms used in the processing of chemical descriptors: methods for descriptor normalization (range scaling and autoscaling); exclusion of descriptors with low variance; pairwise correlation analysis; PCA; and UFS, and in Section 6.6, we have considered stochastic cluster analysis, which is an important algorithm for dividing a large dataset into smaller clusters, finding small clusters of outliers, and dividing a dataset into training and test sets.

In Section 6.7, we have addressed the problem of detecting and removing structural (leverage) and activity outliers. We have demonstrated that a widely used approach of detecting structural outliers using leverage values is insufficient; thus, we have introduced a method based on distances to nearest neighbors. We have also considered a possible way to detect activity outliers prior to QSAR studies. The algorithm is also based on distances between compounds and employs the Dixon's and Grubb's tests. Then, in Section 6.8, we have considered the problem of preprocessing the imbalanced datasets for both classification and category QSAR modeling. We have pointed out that training, test, and external evaluation sets should be separately generated for each class or category and then combined. We have also noticed that in many cases, points representing different classes within the dataset may occupy partially different areas in the descriptor space and that areas where points of only one class are present should be excluded when one develops a QSAR model. We have also mentioned such approaches as oversampling and undersampling but did not consider them in detail.

Then, in Section 6.9, we have addressed a problem of model validation, briefly considered the importance of dividing a dataset into external evaluation and modeling sets and then dividing modeling sets into training and test sets, and discussed the role of external evaluation sets in the assessment of QSAR model performance in virtual screening. Finally, in Section 6.10, we have proposed several conditions that should be satisfied by training and test sets. We have described several algorithms for the division of a modeling set into training and test sets and showed that our algorithms based on the sphere-exclusion approach satisfy these conditions better than some alternative techniques.

In summary, in this chapter we have introduced critical procedures that should be used to preprocess the experimental datasets prior to building QSAR models; the approaches used for model development and validation are the subject of the next chapter. We will discuss different target functions and measures of the prediction accuracy, approaches to model validation, model AD, consensus prediction, and the use of QSAR models in virtual screening. We stress that throughout both chapters we emphasize that the integration of multiple individual components of the unified QSAR modeling workflow is absolutely necessary for achieving rigorously validated and truly predictive QSAR models.

REFERENCES

1. PubChem. http://pubchem.ncbi.nlm.nih.gov/. 2008.
2. Oprea, T. and Tropsha, A., Target, chemical and bioactivity databases—integration is key. *Drug Discov. Today* 2006, *3*, 357–365.
3. Golbraikh, A. and Tropsha, A., Predictive QSAR modeling based on diversity sampling of experimental datasets for the training and test set selection. *Mol. Divers.* 2002, *5*, 231–243.
4. Golbraikh, A., Shen, M., Xiao, Z., Xiao, Y. D., Lee, K. H., and Tropsha, A., Rational selection of training and test sets for the development of validated QSAR models. *J. Comput. Aided Mol. Des.* 2003, *17*, 241–253.
5. Irwin, J. J. and Shoichet, B. K., ZINC—a free database of commercially available compounds for virtual screening. *J. Chem. Inf. Model.* 2005, *45*, 177–182.
6. Tropsha, A., Application of predictive QSAR models to database mining. In T. Oprea (Ed.), *Cheminformatics in Drug Discovery.* Wiley-VCH, Weinheim, Germany, 2005.

7. Tropsha, A., Predictive QSAR (quantitative structure activity relationships) modeling. In: J. Mason (Ed.), *Comprehensive Medicinal Chemistry II.* V. 4 (Computer-Aided Drug Design). Elsevier, Oxford, UK, pp. 149–165, 2006.
8. Tropsha, A., Gramatica, P., and Gombar, V. K., The importance of being earnest: Validation is the absolute essential for successful application and interpretation of QSPR models. *Quant. Struct. Act. Relat. Comb. Sci.* 2003, *22*, 69–77.
9. Tropsha, A. and Golbraikh, A., Predictive QSAR modeling workflow, model applicability domains, and virtual screening. *Curr. Pharm. Des.* 2007, *13*, 3494–3504.
10. Cho, S. J., Zheng, W., and Tropsha, A., Rational combinatorial library design. 2. Rational design of targeted combinatorial peptide libraries using chemical similarity probe and the inverse QSAR approaches. *J. Chem. Inf. Comput. Sci.* 1998, *38*, 259–268.
11. Tropsha, A., Cho, S. J., and Zheng, W., "New Tricks for an Old Dog": Development and application of novel QSAR methods for rational design of combinatorial chemical libraries and database mining. In: A. L. Parrill and M. R. Reddy (Eds), *Rational Drug Design: Novel Methodology and Practical Applications.* American Chemical Society, Washington, pp. 198–211, 1999.
12. Gussio, R., Pattabiraman, N., Kellogg, G. E., and Zaharevitz, D. W., Use of 3D QSAR methodology for data mining the National Cancer Institute repository of small molecules: Application to HIV-1 reverse transcriptase inhibition. *Methods* 1998, *14*, 255–263.
13. Shen, M., Beguin, C., Golbraikh, A., Stables, J. P., Kohn, H., and Tropsha, A., Application of predictive QSAR models to database mining: Identification and experimental validation of novel anticonvulsant compounds. *J. Med. Chem.* 2004, *47*, 2356–2364.
14. Medina-Franco, J. L., Golbraikh, A., Oloff, S., Castillo, R., and Tropsha, A., Quantitative structure–activity relationship analysis of pyridinone HIV-1 reverse transcriptase inhibitors using the k nearest neighbor method and QSAR-based database mining. *J. Comput. Aided Mol. Des.* 2005, *19*, 229–242.
15. Oloff, S., Mailman, R. B., and Tropsha, A., Application of validated QSAR models of D1 dopaminergic antagonists for database mining. *J. Med. Chem.* 2005, *48*, 7322–7332.
16. Zhang, S., Wei, L., Bastow, K., Zheng, W., Brossi, A., Lee, K. H., and Tropsha, A., Antitumor agents 252. Application of validated QSAR models to database mining: Discovery of novel tylophorine derivatives as potential anticancer agents. *J. Comput. Aided Mol. Des.* 2007, *21*, 97–112.
17. Hsieh, J. H., Wang, X. S., Teotico, D., Golbraikh, A., and Tropsha, A., Differentiation of AmpC beta-lactamase binders vs. decoys using classification *K*NN QSAR modeling and application of the QSAR classifier to virtual screening. *J. Comput. Aided Mol. Des.* 2008, *22*, 593–609.
18. Tang, H., Wang, X. S., Huang, X. P., Roth, B. L., Butler, K. V., Kozikowski, A. P., Jung, M., and Tropsha, A., Novel inhibitors of human histone deacetylase (HDAC) identified by QSAR modeling of known inhibitors, virtual screening, and experimental validation. *J. Chem. Inf. Model.* 2009, 49, 461–476.
19. Tropsha, A. and Zheng, W., Identification of the descriptor pharmacophores using variable selection QSAR: Applications to database mining. *Curr. Pharm. Des.* 2001, *7*, 599–612.
20. Hoffman, B., Cho, S. J., Zheng, W., Wyrick, S., Nichols, D. E., Mailman, R. B., and Tropsha, A., Quantitative structure–activity relationship modeling of dopamine D(1) antagonists using comparative molecular field analysis, genetic algorithms-partial least-squares, and K nearest neighbor methods. *J Med. Chem.* 1999, *42*, 3217–3226.

21. Zheng, W. and Tropsha, A., Novel variable selection quantitative structure–property relationship approach based on the K-nearest-neighbor principle. *J Chem. Inf. Comput. Sci.* 2000, *40*, 185–194.
22. Elkan, C., The foundations of cost-sensitive learning. *Proceedings of the 17th International Joint Conference on Artificial Intelligence*, pp. 973–978, 2001.
23. Chen, C., Liaw, A., and Breiman, L., *Using Random Forest to Learn Imbalanced Data*, p. 666. Department of Statistics, University of California, Berkeley, 2004.
24. Young, D. and Martin, T., Are the chemical structures in your QSAR correct? *Qsar & Combinatorial Science* 2008, *27*, 1337–1345.
25. Chemical Computig Group. Molecular Operating Environment (MOE), http://www.chemcomp.com/. 2008.
26. ChemAxon, ChemAxon http://www.chemaxon.com, 2008.
27. OpenEye, OpenBabel, 2008.
28. Wiener, H. J., Structural determination of paraffin boiling points. *J. Am. Chem. Soc.* 1947, *15*, 17–20.
29. Platt, J. R., Influence of neighbor bonds on additive bond properties in paraffins. *J. Chem. Phys.* 1947, *15*, 419–420.
30. Todeschini, R. and Consonni, V., *Handbook of Molecular Descriptors*. Wiley-VCH, Weinheim, Germany, 2000.
31. Todeschini, R. and Consonni, V., Dragon, http://www.talete.mi.it/help/dragon_help/index.html?IntroducingDRAGON, 2007.
32. EduSoft, MolconnZ http://www.edusoft-lc.com/, 2007.
33. Golbraikh, A., Bonchev, D., and Tropsha, A., Novel chirality descriptors derived from molecular topology. *J Chem. Inf. Comput. Sci.* 2001, *41*, 147–158.
34. Golbraikh, A., Bonchev, D., and Tropsha, A., Novel ZE-isomerism descriptors derived from molecular topology and their application to QSAR analysis. *J. Chem. Inf. Comput. Sci.* 2002, *42*, 769–787.
35. Golbraikh, A. and Tropsha, A., QSAR modeling using chirality descriptors derived from molecular topology. *J Chem. Inf. Comput. Sci.* 2003, *43*, 144–154.
36. Carhart, R. E., Smith, D. H., and Venkataraghavan, R., Atom pairs as molecular features in structure–activity studies: Definition and applications. *J. Chem. Inf. Comput. Sci.* 1985, *25*, 64–73.
37. Kovatcheva, A., Golbraikh, A., Oloff, S., Feng, J., Zheng, W., and Tropsha, A., QSAR modeling of datasets with enantioselective compounds using chirality sensitive molecular descriptors. *SAR QSAR Environ. Res.* 2005, *16*, 93–102.
38. Marshall, G. R. and Cramer, R. D., III Three-dimensional structure–activity relationships. *Trends Pharmacol. Sci.* 1988, *9*, 285–289.
39. Klebe, G., Comparative molecular similarity indices: CoMSI. In H. Kubinyi, G. Folkers, and Y. Martin (Eds), *3D QSAR in Drug Design*, Vol. 3. Kluwer Academic Publisher, Great Britain, 1998.
40. Kubinyi, H., Hamprecht, F. A., and Mietzner, T., Three-dimensional quantitative similarity-activity relationships (3D QSiAR) from SEAL similarity matrices. *J Med. Chem.* 1998, *41*, 2553–2564.
41. Robinson, D. D., Winn, P. J., Lyne, P. D., and Richards, W. G., Self-organizing molecular field analysis: A tool for structure–activity studies. *J Med. Chem.* 1999, *42*, 573–583.
42. Tripos, Sybyl http://www.tripos.com, 2008.
43. Accelrys, Catalyst http://www.accelrys.com, 2008.
44. Inte:Ligand, LigandScout http://www.inteligand.com/ligandscout/, 2008.
45. Bures, M. G. and Martin, Y. C., Computational methods in molecular diversity and combinatorial chemistry. *Curr. Opin. Chem. Biol.* 1998, *2*, 376–380.

46. Cherkasov, A., An updated steroid benchmark set and its application in the discovery of novel nanomolar ligands of sex hormone-binding globulin. *J. Med. Chem.* 2008, *51*, 2047–2056.
47. Crivori, P., Cruciani, G., Carrupt, P. A., and Testa, B., Predicting blood–brain barrier permeation from three-dimensional molecular structure. *J. Med. Chem.* 2000, *43*, 2204–2216.
48. Cruciani, G., Pastor, M., and Guba, W., VolSurf: A new tool for the pharmacokinetic optimization of lead compounds 1. *Eur. J Pharm. Sci.* 2000, *11*(Suppl 2), S29–S39.
49. Pastor, M., Cruciani, G., McLay, I., Pickett, S., and Clementi, S., GRid-INdependent Descriptors (GRIND): A novel class of alignment-independent three-dimensional molecular descriptors. *J Med. Chem.* 2000, *43*, 3233–3243.
50. McGregor, M. J. and Pallai, P. V., Clustering of large databases of compounds: Using the MDL "Keys" as structural descriptors. *J. Chem. Inf. Comput. Sci.* 1997, *37*, 443–448.
51. Waller, C. L., A comparative QSAR study using CoMFA, HQSAR, and FRED/SKEYS paradigms for estrogen receptor binding affinities of structurally diverse compounds. *J. Chem. Inf. Comput. Sci.* 2004, *44*, 758–765.
52. Huan, J., Bandyopadhyay, D., Prins, J., Snoeyink, J., Tropsha, A., and Wang, W., Distance-based identification of structure motifs in proteins using constrained frequent subgraph mining. *Comput. Syst. Bioinform. Conf.* 2006, *47*, 227–238.
53. Horvath, D., Bonachera, F., Solov'ev, V., Gaudin, C., and Varnek, A., Stochastic versus stepwise strategies for quantitative structure–activity relationship generation—how much effort may the mining for successful QSAR models take? *J. Chem. Inf. Model.* 2007, *47*, 927–939.
54. Hong, H., Xie, Q., Ge, W., Qian, F., Fang, H., Shi, L., Su, Z., Perkins, R., and Tong, W., Mold(2), molecular descriptors from 2D structures for chemoinformatics and toxicoinformatics. *J. Chem. Inf. Model.* 2008, *48*, 1337–1344.
55. Schroeter, T. S., Schwaighofer, A., Mika, S., Ter, L. A., Suelzle, D., Ganzer, U., Heinrich, N., and Muller, K. R., Estimating the domain of applicability for machine learning QSAR models: A study on aqueous solubility of drug discovery molecules. *J. Comput. Aided Mol. Des.* 2007, *21*, 485–498.
56. Willett, P., Barnard, J. M., and Owns, G. M., Chemical similarity searching. *J. Chem. Inf. Comput. Sci.* 1998, *38*, 983–996.
57. Sachs, L., *Applied Statistics: A Handbook of Techniques*. Springer, New York, 1984.
58. Adams, M. J., *Chemometrics in Analytical Spectroscopy*. The Royal Society of Chemistry, Cambridge, 2004.
59. Whitley, D. C., Ford, M. G., and Livingstone, D. J., Unsupervised forward selection: A method for eliminating redundant variables. *J. Chem. Inf. Comput. Sci.* 2000, *40*, 1160–1168.
60. Zhang, S., Golbraikh, A., Oloff, S., Kohn, H., and Tropsha, A., A novel automated lazy learning QSAR (ALL-QSAR) approach: Method development, applications, and virtual screening of chemical databases using validated ALL-QSAR models. *J. Chem. Inf. Model.* 2006, *46*, 1984–1995.
61. Reynolds, C. H., Tropsha, A., Pfahler, L. B., Druker, R., Chakravorty, S., Ethiraj, G., and Zheng, W., Diversity and coverage of structural sublibraries selected using the SAGE and SCA algorithms. *J. Chem. Inf. Comput. Sci.* 2001, *41*, 1470–1477.
62. Reynolds, C. H., Druker, R., and Pfahler, L. B., Lead discovery using stochastic cluster analysis (SCA): A new method for clustering. *J. Chem. Inf. Comput. Sci.* 1998, *38*, 305–312.
63. Maggiora, G. M., On outliers and activity cliffs—why QSAR often disappoints. *J. Chem. Inf. Model.* 2006, *46*, 1535.

64. Zhang, L., Zhu, H., Oprea, T. I., Golbraikh, A., and Tropsha, A., QSAR modeling of the blood–brain barrier permeability for diverse organic compounds. *Pharm. Res.* 2008, *25*, 1902–1914.
65. Zhu, H., Ye, L., Richard, A., Golbraikh, A., Wright, F. A., Rusyn, I., and Tropsha A. A novel two-step hierarchical quantitative structure-activity relationship modeling work flow for predicting acute toxicity of chemicals in rodents. *Environ. Health Perspect.* 2009, *117*, 1257–1264.
66. Dixon, W. T. Processing data for outliers. *Biometrics*, 1953, *9*, 74–89.
67. Fallon, A. and Spada, C., Detection and accommodation of outliers in normally distributed data sets, http://ewr.cee.vt.edu/environmental/teach/smprimer/outlier/outlier.html, 1997.
68. Environmental Protection Agency. Statistical training course for ground-water monitoring data analysis, EPA/530-R-93-003, Office of Solid Waste, Washington, DC, 1992.
69. Taylor, J. K. Quality assurance of chemical measurements, Lewis Publishers, Chelsea, MI, 1987.
70. Kanji, G. K., *100 Statistical Tests*. Sage, 1993.
71. Yen, S.-J. and Lee, Y.-S., Under-sampling approaches for improving prediction of the minority class in an imbalanced dataset. *Lecture Notes in Control and Information Sciences* 2006, *344*, 731–740.
72. Kubat, M. and Matwin, S., Addressing the curse of imbalanced training sets: One sided selection. *Proceedings of the 14th International Conference on Machine Learning*, San Francisco, CA, Morgan Kaufmann, 1997.
73. Japkowicz, N., Learning from imbalanced datasets: A comparison of various strategies. AAAI Workshop, Learning From Imbalanced Datasets, Papers From The AAAI Workshop, AAAI Press, Menlo Park, CA, 2000.
74. Golbraikh, A. and Tropsha, A., Beware of Q2! *J. Mol. Graph. Model.* 2002, *20*, 269–276.
75. Organisation for Economic and Co-operation Development (OECD), Quantitative Structure–Activity Relationships [(Q)SARs] Project, http://www.oecd.org/document/23/0,3343, en_2649_34365_33957015_1_1_1_1,00.html, 2008.
76. Harju, M., Hamers, T., Kamstra, J. H., Sonneveld, E., Boon, J. P., Tysklind, M., and Andersson, P. L., Quantitative structure–activity relationship modeling on *in vitro* endocrine effects and metabolic stability involving 26 selected brominated flame retardants. *Environ. Toxicol. Chem.* 2007, *26*, 816–826.
77. Sharma, D., Narasimhan, B., Kumar, P., and Jalbout, A., Synthesis and QSAR evaluation of 2-(substituted phenyl)-1H-benzimidazoles and [2-(substituted phenyl)-benzimidazol-1-Yl]-pyridin-3-Yl-methanones. *Eur. J. Med. Chem.* 2009, *44*, 1119–1127.
78. Zvinavashe, E., van den, B. H., Soffers, A. E., Vervoort, J., Freidig, A., Murk, A. J., and Rietjens, I. M., QSAR models for predicting *in vivo* aquatic toxicity of chlorinated alkanes to fish. *Chem. Res. Toxicol.* 2008, *21*, 739–745.
79. Padmanabhan, J., Parthasarathi, R., Subramanian, V., and Chattaraj, P. K., Group philicity and electrophilicity as possible descriptors for modeling ecotoxicity applied to chlorophenols. *Chem. Res. Toxicol.* 2006, *19*, 356–364.
80. Song, M. and Clark, M., Development and evaluation of an in silico model for HERG binding. *J. Chem. Inf. Model.* 2006, *46*, 392–400.
81. Iyer, M., Zheng, T., Hopfinger, A. J., and Tseng, Y. J., QSAR analyses of skin penetration enhancers. *J. Chem. Inf. Model.* 2007, *47*, 1130–1149.
82. Kovatcheva, A., Golbraikh, A., Oloff, S., Xiao, Y. D., Zheng, W., Wolschann, P., Buchbauer, G., and Tropsha, A., Combinatorial QSAR of ambergris fragrance compounds. *J. Chem. Inf. Comput. Sci.* 2004, *44*, 582–595.
83. de Cerqueira, L. P., Golbraikh, A., Oloff, S., Xiao, Y., and Tropsha, A., Combinatorial QSAR modeling of P-glycoprotein substrates. *J. Chem. Inf. Model.* 2006, *46*, 1245–1254.

84. Pandey, G. and Saxena, A. K., 3D QSAR studies on protein tyrosine phosphatase 1B inhibitors: Comparison of the quality and predictivity among 3D QSAR models obtained from different conformer-based alignments. *J. Chem. Inf. Model.* 2006, *46*, 2579–2590.
85. Roberts, D. W., Aptula, A. O., and Patlewicz, G., Mechanistic applicability domains for non-animal based prediction of toxicological endpoints. QSAR analysis of the Schiff base applicability domain for skin sensitization. *Chem. Res. Toxicol.* 2006, *19*, 1228–1233.
86. Ferré, J. and Rius, F. X., Selection of the best calibration sample subset for multivariate regression. *Anal. Chem.* 1996, *68*, 1565–1571.
87. Ferré, J. and Rius, F. X., Constructing D-optimal designs from a list of candidate samples. *Trends Anal. Chem.* 1997, *16*, 70–73.
88. Maesschalck, R. De., Estienne, F., Verdú-Andrés, J., Candolfi, A., Centner, V., Despagne, F., Jouan-Rimbaud, D., Walczak, B., Massart, D. L., Jong, S., de Noord, O. E. D., Puel, C., and Vandeginste, B. M. G., PCR tutorial. 10. Selection and representativity of the calibration sample subset, http://www.vub.ac.be/fabi/multi/pcr/chaps/chap10.html#fin, 2008.
89. Kennard, R. W. and Stone, L. A., Computer-aided design of experiments. *Technometrics* 1969, *11*, 137–148.

7 Predictive Quantitative Structure–Activity Relationships Modeling

Development and Validation of QSAR Models

Alexander Tropsha and Alexander Golbraikh

CONTENTS

In this chapter, we continue to discuss the general framework of quantitative structure–activity relationships (QSAR) modeling. In the previous chapter, we have addressed the issue of data preparation for QSAR studies. The main topic of this chapter is the general principles of QSAR model development and validation irrespective of specifics of any particular QSAR modeling routine. We introduce the concept of combinatorial QSAR modeling, which consists of building QSAR models for all combinations of descriptor types and optimization procedures. We classify QSAR approaches based on the response variable, which can be continuous (i.e., take multiple values spread over a certain interval), represent a category or rank of activity or property (e.g., very active, active, moderately active, and inactive), or a class of compounds (e.g., ligands of different receptors). For each type of the response variable, we introduce target functions that should be optimized by a QSAR procedure and criteria of model accuracy. Particular attention is paid to imbalanced datasets, in which the counts of compounds belonging to different categories or classes are significantly different. We consider different validation procedures including cross-validation (which is included in model training), prediction for test sets (i.e., compounds that were not used in model training), and Y-randomization test (i.e., building and evaluating models with randomized activities of the response variable). We introduce a concept of model stability that can be established by resampling of training and test sets. We discuss the advantages and disadvantages of different definitions of applicability domains (AD) of QSAR models. Finally, consensus prediction of external evaluation sets by all predictive models is considered as a test of the applicability of QSAR models to chemical database mining and virtual screening. We emphasize that the integration of all the steps of QSAR modeling considered in both the previous chapter and this chapter in a unified workflow is critical for the development of validated and externally predictive QSAR models.

7.1 INTRODUCTION: COMBINATORIAL QSAR MODELING

Different descriptor collections can be combined with different optimization methods. For example, models built with *k*NN QSAR and Dragon descriptors, or Chirality descriptors, or MolconnZ descriptors, and so on are developed. For example, for seven descriptor collections and four methods, the total number of different QSAR studies would be 28 [1].

In the majority of QSAR studies, models are typically generated with a single modeling technique. To achieve QSAR models of the highest quality, that is, with high internal and, most importantly, external accuracies, we shall rely on the combinatorial QSAR approach (combi-QSAR), which explores all possible combinations of various collections of descriptors and optimization methods along with external model validation (Figure 7.1). The chief hypothesis of the combi-QSAR approach that we introduced and employed in recent studies [2–4] is that if an implicit structure–activity relationship exists for a given dataset, it can be formally manifested via a variety of QSAR models obtained with different descriptors and optimization protocols. Our experience indicates that there is no universal QSAR method that is guaranteed to give the best results for any dataset. Thus we believe that multiple alternative QSAR models should be developed (as opposed to a single model using some favorite QSAR method) for each dataset to identify the most successful technique in the context of the given dataset. Since QSAR modeling is relatively fast, these alternative models could be explored simultaneously when making predictions for external datasets. The consensus predictions of biological activity for novel test set compounds on the basis of several QSAR models, especially when they converge, are more reliable and provide better justification for the experimental exploration of hits.

Our current approach to combi-QSAR modeling is summarized on the workflow diagram (Figure 7.1). To achieve QSAR models of the highest internal and, most importantly, external accuracy, the combi-QSAR approach explores all possible binary combinations of various descriptor types and optimization methods along with external model validation. Each combination of descriptor sets and optimization techniques is likely to capture certain unique aspects of the structure–activity relationship. Since our ultimate goal is to use the resulting models as reliable activity

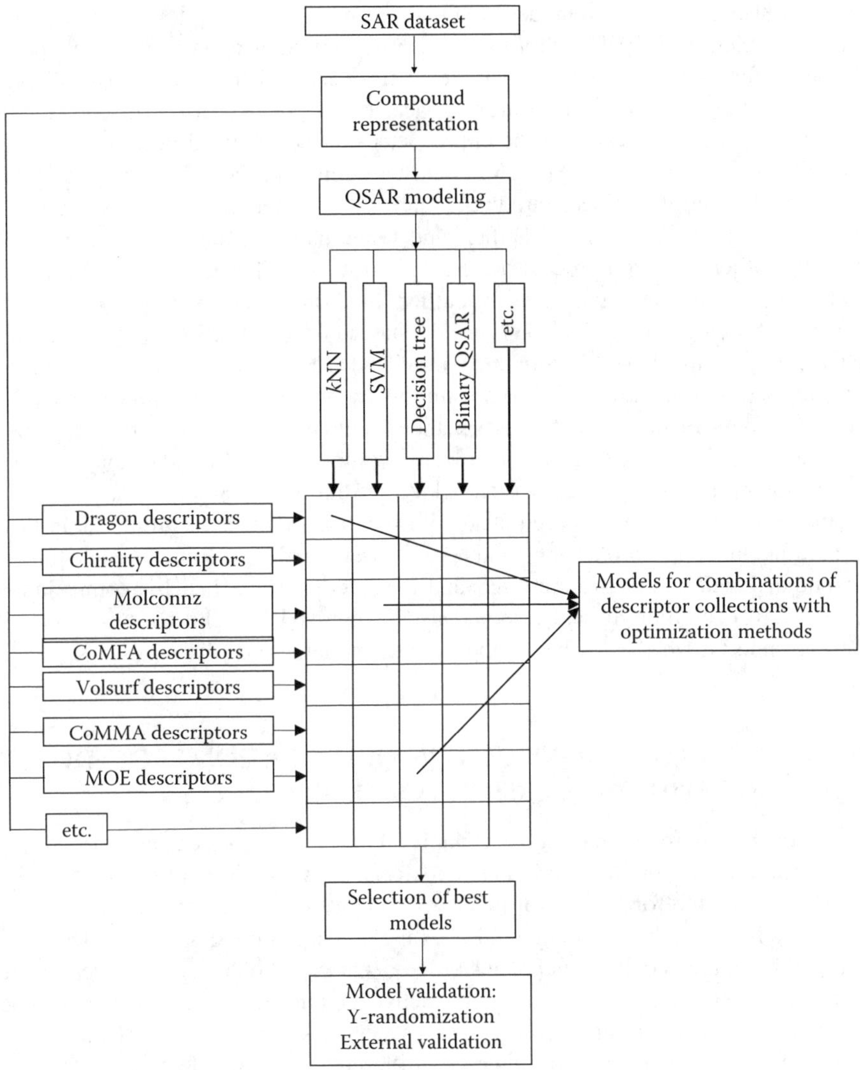

FIGURE 7.1 Combinatorial QSAR modeling workflow.

(property) predictors, application of different combinations of modeling techniques and descriptor sets will increase our chances for success. We typically employ different types of descriptors and modeling techniques, which are described in detail in our recent publications on the implementation of the combi-QSAR strategy [1,3,4].

What if there are outliers in the model? *Outliers in the model should be considered as a serious problem.* In general, there should be an important reason to delete these outliers. For example, a compound could have some biological property that makes it completely different from other compounds. Even the elimination of leverage and activity outliers prior to QSAR modeling may not remove all outliers. Indeed, the order of distances in the entire descriptor space and in the descriptor subspace determined by the model built with a variable selection QSAR approach (model space) can be different, and compounds that are far from each other in the entire descriptor space could become close in the model space. The opposite is also true: compounds that are close to each other in the entire descriptor space could become relatively distant in the model space. So there could be outliers in the model space as well. Moreover, if multiple models are built, some of them may have outliers and some not, some may have one set of outliers and some may have another set, and so on. Ideally, outlier detection procedures should be run again for each model. However, in this case, only compounds that are outliers in all models (probably those that act via unique biological mechanisms) can be removed from the entire modeling set. We should also distinguish outliers in the training and test sets. If there are outliers in the training set, they should be removed from this training set and the model should be rebuilt. If there are outliers in the test set, they should be removed from these test sets, and new statistics characterizing the prediction of this test set should be obtained. All such outliers are activity outliers because leverage outliers are automatically removed from the test sets since they are outside of the model AD (see Section 7.5). In fact, if multiple divisions into training and test sets were made, the same possible outlier can be included in some training sets and in some test sets. If such a compound is removed from all training sets, it should also be removed from the test sets. In fact, it is very important to check whether there are no model outliers.

7.2 TARGET FUNCTIONS USED IN OPTIMIZATION PROCEDURES AND VALIDATION CRITERIA OF QSAR MODELS

Based on the nature of the response variable, QSAR approaches can be grouped into classification, category, or continuous QSAR. Classes are different from categories in a sense that the former cannot be ordered in any scientifically meaningful way, while the latter can be rank ordered. For example, classes of compounds interacting with different receptors cannot be rank ordered. On the other hand, categories can be defined as very active, active, moderately active, and inactive so a rank order can be introduced. If there are just two classes (or categories), the same statistical criteria are used to validate the models. Otherwise, different criteria should be used. Thus, in general, for each group of models, different validation criteria are used. Of course, prior to predicting the properties of compounds that form the test set, the AD for the corresponding training set should be defined (see Section 7.5).

Target functions and validation criteria for continuous QSAR: We suggested that the following validation criteria should be used for continuous QSAR models [5]: (i) leave-one-out (LOO) cross-validation q^2 (which is also used as the target function, that is, it is optimized by the QSAR modeling procedure); (ii) square of the correlation coefficient R (R^2) between the predicted and observed activities; (iii) coefficients of determination for regression lines through the origin (predicted versus observed activities R_0^2 and observed versus predicted activities $R_0^{\prime 2}$); (iv) slopes k and k' of regression lines (predicted versus observed activities and observed versus predicted activities) through the origin. These criteria are calculated according to the following formulas:

$$q^2 = 1 - \frac{\sum_{i=1}^{N}(y_i - \tilde{y}_i)^2}{\sum_{i=1}^{N}(y_i - \bar{y})^2}, \tag{7.1}$$

$$R = \frac{\sum(y_i - \bar{y})(\tilde{y}_i - \bar{\tilde{y}})}{\sqrt{\sum(y_i - \bar{y})^2 \sum(\tilde{y}_i - \bar{\tilde{y}})^2}}, \tag{7.2}$$

$$R_0'^2 = 1 - \frac{\sum(\tilde{y}_i - \tilde{y}_i^{r0})^2}{\sum(\tilde{y}_i - \bar{\tilde{y}})^2} \quad \text{(predicted versus observed)}, \tag{7.3a}$$

$$R_0^2 = 1 - \frac{\sum(y_i - y_i^{r0})^2}{\sum(y_i - \bar{y})^2} \quad \text{(observed versus predicted)}, \tag{7.3b}$$

$$k = \frac{\sum y_i \tilde{y}_i}{\sum y_i^2} \quad \text{(predicted versus observed)}, \tag{7.4a}$$

$$k' = \frac{\sum y_i \tilde{y}_i}{\sum \tilde{y}_i^2} \quad \text{(observed versus predicted)}, \tag{7.4b}$$

where y_i and $\tilde{y}_i$ are the observed and predicted activities, R_0^2 and $R_0'^2$ are the coefficients of determination for regressions through the origin for predicted versus observed and observed versus predicted activities, respectively, k and k' are the corresponding slopes, and $\tilde{y}^{r0} = ky$ and $y^{r0} = k'\tilde{y}$ are the regressions through the origin for predicted versus observed and observed versus predicted activities. In our studies, we consider models acceptable, if they have (i) $q^2 > 0.5$; (ii) $R^2 > 0.6$; (iii) $(R^2 - R_0^2)/R^2 < 0.1$ and $0.85 \leq k \leq 1.15$ or $(R^2 - R_0'^2)/R^2 < 0.1$ and $0.85 \leq k' \leq 1.15$; and (iv) $|R_0^2 - R_0'^2| < 0.3$. Sometimes, stricter criteria are used.

In some papers, other criteria are used. For example, sometimes standard error of prediction (SEP) is used instead of (or together with) R^2. SEP itself makes no sense until we compare it with the standard deviation for activities of the test set, which brings us back to the correlation coefficients. If used, mean absolute error (MAE) should be compared with the mean absolute deviation from the mean. Sometimes, especially in the Hansch QSAR, F-ratio is calculated. F-ratio is the variance explained by the model divided by the unexplained variance. It is believed that the higher the F-ratio, the better the model. We believe that when the F-ratio is used, it must always be accompanied by the corresponding p-value.

Frequently, especially for linear models such as developed with multiple linear regression (MLR) or partial least squares (PLS), the adjusted R^2 is used:

$$R_{\text{adj}}^2 = 1 - \left(1 - R^2\right)\frac{n-1}{n-c-1}, \tag{7.5}$$

where n is the number of compounds in the dataset and c is the number of variables (descriptors) included in the regression equation. It should be noted that $R_{\text{adj}}^2 \leq R^2$. The higher the number of explanatory variables c, the lower the R_{adj}^2 value. R_{adj}^2 is particularly important for linear QSAR models developed with variable selection. R^2_{adj} is not a good criterion for variable selection *k*NN QSAR models, since contrary to regression methods, in the *k*NN algorithm descriptors are just selected or not selected, that is, their weights are either zero or one. As a result, a much larger set of descriptors is selected by the *k*NN procedure than, for example, by stepwise regression.

Target functions and validation criteria for classification QSAR models: We consider a classification QSAR model predictive, if the prediction accuracy characterized by the correct classification rate (CCR) for each class is sufficiently large:

$$\mathrm{CCR_{class}} = \frac{N^{\mathrm{corr}}_{\mathrm{class}}}{N^{\mathrm{total}}_{\mathrm{class}}}, \tag{7.6}$$

and the *p*-value for this $\mathrm{CCR_{class}}$ value is not higher than a predefined threshold (in the case of two classes, the $\mathrm{CCR_{class}}$ threshold should not be lower than 0.65–0.70, and for any number of classes, the *p*-value should not be higher than 0.05 for each class). For example, in Ref. [6], binary classification QSAR model has been built for a training set of 105 chemicals activating the estrogen gene. The test set included 12 compounds, eight of them active and four inactive. The prediction accuracy for the inactive class was 75% (three compounds out of four were predicted correctly). The *p*-value of predicting at least three out of four compounds correctly was 0.31 [there is a probability of $(0.5)^4$ of predicting all four compounds correctly and $4 \times (0.5)^4$ for predicting exactly three compounds correctly; by adding both values, we get *p*-value = 0.31]. In fact, even if all four compounds were predicted correctly, the *p*-value would be 6.25×10^{-2}, which is higher than the threshold value of 0.05. If a class contains only four compounds, prediction accuracy cannot be statistically significant, that is, the null hypothesis H_0 that the model predicts not better than random cannot be rejected with the significance level of 0.95. Thus, at least five compounds of each class should be included in the test set. In this case, if all five compounds are predicted correctly, the *p*-value would be 3.13×10^{-2}, which is statistically significant. However, if one compound out of five would be predicted incorrectly, the *p*-value would be 0.19, which makes the prediction statistically insignificant. In Ref. [7], we built a binary QSAR model for a dataset of AmpC β-lactamase binders versus nonbinders. The test set included 10 compounds, five for each class. All 10 compounds were predicted correctly. It was sufficient to consider the developed model as acceptable and statistically significant.

$\mathrm{CCR_{class}}$ values corresponding to a given *p*-value depend on the number of classes and the number of compounds in the class. For example, we have found recently that in the case of two classes the following assertions are true. If a class includes less than 23 compounds (of course, it should be higher than five compounds; see above), the maximum $\mathrm{CCR_{class}}$ corresponding to the *p*-value of 0.05 is always higher than 0.70. At the same time, if it includes 28 or more compounds, the $\mathrm{CCR_{class}}$ of 0.70 always means that the *p*-value is lower than 0.05. Thus, we have established that in the case of two classes, if the number of compounds in the class is lower than 23, then the *p*-value should be used to accept the model, but if the number of compounds in the class is at least 28, then the threshold 0.70 for $\mathrm{CCR_{class}}$ values should be used. Formally, for each number of compounds between 24 and 27, one of the two criteria should be used based on the maximum error. It is due to the discrete nature of the error distribution. In practice, for these counts of compounds, the minimum CCR_{class} is very close to 0.70, so any criterion can be used. Similar rules can be established for any number of classes and categories and any p-value.

In some papers (see, e.g., Ref. [8]) as well as some QSAR software, the authors use the target function and the measure of classification accuracy defined as

$$CCR = \frac{N(\text{corr})}{N(\text{total})}, \tag{7.7}$$

where N(correct) and N(total) are the number of compounds in the dataset classified correctly and the total number of compounds in the dataset, respectively. This accuracy measure is correct for balanced datasets only, that is, for datasets including equal numbers of compounds of each class. Otherwise, this measure of accuracy is incorrect and can lead to a wrong conclusion about the predictive power of a model. Suppose that we have a dataset with 80 compounds of class 1 and 20 compounds of class 2, and we have "developed" a model that assigns all compounds to class 1. Then the classification accuracy according to the above criterion would be 80%, and if it were the only criterion of accuracy used, a wrong conclusion would be made that the model is highly predictive.

For the classification QSAR with K classes, we shall use the following criterion:

$$CCR = \frac{1}{K}\sum_{i=1}^{K} CCR_i = \frac{1}{K}\sum_{i=1}^{K} \frac{N_k^{\text{corr}}}{N_k^{\text{total}}}, \tag{7.8}$$

along with the CCR for each class (see Formula 7.6). Criterion 7.8 is correct for both balanced and imbalanced (biased) datasets (i.e., when the number of compounds of each class is different). For imbalanced datasets, formula N(corr)/N(total), where N(corr) and N(total) are the number of compounds predicted correctly and the total number of compounds in the dataset, is incorrect. Another alternative to Formula 7.7 could include weights for each class.

$$CCR = \frac{\sum_{i=1}^{K} w_i N_i^{\text{corr}}}{\sum_{i=1}^{K} w_i N_i^{\text{total}}}, \quad \sum_{i=1}^{K} w_i = 1, \tag{7.9}$$

with smaller weights for larger classes. For two classes and weights inversely proportional to the sizes of classes, this formula is identical to Formula 7.8.

We should also be aware that the p-value for prediction for each class does not exceed the predefined threshold (usually 0.05; see above). This is particularly important for small test sets. The same CCR definitions are used as target functions in model optimization by the cross-validation procedure. However, for model optimization, some penalty terms can be subtracted from the target function. These additional terms penalize the target function, if CCRs for different classes are different (Formula 7.10)

$$CCR = \frac{1}{K}\left[\sum_{i=1}^{K} CCR_i - \sum_{i=1}^{K-1}\sum_{j=i+1}^{K} \alpha_{ij}\left|CCR_i - CCR_j\right|\right]. \tag{7.10}$$

Thus, we want to make prediction accuracies for different classes close to each other. Other penalizing schemes are possible; for example, penalty terms are added only if the differences in the CCR values exceed some threshold. If the classes of the dataset under study have significantly different sizes (a dataset is imbalanced or biased), additional modification of Formula 7.9 could be necessary. Usually, in this case, CCR values for larger classes are higher than for smaller ones, so weights w_i for classes are introduced (Formula 7.11).

$$\mathrm{CCR} = \sum_{i=1}^{K} w_i \mathrm{CCR}_i - \frac{1}{K}\sum_{i=1}^{K-1}\sum_{j=i+1}^{K} \alpha_{ij} \left|\mathrm{CCR}_i - \mathrm{CCR}_j\right|, \quad \sum_{i=1}^{K} w_i = 1. \tag{7.11}$$

Higher weights are given for smaller classes. The sum of the weights is equal to one.

Target functions and validation criteria for category QSAR models: Category QSAR with more than two classes should use target functions and validation criteria other than those used in classification QSAR. These target functions and validation criteria should consider errors as differences between predicted and observed categories, or increasing functions of these differences. The total error of prediction over all compounds is the sum of all errors of predictions for individual compounds. Let n_{ij} be the number of compounds of category i assigned by a model to category j (i, $j = 1, \ldots, K$). Then the total error is calculated as follows (Formula 7.12):

$$E = \sum_{i=1}^{K}\sum_{j=1}^{K} n_{ij} f\left(|i-j|\right), \tag{7.12}$$

where $f(|i-j|)$ is the increasing function of errors and $f(0) = 0$. In the case of biased datasets, it would be important to normalize the errors for compounds of category i on the number of compounds in this category:

$$E = \sum_{i=1}^{K} \frac{1}{N_i} \sum_{j=1}^{K} n_{ij} f\left(|i-j|\right), \tag{7.13}$$

where N_i is the number of compounds of category i. Finally, it is possible to use different weights in the calculation of the error: larger weights should be given for errors for categories including a smaller number of compounds to make the total error E more sensitive to errors for smaller categories (Formula 7.14).

$$E = K\sum_{i=1}^{K} \frac{w_i}{N_i} \sum_{j=1}^{K} n_{ij} f\left(|i-j|\right), \quad \sum_{i=1}^{K} w_i = 1. \tag{7.14}$$

The prediction accuracy can be calculated as

$$A = 1 - \frac{E}{E_{\max}}. \tag{7.15a}$$

According to this definition, $0 \leq A \leq 1$. If weights are not used, the maximum possible error E_{max} can be calculated as follows (if errors are not normalized):

$$E_{max} = \sum_{i=1}^{[K/2]} N_i f(|K - i|) + \sum_{i=[K/2+1]}^{K} N_i f(|i - 1|), \tag{7.16a}$$

or as follows (if errors are normalized):

$$E_{max} = \sum_{i=1}^{[K/2]} f(|K - i|) + \sum_{i=[K/2+1]}^{K} f(|i - 1|). \tag{7.16b}$$

If weights are used, the maximum possible error E_{max} can be calculated as follows (if errors are not normalized):

$$E_{max} = K \sum_{i=1}^{K} \max_j \left[w_i N_i f(|i - j|) \right] \quad j = 1, 2, \ldots, K, \tag{7.16c}$$

or as follows (if errors are normalized):

$$E_{max} = K \sum_{i=1}^{K} \max_j \left[w_i f \left(|i - j|\right) \right], \quad j = 1, 2, , \ldots K. \tag{7.16d}$$

An alternative definition of the accuracy is as follows:

$$A' = 1 - \frac{E}{E_{exp}}, \tag{7.15b}$$

where E_{exp} is the expected error. If weights are not used, the expected error E_{exp} can be calculated as follows (if errors are not normalized):

$$E_{exp} = \frac{1}{K} \sum_{i=1}^{K} N_i \sum_{j=1}^{K} f(|i - j|), \tag{7.17a}$$

or as follows (if errors are normalized):

$$E_{exp} = \frac{1}{K} \sum_{i=1}^{K} \sum_{j=1}^{K} f(|i - j|). \tag{7.17b}$$

If weights are used, the maximum possible error E_{max} can be calculated as follows (if errors are not normalized):

$$E_{exp} = \sum_{i=1}^{K} w_i N_i \sum_{j=1}^{K} f(|i - j|), \tag{7.17c}$$

or as follows (if errors are normalized):

$$E_{\text{exp}} = \sum_{i=1}^{K} w_i \sum_{j=1}^{K} f(|i-j|). \tag{7.17d}$$

Prediction statistics for category QSAR models should also be combined with the corresponding p-values.

Threshold moving: Finally, when predicting a compound by a biased dataset, threshold moving can be used. For example, if a compound's predicted class or category is 1.6, by rounding it is assigned to class 2. However, if a model is built for a biased dataset with class or category 1 being much smaller than class or category 2, it can be assigned to class or category 1; if the threshold between categories is moved from 1.5 to, say, 1.7, then all compounds with predicted class or category lower than 1.7 are assigned to class or category 1, otherwise they are assigned to class or category 2.

7.3 VALIDATION OF QSAR MODELS: Y-RANDOMIZATION

To establish model robustness, the Y-randomization (randomization of the response variable) test should be used. This test consists of repeating all the calculations with scrambled activities of the training set. Ideally, calculations should be repeated at least five or 10 times. The goal of this procedure is to establish whether models built with real activities of the training set have good statistics not due to overfitting or chance correlation. If all models built with randomized activities of the training set have statistically significantly lower predictive power for the training or the test set, then the models built with real activities of the training set are reliable. Using different parameters of a model development procedure, multiple QSAR models are built that have acceptable statistics. Suppose that the number of these models is m. The Y-randomization test can also give n models with acceptable statistics. For acceptance of models developed with real activities of the training set, the condition $n \ll m$ should be satisfied. In Refs. [2,3], we have introduced the measure of robustness $R = 1 - n/m$. If $R > 0.9$, the models are considered robust and their high predictive accuracy cannot be explained by chance correlation or overfitting. The Y-randomization test is particularly important for small datasets. Unfortunately, in many publications on QSAR studies, the Y-randomization test is not performed but all QSAR practitioners must be strongly encouraged to use this simple procedure.

7.4 VALIDATION OF QSAR MODELS: TRAINING AND TEST SET RESAMPLING. STABILITY OF QSAR MODELS

If a model has a high predictive accuracy for one division into the training and test sets, it must be tested using other divisions. It is particularly important for small datasets, when division into three sets (training, test, and independent validation) is impossible, since the smaller is the number of objects for building models, the worse are the chances of building a truly predictive model. We are proposing here a new method, namely training and test set resampling, a nonparametric technique that could be used to estimate statistics and confidence intervals for a population of objects when only a representative subset of the population is available (a dataset used to build models). "Resampling" means multiple random divisions of a dataset into training and test sets of the same size as were used for building models. The training sets are used for the calculation of internal prediction accuracy, such as cross-validation q^2 for continuous problems, or CCR and/or A for classification or category QSAR (see Formulas 7.8 through 7.18). The corresponding test sets are used for the calculation of the correlation coefficient between predicted and observed response variables, and coefficients of determination and slopes of regression lines of predicted versus observed and observed versus predicted response variables through the origin for continuous problems. In case of classification or category response variable, test sets are used for estimation of the total classification accuracy as well as classification accuracy for each class or category. Prior to prediction of compounds from the test set, the AD for the corresponding training set should be defined (see Section 7.5). Prediction should be made only for those compounds of the test sets that are within the ADs of the training sets. We argue that predictive models should have similar statistics to those obtained with the initial training and test sets. Large differences between model statistics will evidence that the model is unstable. Average statistics values obtained using the training and test set resampling approach are expected to be better estimates of the population statistics than those obtained with the initial training and test sets. It will be also possible to estimate confidence intervals of the model statistics, which are important characteristics of the model stability.

A similar method of validation, which is used in QSAR and other data analysis areas, is bootstrapping [9–11]. Like the resampling of training and test sets, bootstrapping is a nonparametric approach to obtain estimates of statistics and confidence intervals for a population of objects when only a representative subset of the population is available. Bootstrapping consists of choosing N objects with returns from a dataset of N objects. Due to returns of the selected objects to the initial dataset, some objects will be included in the bootstrapped datasets several times, while others will not be included at all. It has been shown that if the procedure is repeated many times (about 1000 times or more), average bootstrapped statistics are good estimates of population statistics. Bootstrapping can be used separately for training and test sets. Selecting the same compounds several times into the training sets is unacceptable for some QSAR methodologies like kNN. On the other hand, training and test set resampling is free from this disadvantage, because in different realizations it will include the same objects in the training or in the test set. Thus, after many realizations, both training and test sets will be represented by all objects included in the dataset. To obtain population statistics estimates, we shall use the same approaches as used for bootstrapping. They are described elsewhere [10–12].

The authors of a recent publication [13] assert that cross-validation and bootstrapping are not reliable in estimating the true predictive power of a classifier, if a dataset includes less than about 1000 objects, and suggest the Bayesian confidence intervals should be used instead. Cross-validation and bootstrapping are particularly unreliable for small datasets (including less than about 100 compounds). But 95% Bayesian confidence intervals for these datasets are very wide [13]. The authors show that the Repeated Independent Design and Test [14] introduced earlier does not solve the problem of cross-validation and bootstrapping. The procedure consists of splitting a dataset into two parts (which are called design bag and test bag) and repeated selection of objects from these bags into training and test sets with replacement.

7.5 APPLICABILITY DOMAINS OF QSAR MODELS

Formally, a QSAR model can predict the target property for any compound for which chemical descriptors can be calculated. However, if a compound is highly dissimilar from all compounds of the modeling set, reliable prediction of its activity is unlikely to be realized. A concept of AD was developed and used to avoid such an unjustified extrapolation in activity prediction. Classification of existing definitions of AD was given in recent publications by Jaworska and colleagues [15,16]. Here, we will follow this classification [paragraphs (i) through (v)].

i. *Descriptor-range based:* AD is defined as a hyperparallelepiped in the descriptor space in which representative points are distributed [6,17,18]. Dimensionality of the hyperparallelepiped is equal to the number of descriptors, and the size of each dimension is defined by the minimum and maximum values of the corresponding descriptor or it stretches beyond these limits to some extent up to predefined thresholds. The *drawbacks* of this definition are as follows. Generally, the representative points are distributed not in the entire hyperparallelepiped, but only in a small part of it. For example, in many cases, principal component analysis (PCA) shows that a small number of PCs explains 90–95% of the variance of descriptor values; this means that the representative points are predominantly distributed along a hyperplane of much lower dimensionality than that of the entire descriptor space. It is as if, in the three-dimensional hypercube, the points would be concentrated around one of its diagonals, while most of the space in the cube would be empty. Another possibility is that there are structural outliers in the dataset, which were not removed prior to QSAR studies. Even one such outlier can enormously increase the size of the hypercube, and the area around the outlier will contain no other points. Consequently, for many compounds within the hypercube, prediction will be unreliable.
ii. *Geometric methods: convex hull AD:* AD is defined as a convex hull of points in the multidimensional descriptor space [19]. The *drawbacks* of the convex hull AD are the same as those for the descriptor-range-based AD: the representative points might be distributed not in the entire convex hull, but only in a small part of it. Outliers not removed prior to QSAR modeling may significantly increase the volume of the convex hull. Besides, in many cases, especially for small datasets, the number of descriptors exceeds the number of compounds. In this case, the convex hull is not unique.
iii. *Leverage based:* Leverage for a compound is defined as the corresponding diagonal element of the hat matrix [20]. A compound is defined as outside of the AD if its leverage L is higher than 3 K/N, where K is the number

of descriptors and N is the number of compounds. The *drawbacks* of the leverage-based AD are as follows. (a) for each external compound, it is necessary to recalculate leverage; (b) if there are cavities in the representative point distribution area, a compound under prediction in this area will be considered to be within the AD (see Figure 6.2).

iv. *Probability density distribution based:* It is believed that more reliable predictions are in the areas of higher density of representative points than in sparse density. *Drawbacks*: (a) there is a problem of estimating the local density distribution function of points. For example, if a query point is close to the border of the area occupied by representative points, the probability density function can be underestimated; (b) in fact, prediction will be more precise, if nearest neighbors of the query compound have similar activities. It is not necessarily true in high density distribution areas. Moreover, in these areas, activity outliers and even activity cliffs are more likely to be encountered; hence, if they are not taken care of prior to or during QSAR studies, they might pose serious problems. *Bayesian classifiers* belong to this type of algorithms. Let the density distribution functions at point x for each class i be $p(x \mid i)$, and the *a priori* probability for class i be $P(i)$. Then the point x is assigned to the class for which $p(x|i)\ P(i)$ is the highest (the approach is adapted from a lecture on Pattern Recognition [21] and generalized). Another approach, the R-NN *curve approach*, was developed by Guha et al. [22]. The number of neighbors of each compound in the descriptor space is considered as a function of distance to this compound. Low values of this function for short distances to this compound mean that the distribution density of compounds in the neighborhood of this compound is low; hence, this compound can be considered as an outlier.

v. *Ensemble based:* This method was applied to water solubility data [23]. It consists of the following steps. (a) For a query compound, find nearest neighbors with known activities in the database. If the similarity to all compounds in the database is below a predefined threshold, a compound is considered to be outside of the AD. Atom-centered fragments were used as descriptors and a modified Dice coefficient was used as a similarity measure. (b) Predict these compounds by multiple models. In the paper, seven models for assessing the compounds' solubility built by other authors were considered. (c) Select a model that predicted these compounds most accurately. (d) Predict the query compound by this model. *Drawbacks and criticism*: A large amount of data should be available. The authors used a database of more than 1800 compounds with known water solubility. For most of the targets, so much data are not available. If models were built by the same authors, a question would arise why the database compounds were not used in model building, which would increase the predictive accuracy of their models.

vi. *Distance-based AD:* In our studies, the AD is defined as the Euclidean distance threshold D_{cutoff} between a compound under prediction and its closest nearest neighbor of the training set. It is calculated as follows:

$$D_{\text{cutoff}} = \langle D \rangle + Zs. \tag{7.18}$$

Here, $\langle D \rangle$ is the average Euclidean distance between each compound and its k nearest neighbors in the training set (where k is the parameter optimized in the course of QSAR modeling, and the distances are calculated using descriptors selected by the optimized model only), s is the standard deviation of these Euclidean distances, and Z is an arbitrary parameter to control the significance level [2,3,7,24]. We set the default value of this parameter Z at 0.5, which formally places the allowed distance threshold at the mean plus one-half of the standard deviation. We also define the AD in the entire descriptor space. In this case, the same Formula 7.18 is used, $k = 1$, $Z = 0.5$, and Euclidean distances are calculated using all descriptors. Thus, if the distance of the external compound from its nearest neighbor in the training set within either the entire descriptor space or the selected descriptor space exceeds these thresholds, the prediction is not made. We have also investigated changes of predictive power by changing the values of Z-cutoff. We have found that in general, predictive power decreases while Z-cutoff value increases (unpublished observations). Instead of Euclidean distances, other distances and similarity measures can be used.

vii. Consensus prediction value of the activity of a query compound by an ensemble of QSAR models is calculated as the average over all prediction values. For classification and category QSAR, the average prediction value is rounded to the closest integer (which is a class or category number); in the case of imbalanced datasets, rounding can be performed using the moving threshold (see Section 7.2). Predicted average classes or categories (before rounding) that are closer to the nearest integers are considered more reliable. For example, before rounding, one compound has the prediction value of 1.2, but the other has 1.4; hence, both compounds are predicted to belong to class 1. Prediction for the first compound is considered more reliable. Using these prediction values, AD can be defined by a threshold of the absolute difference between the predicted and the rounded predicted activity (unpublished observations).

Another recent publication considering problems of AD definitions and prediction accuracy [25] highlights the following methods [(viii) through (xi)].

viii. Target property should influence the AD, for instance, Ref. [26] recommends using different weights w_n $(n = 1, \ldots, N)$ for descriptors in the calculation of distances between compounds, that is, the distance between compounds i and j is calculated as

$$D_{ij} = \sqrt{\sum_{n=1}^{N} w_n (X_{in} - X_{jn})^2}. \qquad (7.19)$$

The weights in Formula 7.19 could be proportional to the coefficients with which the corresponding descriptors are included in the QSAR model. AD is based on the distances defined by Formula 7.19 rather than the Euclidean distances. Descriptors should be autoscaled.

ix. The distance of a query compound i to the model $D_{i,\text{model}}$ is defined as the smallest Mahalanobis distance between this compound and all compounds of the training set [27]. It has been shown that for an ensemble of the Bayesian regularized neural networks, these distances correlate with the errors of predictions for query compounds.

x. It has been shown that for an ensemble of neural network models, the smaller standard deviation of predicted activities of query compounds corresponds to smaller errors of prediction [28].

xi. Distance to a model and standard deviation were combined in one measure called combined distance, which is the product of the standard deviation and the distance to the model [29]. Different definitions of distances to a model are considered in Ref. [30].

7.6 CONSENSUS PREDICTION

The consensus prediction approach is based on the central limit theorem [31], which can be formulated as follows. If $X_1, X_2, \ldots, X_n$ is a sequence of random variables from the same distribution with mean μ and variance σ^2, then for sufficiently large n, the average $\langle X \rangle$ has approximately normal distribution with mean μ and variance σ^2/n. Thus, ideally, if we use top predictive models with identical predictive power, consensus prediction of a compound's activity is expected to give better (or at least more stable) prediction (or at least more stable) accuracy than each of the individual models. Now let us include more and more predictive models in our consensus prediction. As soon as all these models have comparable predictive power, the accuracy of prediction in terms of the variance will continue to increase. But when we start to include models with lower predictive power, the error of consensus prediction starts to increase and the prediction accuracy and precision typically go down. The consensus prediction of biological activity for an external compound on the basis of several QSAR models is more reliable and provides better justification for the experimental exploration of hits.

External evaluation set compounds are predicted by models that have passed all validation criteria described in Sections 7.3 and 7.4. Each model has its own AD, and each compound is predicted by all models for which it is within the corresponding AD. Actually, each external compound should be within the AD of the training set within the entire descriptor space as well; see Section 7.5 (vi). A useful parameter for consensus prediction is the minimum number (or percentage) of models for which a compound is within the AD; it is defined by the user. If the compound is found within the AD of a relatively small number of models, it is considered to be outside of the AD. Prediction value is the average of predictions by all these models. If the compound is predicted by more than one model, the standard deviation of all predictions by these models is also calculated. For classification and category QSAR, the average prediction value is rounded to the closest integer (which is a class or category number); in the case of imbalanced datasets, rounding can be done using the moving threshold (see Section 7.2).

Predicted average classes or categories (before rounding), which are closer to the nearest integers, are considered more reliable [24]. Using these prediction values, the

AD can be defined by a threshold of the absolute difference between the predicted and the rounded predicted activity. For classification and category QSAR, the same prediction accuracy criteria are used as for the training and test sets (see Formulas 7.8 through 7.17). The situation is more complex for the continuous QSAR. In this case, if the range of activities of the external evaluation set is comparable to that for the modeling set, criteria 7.1 through 7.4 are used. If calculations with multiple external evaluation sets are carried out, and all external evaluation sets are obtained in the manner of external leave-group-out cross-validation, criteria 7.1 through 7.4 can be used as well. Sometimes, however, the external evaluation set may have a much smaller range of activities than the modeling set; hence, it may not be possible to obtain a sufficiently large R^2 value (and other acceptable statistical characteristics) for it. In this case, we recommend using the MAE or the SEP in the following way (Figure 7.2).

i. Build a plot of MAE or SEP against R^2 for the prediction of the test sets by models used in consensus prediction.
ii. On the plot, build straight lines for the threshold R^2 value (see Figure 7.2) and the corresponding MAE or SEP value for the external evaluation set. These lines are parallel to coordinate axes.
iii. If there are many models (e.g., more than 10% of the total number of models used in consensus prediction) in the upper right quadrant defined by the lines built on step (ii), then the prediction of the external evaluation set is acceptable; otherwise it is unacceptable. Actually, the greater the number of models in this area, the better the prediction of the external evaluation set.

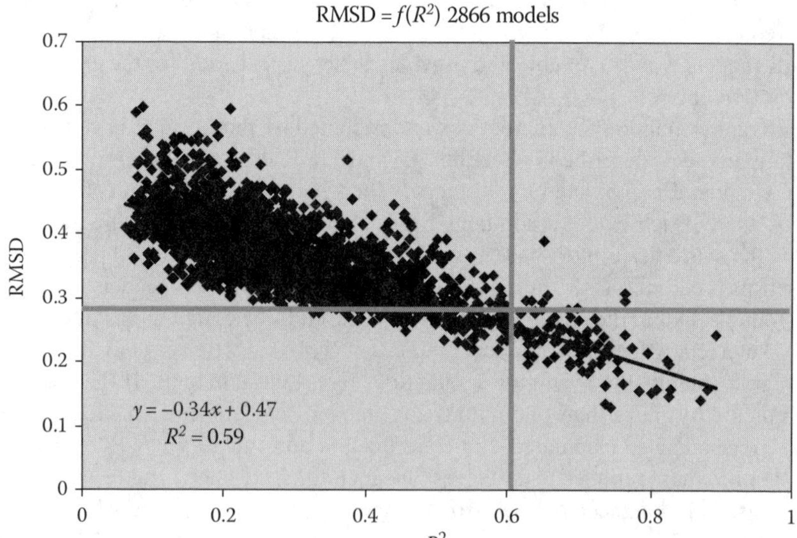

FIGURE 7.2 Prediction power for the external evaluation set.

We have used consensus prediction in many studies [2,3,24,32–35] and have shown that, in most cases, it gives better prediction and coverage than most of the individual predictive models. Thus, we recommend using consensus prediction for virtual screening of chemical databases and combinatorial libraries for finding new lead compounds for drug discovery.

About 2866 models were built for a dataset of anticonvulsants. The dataset consisted of 91 compounds; 15 compounds were randomly selected as the external evaluation set, and a modeling set of 76 compounds was divided into training and test sets using the sphere-exclusion algorithm. Consensus prediction of the external evaluation set by acceptable models (for all these models, LOO cross-validation $q^2 \geq 0.5$ and $R^2 \geq 0.6$ for the test set; other statistics [see Formulas 7.2 through 7.4] were also acceptable) gave SEP $= 0.286$. Is this prediction acceptable? Each dot on the plot corresponds to prediction of a test set by one model built using the corresponding training set. Points corresponding to acceptable models are on the right side of the vertical thick line. The horizontal red line corresponds to SEP $= 0.286$ for the external evaluation set. There are only seven points in the top right quadrant, while there are 110 models with $R^2 \geq 0.6$. Hence, consensus prediction of the external evaluation set is poor.

7.7 CONCLUDING REMARKS

In both Chapters 6 and 7, we have discussed the need for developing reliable QSAR models that could be employed for accurate activity (or property) prediction of external molecules. We have considered some of the most important components of the QSAR modeling workflow. Particular attention was paid to the issue of model validation. It is important to recognize that all the steps of the workflow described in these two chapters should be carried out; skipping any of these steps may lead to models with poor prediction for external compounds, that is, those that were not used in model building or selection, which will essentially undermine the entire exercise. For example, the presence of structural and/or activity outliers in the modeling set may lead not only to poor models, but (what is even worse) to models with overestimated q^2 and R^2 (Figure 7.3). Thus, we cannot agree with the following statement [36]: "Outliers are compounds which are predicted incorrectly—with large residual." Too small training sets may also result in the impossibility to build predictive models, or (what is even worse) in models with delusively good statistics. Too small test sets can lead to statistically insignificant results, because the null hypothesis that the model predicts no better than random selection cannot be rejected with a reasonable significance level. Incorrect target functions for biased datasets can give overestimated (or sometimes underestimated) prediction accuracy. Without rigorous validation of QSAR models including making predictions for compounds in the external evaluation set, there is no empirical evidence that the models are externally predictive at all and that they can be used in virtual screening. Without establishing the model AD, activities of compounds with no similarity to those for which the model was built will be predicted, which makes no sense at all.

The entire process of combi-QSAR modeling can be almost entirely automated. In this scenario, ideally, a user can be given an opportunity to select descriptor collections and model development procedures. Then descriptor matrices (Table 6.1) are

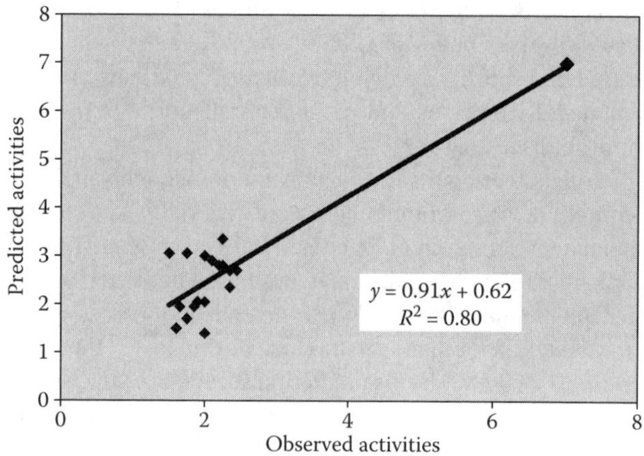

FIGURE 7.3 A typical example when one outlier increases the prediction accuracy. Without the outlier, $R^2 = 0.13$.

obtained (see Section 6.4) and each descriptor matrix is processed by some algorithms (depending on the user's decision and types of descriptors and optimization procedures selected) described in Section 6.5 and possibly other procedures not mentioned here. If necessary, a diverse subset of compounds is selected from the entire modeling set (see Section 6.6). Possible structural and activity outliers are removed from the modeling set (see Section 6.7). Additional procedures may be required for classification and category QSAR modeling, especially, if a dataset is imbalanced or large (see Section 6.8). Each dataset should be divided into modeling and external evaluation sets (see Sections 6.9 and 6.10). For each descriptor set, splitting a modeling set into multiple training and test sets is carried out using a sphere-exclusion algorithm (see Section 6.10). Then the models are developed (see Section 7.1) for all combinations of pairs (descriptor collection, method); appropriate target function and criteria of predictivity are applied (see Section 7.2) based on the nature of the response variable (continuous, classification, or category). The models with acceptable statistics for both training and test sets (for the test set prediction, the respective ADs are used—see Section 7.5), if any, should be rigorously validated by procedures described in Section 6.9 and Sections 7.3 and 7.4, and some other procedures and their ADs should be established (see Section 7.5). Finally, the results of QSAR modeling will be ready to be used for chemical database or virtual library mining.

With more than 40 years of history behind it, QSAR modeling is a well-established research field that (as perhaps with any scientific area) had its ups and downs. There were several recent publications that criticized the current state of the field. Thus, a recent editorial published by the leading chemoinformatics Journal of Chemical Information and Modeling (JCIM; also reproduced by the Journal of Medicinal Chemistry) introduced severe limitations on the level and quality of QSAR papers to be considered acceptable [37]. Another recent editorial opinion by G. Maggiora [38] outlined limitations and some reasons for failures of QSAR modeling that relate to the so-called activity cliffs. In another recent important paper, T. Stouch addressed the question as to why *in silico* ADME/Tox models fail [39]. These examples naturally lead to an important and perhaps critical question as to whether there is any room for further advancement of the field via innovative methodologies and important applications.

Our previous and ongoing research in the area of QSAR suggests that the answer is a resounding "yes." We believe strongly that many examples of low impact QSAR research are due to frequent exploration of datasets of limited size with little attention paid to model *external* validation. This limitation leads to models having questionable "mechanistic" explanatory power but perhaps little if any forecasting ability outside of the training sets used for model development. We believe that the latter ability along with the capabilities of QSAR models to explore chemically diverse datasets with complex biological properties should become the chief focus of QSAR studies. This focus requires the re-evaluation of the success criteria for the modeling as well as the development of novel chemical data mining algorithms and model validation approaches. In fact, we think that the most interesting era in QSAR modeling is just beginning with the rapid growth of the experimental SAR data space [40].

In the last 15 years, innovative technologies that enable rapid synthesis and high-throughput screening of large libraries of compounds have been adopted in almost all major pharmaceutical and biotech companies. As a result, there has been a huge increase in the number of compounds available on a routine basis to quickly screen for novel drug candidates against new targets or pathways. In contrast, such technologies have rarely become available to the academic research community, thus limiting its ability to conduct large-scale chemical genetics or chemical genomics research. The NIH Molecular Libraries Roadmap Initiative has changed this situation by forming the national Molecular Library Screening Centers Network (MLSCN) [41] with the results of screening assays made publicly available via PubChem [42]. These efforts have already led to the unprecedented growth of *available* databases of biologically tested compounds (cf. our recent review where we list about 20 available databases of compounds with known bioactivity [40]). This growth creates new challenges for QSAR modeling such as developing novel approaches for the analysis and visualization of large databases of screening data, novel biologically relevant chemical diversity or similarity measures, and novel tools for virtual screening of compound libraries to ensure high expected hit rates. Due to the significant increase in recent years of the number of publicly available datasets of biologically active compounds and the critical need to improve the hit rate of experimental compound screening, there is a strong need in developing widely accessible and reliable computational QSAR modeling techniques and specific end-point predictors.

REFERENCES

1. Kovatcheva, A., Golbraikh, A., Oloff, S., Xiao, Y. D., Zheng, W., Wolschann, P., Buchbauer, G., and Tropsha, A., Combinatorial QSAR of ambergris fragrance compounds. *J. Chem. Inf. Comput. Sci.* 2004, *44*, 582–595.
2. Kovatcheva, A., Golbraikh, A., Oloff, S., Feng, J., Zheng, W., and Tropsha, A., QSAR modeling of datasets with enantioselective compounds using chirality sensitive molecular descriptors. *SAR QSAR Environ. Res.* 2005, *16*, 93–102.

3. de Cerqueira, L. P., Golbraikh, A., Oloff, S., Xiao, Y., and Tropsha, A., combinatorial QSAR modeling of P-glycoprotein substrates. *J. Chem. Inf. Model.* 2006, *46*, 1245–1254.
4. Wang, X. S., Tang, H., Golbraikh, A., and Tropsha, A., Combinatorial QSAR modeling of specificity and subtype selectivity of ligands binding to serotonin receptors 5HT1E and 5HT1F. *J. Chem. Inf. Model.* 2008, *48*, 997–1013.
5. Golbraikh, A. and Tropsha, A., Beware of Q2! *J. Mol. Graph. Model.* 2002, *20*, 269–276.
6. Saliner, A. G., Netzeva, T. I., and Worth, A.P., Prediction of estrogenicity: Validation of a classification model. *SAR QSAR. Environ. Res.* 2006, *17*, 195–223.
7. Hsieh, J. H., Wang, X. S., Teotico, D., Golbraikh, A., and Tropsha, A., Differentiation of AmpC beta-lactamase binders vs. decoys using classification *k*NN QSAR modeling and application of the QSAR classifier to virtual screening. *J. Comput. Aided Mol. Des.* 2008, *22*, 593–609.
8. Li, Y., Pan, D., Liu, J., Kern, P. S., Gerberick, G. F., Hopfinger, A. J., and Tseng, Y. J., Categorical QSAR models for skin sensitization based upon local lymph node assay classification measures part 2: 4D-Fingerprint three-state and two-2-state logistic regression models. *Toxicol. Sci.* 2007, *99*, 532–544.
9. Efron, B., Bootstrap methods: Another look at the jackknife. *The Annals of Statistics* 1979, *7*, 1–26.
10. Mooney, C. Z. and Duval, R. D., *Bootstrapping. A Non-Parametric Approach to Statistical Inference*. Sage, Newbury Park, CA, 1993.
11. DiCiccio, T. J. and Efron, B., Bootstrap confidence intervals. *Statist. Sci.* 1996, *11*, 189–228.
12. Efron, B., Better bootstrap confidence intervals. *J. Amer. Statist. Assoc.* 1987, *82*, 171–220.
13. Isaksson, A., Wallman, M., Göransson, H., and Gustafsson, M. G., Cross-validation and bootstrapping are unreliable in small sample classification. *Pattern Recogn. Lett.* 2008, *29*, 1960–1965.
14. Wickenberg-Bolin, U., Goransson, H., Fryknas, M., Gustafsson, M. G., and Isaksson, A., Improved variance estimation of classification performance via reduction of bias caused by small sample size. *BMC Bioinform.* 2006, 7, 127.
15. Jaworska, J., Nikolova-Jeliazkova, N., and Aldenberg, T., QSAR applicabilty domain estimation by projection of the training set descriptor space: A review. *Altern. Lab Anim.* 2005, *33*, 445–459.
16. Jaworska, J. and Nikolova-Jeliazkova, N., Review of methods to assess a QSAR applicability domain, http://ambit.acad.bg/nina/publications/2004/AppDomain_qsar04.ppt, 2008.
17. Nikolova-Jeliazkova, N. and Jaworska, J., An approach to determining applicability domains for QSAR group contribution models: An analysis of SRC KOWWIN. *Altern. Lab Anim.* 2005, *33*, 461–470.
18. Netzeva, T. I., Gallegos, S. A., and Worth, A. P., Comparison of the applicability domain of a quantitative structure–activity relationship for estrogenicity with a large chemical inventory. *Environ. Toxicol. Chem.* 2006, *25*, 1223–1230.
19. Fechner, N., Hinselmann, G., Schmiedl, C., and Zell, A., Estimating the applicability domain of Kernel-based QSPR models using classical descriptor vectors. pdf. *Chem. Central J.* 2008, *2*(Suppl. 1), 2.
20. Afantitis, A., Melagraki, G., Sarimveis, H., Koutentis, P. A., Markopoulos, J., and Igglessi-Markopoulou, O., A Novel QSAR model for predicting induction of apoptosis by 4-Aryl-4H-chromenes. *Bioorg. Med. Chem.* 2006, *14*, 6686–6694.
21. Govindraju, V., Pattern Recognition Review (1)—Definitions, Bayesian Classifiers, http://www.cedar.buffalo.edu/~govind/CSE666/fall2007/Pattern_recognition_lecture_slides_1.pdf, 2008.

22. Guha, R., Dutta, D., Jurs, P. C., and Chen, T., *R*–NN curves: An intuitive approach to outlier detection using a distance based method. *J. Chem. Inf. Model.* 2006, *46*, 1713–1722.
23. Kuhne, R., Ebert, R. U., and Schuurmann, G., Model selection based on structural similarity-method description and application to water solubility prediction. *J. Chem. Inf. Model.* 2006, *46*, 636–641.
24. Zhang, L., Zhu, H., Oprea, T. I., Golbraikh, A., and Tropsha, A., QSAR modeling of the blood–brain barrier permeability for diverse organic compounds. *Pharm. Res.* 2008, *25*, 1902–1914.
25. Tetko, I. V., Bruneau, P., Mewes, H. W., Rohrer, D. C., and Poda, G. I., Can we estimate the accuracy of ADME-tox predictions? *Drug Discov. Today* 2006, *11*, 700–707.
26. Netzeva, T. I., Worth, A., Aldenberg, T., Benigni, R., Cronin, M. T., Gramatica, P., Jaworska, J. S., et al., Current status of methods for defining the applicability domain of (Quantitative) structure–activity relationships. The Report and Recommendations of ECVAM Workshop 52. *Altern. Lab Anim.* 2005, *33*, 155–173.
27. Manallack, D. T., Tehan, B. G., Gancia, E., Hudson, B. D., Ford, M. G., Livingstone, D. J., Whitley, D. C., and Pitt, W. R., A consensus neural network-based technique for discriminating soluble and poorly soluble compounds. *J. Chem. Inf. Comput. Sci.* 2003, *43*, 674–679.
28. Bruneau, P. and McElroy, N. R., LogD7.4 Modeling using Bayesian regularized neural networks. Assessment and correction of the errors of prediction. *J. Chem. Inf. Model.* 2006, *46*, 1379–1387.
29. Bruneau, P., Search for predictive generic model of aqueous solubility using Bayesian neural nets. *J. Chem. Inf. Comput. Sci.* 2001, *41*, 1605–1616.
30. Tetko, I. V., Sushko, I., Pandey, A. K., Zhu, H., Tropsha, A., Papa, E., Oberg, T., Todeschini, R., Fourches, D., and Varnek, A., Critical assessment of QSAR models of environmental toxicity against tetrahymena pyriformis: Focusing on applicability domain and overfitting by variable selection 1. *J. Chem. Inf. Model.* 2008, *48*, 1733–1746.
31. Sachs, L., *Applied Statistics: A Handbook of Techniques*. Springer, New York, 1984.
32. Shen, M., Beguin, C., Golbraikh, A., Stables, J. P., Kohn, H., and Tropsha, A., Application of predictive QSAR models to database mining: Identification and experimental validation of novel anticonvulsant compounds. *J. Med. Chem.* 2004, *47*, 2356–2364.
33. Votano, J. R., Parham, M., Hall, L. H., Kier, L. B., Oloff, S., Tropsha, A., Xie, Q., and Tong, W., Three new consensus QSAR models for the prediction of Ames genotoxicity. *Mutagenesis* 2004, *19*, 365–377.
34. Zhang, S., Wei, L., Bastow, K., Zheng, W., Brossi, A., Lee, K. H., and Tropsha, A., Antitumor agents 252. Application of validated QSAR models to database mining: Discovery of novel tylophorine derivatives as potential anticancer agents. *J. Comput. Aided Mol. Des.* 2007, *21*, 97–112.
35. Zhu, H., Tropsha, A., Fourches, D., Varnek, A., Papa, E., Gramatica, P., Oberg, T., Dao, P., Cherkasov, A., and Tetko, I. V., Combinatorial QSAR modeling of chemical toxicants tested against tetrahymena pyriformis. *J. Chem. Inf. Model.* 2008, *48*, 766–784.
36. Vasanthanathan, P., Lakshmi, M., Arockia, B. M., Gupta, A. K., and Kaskhedikar, S. G., QSAR study of 3-phenyl-5-acyloxymethyl-2H,5H-furan-2-ones as antifungal agents: The dominant role of electronic parameter. *Chem. Pharm. Bull. (Tokyo)* 2006, *54*, 583–587.
37. Jorgensen, W. L. and Tirado-Rives, J., QSAR/QSPR and proprietary data. *J. Chem. Inf. Model.* 2006, *46*, 937.
38. Maggiora, G. M., On outliers and activity cliffs—why QSAR often disappoints. *J. Chem. Inf. Model.* 2006, *46*, 1535.

39. Stouch, T. R., Kenyon, J. R., Johnson, S. R., Chen, X. Q., Doweyko, A., and Li, Y., *In silico* ADME/Tox: Why models fail. *J. Comput. Aided Mol. Des.* 2003, *17*, 83–92.
40. Oprea, T. and Tropsha, A., Target, chemical and bioactivity databases—integration is key. *Drug Discov. Today* 2006, *3*, 357–365.
41. Austin, C. P., Brady, L. S., Insel, T. R., and Collins, F. S., NIH molecular libraries initiative. *Science* 2004, *306*, 1138–1139.
42. PubChem, http://pubchem.ncbi.nlm.nih.gov/, 2009.

8 Structure Enumeration and Sampling

Markus Meringer

CONTENTS

Chemical structure enumeration and sampling have been studied by mathematicians, computer scientists, and chemists for quite a long time. Given a molecular formula plus, optionally, a list of structural constraints, the typical questions are: (1) How many isomers exist? (2) Which are they? And, especially if (2) cannot be answered completely: (3) How to get a sample?

In this chapter, we describe algorithms for solving these problems. The techniques are based on the representation of chemical compounds as molecular graphs (see Chapter 2), that is, they are mainly applied to constitutional isomers. The major problem is that *in silico* molecular graphs have to be represented as labeled structures, while in chemical compounds, the atoms are not labeled. The mathematical concept for approaching this problem is to consider orbits of labeled molecular graphs under the operation of the symmetric group. We have to solve the so-called *isomorphism problem*.

According to our introductory questions, we distinguish several disciplines: counting, enumerating, and sampling isomers. While counting only delivers the number of isomers, the remaining disciplines refer to constructive methods. Enumeration typically encompasses exhaustive and non-redundant methods, while sampling typically lacks these characteristics. However, sampling methods are sometimes better suited to solve real-world problems.

There is a wide range of applications where counting, enumeration, and sampling techniques are helpful or even essential. Some of these applications are closely linked to other chapters of this book. Counting techniques deliver pure *chemical information*; they can help to estimate or even determine sizes of chemical databases or compound libraries obtained from combinatorial chemistry.

Constructive methods are essential to *structure elucidation* systems (see Chapter 9). They are used to generate structures that fulfill structural restrictions obtained from spectroscopy in a pregeneration step, while in a postgeneration step, virtual spectra of the generated structures can be computed and compared with the measured data in order to determine which of them achieves the best fit.

Other applications use structure enumeration algorithms in order to produce candidate structures for *virtual screening* (see Chapter 5). *Structure–activity* and *structure–property relationships*, as introduced in Chapter 6, can be used in combination with structure enumeration or sampling as rudimentary approaches toward inverse QSAR (see Chapter 10) and *de novo design* algorithms often have their roots in conventional structure generation.

The nonquantitative aspects of reaction network generation (see Chapter 11) are also based on methods similar to those used for isomer enumeration.

8.1 ISOMER COUNTING

Counting means that only the number of structures is calculated and the structures themselves are not produced by the algorithm. The most powerful counting technique available to chemists is Pólya's theorem [1], see also Refs. [2,3]. There are, of course, various predecessors, for example a paper by Lunn and Senior [4], who were the first to note that group theory plays a role, and a paper by Redfield [5] that contained even better results. However, Pólya's paper gave rise to the development of a whole theory that is nowadays called *Pólya's Theory of Counting*. Typical applications are counting of *permutational isomers* and *acyclic compounds*.

8.1.1 Counting Permutational Isomers

Pólya's approach to the enumeration of molecules with a given molecular formula is to subdivide the molecule in question into a *skeleton* and a set of *univalent substituents*. It leads to the following challenge: Evaluate the set of essentially different distributions of the substituents over the sites of the skeleton with respect to the given *symmetry group* of the skeleton.

In mathematical terms, the symmetry group G acts on the set of mappings m^n from the n sites of the skeleton onto the m available substituents. The set of orbits under this group operation, $m^n//G$, is in a one-to-one relation with the different constitutions.

If the skeleton shows no symmetries, that is, if G is of order 1, then it is clear that there are m^n different substitutions. Note that m^n has two different meanings, one for denoting the set of mappings and another for its cardinality. If the order of G is larger than 1, the situation is more interesting.

The resulting isomers are called *permutational* or *substitutional* isomers. For example the 22 permutational isomers of dioxin (tetrachlorodibenzo-*p*-dioxin) are the essentially different distributions of four hydrogen and four chlorine atoms over the eight sites of the skeleton depicted on the left:

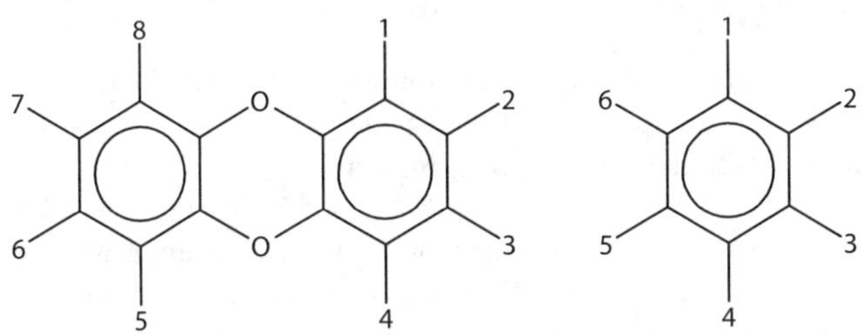

Counting these isomers is described in detail in Kerber's comprehensive book [6] on finite group actions. In this section we will discuss the example of permutational isomers of dichlorobenzene, which are based on the benzene skeleton sketched on the right.

First we will try to describe Pólya's approach in general. The symmetry group G of the skeleton with respect to the n binding sites is required as input. The procedure works with the topological symmetry group as well as with the geometrical symmetry group. Of course, the results might differ. For ways to compute a molecule's symmetry group, see Chapter 2. A suitable data structure for representing groups is described by Sims [7].

The reader should be familiar with the cycle notation of permutations, which is briefly described in Example 8.1.1. At this point, it is useful to know that every permutation has a unique decomposition into disjoint cycles. For more details of permutations and cycles, the reader is referred to Ref. [6].

The *cycle index* of a permutation $g \in G$ is a monomial in variables z_k. It is defined as

$$Z(g) = \prod_{k=1}^{n} {z_k}^{c_k(g)}, \tag{8.1}$$

where $c_k(g)$ is the number of cycles of g having length k. The cycle index $Z(G)$ of G is the averaged sum of cycle indices of group elements:

$$Z(G) = \frac{1}{|G|} \sum_{g \in G} \prod_{k=1}^{n} {z_k}^{c_k(g)}. \qquad (8.2)$$

In order to obtain a *counting series* from the cycle index, a so-called *generating function* has to be inserted. Having $m > 1$ different chemical elements to choose from, a suitable generating function is $y_1 + \cdots + y_m$. Inserting the generation function into the cycle index means that every occurrence of z_k in $Z(G)$ is replaced by ${y_1}^k + \cdots + {y_m}^k$. If $m = 2$ the easier generation function $1 + x$ can be used instead, and during insertion, z_k is replaced by $1 + x^k$.

The coefficient of the monomial $\prod_{i=1}^{m} y_i^{j_i}$ of the counting series equals the number of isomers with j_i substituents of type i. In the case where we have just two different elements to substitute, say H and Cl, the coefficient of x^j equals the number of isomers with j hydrogen atoms. To summarize, we can formulate the following.

ALGORITHM 8.1.1 COUNTING PERMUTATIONAL ISOMERS BY PÓLYA'S THEOREM

1. Calculate the cycle index of the skeleton with respect to the binding sites
2. Insert the generation function into the cycle index
3. Evaluate the coefficients of the counting series

This formal description of how to solve the counting problem in general needs to be illustrated by an example. Counting $C_6H_4Cl_2$ constitutions with a benzene skeleton will be explained step by step.

Example 8.1.1: Counting Isomers of Dichlorobenzene

Below we see the benzene skeleton together with its symmetry axes (1), ..., (6).

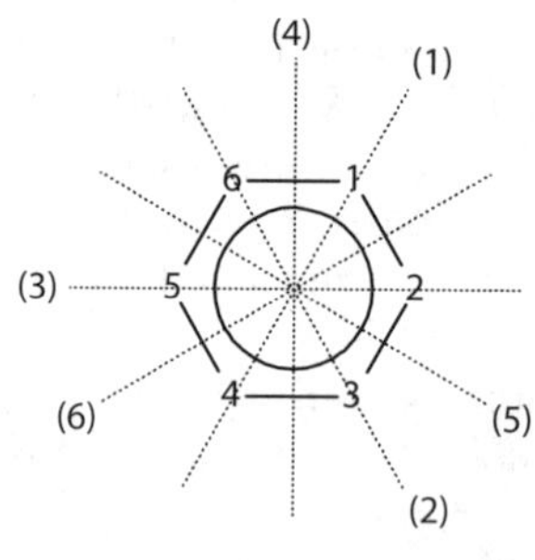

Besides six axis reflections, the benzene skeleton allows some more symmetry operations, namely five proper rotations. Table 8.1 lists all the symmetry operations and the according permutations of benzene's symmetry group D_{6h}. E denotes the identity, $C_i^{+/-}$ represents a rotation by $360/i$ degrees, where the sign describes the direction of the rotation, and $\sigma_v^{(j)}$ represents a reflection at axis (j).

TABLE 8.1
Permutations of the Automorphism Group D_{6h} of Benzene

Operation	List Representation	Cycle Representation	Cycle Index
E	[1 2 3 4 5 6]	(1)(2)(3)(4)(5)(6)	$z_1{}^6$
C_6^+	[2 3 4 5 6 1]	(1 2 3 4 5 6)	$z_6{}^1$
C_6^-	[6 1 2 3 4 5]	(6 5 4 3 2 1)	$z_6{}^1$
C_3^+	[3 4 5 6 1 2]	(1 3 5)(2 4 6)	$z_3{}^2$
C_3^-	[5 6 1 2 3 4]	(5 3 1)(6 4 2)	$z_3{}^2$
C_2	[4 5 6 1 2 3]	(1 4)(2 5)(3 6)	$z_2{}^3$
$\sigma_v^{(1)}$	[1 6 5 4 3 2]	(1)(4)(2 6)(3 5)	$z_1{}^2z_2{}^2$
$\sigma_v^{(2)}$	[5 4 3 2 1 6]	(3)(6)(1 5)(2 4)	$z_1{}^2z_2{}^2$
$\sigma_v^{(3)}$	[3 2 1 6 5 4]	(2)(5)(1 3)(4 6)	$z_1{}^2z_2{}^2$
$\sigma_v^{(4)}$	[6 5 4 3 2 1]	(1 6)(2 5)(3 4)	$z_2{}^3$
$\sigma_v^{(5)}$	[4 3 2 1 6 5]	(1 4)(2 3)(5 6)	$z_2{}^3$
$\sigma_v^{(6)}$	[2 1 6 5 4 3]	(1 2)(3 6)(4 5)	$z_2{}^3$

Permutations are given in two notations. The list representation might appear more straightforward to the reader, because for a permutation π the ith component in the list simply defines the image of i, that is the list representation of π is $[\pi(1), \ldots, \pi(n)]$.

The cycle notation is a little more difficult to understand, but gives direct access to the cycle index, which is needed to compute counting series. The cycle representation consists of 1 to n cycles, which are enclosed by round brackets. Each cycle itself consists of 1 to n elements. Cycles of only one element (i) show that i is fixed under the permutation, that is, $\pi(i) = i$. Sometimes such cycles are even suppressed in cycle notations. Cycles $(i_1, \ldots, i_l)$ with more than one element indicate that i_1 is mapped onto i_2, i_2 is mapped onto i_3, and so on. The length l of the cycle is determined by the minimum number of applications of π that map i again onto i, that is, $l = \min\{h : \pi^h(i) = i\}$. In particular, this means that the last element of the cycle, i_l, is mapped onto i_1, and generally the cycle of i_1 can be written as $[i_1, \pi(i_1), \pi^2(i_1), \ldots, \pi^{l-1}(i_1)]$.

Finally, the cycle indices of the elements of the automorphism group can be derived directly from the cycle notations using Equation 8.1. This results in the cycle index

$$Z(D_{6h}) = \tfrac{1}{12}(z_1{}^6 + 4z_2{}^3 + 2z_3{}^2 + 2z_6{}^1 + 3z_1{}^2z_2{}^2)$$

for benzene's automorphism group D_{6h}. Inserting the generating function $1 + x$ leads to the counting series

$$\begin{aligned} C(D_{6h}) &= \tfrac{1}{12}((1+x)^6 + 4(1+x^2)^3 + 2(1+x^3)^2 \\ &\quad + 2(1+x^6) + 3(1+x)^2(1+x^2)^2) \\ &= 1 + x + 3x^2 + 3x^3 + 3x^4 + x^5 + x^6. \end{aligned}$$

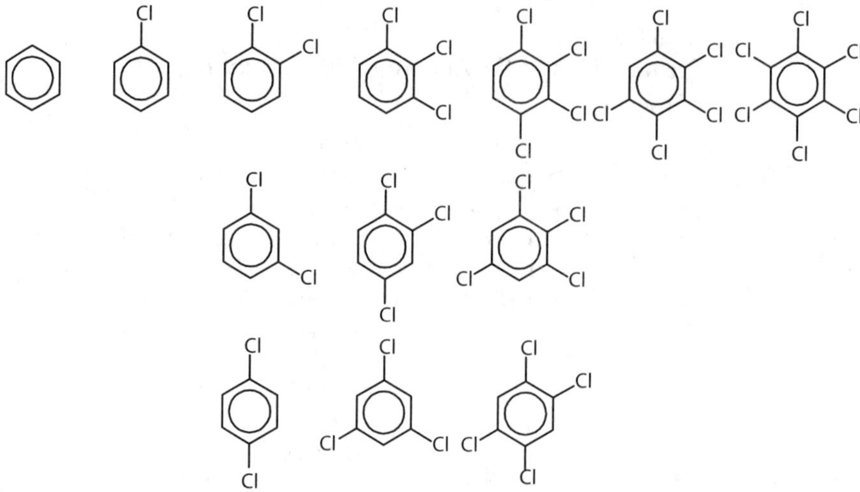

FIGURE 8.1 Thirteen different substitutions of a benzene skeleton with H and C.

The coefficient of x^i in the counting series indicates the number of isomers with i hydrogen and $n - i$ chlorine atoms. Thus we obtain the number of isomers of benzene (one isomer according to coefficient 1 of x^0), chlorobenzene (one isomer according to coefficient 1 of x^1), dichlorobenzene (three isomers according to coefficient 3 of x^2), and so on. This sums up to 13 different substitutions of the benzene skeleton with H and Cl. These 13 compounds are shown in Figure 8.1.

But note that a counting series itself gives no hint of the structures of the counted isomers, that is, as soon as there is more than one isomer found, Pólya's theorem does not show how to attach the substituents to the skeleton in order to obtain all isomers. For this purpose we need constructive methods based on the principles of double cosets developed by Ruch, Klein and others [8–10].

Example 8.1.2: Cycle Indices and Counting Series

In the following we list cycle indices of several benzenoid hydrocarbons, together with their counting series obtained by substituting $1 + x$.

- *Naphthalene:* $Z(D_{2h}) = \frac{1}{4}({z_1}^8 + 3{z_2}^4)$, $C(x) = 1 + 2x + 10x^2 + 14x^3 + 22x^4 + 14x^5 + 10x^6 + 2x^7 + x^8$.
- *Anthracene:* $Z(D_{2h}) = \frac{1}{12}({z_1}^{10} + {z_1}^2{z_2}^4 + 2{z_2}^5)$, $C(x) = 1 + 3x + 15x^2 + 32x^3 + 60x^4 + 66x^5 + 60x^6 + 32x^7 + 15x^8 + 3x^9 + x^{10}$.
- *Phenanthrene:* $Z(C_{2v}) = \frac{1}{2}({z_1}^{10} + {z_2}^5)$, $C(x) = 1 + 5x + 25x^2 + 60x^3 + 110x^4 + 126x^5 + 110x^6 + 60x^7 + 25x^8 + 5x^9 + x^{10}$.
- *Tetracene:* $Z(D_{2h}) = \frac{1}{4}({z_1}^{12} + 2{z_2}^6)$, $C(x) = 1 + 3x + 21x^2 + 55x^3 + 135x^4 + 198x^5 + 236x^6 + 198x^7 + 125x^8 + 55x^9 + 21x^{10} + 3x^{11} + x^{12}$.
- *Triphenylene:* $Z(D_{3h}) = \frac{1}{6}({z_1}^2 + 2{z_2}^6 + 2{z_3}^4)$, $C(x) = 1 + 2x + 14x^2 + 38x^3 + 90x^4 + 132x^5 + 166x^6 + 132x^7 + 90x^8 + 38x^9 + 14x^{10} + 2x^{11} + x^{12}$.

We obtain the same cycle index for naphthalene as for the introductory sample of dioxin. We see that the monomial x^4 has the coefficient 22, that is, there are 22 isomers of dioxin.

It follows from Formulas 8.1 and 8.2 that the total number of different substitutions with respect to G can be computed as

$$|m^n//G| = \frac{1}{|G|}\sum_{g\in G} m^{c(g)}, \tag{8.3}$$

where $c(g)$ denotes the number of cycles of g. We will use this later in Section 8.4.2 for counting constituents of combinatorial libraries, which is closely related to counting permutational isomers.

Van Almsick et al. [11] developed a software tool that calculates the number of permutational isomers using Pólya's approach. Computer algebra systems, such as commercial implementations Mathematica and Maple, or the open source system SYMMETRICA [12] are also able to conduct computations following Pólya's theory, but without any special adaptions to chemistry. The computation of numbers of permutational isomers using SYMMETRICA is available online at symmetrica.uni-bayreuth.de/perm_iso.html (accessibility checked on December 2009).

8.1.2 Counting Isomers of Acyclic Structures and Other Compound Classes

Besides permutational isomers, counting series for several other compound classes have been discovered in the past. However, in contrast to permutational isomers, these cannot be produced using a well-defined algorithm. They were rather individual ideas that led to these counting series. Counting series are known especially for the most prominent acyclic compound classes. Most of them were derived by applying Pólya's theorem in a recursive manner; that is, counting series themselves were used as generating functions.

Alkyl groups have the form $-C_nH_{2n+1}$. They can be interpreted as rooted trees on n nodes, where the root is the carbon atom with the free valence. Let $A_n(x)$ denote the counting series for alkyl groups having n atoms. There is a recursive formula

$$A_n(x) = 1 + \frac{1}{6}x[A_{n-1}{}^3(x) + 3A_{n-1}(x)A_{n-1}(x^2) + 2A_{n-1}(x^3)] \tag{8.4}$$

starting with $A_0(x) = 1$. For $n \to \infty$, the counting series for alkyl groups is often written as

$$A(x) = \sum_{n=0}^{\infty} A_n x^n$$

with certain coefficients calculated from the recursive Equation 8.4. The first terms are

$$A(x) = 1 + x + x^2 + 2x^3 + 4x^4 + 8x^5 + 17x^6 + 39x^7 + 89x^8 + 211x^9 + 507x^{10} + \ldots$$

Based on this recursive approach, several counting series for other acyclic compound classes have been formulated by Read [13]:

- *Primary alcohols:* R–CH_2–OH with an alkyl group R: $xA(x)$
- *Secondary alcohols:* R^1–CH(R^2)–OH with two alkyl groups R^1 and R^2: $\frac{1}{2}x[A^2(x) - 2A(x) + A(x^2)]$
- *Tertiary alcohols:* R^1–C(R^2)(R^3)–OH with alkyl groups R^1, R^2 and R^3: $\frac{1}{6}x[A^3(x) - 3A^2(x) + 3A(x)A(x^2) - 3A(x^2) + 2A(x^3)]$
- *Aldehydes and ketones:* R^1–C(=O)–R^2 with alkyl groups or hydrogen atoms R^1 and R^2: $\frac{1}{2}x[A^2(x) + A(x^2)]$
- *Alkynes:* R^1–C≡C–R^2 with alkyl groups or hydrogen atoms R^1 and R^2: $\frac{1}{2}x^2[A^2(x) + A(x^2)]$
- *Esters:* R^1–C(=O)–O–R^2 with alkyl groups R^1 and R^2 where R^1 can also be a hydrogen atom: $xA(x)[A(x) - 1]$

Perhaps the most important counting series for acyclic compounds is the one for alkanes, that is, compounds with formula C_nH_{2n+2}. It has been determined as

$$a(x) = \frac{1}{24}x[A^4(x) + 6A^2(x)A(x^2) + 3A^2(x^2) + 8A(x)A(x^3) + 6A(x^4)] - \frac{1}{2}[(A(x))^2 - A(x^2) + 1],$$

and the first terms are

$$a(x) = 1 + x + x^2 + x^3 + 2x^4 + 3x^5 + 5x^6 + 9x^7 + 18x^8 + 35x^9 + 75x^{10} + \ldots$$

Other compound classes for which several counting series are known are benzenoids and polyhex hydrocarbons. The review of Faulon et al. [14] offers an extensive overview of these counting series and on how they were deduced.

Although there is a counting series known for simple graphs on n nodes, no general counting series for molecular graphs a with given molecular formula has been found yet. An approach for counting cubic graphs is presented [15]. The relationship between cubic and molecular graphs might not be very obvious at first sight, but will become clearer in Section 8.2.1. Recently, another small step towards a more universal counting series was found for hydroxyl ethers [16], that is, isomers with molecular formula $C_iH_{2i+2}O_j$.

Up to now, the only way to calculate the number of isomers belonging to an arbitrary molecular formula is to use structure generators. Structure generators not only calculate the number of isomers, but also deliver the structures themselves as output. On the other hand, counting series always a good choice to prove the correctness of new structure generator results. In the next section, we will get to know the algorithmic concepts underlying past and present structure generators.

8.2 ISOMER ENUMERATION: DETERMINISTIC STRUCTURE GENERATION

The construction of all constitutional isomers having the same molecular formula has a long history, which will and cannot be reported in detail here. Just as the representation of chemical compounds as graphs was one of the roots of graph theory, their generation was one of the challenges for the development of construction algorithms for computers.

A prominent starting point is the well-known DENDRAL system [17], the development of which began already in the middle and late sixties of the last century. DENDRAL (short for DENDRIC ALgorithm) was developed for the automated structure elucidation of organic compounds by mass spectrometry (MS). For that purpose DENDRAL was endowed with an isomer generator that was able to process structural constraints obtained from MS (especially the molecular formula) and from other spectroscopic methods, in particular nuclear magnetic resonance (NMR) spectroscopy.

DENDRAL is described in many computer science books as the first expert system. Moreover, it can be considered as one of the roots of chemoinformatics. Interestingly, even the NASA was among the founders of this pioneering project, with the ambitious intention to supply future Mars missions with such software, to enable analysis and interpretation of MS samples onboard a space probe and to broadcast only identified structural formulas back to the earth instead of huge gas chromatography/mass spectrometry (GC/MS) data sets.

8.2.1 Early Cyclic and Acyclic Structure Generators

At first, only acyclic structures could be constructed until there was a breakthrough in the early 1970s when a decomposition of the given molecular formula into those of cyclic substructures was found. Cyclic substructures had to be combined by bridges to get molecules with the prescribed molecular formula. All possible decompositions of this kind could be determined by appropriate mathematical theorems prior to constructing these cyclic substructures.

8.2.1.1 Acyclic Structure Generators

Henze and Blair [18] used the fact that a unique centroid can be found in any chemical tree for the enumeration of alkanes as early as the 1930s. The unique *centroid* is the starting point for a canonical labeling of the tree, following simple rules of precedence of the constituent radicals according to their composition and topological structure. An unambiguous notational system was established by Lederberg [19].

However, the existence of a unique (bi)centroid in a tree on n nodes had already been formulated a century earlier in Jordan's theorem [20]:

- For odd $n = 2k + 1$, there exists a unique node, called *centroid*, such that all incident subtrees have at most k nodes
- For even $n = 2k$ there exists either
  - a unique node such that all incident subtrees have less than k nodes, or
  - a unique edge, called *bicentroid* or centroid edge, such that the incident subtrees have exactly k nodes

This theorem shows a recursive way to generate trees. A tree on $n = 2k + 1$ nodes is composed from one node (the centroid, having degree d) and d rooted trees with less than k nodes in such a way that the sum of nodes is $n - 1$. In terms of acyclic chemical graphs (without multiple bonds) one will have to loop over all different atoms as centroid, partition the remaining atoms into subsets according to the centroid's valency, and then iterate this procedure on the subsets (with the small difference that now rooted trees have to be built). The iteration ends when no more partitioning is possible. The case of odd numbers of nodes can be processed similarly, with the variation that two atoms have to be chosen for a bicentroid. Reference [21] offers a pseudo code for such an algorithm applied to alkanes.

An implementation with respect to general chemical trees was part of the DENDRAL system. Results of this acyclic generator have been published in Ref. [22].

8.2.1.2 Cyclic Structure Generator

Approaching the challenge of cyclic structures, Lederberg introduced a series of simplification steps that finally (apart from certain exceptions) showed a mapping from cyclic structures on certain classes of cubic graphs [23].

These initial ideas developed into a structure generator described by Masinter et al. [24,25], which was the first generator that covered both acyclic and cyclic structures. The fundamental ideas of this structure generator will be described below. First, some terminology is required.

Chapter 2 introduced molecular graphs as representations of chemical compounds. In Figure 8.2 we see such a representation **1** of a substituted piperazine. The *chemical graph* **2** ignores hydrogen. The symbol U is used in the composition to denote the number of unsaturations. The number of unsaturations u is computed from the molecular formula as follows:

$$u = \frac{1}{2}\left(2 + \sum_{i=1}^{k}(i-2)a_i\right), \tag{8.5}$$

where a_i denotes the number of atoms of valence i and k is the maximum valence of the composition.

An atom of a chemical graph is called *cyclic* if it lies on a cycle (or ring); otherwise it is called *acyclic*. This way, a chemical graph can be separated into cyclic and acyclic

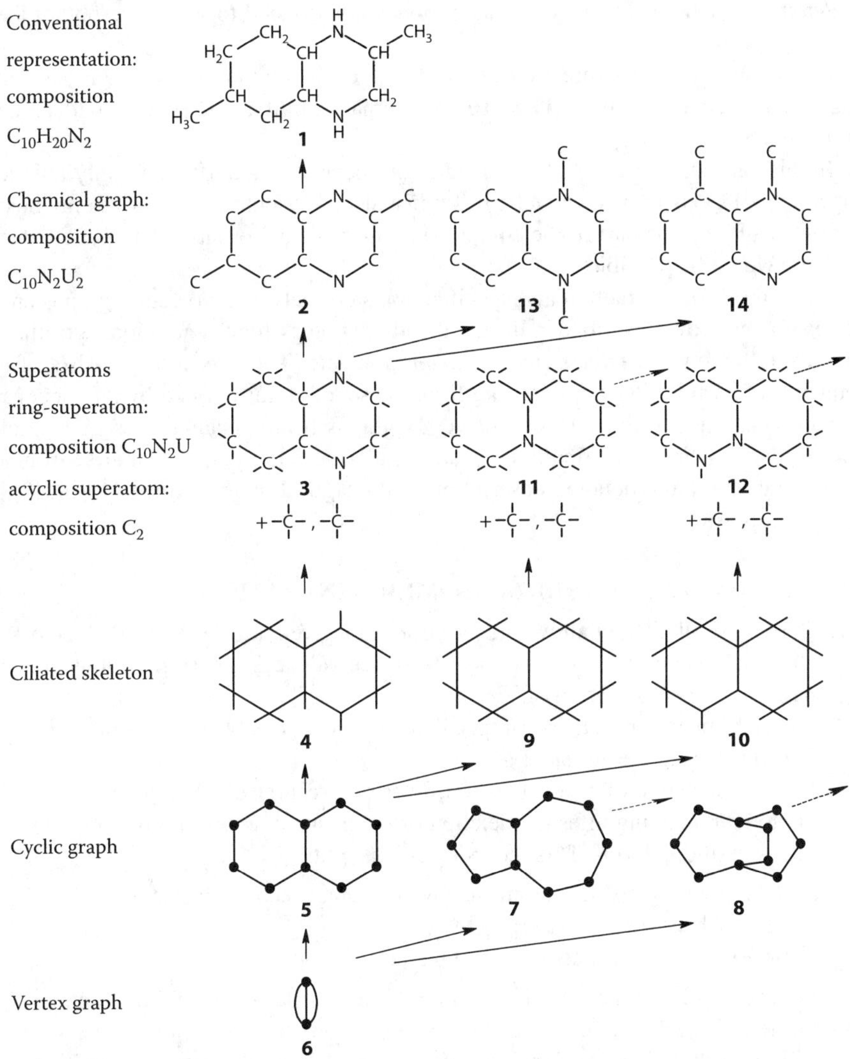

FIGURE 8.2 Examples of abstraction and refinement steps used in the generation of cyclic structures.

parts. Connected components of the chemical graph induced by the cyclic atoms are called *superatoms*. Graph-theoretically, a superatom is a connected isthmus-free multigraph (short cif-graph), that is, with no edge whose deletion would disconnect the graph. The number of free valences of the superatom is determined by the number of connections to atoms outside the superatom. The chemical graph **2** is composed of the superatom **3**, having 16 free valencies, and two acyclic carbon atoms.

The *ciliated skeleton* **4** is obtained from **3** by stripping the element symbols. A further step of abstraction is the deletion of free valences, resulting in the *cyclic* *skeleton* **5**. Finally, if chains of bivalent nodes are reduced to edges we obtain the *vertex graph* **6**.

Going into the reverse direction, starting from **6**, two alternative cyclic graphs that can be obtained are **7** and **8**. **9** and **10** are alternative cilitated skeletons that can be built from **5**.

In this particular example, the valencies of nodes in **9** and **10** allow only unique superatoms **11** and **12**, respectively. If, for instance, 4-valent sulfur or 3-valent phosphorus would also be part of the composition, more than one superatom per ciliated skeleton would be possible.

The scheme of abstraction and specification steps between molecular graphs and vertex graphs above described indicates already a strategy for a generation algorithm, roughly following the *Divide and Conquer* principle. The algorithm consists of a sequence of partitioning steps starting from the set of atoms defined by a molecular formula that leads to the selection of vertex graphs from a catalog. A sequence of consecutive labeling steps finally reconstructs all molecular graphs that arise from a vertex graph. A more detailed description of the algorithm as outlined in Ref. [25] reads as follows:

ALGORITHM 8.2.1 DENDRAL ISOMER GENERATION

1. Determine all distinct allowable partitions of a given degree sequence V into atoms and superatom sets with assigned free valences. These partitions are based on the unsaturation of V.
2. For each superatom set, determine all the distinct allowable allocations of the free valences to the atoms of the set.
3. For each such free valence allocation, determine recursively the allowable sets of atoms remaining after the deletion of the bivalent atoms and the pruning of any resulting loops. This recursion is done until
    a. the remaining bivalent atoms in any cif-graph based on the set must all be on edges, or
    b. one of two special cases is encountered.
4. For each such set of atoms, if condition (a) terminates the recursion, look up in the catalog all the cif-graphs based on the nonbivalent atoms in the set, and for each such graph, label the edges with the bivalent atoms. If condition (b) terminates the recursion, directly write down the allowable graphs.
5. For each such graph, recursively label the atoms with loops and the loops and edges with bivalent atoms.
6. For each graph so obtained, label the atoms with the free valences.
7. For each set of atoms and superatoms obtained as above, use the tree generator to construct all the nonisomorphic connected multigraphs based on these atoms and superatoms.

This algorithm uses several subroutines that cannot be described in detail here. The *superatom partitioner* (step 1), the *free valence partitioner* (2), the *loop-bivalent partitioner* (3) with the definition of the special cases (b), the look-up routine from the catalog (4), the *loop-bivalent labeler* (5) and the *free-valence labeler* are subject

TABLE 8.2
Allowed Partitions of C_6U_3 into Superatom Pots and the Remaining Pot

		Superatom Pot			
Partition	Superatom Pots	1	2	3	Remaining Pot
1	1	C_6U_3			
2	1	C_5U_3			C
3	1	C_4U_3			C_2
4	1	C_3U_3			C_3
5	2	C_4U_2	C_2U		
6	2	C_3U_2	C_2U		C
7	2	C_2U_2	C_2U		C_2
8	2	C_4U	C_2U_2		
9	2	C_3U	C_2U_2		C
10	2	C_3U_2	C_3U		
11	3	C_2U	C_2U	C_2U	

to Ref. [25] and the references cited therein. Especially for the labeling steps, see Refs. [26–28].

Example 8.2.1: Superatom Partitioner

Step (1) of Algorithm 8.2.1.2 will be illustrated here. Table 8.2 shows the results of the superatom partitioner for C_6H_8. Firstly, hydrogens are replaced by unsaturations U. According to Equation 8.5, C_6H_8 has $u = 3$ unsaturations. A total of 11 allowed partitions of up to three superatoms are obtained.

According to the terminology introduced in Figure 8.2, the results of step (5) are cyclic graphs, at step (6) cilitated skeletons are obtained, and step (7) delivers chemical graphs. Step (7) is also described in Ref. [24]. However, some words on the treatment of superatoms are appropriate.

In the final step, superatoms require some special treatment in the tree generator. If a superatom A has k free valences, then in forming molecular structures that include A, A behaves differently from an atom of valence k. The difference in forming structures including A and those including an atom of valence k is the following: the k free valences on an atom of valence k are, as edge endpoints in a graph, indistinguishable, that is, the free valences on the atom admit as symmetry group the group S_k, the full permutation group on k objects. However, the k free valences on the superatom A are usually distinguishable from a symmetry viewpoint, so the free valences on A will, in general, admit only a subgroup of S_k.

The structure generator outlined here has become popular under the name CONGEN (short for CONstrained GENerator) and was used within the DENDRAL project until it was finally replaced by the advanced generator GENOA [29] (short for GENeration with Overlapping Atoms).

From today's point of view it is remarkable that a project like DENDRAL could be successfully realized. Computers were slow at that time and extremely limited in memory. Programming languages were still on a low level and software engineering was hardly recognized as a new technological discipline. However, mathematically it was state-of-the-art. But the various partitioning and labeling steps implicate a problem: it is difficult to process structural constraints efficiently. Efficiency means that constraints can already be tested during structure generation, help us to reduce intermediate results and speed up the enumeration process. Among others, this feature will be the subject of the next subsection.

8.2.2 Orderly Generation

There was a development by Read [30] and Faradzev [31,32] who both presented the technique of orderly generation independently in 1978. In this technique an artificial ordering is imposed on the set of graphs that are to be generated, such that the smallest representative of a given isomorphism type always contains a subgraph that is the smallest representative of its isomorphism type. Thus, only the smallest representatives have to be extended and the results have to be tested for being the smallest again.

This approach allowed avoidance of pairwise isomorphism testing and keeping long lists of graphs in memory for comparison. An advantage compared with the DENDRAL generators is that orderly generation does not require any catalog of elemental graphs.

8.2.2.1 Enumerating Labeled Graphs

The principles underlying orderly generation are best explained using simple graphs. Let γ and γ' be simple graphs on n nodes. Nodes are labeled with numbers from 1 to n. There is an order on edges of such graphs defined as follows: for edges $e = (x, y)$, $e' = (x', y')$ with $x < y$, $x' < y'$, e is less than e', if and only if $x < x'$, or $x = x'$ and $y < y'$. This can be expressed more precisely in mathematical terms:

$$e < e' \quad :\Longleftrightarrow \quad x < x' \ \vee \ (x = x' \ \wedge \ y < y').$$

This induces a *lexicographical order* on the set of graphs on n nodes. Let $e_1, \dots, e_t$ be the edges of γ and $e'_1, \dots, e'_{t'}$ those of γ' sorted in the above order, that is, $e_1 < \cdots < e_t$ and $e'_1 < \cdots < e'_{t'}$. Then γ is less than γ', if and only if there exists an index i with $e_i < e'_i$ and $e_j = e'_j$ for all $j < i$, or $t < t'$ and $e_j = e'_j$ for all $j \leq t$. Again, this can be expressed more conveniently using mathematical notation:

$$\gamma < \gamma' \quad :\Longleftrightarrow \quad (\exists i < \min\{t, t'\} : e_i < e'_i \wedge \forall j < i : e_j = e'_j)$$
$$\vee \ (t < t' \wedge \forall j \leq t : e_j = e'_j).$$

As a first application, this order shows a way to construct labeled structures. We can define an algorithm that constructs labeled simple graphs according to this order.

ALGORITHM 8.2.2 LABELED ENUMERATION (γ)

1. Output γ
2. For each edge $e > \max\{e' \in \gamma\}$ do in ascending order of e

   Call *Labeled Enumeration* $(\gamma \cup \{e\})$

Example 8.2.2: Labeled Graphs on Three Nodes

Let us have a brief look at the minimalistic example of $n = 3$ nodes. Figure 8.3 shows the way edges are inserted during recursive calls of *Labeled Enumeration*. During the first call with the empty graph {} edges $(1, 2)$, $(1, 3)$, and $(2, 3)$ are used for augmentation. In the second call with graph $\{(1, 2)\}$ as the argument, edges $(1, 3)$ and $(2, 3)$ are considered, and so on. Thus graphs are written to the output in the following order:

$$\{\}, \{(1,2)\}, \{(1,2),(1,3)\}, \{(1,2),(1,3),(2,3)\}, \{(1,2),(2,3)\},$$
$$\{(1,3)\}, \{(1,3),(2,3)\}, \{(2,3)\}.$$

It is easy to check that this is the lexicographical order as introduced above.

8.2.2.2 Enumerating Unlabeled Graphs

Beyond the construction sequence the ordering on the set of graphs provides a canonical form. Selecting the minimal orbit representative shows a way to avoid producing isomorphic duplicates. A graph γ is defined as *canonical* if it is *minimal* in its orbit. In mathematical terms:

$$\forall \pi \in S_n : \gamma \leq \gamma^{\pi}.$$

Algorithm 8.2.2 can be upgraded to generate minimal orbit representatives only by modifying step (1):

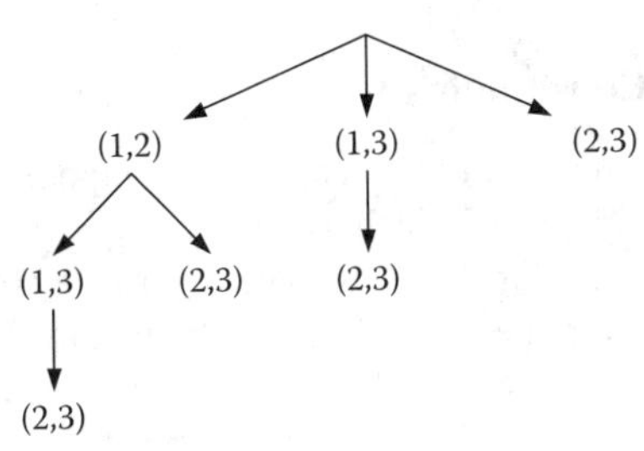

FIGURE 8.3 Generating tree for labeled graphs on three nodes.

ALGORITHM 8.2.3 UNLABELED ENUMERATION (γ)

1. If γ is minimal in its orbit, then

    Output γ

2. For each edge $e > \max\{e' \in \gamma\}$ do in ascending order of e

    Call *Unlabeled Enumeration* ($\gamma \cup \{e\}$)

However, this algorithm has to check all of the $2^{n(n-1)/2}$ labeled graphs on n nodes for canonicity. The main finding of Read [30] and Faradzev [31,32] was that every minimal orbit representative with q edges has a minimal subgraph with $q-1$ edges. Thus, nonminimal intermediates do not have to be considered for further augmentation. Using this knowledge, Algorithm 8.2.3 can be improved to the following.

ALGORITHM 8.2.4 ORDERLY ENUMERATION (γ)

1. If γ is not minimal in its orbit, then

    Return

2. Output γ
3. For each edge $e > \max\{e' \in \gamma\}$ do in ascending order of e

    Call *Orderly Enumeration* ($\gamma \cup \{e\}$)

Example 8.2.3: Unlabeled Graphs on Three Nodes

Continuing Example 8.2.2, we notice that there are four unlabeled graphs on three nodes. They have zero to three edges. The minimal orbit representatives are

$$\{\}, \{(1,2)\}, \{(1,2), (1,3)\}, \{(1,2), (1,3), (2,3)\}.$$

Comparing Algorithms 8.2.3 and 8.2.4, one canonicity test could be saved using the latter: graph $\{(1,3)\}$ would be recognized as nonminimal, and its augmentation $\{(1,3), (2,3)\}$ would not have to be considered. Of course, for increasing n the improvement in Algorithm 8.2.4 leads to much bigger gains in speed.

8.2.2.3 Introducing Constraints

Typically, one is not interested in enumerating all graphs, but just certain subsets, often denoted as *classes of graphs*. Such a class of graphs is characterized by one or more *constraints*, or *restrictions*. In mathematical terms a constraint is a mapping R from the set of graphs on n nodes onto the set of boolean values $\{true, false\}$, which is symmetry invariant:

$$\forall \pi \in S_n : R(\gamma) = R(\gamma^{\pi}).$$

A graph γ fulfills R, if $R(\gamma) = true$. Otherwise γ violates the constraint. A constraint R is called *consistent* if the violation of a graph γ to R implies that every augmentation γ' of γ violates R:

$$R(\gamma) = false \wedge \gamma \subset \gamma' \quad \Longrightarrow \quad R(\gamma') = false.$$

Examples of consistent constraints are an upper number of edges, a minimal cycle size or graph-theoretical planarity. On the other hand, the presence or absence of a certain subgraph or a maximum ring size are examples for inconsistent constraints (the precise definition of these terms would require a section on its own).

Consistent constraints can be incorporated into generating algorithms in a way that structure enumeration is accelerated. Such restrictions can be checked after each insertion of a new edge and can help to prune the generating tree. Inconsistent constraints are more problematic. Testing these constraints is only useful when a graph is completed. Completeness itself is also described by constraints. As to generating constitutional isomers, completeness is typically defined by a given degree sequence.

ALGORITHM 8.2.5 ORDERLY ENUMERATION WITH CONSTRAINTS (γ)

1. If γ is not minimal in its orbit then
   Return
2. If γ violates any consistent constraint then
   Return
3. If γ fulfills all inconsistent constraints then
   Output γ
4. For each edge $e > \max\{e' \in \gamma\}$ do in ascending order of e
   Call *Orderly Enumeration With Constraints* ($\gamma \cup \{e\}$)

8.2.2.4 Variations and Refinements

There are several variations and refinements possible that might, depending on the type of constraints, lead to a considerable speedup.

- Testing completeness is typically cheaper than other constraints like presence and absence of substructures. Thus these more expensive inconsistent constraints should be tested after completeness has been confirmed.
- Testing inconsistent constraints is often cheaper than testing canonicity. Thus it can be useful to process step (2) before step (1).

In general the sequence of tests is affected by two strategies:

- Process cheap tests first, that is, tests that consume least computation time
- Process selective tests first, that is, tests that eliminate most intermediates

Those tests that fulfill both criteria should surely be processed first, and those that fulfill none of them should be executed last. However, for expensive tests that are very selective and for cheap tests with low selectivity, one has to find a trade-off.

Going back to Algorithm 8.2.5, step (2) is often replaced by a cheaper criterion that only tests a necessary condition for canonicity, the so-called *semicanonicity*. Without going into details, this criterion only checks for transpositions τ if $\gamma \leq \gamma^{\tau}$. For a more detailed description, see Ref. [33] or [34]. The full canonicity test will be delayed until the graph is completed.

If some candidate solution then turns out not to be canonical, a so-called *learning criterion* provides a necessary condition for the canonicity of the lexicographic successors. The earliest extension step is determined where nonminimality could have been detected in the generation procedure. Applying this criterion will further prune the generating tree. Details this criterion can also be found in Refs. [33] and [34].

8.2.2.5 From Simple Graphs to Molecular Graphs

Now that we have learnt the principles of orderly generation, it is about time to adapt them to molecular graphs. In contrast to simple graphs, edges of molecular graphs have a *bond multiplicity* (or *bond order*). It is convenient to use the lexicographical order on the adjacency matrix (or equivalently on the connectivity stack) as a construction sequence. Objects with maximal connectivity stack are defined as canonical orbit representatives. This definition of canonicity is backward compatible in the following sense: a minimal simple graph as defined in Section 8.2.2.2 has the maximum connectivity stack in its orbit and vice versa.

Nodes of molecular graphs are colored by *element symbols*. Hydrogen atoms are typically treated implicitly, that is, they are not represented by nodes, but instead each non-hydrogen atom has a *hydrogen count* as an attribute. Further attributes of atoms are the sum of *remaining valencies*, that is, those not bonded to hydrogen, *charges*, and *unpaired electrons*. These attributes impose invariants on the set of atoms. Additionally, the *bond order distribution* of bonds incident with an atom can be used as invariant.

The combination of these attributes defines the atom state. Before starting to fill the adjacency matrix A, the atom states are assigned to rows (and columns) of A. If the number of atoms of each state cannot be deduced directly from the input, all possible distributions of atom states are generated and filling the adjacency matrix is repeated for each atom state distribution.

The assignment of atom states to rows and columns of the adjacency matrix introduces a block structure as depicted in Figure 8.4. Each block belongs to one of the t different atom types; λ_r equals the number of atoms of a state r.

As a first gain of this block structure no longer all $n!$ permutations of the full symmetric group S_n have to be checked during the canonicity test. Only the $\prod_{i=1}^{t} \lambda_i!$ permutations that respect the block structure have to be considered. This reduces the computational costs for canonicity testing immensely.

Algorithm 8.2.6 is taken from Ref. [33] and shows how the structure generator underlying MOLGEN (short for MOLecular structure GENerator), version 3.5 [35,36] fills the adjacency matrix. Filling matrix blocks (steps 3 and 4) is iterated

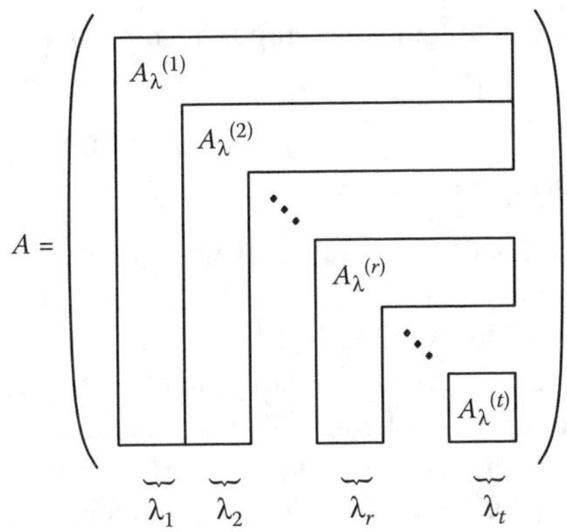

FIGURE 8.4 Adjacency matrix with block structure as used in Algorithm 8.2.6.

with testing canonicity for matrix blocks (step 5). For canonicity testing of block r, only permutations from the formerly calculated automorphism group Aut^{r-1} of blocks $1, \ldots, r-1$ have to be taken into account.

ALGORITHM 8.2.6 MOLGEN ORDERLY ENUMERATION

1. Start: set $r := 0$ and goto (3).
2. Stop criterion: if $r = 0$ stop; else goto (4).
3. Maximum filling: fill block $A^{(r)}$ (depending on $A^{(1)}, \ldots, A^{(n-1)}$) in a lexicographically *maximal* manner so that $A^{(r)}$ fulfills the desired matrix properties (regarding atom states and consistent constraints).
   If no such filling exists, then set $r := r - 1$ and goto (2); else goto (5).
4. Next smaller filling: fill block $A^{(r)}$ (depending on $A^{(1)}, \ldots, A^{(n-1)}$) in a lexicographically *next smaller* manner so that $A^{(r)}$ fulfills the desired matrix properties (regarding atom states and consistent constraints).
   If no such filling exists then set $r := r - 1$ and goto (2); else goto (5).
5. Test canonicity: if $\forall \pi \in Aut^{(r-1)}(A) : A^{(r)} \geq A^{(r)}\pi$, then
   if $r = t$ (canonical matrix complete) then
      a. if constraints are fulfilled, then output A.
      b. goto (4)
   else determine $Aut^{(r)}(A)$, set $r := r + 1$ and goto (3).
   else goto (4).

This algorithm uses two subroutines, the filling of a matrix block and the canonicity test of a matrix block. Filling a matrix block is called in two different situations: In step (3), block $A^{(r)}$ is initially filled in a maximal manner. When step (4) is called, block $A^{(r)}$ had already been filled before, and now the next smaller filling is produced. Due to their huge technical overhead, these subroutines will not be described in detail here. The reader is referred to the original publication [33]. However, this book comprises the principles of these subroutines. Canonical labeling has been introduced in Chapter 2. Filling a matrix block is done in a lexicographically descending order, which is similar to constructing labeled graphs as introduced at the beginning of this subsection.

8.2.3 Beyond Orderly Generation

Of course, other principles can be combined with orderly generation. For instance the above-mentioned MOLGEN 3.5 allows definition of *macroatoms*. These are substructures that are treated during orderly generation as a special atom type and are expanded whenever a canonical matrix is complete. Double coset representatives are used to avoid isomorphic duplicates. This principle is already known from the construction of permutational isomers and from the treatment of superatoms during tree generation in the DENDRAL generator. In mathematics, this method of joining partial structures without producing isomorphic duplicates is known as *gluing lemma* [37,38].

References [37,38] also describe the *principle of homomorphisms*. A homomorphism is a simplification of a structure, which maps isomorphic objects onto isomorphic simplified ones. The simplification from molecular graphs to multigraphs by removing element symbols, or from multigraphs to simple graphs by forgetting bond multiplicities, are examples of homomorphisms. Indeed, the DENDRAL strategy already relied on these simplification steps, only the general principle had not been worked out. In Ref. [39], this approach of simplifying by homomorphisms has been pushed to an extreme by constructing graphs with a prescribed degree sequence from regular graphs as the most simple graphs. It turned out that for huge numbers of nodes n, such a generator is much faster than orderly generation only. However, for small n, which still allow generation of full lists of graphs, the generator accelerated by homomorphisms was not able to keep up with ordinary orderly generation.

Another variation of orderly generation is also worth mentioning: McKay's *enumeration by canonical construction path* [40] restricts extensions to those structures where the new edges are taken from a certain orbit of the automorphism group.

Speed plays an important role in structure enumeration, but only a few theoretical results about the computational complexity are known. Goldberg's work [41] proves that the results in orderly enumeration can be computed with polynomial delay and a paper of Luks [42] shows that isomorphism testing of molecular graphs can be done in polynomial time.

A new approach named *constrained generation* [43] pays attention to the fact that isomer generators in structure elucidation typically aim at small numbers of solutions. For this reason, the ability to generate labeled structures that fulfill long lists of constraints becomes more important than efficient isomorphism avoidance. This generator has no fixed sequence of filling the adjacency matrix. Instead, a heuristic method has to decide which alternative makes best use of the actual constraints. It only has to be guaranteed that each isomorphism type is constructed at least once. Its canonical representation is then stored in a hash table. If it is new, it will be written to the output, otherwise it is a duplicate. Although giving up all the expertise from orderly generation, gluing lemma and homomorphism principle looks like a step backwards, this approach, implemented in MOLGEN 4.0 [44], currently appears to be the best suited solution for application in structure elucidation. It is being used in chemical and pharmaceutical companies (where results typically are not disclosed to the public domain), as well as in public research institutions (see, e.g. Ref. [45]).

Of course not all generation algorithms and implementations can be discussed in detail here. At least the most popular ones such as CHEMICS [46], ASSEMBLE [47,48], as well as Refs. [49–51] are worth being cited. Number 27 of the journal MATCH is completely devoted to this topic. Faulon's review [14] also contains a large section about this topic. Free online access to MOLGEN 3.5 and the new MOLGEN 5.0 are available at unimolis.uni-bayreuth.de/molgen and molgen.de, respectively (accessibility checked in December 2009).

8.3 ISOMER SAMPLING: STOCHASTIC STRUCTURE GENERATION

Due to the *combinatorial explosion* of numbers of constitutions with increasing numbers of atoms, it is often impossible to generate all molecular graphs belonging to a given molecular formula. Alternative methods are required, especially if no structural constraints are available, for instance if statistical statements on structural or physico-chemical properties of isomers of a certain molecular formula have to be made. Sampling techniques help to tackle such problems. A frequent requirement is a uniform probability distribution for all isomorphism types.

8.3.1 Uniformly Distributed Random Sampling

To explain the basic principle of uniformly distributed random sampling, we will again start with simple graphs. Labeled simple graphs on n nodes can be sampled with uniform distribution by simply choosing each pair of nodes with probability 0.5 as an edge.

However, the different isomorphism types have different numbers of labeled structures; thus other methods are required for uniformly distributed random sampling for unlabeled structures. Dixon and Wilf [52] solved the problem as follows.

Firstly, they choose a permutation at random from S_n and next a graph is constructed at random that is fixed by this permutation. The details of this procedure are described below.

ALGORITHM 8.3.1 SAMPLING UNLABELED GRAPHS UNIFORMLY AT RANDOM

1. Select a permutation $\pi \in S_n$ at random
2. Compute $\pi^* \in S_{\binom{n}{2}}$ corresponding to π
3. For each cycle of π^* select a boolean value at random
4. Output the graph composed by edges of cycles with value *true*

The operation of S_n on the nodes of graphs induces an operation of $S_{\binom{n}{2}}$ on the edges of graphs. $\pi^* \in S_{\binom{n}{2}}$ in step (2) is defined as

$$\pi^*((i,j)) := (\pi(i), \pi(j)).$$

This is the key to generate a random graph fixed by π. A graph constructed this way is drawn randomly with uniform distribution from all unlabeled graphs on n nodes.

Example 8.3.1: Unlabeled Simple Graphs on Six Nodes

Having selected $\pi = [3\ 4\ 5\ 6\ 1\ 2] = (1\ 3\ 5)(2\ 4\ 6)$ at random, the corresponding permutation in $S_{\binom{n}{2}}$ is

$$\pi^* = ((1,2)\ (3,4)\ (5,6))((1,3)\ (3,5)\ (1,5))$$
$$((1,4)\ (3,6)\ (2,5))((1,6)\ (2,3)\ (4,5))((2,4)\ (4,6)\ (2,6)).$$

Random values *true* for the cycles

$$((1,2)\ (3,4)\ (5,6)) \text{ and } ((1,6)\ (2,3)\ (4,5))$$

would lead to the graph with edgeset

$$\{(1,2),(1,6),(2,3),(3,4),(4,5),(5,6)\},$$

the cycle graph on six nodes.

The Dixon–Wilf technique was later expanded by Wormald [53] to sample regular graphs. An extension to molecular graphs with given molecular formula was developed by Goldberg and Jerum [54]. Their algorithm is a two-step procedure. First, a core structure that does not contain vertices of degree one or two is sampled using a Dixon–Wilf–Wormald's type algorithm. Then, the core is extended by adding trees and chains of trees (vertices of degree one or two). This strategy is similar to the processing in DENDRAL, where once cyclic substructures were generated, and connections representing the acyclic parts are added afterwards (see Section 8.2.1).

8.3.2 Monte Carlo and Simulated Annealing

Uniformly distributed random sampling is appropriate to calculate average properties of compounds from specific compound classes, but is rather time-consuming when used to search for the best compounds matching target properties or experimental data. In such an instance, optimization methods such as Monte Carlo (MC) and simulated annealing (SA) or genetic algorithms (GA) are more suitable.

Here, structures are optimized with respect to a certain target property. Any mapping from the constitutional space onto real numbers can be used as target property. Of course this mapping must be invariant with respect to atom numbering. Topological indices, group contribution calculations or potential energy are prominent examples used by Faulon [55].

Algorithm 8.3.2 is extracted from Ref. [55] and outlines the principle of MC/SA. In each annealing step a molecular graph, represented by its adjacency matrix A, is given a small displacement in order to obtain a new structure, represented by A'.

If the new structure is better with respect to the target property, it is accepted for the next annealing step. Otherwise it could still be accepted depending on a random decision guided by a so-called annealing schedule. The coefficient kT calculated in g is typically dependent on an initial coefficient, the current step number and the total number of scheduled annealing steps. Indeed the annealing schedule is the only difference between SA and MC algorithms. This procedure is repeated until a given number of annealing steps was carried out.

ALGORITHM 8.3.2 SIMULATED ANNEALING

1. Generate an initial A using a deterministic technique
2. For each SA step
    a. Choose four distinct atoms x_1, y_1, x_2, y_2 randomly
    b. Set $A' :=$ *Displacement* (x_1, y_1, x_2, y_2)
    c. If A' does not meet the chemical constraints goto (a)
    d. Compute the cost function $e(A')$
    e. $\Delta e := e(A') - e(A)$
    f. $RN :=$ random number between 0 and 1
    g. Compute the coefficient kT according to the annealing schedule
    h. If $\Delta e < 0$ or $RN < exp(-\Delta e/kT)$ then $A := A'$ and Output A

Subroutine *Displacement* (x_1, y_1, x_2, y_2)

1. Initialize $A' := A$,
   $a_{11} := A(x_1, y_1)$, $a_{12} := A(x_1, y_2)$,
   $a_{21} := A(x_2, y_1)$, $a_{22} := A(x_2, y_2)$.
2. Choose $b_{11} \neq a_{11}$ at random so that
   $b_{11} \geq \max(0, a_{11} - a_{22}, a_{11} + a_{12} - 3, a_{11} + a_{21} - 3)$ and
   $b_{11} \leq \min(3, a_{11} + a_{12}, a_{11} + a_{21}, a_{11} - a_{22} + 3)$.
3. Set $A'(x_1, y_1) := b_{11}$,
   $A'(x_1, y_2) := a_{11} + a_{12} - b_{11}$,
   $A'(x_2, y_1) := a_{11} + a_{21} - b_{11}$,
   $A'(x_2, y_2) := a_{22} - a_{11} + b_{11}$.
4. Return A'.

The crucial step in this procedure is the random displacement, which can be regarded as a transformation of a molecular graph in such a way that another isomer is obtained. Random displacements are implemented by modifying bond orders [56]. This includes creation of bonds in case a bond order is changed from zero to a positive value and deletion of bonds if the bond order is set to zero during the modification. Because isomers must have the same total number of bonds, when a bond order is increased, another bond order must be decreased. Hence, such a transformation implies the selection of at least two bonds or four atoms.

The *bond order switch* is described in subroutine *Displacement* of Algorithm 8.3.2. Numbers of randomly selected atoms are parameter values for this subroutine. The inequations in step (2) reflect the fact that bond orders range from zero to three. The new bond orders assigned in step (3) maintain the atom's valencies. In Ref. [55] it has been shown by computer experiments that all possible constitutional isomers of a given molecular formula can be reached using this bond order switch.

Example 8.3.2: Bond Order Switch

Figure 8.5 shows several examples of such random displacements. The new bond orders assigned to (x_1, y_1) are sketched by arrows labeled with the change in the adjacency matrix. In the upper two bond switches, a bond between x_1 and y_1 is deleted and created in reverse direction. In the third and fourth bond switches the bond order changes from two to one and vice versa. The lower bond switch shows a change from triple to double bond between x_1 and y_1.

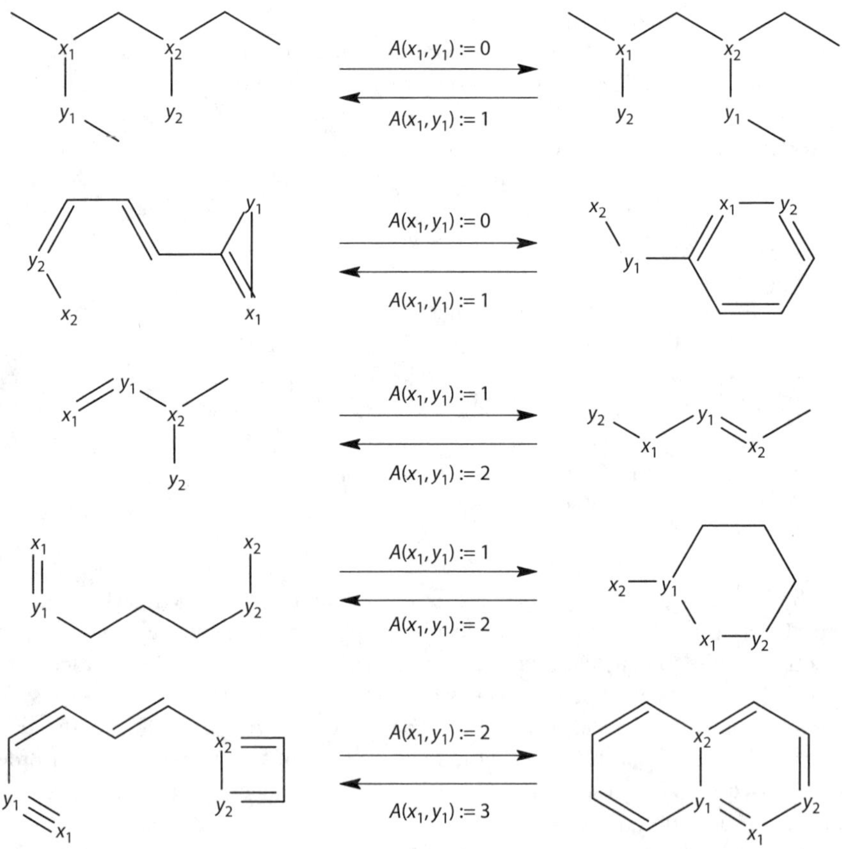

FIGURE 8.5 Examples of random displacements as used in Monte Carlo and simulated annealing algorithms.

8.3.3 Genetic Algorithms

Another type of stochastic structure generators is based on the technique of GA. GAs try to simulate principles from biological evolution, such as *inheritance*, *mutation*, *crossover* (or *recombination*), and *selection*. Except for recombination, these principles have already been used in MC/SA algorithms. However, terminology is taken from biology's evolutionary theory.

A fitness function serves for the selection of structures that fit a problem-specific target property well. The connectivity stack can be used as *genetic code*. The two types of structure manipulations, mutation and recombination, can be seen as operations on the genetic code. Mutations can be defined like random displacements known from MC/SA. Crossover involves two parent structures and at positions where their genetic codes differ a random decision determines which parent's information should be passed to the child structure.

Algorithm 8.3.3 shows the construction of a new generation of structures as described in Meiler's work [57,58]. This study was devoted to structure elucidation of small organic compounds by means of ^{13}C NMR spectra. The root-mean-square deviation between the experimental chemical shifts and the predicted chemical shifts obtained by an artificial neural network served as fitness function.

ALGORITHM 8.3.3 GENETIC ALGORITHM (CONSTRUCTION OF A NEW GENERATION)

1. Set $i := 0$
2. While the number of populations $i < n$
   a. Set $j := 0$
   b. While the number of molecules $j < m$
      i. Select first parent structure at random according to a probability based on the fitness function
      ii. If random decision for recombination is *true* then Select second parent and perform recombination else goto (iv)
      iii. If random decision for mutation is *false* then goto (v).
      iv. Perform mutation
      v. If molecule is new then increase j by 1
   c. Calculate fitting values of the molecules of the child population
   d. Replace the l worst molecules of the child population by the l best parents
   e. Increase i by 1

Each generation consists of a predefined number of populations n, where each population comprises m molecules. In step (i) the first parent for recombination is chosen at random. The probability distribution for this random selection is based on the values of the fitness function in the parent population. This guarantees that better structures have higher probability to hand down their genetic information to child structures. In the next step it is decided randomly if recombination or mutation should take place. In the case of recombination a second parent is chosen, and the recombination is carried out. Otherwise the parent structure is directly passed to mutation in step (iv). After recombination, the child structure can also still be subject to mutation, again based on a random decision. After these structure manipulation steps, fitting values of the child population are calculated. Step (d) finally prevents losing good solutions already obtained in the parent population, via the user-parameter l.

While mutation was already known from the previous subsection, recombination is a special feature of the GA. From the chemoinformatics point of view it is interesting how this is applied to constitutional isomers. For pairs of atoms (x_i, y_i) that have the same bond order in both parent structures A and B, the child structure C also obtains this bond order for (x_i, y_i). If the bond orders differ in the parent structures, the corresponding bond order in the child structure is selected randomly from one of the parent structures.

Example 8.3.3: Recombination

Figure 8.6 depicts two parent structures *A* (left) and *B* (right) and a resulting child structure *C* (bottom). The parts contributed by the parent structures are highlighted in gray. However, the process of recombination can be better understood when looking at the genetic codes of the involved structures. Remember, the genetic codes are the upper left triangles of the adjacency matrices written in one row. The different rows of the adjacency matrices are separated by blanks in

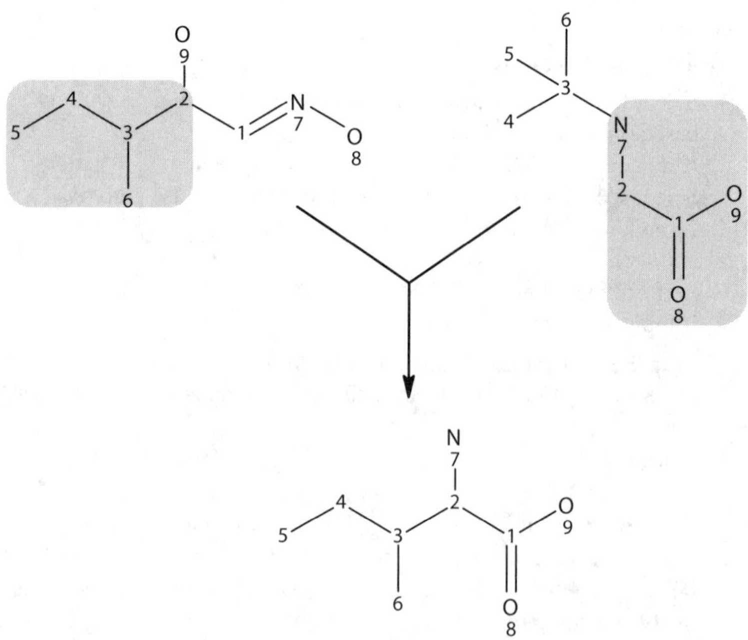

FIGURE 8.6 Example of a recombination as used in genetic algorithms.

the following representation:

A:	10000200	1000001	101000	10000	0000	000	10	0
	↓↓↓	↓ ↓					↓	
C:	10000021	1000100	101000	10000	0000	000	00	0
		↑	↑ ↑	↑				
B:	10000021	0000100	111100	00000	0000	000	00	0

We see that the genetic codes of the parent structures differ in 10 positions. Arrows mark the positions in the parent structures that have to be changed in order to obtain the child structure.

But in contrast to the bond order switch introduced in Section 8.3.2, it could happen that the child structure is not necessarily chemically valid. Structures have to be checked after recombination and further bond orders might have to be changed. Another approach for recombination, which avoids this deficiency, is proposed in Ref. [59].

The introduced MC/SA algorithm and the GA do not intend to avoid isomorphic duplicates. However, since these algorithms typically aim at producing small numbers of constitutions that fit the target property well, they could easily be upgraded to avoid duplicates by using a canonical labeling algorithm and a hash map.

8.4 BEYOND ISOMER ENUMERATION

Beyond generation of isomers, the methods described in the first two sections of this chapter can be applied to generate extensive parts of the chemical space or to count and construct combinatorial libraries. In the following we outline the adaptations in algorithms required for these purposes.

8.4.1 Virtual Chemical Space

There are several reasons to generate large parts of the chemical space by means of computers. In Refs. [60,61] the authors generate and examine a virtual chemical space of small molecules (up to 11 non-hydrogen atoms, elements C, H, N, O, F) with respect to ring systems, stereochemistry, compound classes, physico-chemical properties, as well as drug- and lead-likeness.

The algorithm starts with the construction of simple graphs, subsequently introduces multiple bonds and element symbols, and ends with the generation of stereoisomers. More precisely the algorithm's pseudo code looks as follows:

ALGORITHM 8.4.1 VIRTUAL CHEMICAL UNIVERSE UP TO 11 ATOMS

1. Generate all simple connected graphs on up to 11 vertices with vertex valency up to 4 (corresponding to saturated hydrocarbons).
2. Selection of graphs with moderate ring strain using topological criteria and molecular mechanics that eliminate
   a. graphs containing one or more nodes in two small (three- or four-membered) rings,

   b. graphs containing a ring system with a tetravalent bridgehead in a small ring,
   c. all graph-theoretically non-planar graphs,
   d. graphs containing highly distorted centers not identified by topology, but with an adapted MM2 force field.
3. Introduction of multiple bonds and elements C, H, N, O, F:
   a. Determine the symmetry of the simple graphs.
   b. Introduce double and triple bonds combinatorially with respect to symmetry in order to avoid duplicates. Bridgehead double bonds, triple bonds in rings smaller than nine and allenes are excluded as considered potentially problematic for synthesis.
   c. Determine symmetry of the multigraphs.
   d. Introduce element symbols combinatorially with respect to symmetry and under consideration of valency rules.
4. Filtering for chemical stability, tautomeric and aromatic duplicates:
   a. Structures with unstable functional groups are identified by substructure search and removed.
   b. Tautomeric and aromatic duplicates are removed.
5. Stereoisomer generation.

Step (1) was executed using GENG which is part of the freely available NAUTY system [40,62]. This resulted in 843,335 simple, connected graphs. The three topological selection steps (2a) through (2c) reduced the number of graphs from 843,335 to 16,009, the molecular-mechanics-based procedure eliminated another 283 graphs, leaving a final set of 15,726 graphs. The introduction of multiple bonds in Steps (3a) and (3b) resulted in 276,220 multigraphs, and the introduction of element symbols in steps (3c) and (3d) led to 1,720,329,902 molecular graphs. For the symmetry perception in steps (3a) and (3c) an algorithm described in Ref. [63] was used, which is based on the methods of Ref. [64], but introduces additional atomic invariants. Note that we see two typical applications of the homomorphism principle here. After removing unstable structures in step (4a), 27,681,431 structures remained. Keeping only the most probable tautomer in step (4b) resulted in 26,434,571 structures. Stereoisomers in step (5) were constructed by an implementation of Ref. [65] and led to 110,979,507 configurations.

Another approach to generate molecular structures as SMILES has been proposed in Ref. [66]. The software GENSMI is based on the commercial toolkit of Daylight Chemical Information Systems, Inc. The aim of this project was to provide structures for the search of new drug candidates that involve virtual screening by evaluating protein–ligand interactions.

A study to point out the discrepancy between compounds registered in structural databases and the number of mathematically possible compounds has been published in Ref. [67]. Mathematically possible compounds consisting of C, H, N, O and having a mass $\leq$ 150 Da were generated exhaustively and were compared with the Beilstein registry and the NIST '98 mass spectral library. As expected, it turned out that the spectral library contains only a small fraction of the compounds in the Beilstein registry, which itself represents only an even smaller fraction of the mathematically possible compounds (ratio 1:11:404976). This result emphasizes the need for structure generation software in the fields of structure elucidation and drug discovery.

The algorithm used for the generation of the chemical space in Ref. [67] is mainly based on the structure generator MOLGEN 3.5 [35,36], which was fed with all possible molecular formulas:

ALGORITHM 8.4.2 MOLECULES IN SILICO UP TO 150 Da

For each mass m between 1 and 150 do
  for each graphical molecular formula f with mass m do
    generate all molecular graphs with this molecular formula f using MOLGEN.

The program run resulted in 1405 valid molecular formulas. In this context a molecular formula is denoted as *valid*, if it belongs to at least one connected molecular graph of an organic compounds (i.e., at least one C), and where the involved elements appear with standard valencies (4, 1, 3, 2 for C, H, N, O, respectively). Structure generation finally resulted in 3,699,858,517 nonisomorphic molecular graphs. Detailed tables that list molecular formulas by mass and constitutions by molecular formula, as well as the numbers of structures in the above mentioned databases, are included in Ref. [68].

8.4.2 Combinatorial Libraries

During the past decades, combinatorial chemistry has become an appropriate method to synthesize huge libraries of new compounds for biochemical screening. Especially with the development of high-throughput screening technology and better bioassays, this method has gained much attraction in the drug development workflow. In order to reduce costs it is useful to plan such experiments using computer programs. Depending on the stage in the drug discovery process, combinatorial chemistry experiments may follow different strategies. In early stages, say during lead discovery, one may want to produce libraries with a high structural diversity. In later stages, for instance during lead optimization, pharmaceutical and medicinal chemists are rather interested in focused libraries.

For any of these purposes it is useful to generate the library compounds at first *in silico* and apply appropriate tools in order to calculate parameters representing diversity, or virtual screening methods in order to optimize the experiment into a direction where most promising hits can be expected.

8.4.2.1 Counting Combinatorial Libraries

Analogously to permutational isomers we will at first show how to calculate sizes of combinatorial libraries. Most combinatorial chemistry experiments can be reduced to the situation where building blocks from a pool of substituents are connected to a central molecule, a so-called core structure with n active sites. Even reactions with multiple reaction steps, as for instance Ugi's seven-component reaction, can be processed this way.

If the core structure shows no symmetry with respect to the active sites, the size of the library is simply the product $\prod_{i=1}^{n} a_i$, where a_i denotes the numbers of possible substituents for active site i. However, if the central molecule shows symmetries, the situation is more complicated. But it can be solved with the methods from Section 8.1.1, as illustrated in Example 8.4.1.

Example 8.4.1: Amidation of Benzene Trisacetylcychloride

As an example, we consider the exhaustive amidation of benzene trisacetylcychloride as a central molecule:

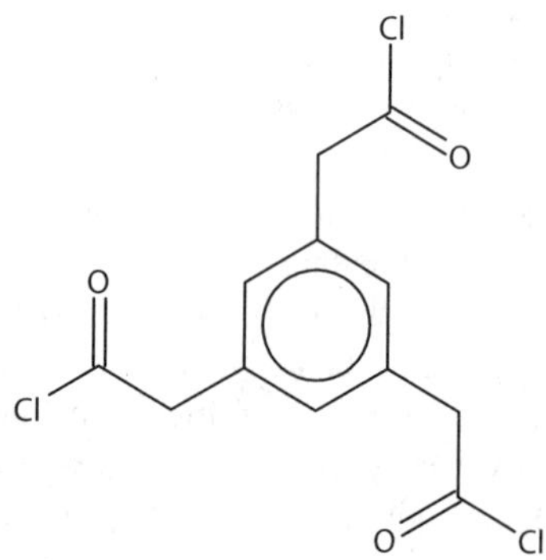

Different amino acids are attached to this central molecule as shown below. An acyl chloride group reacts with an amino group in α position to the carboxyl group:

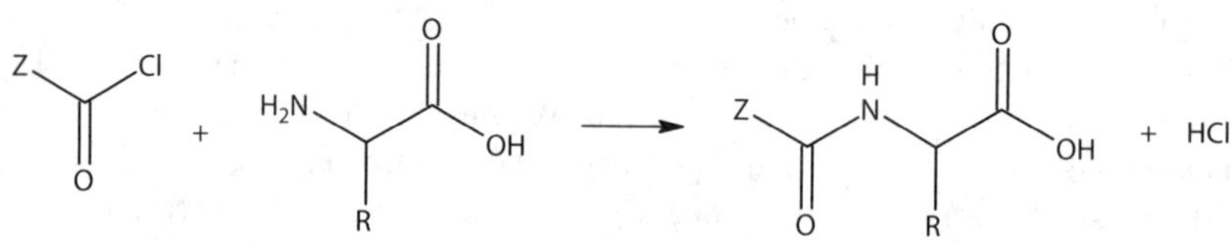

Altogether there are m^3 possible attachments of m amino acid molecules to the central molecule. But with respect to the central molecule's automorphism group, the essentially different attachments are obtained as orbits of the operation of the automorphism group applied to the set of m^3 mappings.

We face a similar situation as in Section 8.1.1. The topological automorphism group of the central molecule D_{3h} has six permutations, the identity (1)(2)(3), three reflections (1 2)(3), (1 3)(2), (1)(2 3), and two rotations (1 2 3), (1 3 2). According to Equation 8.3, the number of orbits is

$$|m^3//D_{3h}| = \tfrac{1}{6}(m^3 + 3m^2 + 2m).$$

For $m = 20$ amino acids this makes a library size of 1540 compounds, while ignoring symmetry would lead to $20^3 = 8000$ possible attachments. As experienced earlier in Section 8.1.1, we can also apply Pólya's theorem. We obtain the cycle index

$$Z(D_{3h}) = \tfrac{1}{6}({z_1}^3 + 3z_1z_2 + 2z_3),$$

and for instance if we want to examine the situation of three different amino acids as possible substituents, we replace z_k by $\sum_{i=1}^{3} y_i^k$ and obtain

$$C(D_{3h}) = \frac{1}{6}\left(\left(\sum_{i=1}^{3} y_i\right)^3 + \left(\sum_{i=1}^{3} y_i\right)\left(\sum_{i=1}^{3} y_i^2\right) + \sum_{i=1}^{3} y_i^3\right)$$

$$= y_1^3 + y_2^3 + y_3^3 + y_1^2 y_2 + y_1^2 y_3 + y_1 y_2^2 + y_2^2 y_3 + y_1 y_3^2 + y_2 y_3^2 + y_1 y_2 y_3.$$

The coefficient of the monomial $y_1^{j_1} y_2^{j_2} y_3^{j_3}$ gives the number of library molecules where amino acid i has been attached j_i times. In this example every monomial occurs with coefficient 1. Using this knowledge, it is quite easy to construct all the library members by hand:

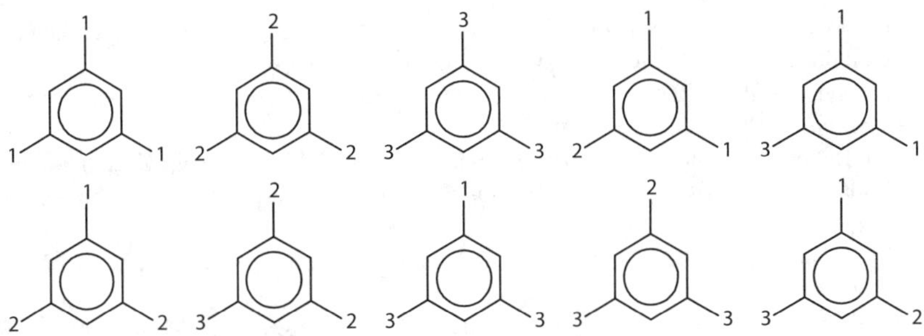

8.4.2.2 Generating Combinatorial Libraries

In general it is not possible to construct library members directly from counting series. In order to solve this problem we need to apply the relationship between permutational isomers and double cosets introduced in Refs. [8,10]. It says that the library members created from j_1 building blocks $M_1, \ldots, j_m$ building blocks M_m are in one-to-one correspondence to the set of double cosets

$$G \setminus S_n / S_{j_1} \oplus \cdots \oplus S_{j_m},$$

where G denotes the automorphism group of the central molecule and $S_n, S_{j_1}, \ldots, S_{j_m}$ the full symmetric groups of order $n, j_1, \ldots, j_m$, respectively. One method to generate double cosets is using subgroup ladders [69]. An implementation MOLGEN-COMB [70], which is specialized to applications in combinatorial chemistry, uses orderly generation for the construction of double cosets.

A completely different approach to generate combinatorial libraries and reaction networks in general is to execute all possible reactions using a backtracking strategy and to filter duplicate products using a canonical labeling algorithm. [71] and [72] are just two references that describe such an approach. Chapter 11 will discuss the generation of reaction networks in more detail.

Many of the well-known molecular modeling packages contain modules to generate combinatorial libraries, for instance CombiLibMaker (package: SYBYL, company: Tripos), LibraryMaker (BenchWare, Tripos), Analog Builder (Cerius^2, Accelrys), Afferent (company: MDL), or Structure Designer (ACD). For more detailed information on the features of these tools, the reader is referred to the product information published by the various companies.

ACKNOWLEDGMENT

The author thanks Emma Schymanski and Christoph Rücker for carefully proofreading the manuscript.

REFERENCES

1. G. Pólya. Kombinatorische anzahlbestimmungen für gruppen, graphen und chemische verbindungen. *Acta Math.*, 68: 145–253, 1937.
2. G. Pólya and R. C. Read. *Combinatorial Enumeration of Groups, Graphs and Chemical Compounds*. Springer, New York, 1987.
3. A. T. Balaban. *Enumeration of isomers*. In D. Bonchev and D. H. Rouvray, (eds). *Chemical Graph Theory. Introduction and Fundamentals*, Abacus Press, Gordon and Breach, New York, 1991, pp. 177–234.
4. A. C. Lunn and J. K. Senior. Isomerism and configuration. *J. Phys. Chem.*, 33: 1027–1079, 1929.
5. J. H. Redfield. The theory of group-reduced distributions. *Am. J. Math.*, 49: 433–455, 1927.
6. A. Kerber. *Algebraic Combinatorics via Finite Group Actions*. B.I. Wissenschaftsverlag, Manheim, Germany, 2nd ed., 1999.
7. C. C. Sims. *Computation with permutation groups*. In S. R. Petrick, (ed.), Proceedings of the Second Symposium on Symbolic and Algebraic Manipulation, New York, pp. 23–28, 1971.
8. E. Ruch, W. Hässelbarth, and B. Richter. Doppelnebenklassen als klassenbegriff und nomenklaturprinzip für Isomere und ihre abzählung. *Theor. Chim. Acta*, 19: 288–300, 1970.
9. W. Hässelbarth, E. Ruch, D. J. Klein, and T. H. Seligman. Bilateral classes. *J. Math. Phys.*, 21: 951–953, 1980.
10. E. Ruch and D. J. Klein. Double cosets in chemistry and physics. *Theor. Chim. Acta*, 63: 447–472, 1983.
11. M. van Almsick, H. Dolhaine, and H. Hönig. Efficient algorithms to enumerate isomers and diamutamers with more than one type of substituent. *J. Chem. Inf. Comput. Sci.*, 40: 956–966, 2000.
12. A. Kerber, A. Kohnert, and A. Lascoux. SYMMETRICA, an object oriented computer-algebra system for the symmetric group. *J. Symb. Comput.*, 14: 195–203, 1992.
13. R. C. Read. The enumeration of locally restricted graphs II. *J. Lond. Math. Soc.*, 35: 344, 1960.
14. J.-L. Faulon, D. Visco Jr., and D. Roe. Enumerating molecules. In K. Lipkowitz (Ed.), *Reviews in Computational Chemistry*, Vol. 21, Wiley–VCH, New York, pp. 209–286, 2005.

15. R. W. Robinson and W. C. Wormald. Numbers of cubic graphs. *J. Graph Theory*, 7: 436–467, 1983.
16. J. Wang, R. Li, and S. Wang. Enumeration of isomers of acyclic saturated hydroxyl ethers. *J. Math. Chem.*, 33: 171–179, 2003.
17. R. K. Lindsay, B. G. Buchanan, E. A. Feigenbaum, and J. Lederberg. *Applications of Artificial Intelligence for Organic Chemistry: The DENDRAL Project*. McGraw-Hill Book, New York, 1980.
18. H. R. Henze and C. M. Blair. The number of structurally isomeric alcohols of the methanol series. *J. Am. Chem. Soc.*, 53: 3077–3085, 1931.
19. J. Lederberg. DENDRAL-64, a system for computer construction, enumeration and notation of organic molecules as tree structures and cyclic graphs. Interim Report. NASA CR-57029, National Aeronautics and Space Administration, 1964.
20. C. Jordan. Sur les assemblages des lignes. *Reine Angew. Math.*, 70: 185–190, 1869.
21. R. Aringhieri, P. Hansen, and F. Malucelli. Chemical trees enumeration algorithms. Technical Report. TR-99-09, Università di Pisa, 1999.
22. J. Lederberg, G. L. Sutherland, B. G. Buchanan, E. A. Feigenbaum, A. V. Robertson, A. M. Duffield, and C. Djerassi. Applications of artificial intelligence for chemical inference. I. Number of possible organic compounds. Acyclic structures containing carbon, hydrogen, oxygen, and nitrogen. *J. Am. Chem. Soc.*, 91: 2973–2976, 1969.
23. J. Lederberg. Topological mapping of organic molecules. *Proc. Natl. Acad. Sci.* USA, 53: 134–139, 1965.
24. L. M. Masinter, N. S. Sridharan, J. Lederberg, and D. H. Smith. Applications of artificial intelligence for chemical inference. XII. Exhaustive generation of cyclic and acyclic isomers. *J. Am. Chem. Soc.*, 96: 7702–7714, 1974.
25. H. Brown and L. Masinter. An algorithm for the construction of the graphs of organic molecules. Technical Report, STAN-CS-73-361, Computer Science Department, Stanford University, 1973.
26. H. Brown, L. Masinter, and L. Hjelmeland. Constructive graph labeling using double cosets. Technical Report, STAN-CS-72-318, Computer Science Department, Stanford University, 1972.
27. H. Brown, L. Hjelmeland, and L. Masinter. Constructive graph labeling using double cosets. *Discr. Math.*, 7: 1–30, 1974.
28. L. M. Masinter, N. S. Sridharan, R. E. Carhart, and D. H. Smith. Applications of artificial intelligence for chemical inference. XIII. Labeling of objects having symmetry. *J. Am. Chem. Soc.*, 96: 7714–7723, 1974.
29. R. E. Carhart, D. H. Smith, N. A. B. Gray, J. G. Nourse, and C. Djerassi. GENOA: A computer program for structure elucidation utilizing overlapping and alternative substructures. *J. Org. Chem.*, 46: 1708–1718, 1981.
30. R. C. Read. *Everyone a winner*, Volume 2 of *Annals of Discrete Mathematics*, North-Holland Publishing Company, Amsterdam, The Netherlands, pp. 107–120, 1978.
31. I. A. Faradzhev. *Generation of Nonisomorphic Graphs with a Given Degree Sequence*, Algorithmic Studies in Combinatorics. NAUKA, Moscow, Russia, pp. 11–19, 1978. In Russian.
32. I. A. Faradzhev. Constructive enumeration of combinatorial objects. Problèmes Combinatoires et Théorie des Graphes, 260: 131–135, 1978. (Colloq. Internat. CNRS, University of Orsay, Orsay 1976). For details see http://www.amazon.com/Problemes-combinatoires-theorie-graphes-internationaux/dp/2222020700
33. R. Grund. Construction of molecular graphs with given hybridizations and non-overlapping fragments. *Bayreuther Math. Schr.*, 49: 1–113, 1995. In German.

34. M. Meringer. Fast generation of regular graphs and construction of cages. *J. Graph Theory*, 30: 137–146, 1999.
35. C. Benecke, R. Grund, R. Hohberger, A. Kerber, R. Laue, and T. Wieland. MOLGEN+, a Generator of connectivity isomers and stereoisomers for molecular structure elucidation. *Anal. Chim. Acta*, 314: 141–147, 1995.
36. C. Benecke, T. Grüner, A. Kerber, R. Laue, and T. Wieland. Molecular structure feneration with MOLGEN, new features and future developments. *Fresenius J. Anal. Chem.*, 358: 23–32, 1997.
37. R. Laue. Construction of combinatorial objects—A tutorial. *Bayreuther Mathematische Schriften*, 43: 53–96, 1993.
38. R. Kerber and A. Laue. Group actions, double cosets, and homomorphisms: Unifying concepts for the constructive theory of discrete structures. *Acta Appl. Math.*, 52: 63–90, 1998.
39. T. Grüner, R. Laue, and M. Meringer. *Algorithms for group actions: homomorphism principle and orderly generation applied to graphs*, volume 28 of *DIMACS series in discrete mathematics and theoretical computer science*, American Mathematical Society, Providence, RI, 113–122, 1996.
40. B. D. McKay. Isomorph-free exhaustive generation. *J. Algorithms*, 26: 306–324, 1998.
41. L. A. Goldberg. Efficient algorithms for listing unlabeled graphs. *J. Algorithms*, 13: 128–143, 1992.
42. E. Luks. Isomorphism of graphs of bounded valence can be tested in polynomial time. *J. Comput. Syst. Sci.*, 25: 42–65, 1982.
43. R. Laue, T. Grüner, M. Meringer, and A. Kerber. Constrained generation of molecular graphs, volume 69 of DIMACS series in discrete mathematics and theoretical computer science: Graphs and discovery, *Am. Math. Soc.*, 319–332, 2005.
44. T. Grüner, A. Kerber, R. Laue, and M. Meringer. MOLGEN 4.0. MATCH Commun. *Math. Comput. Chem.*, 37: 205–208, 1998.
45. E. L. Schymanski, C. Meinert, M. Meringer, and W. Brack. The use of MS classifiers and structure generation to assist in the identification of unknowns in effect-directed analysis. *Anal. Chim. Acta*, 615: 136–147, 2008.
46. K. Funatsu, N. Miyabayaski, and S. Sasaki. Further development of structure generation in the automated structure elucidation system CHEMICS. *J. Chem. Inf. Comput. Sci.*, 28: 18–28, 1988.
47. V. Kvasnicka and J. Pospichal. Canonical indexing and the constructive enumeration of molecular graphs. *J. Chem. Inf. Comput. Sci.*, 30: 99–105, 1990.
48. M. Badertscher, A. Korytko, K.-P. Schulz, M. Madison, M. E. Munk, P. Portman, M. Junghans, P. Fontana, and E. Pretsch. Assemble 2.0: A structure generator. *Chemom. Intel. Lab. Syst.*, 51: 73–79, 2000.
49. M. E. Elyashberg, E. R. Martirosian, Y. Z. Karasev, H. Thiele, and H. Somberg. X-PERT: A user friendly expert system for molecular structure elucidation by spectral methods. *Anal. Chim. Acta*, 337: 265–286, 1997.
50. M. S. Molchanova and N. S. Zefirov. Irredundant generation of isomeric molecular structures with some known fragments. *J. Chem. Inf. Comput. Sci.*, 38: 8–22, 1998.
51. S. G. Molodtsov. The generation of molecular graphs with obligatory, forbidden and desirable fragments. *MATCH Commun. Math. Comput. Chem.*, 37: 157–162, 1998.
52. J. D. Dixon and H. S. Wilf. The random selection of unlabeled graphs. *J. Algorithms*, 4: 205–213, 1983.
53. N. C. Wormald. Generating random unlabeled graphs. *SIAM J. Comput.*, 16: 717–727, 1987.

54. L. A. Goldberg and M. Jerrum. Randomly sampling molecules. *SIAM J. Comput.*, 29: 834–853, 1999.
55. J.-L. Faulon. Stochastic generator of chemical structure. 2. Using simulated annealing to search the space of constitutional isomers. *J. Chem. Inf. Comput. Sci.*, 36: 731–740, 1996.
56. J.-L. Faulon. Isomorphism, automorphism partitioning, and canonical labeling can be solved in polynomial-time for molecular graphs. *J. Chem. Inf. Comput. Sci.*, 38: 432–444, 1998.
57. J. Meiler and M. Will. Automated structure elucidation of organic molecules from 13C NMR spectra using genetic algorithms and neural networks. *J. Chem. Inf. Comput. Sci.*, 41(6): 1535–1546, 2001.
58. M. Will and J. Meiler. Genius: A genetic algorithm for automated structure elucidation from 13C NMR spectra. *J. Am. Chem. Soc.*, 124: 1868–1870, 2002.
59. A. Globus, J. Lawton, and T. Wipke. Automatic molecular design using evolutionary techniques. *Nanotechnology*, 10: 290–299, 1999.
60. T. Fink, H. Bruggesser, and J.-L. Reymond. Virtual exploration of the small-molecule chemical universe below 160 Daltons. *Angew. Chem. Internat. Ed.*, 44(10): 1504–1508, 2005.
61. T. Fink and J. L. Reymond. Virtual exploration of the chemical universe up to 11 atoms of C, N, O, F: Assembly of 26.4 million structures (110.9 million stereoisomers) and analysis for new ring systems, stereochemistry, physicochemical properties, compound classes, and drug discovery. *J. Chem. Inf. Model.*, 47: 342–353, 2007.
62. B. D. McKay. Nauty User's Guide (version 1.5). Technical Report, TR-CS-90-02, Department of Computer Science, Australian National University, 1990.
63. S. Bohanec and M. Perdih. Symmetry of chemical structures: A novel method of graph automorphism group determination. *J. Chem. Inf. Comput. Sci.*, 33: 719–726, 1993.
64. D. Weininger, A. Weininger, and J. L. Weininger. SMILES. 2. Algorithm for generation of unique SMILES notation. *J. Chem. Inf. Comput. Sci.*, 29: 97– 101, 1989.
65. J. G. Nourse, R. E. Carhart, D. H. Smith, and C. Djerassi. Exhaustive generation of stereoisomers for structure elucidation. *J. Am. Chem. Soc.*, 101: 1216–1223, 1979.
66. T. I. Oprea and J. M. Blaney. Cheminformatics approaches to fragment-based lead discovery, volume 34 of methods and principles in medicinal chemistry: Fragment-based approaches in drug discovery, chapter 5, pp. 91–111. Wiley-VCH, Weinheim, Germany, 2006.
67. A. Kerber, R. Laue, M. Meringer, and C. Rücker. Molecules in Silico: Potential versus known organic compounds. *MATCH Commun. Math. Comput. Chem.*, 54: 301–312, 2005.
68. M. Meringer. *Mathematical Models for Combinatorial Chemistry and Molecular Structure Elucidation*. Logos–Verlag, Berlin, 2004. In German.
69. B. Schmalz. Verwendung von untergruppenleitern zur bestimmung von doppelnebenklassen. *Bayreuther Mathematische Schriften*, 31: 109–143, 1993.
70. R. Gugisch, A. Kerber, R. Laue, M. Meringer, and J. Weidinger. MOLGEN–COMB, a software package for combinatorial chemistry. *MATCH Commun. Math. Comput. Chem.*, 41: 189–203, 2000.
71. G. Benkö, C. Flamm, and F. Stadler. A graph–based toy model of chemistry. *J. Chem. Inf. Comput. Sci.*, 43: 1085–1093, 2003.
72. A. Kerber, R. Laue, M. Meringer, and C. Rücker. Molecules in Silico: A graph description of chemical reactions. *J. Chem. Inf. Model.*, 47: 805–817, 2007.

9 Computer-Aided Molecular Design

Inverse Design

Donald P. Visco, Jr.

CONTENTS

9.1 INTRODUCTION

"Molecular design" is a term that has many connotations in a variety of fields. Certainly one can perform experimental work on various compounds and, through an analysis of their properties, propose a new substance with a desired property value. This would be an example of molecular design. However, in this chapter we will focus on the *in silico* approach to molecular design, which goes by the catch-all term "computer-aided molecular design," or CAMD.

At its most basic level, CAMD is the application of computer-implemented algorithms that are utilized to design a molecule for a particular application. Normally, when one considers the term "molecular design," a common thought is in the area of therapeutics. Many researchers in industry and academia alike are involved in the design of drugs and, accordingly, extensive effort has been afforded to developing techniques specific to these types of systems [1–4]. However, while not as visible or attractive as marketing the latest pharmaceutical, CAMD is a popular and useful technique in many other areas, such as for polymers [5,6] or in solvent design [7].

In general, CAMD has a practically infinite solution space wherein to search for candidates. As we shall see in this chapter, when the desired molecules are for biological systems, the solution space is estimated to be at least 10^{60} [8], which is a relatively tiny fraction of the space as a whole. Large search spaces are both a blessing and a curse. With a vast amount of compounds to evaluate, there is more of a possibility to find a higher-quality and/or novel candidate. This could, in turn, lead to a discovery with the potential for great economic impact for a particular company. On the other hand, with such a big "hay stack," enormous time and effort could be spent in a search that leads nowhere and is unproductive. Accordingly, efforts are made *a priori* to limit the search space using techniques such as full-fledged templating [9] or requiring the presence of certain features in a candidate molecule [10].

The two most visible industries using CAMD are the chemical process industry and the pharmaceutical industry. While both industries are solving CAMD problems, they differ in substantial ways. For example, the chemical process industry regularly uses CAMD in the area of solvent design [7]. Solvents are designed to have certain properties for applications in a particular area and outputs of CAMD algorithms are scored based on predicted properties (most often from group-contribution methods). In the pharmaceutical industry, CAMD is often used in a *de novo* approach [11,12]. In its most popular implementation, ligands are built within an active receptor site through a CAMD algorithm, although in reality the term *de novo* has been used loosely to encompass virtually any sort of computational drug design [11]. Hence, while both industries use algorithms that share the common features of computational molecular design and scoring of candidates, the scoring functions used and (ultimately) the algorithms employed to make (or revise) the selected candidates are different. For more details on specific *de novo* design approaches, many good reviews exist [11,12].

This chapter presents general molecular design methods where compounds are designed using structure–activity or structure–property relationships. Chapter 10 focuses on drug design and, in particular, *de novo* drug design. With *de novo* design, compounds are constructed *ab initio* to complement a given target receptor. In contrast to the next chapter, the techniques presented here are not limited to drugs and do not require the knowledge of potential targets. Precisely, we focus on three inverse design techniques that have been used to design compounds for both product design/engineering applications and for a biological-related application during the last decade. We have chosen these three since they present a broad overview of some of the important issues associated with design of molecules and have been used to design compounds for both engineering applications and bio-related applications. The first algorithm presented, a group-based generate and test approach, is that derived from the work of Gani and coworkers during the early 1980s [13,14], but modified since that time and is still an important CAMD approach to this day [15]. The second algorithm is representative of the approach to treat molecular design as an optimization problem solved using a mixed-integer nonlinear approach, popularized by the work of Maranas [6,16]. Here, topological indices are incorporated in conjunction with adjacency matrices and we present a recent implementation based on the work of Camarda and Sunderesan [17]. The third algorithm comes from the work of Faulon and Visco using fragments of molecules (called Signatures) in conjunction with a powerful structure generator to solve the molecular design problem [18,19]. For each of the three approaches, a case study is presented showing two molecular design results from the implementation of each algorithm: one for an engineering application and one for a bio-related application.

9.2 CAMD AND QUANTITATIVE STRUCTURE–ACTIVITY RELATIONSHIP (QSAR)/INVERSE-QSAR (iQSAR)

Evaluating the fitness of molecular candidates derived from CAMD algorithms is regularly performed by using models that are trained on experimental data. QSARs for a particular property of interest are normally how candidates are scored. In this section, we provide an overview of QSARs and CAMD as well as a way to utilize a QSAR in a reverse fashion as an inverse design technique.

9.2.1 QSAR

A QSAR is a quantitative structure–activity (or property) relationship, which purports to describe something about a molecule (its activity against a certain protein, its boiling point, etc.) based on the molecule's structure. It was introduced in the 1960s with the work of Hansch [20] and is still an active area of research [21] with a rich history [22], although its utility as a predictive tool has been called into question [23]. While molecular properties themselves or whole-molecule descriptors can be used as independent variables in a QSAR, a popular approach is to use independent variables based on subparts of the molecule. For example, group-contribution techniques decompose a molecule into smaller groups where each group provides some contribution to a predicted molecular property. Such approaches are well highlighted in *The Properties of Gases and Liquids* [24]. Other techniques examine a 2D graphical representation of a molecule where atoms are nodes and bonds are edges. Here, an operator on some portion of the molecular graph plays the role of independent variable and many of these descriptors exist in the literature today [25]. Note that "QSAR" is a bit of a catch-all term and we use it in this chapter to denote any property of interest and not just biological activity. Further information on QSARs is given in Chapters 6 and 7.

9.2.2 The Origins of CAMD

Computer-aided molecular design has its origins in the early 1980s with the work of Gani and Brignole [13,14]. Here, functional groups derived to estimate activity coefficients of nonelectrolytes in an approach called UNIFAC [26] were used in a generate and test approach for use in solvent selection. This technique, in general, suffered greatly from combinatorial explosion since many nonfeasible structures are generated through the algorithm. Additionally, accounting for steric effects is often not successful using group-contribution techniques [27].

9.2.3 Inverse QSAR

An alternative approach to using prescribed functional groups in a CAMD algorithm is the so-called iQSAR techniques [28]. Here, rather than using a QSAR to score potential molecules created from combining groups, one fixes a desired value (or range of values) and attempts to solve for the set of independent variables (descriptors) that satisfy the QSAR. Once this is done, a molecule (or molecules) is generated (normally from a structure generator) based on those values of the independent variables (if ever possible).

The first complete attempt at iQSAR was reported by Zefirov and coworkers in 1990 in relation to connectivity indices [29]. In this algorithm, a value of the order-1 connectivity index is used as input and is rewritten in terms of the distribution of edge types. Valence-type distributions are determined from this edge type and structures are generated from a structural isomer generation code. Reported degeneracy issues exist that are associated with many structures possessing the same connectivity index. Additionally, since the edge and valence-type distributions do not follow a one-to-one correspondence, structures that are generated from the valence-type distributions are not guaranteed to possess the required value of the connectivity index.

A few years later, Zefirov and coworkers again looked at iQSAR, but this time in relation to another topological index, here the kappa (κ) indices [30]. The algorithm is based on setting a desired number of vertices in a 2D graph of a molecule and partitioning that number into the number of nodes of degrees 1, 2, 3, and 4. From the conditions set on graph existence associated with a particular partition and by fixing a desired value (or range of values) from a QSAR based on the κ indices, bond path equations are determined and act as constraints on the partitions. The partitions that pass are then generated into structures using a structure generator. While the approach is straightforward, its limitations are the use of κ indices with the QSAR. Additionally, there are known degeneracy issues associated with κ indices.

Other iQSAR approaches for different independent variables have been developed as well, such as Kier and Hall's use of molecular connectivity indices [31,32] and Zefirov's use of the Hosoya Index [33] as well as other information topological indices [34].

9.3 GENERAL FEATURES OF CAMD

In CAMD, molecules are made from fragments or groups. However, there is no standard as to what are considered groups. Accordingly, the same algorithms, which utilize different starting fragments, can end up with different optimally predicted solutions.

Computer-aided molecular design can be broadly described in three general steps, which are presented below.

Step 1: Selection of groups or fragments CAMD requires a pool of groups or fragments in order to build molecules. The selection of these groups is not standard and is normally a function of the problem to be solved. In fact, even what is considered a fragment is problem dependent since fragments can be as small as a single atom or contain many atoms [35].

Step 2: Making molecules (by combining groups/modifying candidates) If one is working with groups, they must be merged together to form molecules. However, there are two issues here. First, how the groups are selected is an algorithmic issue. While one can use a technique that exhaustively selects groups (a so-called generate and test paradigm), a variety of constraints can be implemented at this stage such as requiring certain groups to be present (but not exceeding a certain number) [36]. Second, rules must be developed on how the various groups can merge together based on valence arguments (among other user-defined constraints) [17,36,37]. Those molecules that have been deemed structurally feasible are ready to be evaluated for fitness. Note also that candidate molecules can be modified in this step as well (if part of a feedback loop) through stochastic techniques such as genetic algorithms [38,39].

Step 3: Evaluating candidate fitness Depending on the problem to be solved, a single scoring function may be used or many scoring functions may be layered together, to either filter out or rank solutions. While the scoring is normally based on group contribution (through QSARs), other factors such as molecular stability [40] or synthetic feasibility [41] can be used as well. It is also here where, if using a stochastic algorithm, candidates can be modified (via step 2) in an attempt to improve the rating of potential candidates. The best candidates are ranked and the top ones move onto further analysis, through either experimental verification or additional testing.

9.4 GENERATE AND TEST APPROACH OF GANI AND COWORKERS

One of the most popular CAMD algorithms was implemented by Gani and coworkers based on the use of the UNIFAC groups. We describe that algorithm here. In 1991, Gani and coworkers refined their previous approaches [13,14] to present a methodology for CAMD based on the previously identified UNIFAC groups [36]. Working with the UNIFAC groups provided direct access to various parameters for these groups that are needed during property estimation.

Gani created six classes of compounds based on the number of attachments available in a particular group. Class 0 is just molecules themselves (such as methanol), while Class 5 is specific for aromatic groups. In between those classes, the label refers to the number of attachments available for a particular group (re: CH_3 is in class 1, etc.). Within each class, five categories exist that house information on type of attachment, basically encoding chemical feasibility and stability. The lower the category number, the less restrictions placed on a particular group, in general.

While there have been modifications to this approach since its inception such as the use of second-order groups [35], we focus on the broader methodology that implements these ideas, the Hybrid-CAMD algorithm [40,42,43].

9.4.1 Hybrid-CAMD

The Computer Aided Process-Product Engineering Center (CAPEC) of the Technical University of Denmark offers their Integrated Computer-Aided System (ICAS), inside of which is housed a computer-aided molecular design code called ProCAMD (v 3.6). The educational version of ICAS boasts at least 50 academic users worldwide and the CAPEC counts 32 companies from a varying range of industries as industrial consortium members. ProCAMD is based on the Hybrid-CAMD [40,42,43] approach, which combines the generate and test paradigm of previous techniques with inclusion of higher-order information (such as molecular connectivity through topological indices). Owing to its popularity, the algorithm is described below.

The Hybrid-CAMD approach to molecular design comprises three steps: (1) predesign phase, (2) design phase, and (3) postdesign phase. We will describe each step individually below.

9.4.1.1 Predesign Phase

The predesign phase is important since it speaks of the practicality of solving a particular CAMD problem. If equations or data are not available to estimate a particular property, it is identified at this step.

ALGORITHM 9.1 HYBRID-CAMD: PREDESIGN PHASE ALGORITHM

1. Problem specification, including specific information on properties of interest, range of application, and so on
2. List equations/methods (e.g., QSARs) available to predict required properties
3. Based on step 2, select proper groups to be used to build molecules

9.4.1.2 Design Phase

The design phase is a set of four generate-and-test steps (called levels), which have an increasing sophistication of structural information used at each step. In effect, this is basically a set of four filters, which have an increasingly fine (and computationally expensive) threshold. A lot of solutions can be evaluated quickly in lower levels, yet the higher levels provide more stringent tests of fitness, although they take a longer time to evaluate. Such a multistep approach is purported to mitigate the combinatorial explosion that was associated with earlier versions of this type of generate and test technique [40].

9.4.1.2.1 Level 1

Level 1 is a generate-and-test algorithm that finds sets (called vectors) of feasible groups that also pass any design constraints based on these groups. The groups are the UNIFAC first-order groups and are characterized by five classes (based on valency) and five categories (based on chemical feasibility/stability). Sets of feasible groups passing this level are then feasible from a valency/feasibility/stability standpoint as well as from a property standpoint. This algorithm, which is presented below, follows from that given by the authors.

ALGORITHM 9.2 HYBRID-CAMD: DESIGN PHASE ALGORITHM (LEVEL 1)

1. For a given set of building blocks (re: groups), set min/max number of groups in a compound (Kmin and Kmax) and type of compound (acyclic, cyclic or aromatic)
2. For K = Kmin, Kmax
    a. Solve feasibility constraints associated with classes for a given value of K
        i. Solve constraints associated with categories based on chemical feasibility and stability.
        ii. Generate all possible combinations of groups within a particular solution
        iii. Screen each combination against property constraint(s) based on knowledge of number and type of group only
3. K = K + 1
4. If K > Kmax, Stop, else go to step 2

Note that step 2a is simply a sum of the number of partitions of length 4 that can be obtained from a given value of K since the classes reflect the amount of available attachments for that class (re: class 1 has one free attachment, such as –CH3, etc.). Also, if a cyclic compound is chosen, an additional constraint is added to the partition, which is based on the maximum number of rings in a molecule.

9.4.1.2.2 Level 2

Level 2 is the step that takes the vectors that pass through Level 1 and generates 2D structures in a recursive algorithm that also removes duplicate structures. In essence, the algorithm is finding the feasible spanning trees from the base graph, which provides all of the potential connections between groups. Caution is taken to split up nonsymmetric groups that have more than one free connection. A key aspect of the algorithm is an accounting of which groups have free connections available and which do not as the algorithm proceeds. Note that an additional step is required if cyclic compounds are to be generated.

ALGORITHM 9.3 HYBRID-CAMD: DESIGN PHASE ALGORITHM (LEVEL 2)

1. Set list of generated compounds to 0
2. Choose a solution vector, V
3. Choose a starting group from V
4. For all free connections in V
    a. Select a free connection
        i. For all unused groups in V
            1. If connection is allowed between free connection and unused group, make a copy of V with the new connection and add to list of generated compounds
5. If all groups have been used, STOP
6. Remove duplicate solutions
7. If all groups have not been used, yet no free connections exist, remove solution.
8. Goto step 1, if remaining vectors exist

Once the structures have been generated, the potential exists to use additional property estimation techniques to remove those structures that do not fall within constraint limits. In the Hybrid-CAMD approach, the notion of second-order groups [35] is used, which are, in essence, combinations of smaller first-order groups. Here, a pattern matching technique for these second-order groups is used on the adjacency matrix to determine the absence or presence of these second-order groups. This information, along with QSARs available based on second-order groups, can be used to predict the physical properties of the candidate molecule. Note that the adjacency matrix at this level is group based and provides a 2D table of connectivity between groups in the structure.

9.4.1.2.3 Level 3

The third level aims to transform connectivity information between groups to that between atoms and to utilize constraints that are based on atomistic connectivity (such as topological indices). In this level, the groups of the group-based adjacency matrix are expanded atomistically to create an atom-based adjacency matrix. However, there can be a degeneracy associated with isomers of some groups when transforming to an atomic description. Thus, some group-based adjacency matrices will provide more than one atom-based adjacency matrices.

ALGORITHM 9.4 HYBRID-CAMD: DESIGN PHASE ALGORITHM (LEVEL 3)

1. For each group-based adjacency matrix
    a. Expand the group into constituent atoms in the matrix and fill in "1" to establish connectivity within the group

   b. Where degeneracy exists, create new matrices that account for this connectivity
2. Find where original groups connect and identify with a "1" in the new atom-based adjacency matrix.
3. Stop

Determining atomistic-level connectivity opens up the possibility to use many QSARs that have already been developed based on topological indices. Connectivity indices [44,45] and shape indices [46–48], for example, have been used as molecular descriptors in previous QSARs for a variety of physical properties. Accordingly, this information can be used as a further screen of potential candidate solutions, especially where different isomers possess widely different properties.

9.4.1.2.4 Level 4

The final level of the design phase algorithm converts the 2D representation of the molecule to a 3D representation, with an accounting for potential structural isomers. Once the 3D representation comes into being, additional techniques to evaluate the fitness of the structure may be employed.

ALGORITHM 9.5 HYBRID-CAMD: DESIGN PHASE ALGORITHM (LEVEL 4)

1. Choose an atom-based adjacency matrix
2. Select a single-bonded atom, J, and assign its position as the origin
3. Select the atom it is bonded to, K, and a direction
4. Find number and types of bonds K participates in
5. Determine position of K based on bond length and position of J
6. Find other atoms bonded to K and set distance and direction
7. Repeat until all atoms are used
8. For each atom
   a. If chiral centers possible, duplicate structure and make appropriate position swaps for R/S isomers
9. For each double bond
   a. Analyze if possible for Z/E isomers. If so, duplicate structure and make appropriate positional swaps for Z/E isomers

Note that not included in the algorithm above is an additional step to create cyclic structures. If this occurs, an analysis of the possibility for *cis/trans* isomerism is performed as well. Additionally, there is no step in the algorithm for the proper inclusion of both torsional angles and uniformity of bond lengths in a ring.

Once the 3D structure(s) are obtained from the 2D atom-based adjacency matrix, various structure-based analytical techniques can be used involving molecular force fields. This can potentially address issues of torsional angles or the uniformity of bond lengths in a ring. Potential stability issues in a structure that had not been a factor up to this point might be identified here and that structure discarded from the list of potential CAMD candidate solutions.

9.4.1.3 Postdesign Phase

When candidate solutions make it through the design phase, additional analysis is required to determine if they are suitable compounds to solve the problem at hand. Issues such as price, availability, ease of synthesis, and potential environmental impact are just a few of the factors that might preclude a compound for further consideration. Additionally, some design constraints may not have been accessible through modeling and, thus, experimentation is required to determine these properties (or confirm some of the previously predicted properties inside the CAMD algorithm).

9.4.2 Case Study: Chemical Process Industry Application

As previously mentioned, the Hybrid-CAMD algorithm is implemented in the software package ProCAMD, which is a tool within ICAS. Several case studies have been presented and suggested [15]; we describe one here [7,49].

3-Octanol is oxidized to 3-octanone in the solvent dichloromethane. However, while this solvent has many favorable properties, it is not a green solvent (it has a nonzero ozone-depletion potential) and, thus, a replacement is warranted.

The desired list of properties for a replacement solvent includes

- Melting point <250 K
- Boiling point <380 K
- Hildebrand solubility parameter between 17 and 19 $MPa^{1/2}$
- Liquid density at 298 K between 0.95 and 1.05 g/cm^3
- $\log P < 2$
- $\log LC_{50} < 3$
- ODP and GWP should be low
- Solvent–solute solubility should be high
- Solvent–water should be immiscible

Implementing the Hybrid-CAMD algorithm, it was found that 4936 compounds (which include isomers) were generated with 215 of those satisfying all of the property constraints. The authors report that some of these compounds include 2-pentanone, sec-butyl acetate, 2-oxepanone, γ-valerolactone, and 2-ethoxy ethylacetate, with the first compound being the most environment friendly (Figure 9.1).

9.4.3 Case Study: Bio-Related Application

Gani reports on a case study to design active herbicides with the α-choloroacetamideo-chloroacetanilide backbone [15]. Based on a QSAR where $\log P$ was an independent variable, potential candidates were screened based on activity (by first estimating $\log P$) and the most active compound was found. The backbone structure had three substituent points and the solution with the highest activity is presented in Figure 9.2.

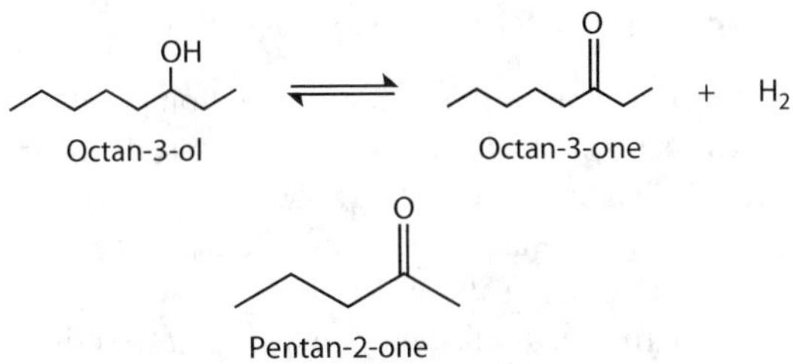

FIGURE 9.1 Using the Hyper CAMD algorithm, pentan-2-one is among the best solvents predicted to replace dichloromethane for the oxidation of octan-3-ol. (From Gani, R., C. Jimenez-Gonzalez, and D.J.C. Constable, *Comput. Chem. Eng.*, 2005, **29**: 1661–1676. With permission. Copyright 2005 Elsevier.)

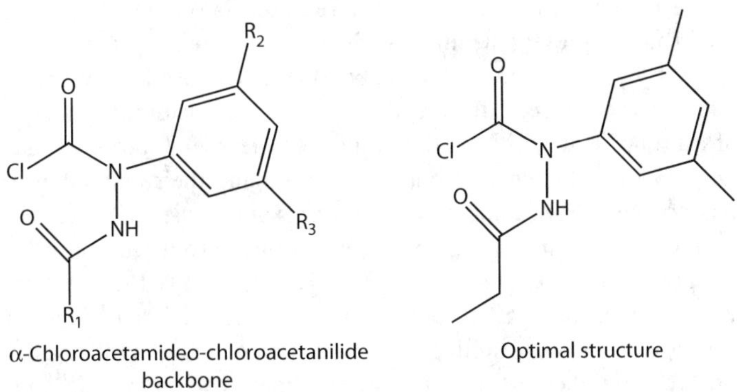

FIGURE 9.2 The template compound and the predicted optimally active structure.

9.5 CAMD AS OPTIMIZATION

Since the goal of CAMD is to build a molecule (or set of molecules) that has a certain property (or range of properties), this can be considered an optimization problem since one can attempt to minimize the difference between the desired and the predicted property. Couched in this fashion, techniques from optimization theory have been brought to bear for use within CAMD. Such approaches were initiated in the 1990s [50,51] and the first optimization attempt to use a connectivity index was reported by 1998 by Raman and Maranas [16].

In general, the CAMD as an optimization approach is to formulate the objective function using the desired properties while incorporating constraints on the structures that can be created. This falls under the wide category of mixed-integer nonlinear programs (MINLP), which are then solved using any of a variety of optimization codes designed for MINLP problems. As an illustration of such an approach, we describe the algorithm behind solving an MILP problem using connectivity indices, although other approaches exist [52,53].

9.5.1 Mixed-Integer Linear Programming Algorithm for CAMD

Camarda and Sunderesan recently solved an optimization problem related to designing valued-added soybean oil products [17]. They used an MILP approach in conjunction with order 0, 1, and 2 (simple and valence) connectivity indices [54]. The key step in their approach was the use of binary variables that allowed them to rewrite the connectivity indices in terms of these variables. In their formulation, molecules were represented by a partitioned adjacency matrix that describes the connectivity of prespecified groups within a molecule (they had chosen 16 groups). The binary variables, in turn, were related to entries in the partitioned adjacency matrix. This, in conjunction with a Glover transformation, created objective function expressions that were linear and, accordingly, easier to solve.

9.5.1.1 Molecular Representation

To represent a molecule, the data structure required is one that identifies whether a particular group is present within a molecule, what other group(s) it is bonded to and the multiplicity of that particular bond. What is used is two sets of binary variables (re: 0 for absence; 1 for presence): W (which determines the presence or absence of a group) and A (which determines connectivity and bond multiplicity). Required *a priori* is a listing of groups to be used and their maximum occurrence number in a molecule. For example [55], we can examine the binary variables W and A for the molecule propane. Here, three basic groups are listed with their allowed multiplicity given parenthetically: CH_3 (3), CH_2 (3), and CH (2). Accordingly, an adjacency matrix can be written, but in this approach it is partitioned to cluster the same groups together, resulting in a structure called a partitioned adjacency matrix (A). We show the partitioned adjacency matrix for propane using the three basic groups with allowed multiplicity for bond multiplicity 1 in Figure 9.3. Note that similar matrices will exist for bond multiplicity 2 and 3, yet those will have zero values for all elements in this instance since propane has no double or triple bonds.

Since propane has two CH_3 groups, the first one is labeled “1,” while the second one is labeled “2.” Such a labeling refers to the row/column in the partitioned adjacency matrix. Likewise, the single CH_2 group in propane is labeled “6.” Note that the other group available, CH, is not present in propane and has zeros for all entries.

Now that the bonding has been accounted for using the partitioned adjacency matrix, the presence and absence of a group is provided in an existence vector, W. In this example, W would be the vector: $W = \{1, 1, 0, 0, 0, 1, 0, 0\}$ where the labeling mentioned previously is retained here. Once the binary variables A and W are written, expressions for the molecular connectivity indices (both simple and valence) can be given in terms of these variables.

9.5.1.2 Constraint Equations

With regard to constraints, molecular feasibility expressions based on the valence of basic groups were written in terms of the binary variables. Additional expressions limit the number of basic groups in a molecule as well as the number and types of rings in a molecule. In order to guarantee that molecules are fully connected, network flow constraints [56] are written in terms of the binary variables using sink and source nodes.

ALGORITHM 9.6 MILP ALGORITHM FOR CAMD

1. Select basic groups
2. Create linear QSARs using connectivity indices for properties of interest
3. Set target values for properties of interest
4. Write objective function in terms of minimizing absolute value of difference between QSAR prediction and target value for all properties
5. Write all constraint equations in terms of binary variables
6. Solve MILP design problem and arrive at optimal solution

The authors also allowed the use of an integer-cut equation as well in the solution to the problem (not listed above in the algorithm) in order to make previously arrived at solutions infeasible within the algorithm. This provided "ranked" solutions rather than just a single optimal solution.

Note that the solutions to this problem are in terms of binary variables which, in turn, are related to the entries of the partitioned adjacency matrix. The authors restrict their solution space through the use of templating, which, in essence, preassigns parts of the partitioned adjacency matrix such that a portion (or most) of the structure is already set prior to solving the problem. This is done because of the complexity of the problem and the computational resources required, even after linearization.

	1	2	3	4	5	6	7	8	
1	0	0	0	0	0	1	0	0	CH_3
2	0	0	0	0	0	1	0	0	
3	0	0	0	0	0	0	0	0	
4	0	0	0	0	0	0	0	0	CH_2
5	0	0	0	0	0	0	0	0	
6	1	1	0	0	0	0	0	0	
7	0	0	0	0	0	0	0	0	CH
8	0	0	0	0	0	0	0	0	

FIGURE 9.3 The partitioned adjacency matrix for propane with bond multiplicity of 1. (Reprinted from Siddhaye, S. et al., *Comput. Chem. Eng.*, 2004, **28**: 425–434. With permission. Copyright 2004 Elsevier.)

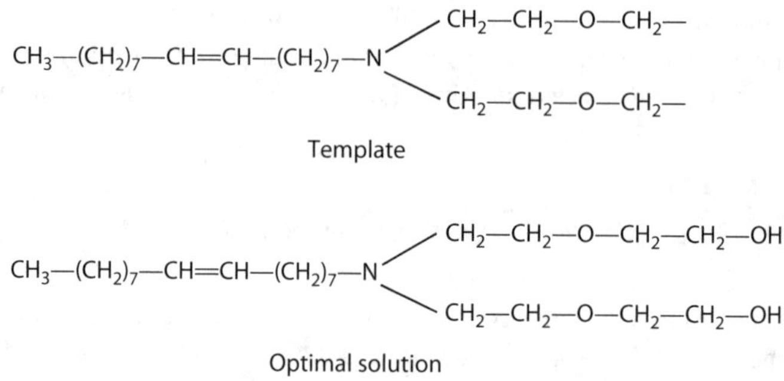

FIGURE 9.4 Starting with a template, an optimal structure was found by the MILP algorithm which is predicted to possess the required properties. (Reprinted from Camarda, K. and P. Sunderesan, *Ind. Eng. Chem. Res.*, 2005, **44**: 4361–4367. With permission. Copyright 2005 American Chemical Society.)

9.5.1.3 Case Study

Chemical process industry application: Camarda and Sunderesan provide a few examples of the implementation of this algorithm as applied to soybean oil products. Here, we describe the design of a fuel additive. The desired properties are as follows:

- Hydrophilic–lipophilic balance, HLB = 8
- Lubricity between 2.0 and 3.75 N/kg
- Critical micelle concentration value between 10^{-1} and 10^{-5} mol/L

The base groups used were C, CH, C=, =CH, CH_2, NH, OH, NO_2, CH_3, =O, O, and N, with normal valency for each implied. QSARs were created using both simple and valence connectivity indices, up to order 2. The problem was solved using CPLEX 6.5 (a mixed-integer optimizer available from ILOG, Inc.) accessed through the General Algebraic Modeling System on a SUN Ultra 10. Using a template (Figure 9.4), the optimal solution was found in approximately 1.5 h.

The optimal structure had a predicted HLB of 7.9, a CMC of 10^{-3} mol/L, and a lubricity of 3.6 N/kg.

9.5.1.4 Case Study: Bio-Related Application

Camarda and coworkers used the MILP formulation for pharmaceutical product design [55]. In one example, starting with a penicillin backbone, their goal was to find the optimum compound that possessed a log P value closest to 0.35 and a melting point closest to 127°C. The base groups used were C, CH, =CH, CH2, NH, OH, CH3, =O, O, and F, with normal valency for each implied. QSARs were created using both simple and valence connectivity indices, up to order 1. The problem was solved using CPLEX 6.5 (a mixed-integer optimizer available from ILOG, Inc.) accessed through the General Algebraic Modeling System on a SUN Ultra 10. The best solution was found in 277 s of CPU time, which is provided in Figure 9.5 along with the template.

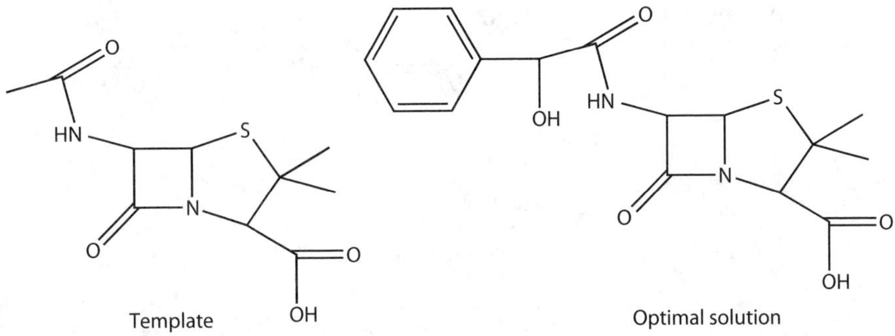

FIGURE 9.5 Starting with a template common to all penicillins, an optimal structure was found by the MILP algorithm which is predicted to possess the required properties. (Reprinted from Siddhaye, S. et al., *Comput. Chem. Eng.*, 2004, **28**: 425–434. With permission. Copyright 2004 Elsevier.)

The optimal structure had a predicted log *P* value of 0.347 and a melting point of 128°C.

9.5.2 CAMD Using Signature

In the mid-1990s, Faulon introduced the Signature molecular descriptor to elucidate structure [57,58], and in the early part of this decade, Faulon and Visco combined Signature with a structure generation code for use in inverse design [19,59,60]. Theirs is an approach that sits somewhere between the CAMD methodology of Gani and that of inverse QSAR. They use a fragmental descriptor, called Signature, and have developed structural constraint and valence equations associated with the presence of these fragments in a molecule. This sets up a series of Diophantine equations that are solved, and then the solutions are filtered through scoring QSARs to arrive at optimal candidates. 2D structures are then generated from these solutions using an in-house code that implements an enumeration algorithm [19,57].

9.5.2.1 What Is Signature?

In the previous two algorithms discussed in this chapter (and, in fact, most CAMD algorithms), predetermined groups are the building blocks to make molecules. These groups can either be self-selected or come from a standard list to facilitate group-contribution techniques, such as the UNIFAC groups [26]. Signature, on the other hand, is a more formalized and systematic approach to generate groups [59].

In order to facilitate a description of CAMD using Signature, we briefly describe the Signature molecular descriptor, although a formal definition is given in Chapter 3. An atomic Signature is a rooted tree of a 2D description of a molecule that spans the local environment around the root. The height of the atomic Signature indicates the path-length from the root atom. Height-0 is just the atom (or root), height-1 is the atom and its neighbors away by one path-length, etc. No backtracking is allowed in the creation of atomic Signatures [59]. In this formalism, a molecule of *n* atoms has *n* atomic Signatures. The molecular Signature of a molecule is then the sum of its

(a)

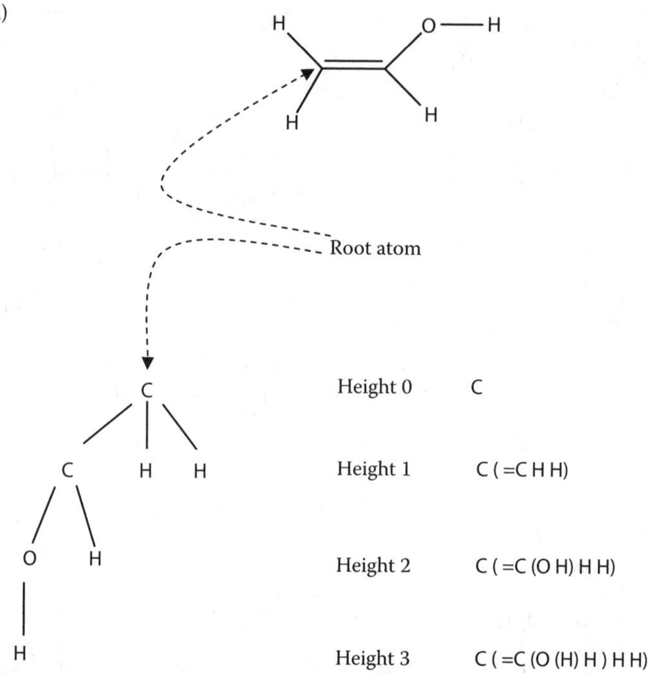

(b)

C(=C H H) + C (=C O H) + O (C H) + 3 H (C) + H (O)

FIGURE 9.6 (a) The atomic Signatures at four heights for one of the carbon atoms in ethenol. (b) The molecular Signature for ethenol at height-1.

atomic Signatures, with coefficients on each atomic Signature signifying the number of occurrences of that atomic Signature in a molecule.

For example, let us consider the 2D graph of ethenol. The dashed arrow identifies one of the carbon atoms for which we will write the atomic signatures of various heights. Since there are seven atoms in ethenol, there are seven atomic Signatures at each height. After writing all of the seven atomic Signatures for ethenol at, say, height-1, their sum would be the molecular Signature at height-1. Note that all seven of the height-1 atomic Signatures for ethenol are not unique; their degeneracies are noted by the coefficients in the molecular Signature. This is shown in Figure 9.6. The types of bonding between atoms in the Signatures are denoted in the atomic Signature itself (as seen in Figure 9.6) and can account for bonding in aromatic rings as well. Also note that various valence states are accounted for through different vertex labeling, such as with nitrogen or phosphorus.

9.5.2.2 Inverse Design Algorithm Using Signature

Inverse design using Signature can be divided into a four-step process: (1) making the atomic Signature database, (2) solving the atomic Signature constraint equations, (3) solution evaluation, and (4) structure generation and analysis [18,61]. We discuss each step in the next section and provide the appropriate algorithm for that step.

9.5.2.2.1 Making the Atomic Signature Database

Unlike the selection of tabulated fragments discussed in the previous two methods, the Signature approach requires the identification of a particular dataset to work from. The normal procedure is to use a dataset for which you have generated a QSAR for a property (or properties) of interest. Once a dataset is selected, a specific atomic Signature height is chosen for the problem. If a small height is used (say height-0), the building blocks are just the elements that appear in the dataset. If a large height is chosen (say height-4), the building blocks are height-4 atomic Signatures, of which there would be many for a given system. A dataset of 100 compounds might produce hundreds of height-4 atomic Signatures. Thus, the selection of an atomic Signature height controls the number and type of fragments used to make the molecules. Traditionally, height-1 or height-2 has been used in inverse design problems using Signature, which provides a convenient trade-off between a too general atomic Signature (low height) and a too specific atomic Signature (high height) [5].

To convert 2D structures into its Signature representation requires the use of a freely available translator code [62]. Structures are input in a common form (such as an MDL mol file) and the molecular Signature of a compound is generated. This information is parsed to develop a list of unique atomic Signatures at a desired height.

ALGORITHM 9.7 INVERSE DESIGN WITH SIGNATURE: GENERATING ATOMIC SIGNATURE DATABASE

1. Select particular dataset of interest
2. Select desired atomic Signature height
3. Run Signature translator code on 2-D structure files
4. Generate list of unique atomic Signatures at pre- selected height

The list of atomic signatures obtained at this step forms the set of fragments from which new and/or novel structures will be generated.

9.5.2.2.2 Solving the Atomic Signature Constraint Equations

In order to make solutions from the atomic Signatures, they must be connected together in a way that satisfies valency and connectivity constraints. The necessary condition to create a connected graph using Signature is called the graphicality equation and there is but one equation per dataset. This expression is based simply on known valences for the various atoms (and their types) in the dataset and is given by the following modulus expression

$$\text{Mod}\left(\sum_{i=2}^{z}(i-2)n_i - n_1 + 2, 0\right) = 0,$$

where z is the maximum number of vertices of atoms in the dataset while n_i is the degree of the root of signature i [18].

The other system constraints on the atomic Signatures are related to the bonding that occurs in a Signature. For example, consider propane, whose atomic Signatures at height-1 (and the degeneracy of occurrence) is shown in Table 9.1.

Here, there are two types of bonds: a carbon bonded to another carbon and a carbon bonded to a hydrogen. Let us examine the carbon–hydrogen bonding first. The first height-1 atomic signature ($x1$) contributes two C−H bonds while the second height-1 atomic signature ($x2$) contributes three C−H bonds. The third atomic signature ($x3$) contributes one H−C bond. Since the number of C−H bonds has to equal the number of H−C bonds, the constraint equation would be: $2x1 + 3x2 - x3 = 0$

When working with bonds between the same atom type, such as carbon–carbon, a similar approach is used. Here, $x1$ contributes 2 carbon–carbon bonds, while $x2$ contributes 1 carbon–carbon bond. However, an equality constraint (as in the C−H bond example) would be unnecessarily strong here, since the atom types are the same. Accordingly, the sum of the contributions from each atomic signature with a C−C bond would need to be an even number. Thus, the constraint equation for the C−C bond would be: $\text{Mod}(2x1 + x2, 2) = 0$. Note, finally, that the graphicality equation here becomes $\text{Mod}(2x1 + 2x2 - x3 + 2, 2) = 0$. The reader can use the occurrence numbers in Table 9.1 to verify the equations [63].

Practically, instead of three equations (as in the example above) to be solved, many (dozens) of equations are the norm for reasonable-sized datasets with a variety of atom types. The constraint equations form a set of Diophantine equations since they involve integer coefficients and require integer solutions. While generic solvers exist for these types of systems [64], the most recent use within the larger CAMD approach with Signature uses an efficient brute force approach that iterates over the range of values of the variables in the dataset [65]. It is efficient since it starts with the least complex variables (fewest iterations) and significantly saves computational time by omitting the portion of the solution space that does not satisfy any single equation.

TABLE 9.1
Height-1 Atomic Signatures for Propane

Signature	Variable	Occurrence
[C]([C][C][H][H])	$x1$	1
[C]([C][H][H][H])	$x2$	2
[H]([C])	$x3$	8

ALGORITHM 9.8 INVERSE DESIGN WITH SIGNATURE: SOLVING THE CONSTRAINT EQUATIONS

1. Generate all constraint equations to form set of Diophantine equations
2. Determine min and max values of occurrence numbers of each atomic signature in the dataset
3. Run efficient brute force algorithm to solve Diophantine equations

The solution to even moderate-sized problems can cause both storage and time concerns. It is not uncommon to generate millions or even billions of solutions [61,63]. One can place scoring functions or other filtering expressions inside the brute force algorithm to mitigate issues of storage, as needed.

9.5.2.2.3 *Solution Evaluation*

Once the solutions have been generated, they need to be evaluated for fitness. At this point (or earlier) a QSAR would be generated based on data available and using the occurrences of the atomic signatures as independent variables. Multiple QSARs can be generated and they would be used (in succession) to identify those candidates that satisfy target values (or ranges). The equations need not be linear. Constraints on the number of allowed rings in the system can be applied at this point as well. Additionally, other input can be used here to score the solutions, such as the Lipinski's Rule of 5 if working on pharmaceuticals [66].

ALGORITHM 9.9 INVERSE DESIGN WITH SIGNATURE: SOLUTION EVALUATION

1. Develop QSARs for properties of interest using atomic signatures as independent variables
2. Screen candidate solutions through QSARs and save those solutions which possess desired predicted property values
3. Set min/max desired cycles in candidate solutions and screen solutions passing through step 2 with cycle filter
4. Screen candidate solutions passing through step 3 with additional filters relevant to problem

9.5.2.2.4 *Structure Generation and Analysis*

Once the solutions (re: a molecular Signature) have passed through the various scoring routines, they are ready to be turned into 2D structures. There is a degeneracy associated with going from a molecular Signature to a 2D structure. At small heights, the degeneracy is quite large, but monotonically decreases with Signature height until there is a unique 2D structure associated with a particular molecular signature. This occurs quite rapidly and most molecular Signatures are basically nondegenerate at height-3 [19,67].

Structure generation is performed using an algorithm developed by Faulon and coworkers [19], based on an earlier isomer enumeration algorithm developed by Faulon [58,68]. The algorithm is iterative and involves starting with a molecular Signature of all atoms and no bonds and then attempts to add bonds in all possible ways to match the target molecular Signature. More details on implementation and examples are provided elsewhere [19].

Once the structures are generated, postprocessing of these candidates can occur to remove those structures that are not reasonable (such as multiple bridges or aromatic rings not following Huckel's rule). Additional filtering can occur based on energy minimization via the use of force fields, if desired.

ALGORITHM 9.10 INVERSE DESIGN WITH SIGNATURE: STRUCTURE GENERATION AND ANALYSIS

1. Submit molecular signatures to structure generation code
2. Evaluate structures based on stability/energetic constraints
   a. Use of Huckel's Rule
   b. Use of intramolecular force field

At this stage, the algorithm is complete and the resulting structures that have passed through postprocessing steps form a focused database of candidate solutions for the problem at hand. Here, these structures may be bought (if available) or synthesized and then evaluated by other means (such as experimentation) in order to gain additional confidence in the solutions.

9.5.2.3 Case Study: Chemical Process Industry Application

Faulon and coworkers have implemented the above algorithm to design novel polymers with desired properties [5]. They used 33 polymers and focused on glass transition temperature, heat capacity, and density. Height-1 atomic signatures were used on the repeat unit of the polymer. This resulted in 63 unique height-1 atomic signatures and, accordingly, 28 constraint equations. Instead of using the brute force approach to solve the constraint equations, they employed an implementation of the Fortenbacher algorithm [64] to arrive at basis vectors for their solutions. From there they calculated linear combinations of the basis vectors and arrived at over 800 million molecular Signatures (with the constraint that no repeat unit could have more than 50 atoms). These solutions were then filtered on a second constraint that no solution could have an occurrence number for a particular atomic Signature above that in the original dataset. Additionally, a normalized Euclidean distance metric was used so that solutions only in the nearest neighborhood of the original dataset were used. Of the more than 800 million solutions exposed to these constraints, only 1327 were acceptable.

When the 1327 molecular signatures were exposed to the three property constraints (313 K for glass transition temperature, 439 J/mol-K for heat capacity, 1.04 g/mol for density; all plus or minus an absolute mean error), 80 solutions were left. From this list, 178 structures were generated. A small number of the structures with some of the predicted target properties are shown in Figure 9.7.

9.5.2.4 Case Study: Bio-Related Application

Recently, Visco and coworkers used the inverse design algorithm with Signature to design novel glucocorticoid receptor ligands with pulmonary selectivity [65]. They

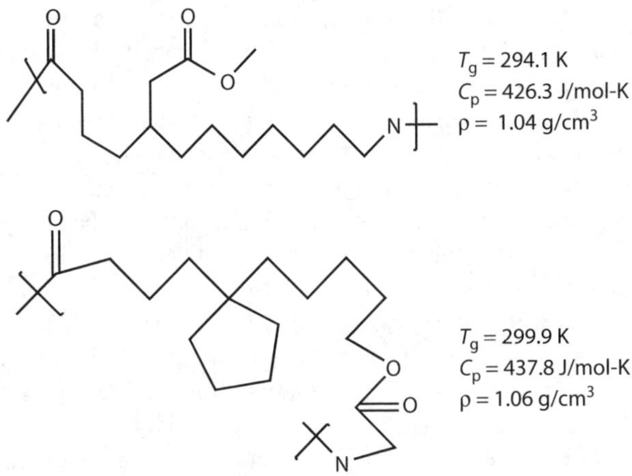

FIGURE 9.7 Two of the 137 polymer repeat units that were identified as having predicted targeted properties by the inverse design algorithm using the Signature molecular descriptor.

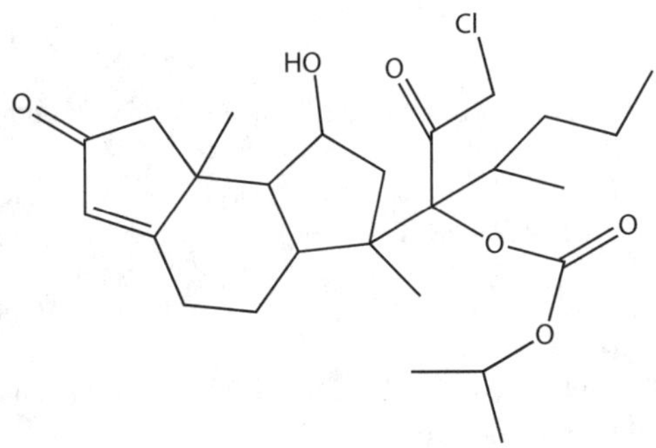

FIGURE 9.8 A compound predicted to be an effective glucocorticoid receptor ligand using the inverse design technique with the Signature molecular descriptor.

explored properties such as high receptor binding affinity, high systemic clearance, high plasma protein binding, and low oral bioavailability using atomic height-2 Signatures designed from previously published data to identify 84 high priority compounds predicted to be at least or more effective than currently available corticosteroids. One such compound developed, which is nonsteroidal in nature, is provided in Figure 9.8.

The optimal structure had a predicted oral bioavailability of 1%, a plasma protein binding of 99.1%, a systemic clearance of 475 L/h, and a relative receptor binding affinity of 1179.45.

9.6 CONCLUDING REMARKS

In this chapter, we examined three CAMD algorithms that are being implemented to solve molecular design problems of both an engineering-related and a biological nature. Each of the techniques has its own strengths and weaknesses. For the Hybrid-CAMD approach, while it does combine information with increasing complexity to mitigate some of the combinatorial explosion, one is still limited (at least in the lowest levels) to the groups selected at the start of the algorithm. Additionally, to design larger molecules, templates are seemingly necessary in order to focus the solution space to an area where higher-quality solutions are expected. For the CAMD as optimization approach, high-quality optimization codes can be brought to bear on a problem formation, thus providing an access to these techniques. However, templating is required to make progress on problems that limit the solution space. Finally, for the CAMD approach with Signature, one is not limited by an initial set of fragments and templating is not required, which can open up nonintuitive areas of the chemical space. However, storage issues and computational time issues can quickly become bottlenecks when the initial dataset molecules become large.

For all three of these techniques, a combination of increased computational and storage resources combined with algorithm refinement will make them viable CAMD techniques for the near future.

REFERENCES

1. Brown, N. and R.A. Lewis, Exploiting QSAR methods in lead optimization. *Curr. Opin. Drug Discov. Devel.*, 2006, **9**(4): 419–424.
2. Clark, D.E., What has computer-aided molecular design ever done for drug discovery? *Expert Opin. Drug Discov.*, 2006, **1**(2): 103–110.
3. de Julian-Ortiz, J.V., Virtual Darwinian drug design: QSAR inverse problem, virtual combinatorial chemistry and computational screening. *Comb. Chem. High Throughput Screen.*, 2001, **4**(3): 295–310.
4. Taft, C.A., V.B. Da Silva, and C. Da Silva, Current topics in computer-aided drug design. *J. Pharma Sci.*, 2008, **97**(3): 1089–1098.
5. Brown, W.M., et al., Designing novel polymers with targeted properties using the signature molecular descriptor. *J. Chem. Inf. Model.*, 2006, **46**: 826–835.
6. Camarda, K.V. and C.D. Maranas, Optimization in polymer design using connectivity. *Ind. Eng. Chem. Res.*, 1999, **38**: 1884–1892.
7. Gani, R., et al., A modern approach to solvent selection. *Chem. Eng.*, 2006, **March**: 30–43.
8. Dobson, C.M., Chemical space and biology. *Nature*, 2004, **432**: 824–828.
9. Garg, S. and L.E.K. Achenie, Mathematical programming assisted drug design for nonclassical antifolates. *Biotechnol. Prog.*, 2001, **17**: 412–418.
10. Lin, B., et al., Computer-aided molecular design using tabu search. *Comput. Chem. Eng.*, 2005, **29**: 337–347.
11. Schneider, G. and U. Fechner, Computer-based *de novo* design of drug-like molecules. *Nat. Rev. Drug Discov.*, 2005, **4**: 649–663.
12. Todorov, N.P., I.L. Alberts, and P.M. Dean, *De novo* design. In: *Comprehensive Medicinal Chemistry II*, J.B. Taylor and D.J. Triggle (eds). Elsevier Ltd.: Oxford, pp. 283–305, 2007.
13. Brignole, E.A., S. Bottini, and R. Gani, A strategy for design and selection of solvents for separation processes. *Fluid Phase Equi.*, 1986, **29**: 125–132.
14. Gani, R. and E.A. Brignole, Molecular design of solvents for liquid extraction based on UNIFAC. *Fluid Phase Equi.*, 1983, **13**: 331–340.

15. Gani, R., Case studies in chemical product design—use of CAMD techniques. In: *Chemical Product Design: Towards a Perspective Through Case Studies*, 23, K.M. Ng, R. Gani, and K. Dam-Johansen (eds). Elsevier: Amsterdam, the Netherlands, pp. 435–458, 2007.
16. Raman, V.S. and C.D. Maranas, Optimization in product design with properties correlated with topological indices. *Comput. Chem. Eng.*, 1998, **22**: 747–763.
17. Camarda, K. and P. Sunderesan, An optimization approach to the design of value-added soybean oil products. *Ind. Eng. Chem. Res.*, 2005, **44**: 4361–4367.
18. Churchwell, C., et al., The signature molecular descriptor. 3. Inverse quantitative structure relationship of ICAM-1 inhibitory peptides. *J. Mol. Graph. Model.*, 2003, **22**: 263–273.
19. Faulon, J.-L., C.J. Churchwell, and D.P. Visco Jr., The signature molecular descriptor. 2. enumerating molecules from their extended valence sequences. *J. Chem. Info. Comput. Sci.*, 2003, **43**: 721–734.
20. Hansch, C., et al., Correlation of biological activity of phenoxyacetic acids with Hammett substituent constants and partition coefficients. *Nature*, 1962, **194**: 178–180.
21. Gong, Z., et al., Quantitative structure–activity relationship study on fish toxicity of substituted benzenes. *QSAR Comb. Sci.*, 2008, **27**: 967–976.
22. Selassie, C.D., History of quantitative structure–activity relationships. In: *Burger's Medicinal Chemistry and Drug Discovery*, D.J. Abraham (ed.). Wiley, Hoboken, NJ, pp. 1–47, 2003.
23. Doweyko, A.M., QSAR: Dead or alive? *J. Comput.-Aided Mol. Des.*, 2008, **22**(2): 81–89.
24. Poling, B., J.M. Prausnitz, and J.P. O'Connell, *Properties of Gases and Liquids* (5th edition). McGraw-Hill Book Company, United States, 2001.
25. Devillers, J. and A.T. Balaban, (eds). *Topological Indices and Related Descriptors in QSAR and QSPR*. Academic Press, Singapore, p. 811, 2000.
26. Fredenslund, A., J. Gmehling, and P. Rasmussen, *Vapor–Liquid Equilibrium Using UNIFAC. A Group-Contribution Method*. Elsevier: Amsterdam, the Netherlands, 1977.
27. Wang, S. and G.W.A. Milne, Graph theory and group contributions in the estimation of boiling points. *J. Chem. Info. Comput. Sci.*, 1994, **34**: 1242–1250.
28. Faulon, J.-L., J. Visco, D.P., and D. Roe, Enumerating molecules. In: *Reviews in Computational Chemistry*, K.B. Lipkowitz, R. Larter, and T.R. Cundari (eds). Wiley-VCH: Hoboken, NJ, pp. 209–286, 2005.
29. Gordeeva, E.V., M.S. Molchanova, and N.S. Zefirov, General methodology and computer program for the exhaustive restoring of chemical structures by molecular connectivity indexes. Solution of the inverse problem in QSAR/QSPR. *Tetrahedron Comput. Methodol.*, 1990, **3**: 389–415.
30. Skvortsova, M.I., et al., Inverse problem in QSAR/QSPR studies for the case of topological indices characterizing molecular shape (Kier indices). *J. Chem. Info. Comput. Sci.*, 1993, **33**: 630–634.
31. Kier, L.B., L.H. Hall, and J.W. Frazer, Design of molecules from quantitative structure–activity relationship models. I. Information transfer between path and vertex degree counts. *J. Chem. Info. Comput. Sci.*, 1993, **33**: 143–147.
32. Kier, L.B., L.H. Hall, and J.W. Frazer, Design of molecules from quantitative structure–activity relationship models. II. Derivation and proof of information transfer relating equations. *J. Chem. Info. Comput. Sci.*, 1993, **33**: 148–152.
33. Skvortsova, M.I., et al., Molecular design of chemical compounds with prescribed properties from QSAR models containing the Hosoya index. Internet electron. *J. Mol. Des.*, 2003, **2**: 70–85.
34. Skvortsova, M.I., et al., Inverse problem in the structure–property relationship analysis for the case of information toplogical indices. *Doklady Chem.*, 1997, **357**: 252–254.

35. Constantinou, L. and R. Gani, New group contribution method for estimating properties of pure compounds. *AIChE J.*, 1994, **40**(10): 1697–1710.
36. Gani, R., B. Nielsen, and A. Fredenslund, A group contribution approach to computer-aided molecular design. *AIChE J.*, 1991, **37**(9): 1318–1332.
37. Chen, B., et al., Application of CAMD in separating hydrocarbons by extractive distillation. *AIChE J.*, 2005, **51**(12): 3114–3121.
38. Patkar, P.R. and V. Venkatasubramanian, Genetic algorithms based CAMD. In: *Computer Aided Molecular Design: Theory and Practice*, L.E.K. Achenie, R. Gani, and V. Venkatasubramanian (eds). Elsevier: Amsterdam, the Netherlands, pp. 95–128, 2003.
39. Venkatasubramanian, V., K. Chan, and J.M. Caruthers, Computer-aided molecular design using genetic algorithms. *Comput. Chem. Eng.*, 1994, **18**(9): 833–844.
40. Harper, P.M., M. Hostrup, and R. Gani, A hybrid CAMD method. In: *Computer Aided Molecular Design*, L.E.K. Achenie, R. Gani, and V. Venkatasubramanian (eds). Elsevier: Amsterdam, pp. 129–165, 2003.
41. Gillett, V.J., et al., SPROUT, HIPPO, and CAESA: Tools for *de novo* structure generation and estimation of synthetic accessibility. *Perspect. Drug Discov. Des.*, 1995, **3**: pp. 34–50.
42. Harper, P.M. and R. Gani, A multi-step and multi-level approach for computer aided molecular design. *Comput. Chem. Eng.*, 2000, **24**: 677–683.
43. Harper, P.M., et al., Computer-aided molecular design with combined molecular modeling and group contribution. *Fluid Phase Equi.*, 1999, **158–160**: 337–347.
44. Hall, L.H. and L.B. Kier, Molecular connectivity chi indices for database analysis and structure–property modeling. In: *Topological Indices and Related Descriptors in QSAR and QSPR*, J. Devillers and A.T. Balaban (eds). Gordon and Breach Science Publishers: Canada, pp. 307–360, 1999.
45. Kier, L.B., et al., Molecular connectivity I: Relation to non-specific local anesthesia. *J. Pharm. Sci.*, 1975, **64**: 1971–1974.
46. Kier, L.B., Indexes of molecular shape from chemical graphs. *Acta Pharm. Jugosl.*, 1986, **36**: 171.
47. Kier, L.B., A shape index from molecular graphs. *Quant. Struct. Act. Relat.*, 1985, **4**: 109.
48. Kier, L.B. and L.H. Hall, *Molecular Structure Description.* Academic Press: San Diego, CA, 1999.
49. Gani, R., C. Jimenez-Gonzalez, and D.J.C. Constable, Method for selection of solvents for promotion of organic reactions. *Comput. Chem. Eng.*, 2005, **29**: 1661–1676.
50. Churi, N. and L.E.K. Achenie, Novel mathematical programming model for computer aided molecular design. *Ind. Eng. Chem. Res.*, 1996, **35**: 3788–3794.
51. Odele, O. and S. Macchietto, Computer aided molecular design: A novel method for optimal solvent selection. *Fluid Phase Equi.*, 1993, **82**: 47–54.
52. Ostrovsky, G.M., L.E.K. Achenie, and M. Sinha, *A reduced dimension branch-and-bound algorithm for molecular design. Comput. Chem. Eng.*, 2003, **27**: 551–567.
53. Sahinidis, N.V., M. Tawarmalani, and M. Yu, Design of alternative refrigerants via global optimization. *AIChE J.*, 2003, **49**(7): 1761–1775.
54. Hall, L.H. and L.B. Kier, The molecular and connectivity chi indexes and kappa shape indexes in structure–property modeling. In: *Reviews in Computational Chemistry*, K.B. Lipkowitz and D.B. Boyd (eds). VCH Publishers: New York, pp. 367–422, 1991.
55. Siddhaye, S., et al., Pharmaceutical product design using combinatorial optimization. *Comput. Chem. Eng.*, 2004, **28**: 425–434.
56. Ahuja, R.K., T.L. Magnanti, and J.B. Orlin, Network flows. In: *Handbooks in Operations Research and Management Science*. G. L. Nemhauser, A. H. G. Rinnooy Kan, and M. J. Todd, (eds). Elsevier: Amsterdam, the Netherlands, pp. 211–360, 1989.

57. Faulon, J.-L., On using graph-equivalent classes for the structure elucidation of large molecules. *J. Chem. Info. Comput. Sci.*, 1992, **32**: 338–348.
58. Faulon, J.-L., Stochastic generator of chemical structure. 1. Application to the structure elucidation of large molecules. *J. Chem. Info. Comput. Sci.*, 1994, **34**: 1204–1218.
59. Faulon, J.-L., D.P. Visco Jr., and R.S. Pophale, The signature molecular descriptor. 1. Extended valence sequences and topological indices. *J. Chem. Info. Comput. Sci.*, 2003, **43**: 707–720.
60. Visco, J.D.P., et al., Developing a methodology for an inverse quantitative structure–activity relationship using the signature molecular descriptor. *J. Mol. Graph. Model.*, 2002, **20**: 429–438.
61. Weis, D., et al., The signature molecular descriptor. 5. The design of hydrofluoroether foam blowing agents using inverse-QSAR. *Ind. Eng. Chem. Res.*, 2005, **44**: 8883–8891.
62. Faulon, J.L., http://www.cs.sandia.gov/~jfaulon/QSAR/downloads.html
63. Weis, D., Inverse QSAR for compound development using the signature descriptor: Application to hydrofluoroethers and γ-secretase inhibitors, in Department of Chemical Engineering. Tennessee Technological University: Cookeville, TN, 2005.
64. Contejean, E. and H. Devie, An efficient incremental algorithm for solving systems of linear diophantine equations. *J. Chem. Inf. Comput. Sci.*, 1994, **113**: 143–172.
65. Jackson, J.D., D.C. Weis, and D.P. Visco, Potential glucocorticoid receptor ligands with pulmonary selectivity using I-QSAR with the signature molecular descriptor. *Chem. Biol. Drug Des.*, 2008, **72**: 540–550.
66. Lipinski, C.A., et al., Experimental and computational approaches to estimate solubility and permeability in drug discovery and development settings. *Adv. Drug Deliv. Rev.*, 1997, **23**: 3–25.
67. Faulon, J.-L., W.M. Brown, and S. Martin, Reverse engineering chemical structures from molecular descriptors: How many solutions? *J. Comput-Aided Mol. Des.*, 2005, **19**(9–10): 637–650.
68. Faulon, J.-L., On using graph-equivalent classes for the structure elucidation of large molecules. *J. Chem. Info. Comput. Sci.*, 1992, **32**: 338–348.

10 Computer-Aided Molecular Design
De Novo Design

Diana C. Roe

CONTENTS

10.1 INTRODUCTION

A vital goal in drug discovery is identifying novel compounds that can serve as a starting point in drug discovery. It is estimated that there are $10^{60}-10^{100}$ [1,2] potential chemical compounds that each have a molecular weight of less than 500 Da. By comparison, PubChem, the largest database of known chemicals, has a little more than 19 million compounds to date [3], covering only a very small percentage of potential chemical space. Even combinatorial libraries, which can range in size up to billions of compounds, do not begin to fully sample the range of all possible compounds. As the full compound space is too vast to search comprehensively, strategies have to be employed to search this space efficiently for the discovery of novel lead drug compounds.

De novo design handles this challenge by building compounds from scratch to complement the target receptor. The guiding principle in this approach is that small molecules that are complementary to the target receptor, both in shape and chemical properties, will have the most specific binding. Resulting compounds also need to be "drug-like" and readily synthetically accessible. In theory, any molecule of chemical space could be constructed using *de novo* approaches. To reduce the search of chemical space to a manageable problem, strong physical constraints must be taken into account at each step during the generation of the lead drug molecule, limiting the chemical space explored to those regions specifically complementary to the target receptor. The advantage of this approach over a virtual screening strategy to identify these compounds is that the search is directed to the relevant regions in chemical space with a far greater range and diversity of potential lead compounds that can be evaluated. Also, since compounds are built within the shape constraints of the target receptor, the structures are generated with optimal conformational geometries for binding. In most virtual screening algorithms these conformations must be sampled and can be missed. The main drawback is that the resulting compounds need to be experimentally synthesized and tested, rather than taken from an in-house inventory or ordered commercially. As our ability to predict binding affinities improves, the trade-off between greater speed of screening and greater diversity of results may drive an increase in use of *de novo* design strategies.

De novo design is inherently combinatoric, as many choices are available at each step in molecule generation, leading to a non-polynomial (NP)-hard problem that cannot be provably solved to the global optimum. Any solution to the problem is going to represent a local optimum. And so the success of this approach depends on well-chosen constraints for the problem and an efficient search strategy. The primary constraints are the geometric and chemical constraints derived from the target receptor or target ligand(s), and internal constraints for the geometry and chemistry of the lead compound being constructed. The shape and chemical constraints include both positive and negative requirements. For example, receptor site points or "hot spots" of interaction must be matched with complementary functional groups in all candidate structures, whereas boundary constraints that define disallowed structural regions must be avoided. These primary constraints are directly handled in the structure generation phase in *de novo* design. The secondary constraints include synthetic accessibility of the final compounds and their predicted adsorption, distribution, metabolism, elimination, and toxicity (ADMET) properties. These constraints are handled both by heuristics employed during structure generation and also as filters on the final set of compounds. While the set of constraints are similar to all *de novo* design algorithms, strategies for generating structures and for searching chemical space to fit these constraints vary considerably among the programs.

One major distinction in *de novo* design programs is whether they are receptor based or ligand based. In receptor-based programs the three-dimensional (3D) structure of the target receptor is known and provides the primary constraints. Ligand-based programs either generate a 3D structural pharmacophore model to generate geometric and chemical constraints that are similar to the receptor-based constraints or they use similarity to a known active ligand or QSAR model as the primary constraint. The features in *de novo* design algorithms can be divided into the following components, and choices at each of these components can distinguish one program from another:

i. *Site points:* Site points represent "hot spots" of interaction with the target receptor. They define the primary geometric and chemical constraints for the structure generation. Some ligand-based *de novo* design programs use similarity to a molecule template instead of site points.
ii. *Molecule building blocks:* These are the units for constructing the structure. They can be atoms, fragments, or templates (generic fragments).
iii. *Structure generation:* The strategy chosen here is one of the ways used to classify *de novo* design programs. Strategies include (a) "grow," which starts from a seed point and grows a structure; (b) "fragment-link," where fragments are placed in site points and linked together; and (c) sampling approaches, where molecules are randomly grown. Implicit in each strategy is rules and geometric constraints to generate chemically reasonable structures.
iv. *Search strategy:* The strategy used to search through the combinatoric set of possible (sub)structures, combined with the structure generation strategy, forms the core of the *de novo* design algorithm. Common search strategies include breadth-first search, depth-first search, evolutionary algorithms (EAs), and Monte Carlo.
v. *Structure evaluation:* Evaluates structures based on primary constraints—geometry and chemistry for binding, or similarity to a known active ligand. By evaluating substructures at every step during structure generation, the choices can be pruned during structure generation.
vi. *ADMET and synthetic accessibility:* These are the secondary constraints that can be taken into consideration as a scoring function during structure generation, as a postfilter after construction. In addition, some programs use heuristics during structure generation to incorporate these constraints.

While there are many variations on the algorithms at each of these component steps, as described below, this review will focus on the structure generation algorithms.

10.2 HISTORICAL APPROACHES TO *DE NOVO* DESIGN

The first *de novo* design programs were receptor based. The field began with programs being developed to describe the binding properties of a target receptor. Initial programs [4–8] devised strategies to identify sites that represented "hot spots" of interaction within a receptor, placed small molecule fragments or skeletons to interact at these sites, and linked them together with generic spacer molecules. Initial ligand-based programs were variations of the receptor-based programs using pharmacophore-derived constraints [9,10]. Later, ligand-based *de novo* programs were developed for cases without site points, and primarily used evolutionary methods to optimize molecules to a QSAR model of activity [11] or by similarity to a known active molecule [12].

Many choices and variations have been tried for each of the different components in *de novo* design as described below.

10.2.1 Identify Interaction Sites

In studying known inhibitors, it is found that there are certain ligand–receptor interactions, deemed "hot spots" that are important for binding and inhibition. These interaction site points can be generated from analyzing the 3D coordinates of a target receptor, or from a 3D pharmacophore model based on a superposition of bioactive ligands. These provide positive primary constraints that are positive (must be matched) during structure generation. The coordinates of the receptor, and excluded volume regions from a pharmacophore model, provide primary constraints that are negative (regions to be avoided) during structure generation. For cases where the 3D structure of the receptor is known, the interaction sites can be taken from a known pharmacophore for that protein, or predicted. The first program to predict "hot spots" was the Goodford GRID [13] program, which created a grid inside a target receptor and calculated the energy of probe atoms placed at each point in the grid to create contours of interaction for different probe types. The contour peaks would represent interaction sites. While several early *de novo* programs used GRID [7,14,15] to identify site points, the first approach to identify site points that was used in a *de novo* design program was HSITE [4], a rule-based method to search for hydrogen bonding sites based on ideal hydrogen bond geometries derived from crystal data. Other early programs expanded the rule-based method by adding lipophilic regions [16]. HIPPO [17] was the first to add covalent bonds, metal-binding sites, and complex hydrogen-bonding patterns. MCSS [18] took a unique approach and used a modified molecular dynamics code to identify hot spots by simulating probe molecules that simultaneously interacted with the protein but not with each other. The lowest energy probes are retained as starting fragments for *de novo* design. Once the interaction sites are identified, most initial programs used a rule-based scheme to place small molecule fragments that interacted with the site points. Other programs use docking codes on fragments to provide initial placements for initial placement of fragments [7,19].

10.2.2 Molecule Building Blocks

The first *de novo* design programs used molecule templates [5,6] as the building blocks, along with programs still in current use [9,20]. Templates are joined to create a 3D molecular graph, termed a "skeleton," whose vertices are labeled solely by hybridization state and edges by bond type. This approach divides structure generation into two steps: primary structure generation of generating a skeleton that fits all geometric constraints, and secondary structure generation [21] of substituting atoms into the graph to fit the chemical constraints such as hydrophobicity and electrostatic properties. This approach collapses the search space by looking at structures with the same geometry simultaneously. In contrast, the atom-based approach starts with real atoms and builds up molecules. It has the theoretical advantage that it allows more diversity in the results, with the corresponding challenge of finding efficient strategies to search through the larger chemical space. Atom-based building blocks have been used successfully in early programs [8,22] but are harder to constrain to reasonable, synthetically accessible, and "drug-like" structures and require larger combinatorial sampling. Atom-based building blocks have become less common in recent algorithms.

Another development was in the choice of fragments and building rules to incorporate synthetic accessibility and "drug-like" heuristics into the structure generation. The first step in this approach was the RECAP [23] procedure, which broke down existing drugs from the Derwent World Drug Index (WDI), according to common retrosynthetic pathways (i.e., to produce a library of reactants). TOPAS [12,24] was the first program to use a library generated in such a way and incorporate the same reaction chemistry into the structure generation, creating 25,000 unique fragments from 11 retrosynthetic pathways.

10.2.3 Structure Generation and Search Strategies

Historically *de novo* design programs have been categorized by their search strategy. The three main categories are (a) fragment-link, (b) grow, and (c) sampling. This section will briefly describe these algorithms as they are further elaborated in the next subsection. The first *de novo* programs used the *fragment-link* approach, where appropriate fragments were placed at key interaction sites and linked together. There were many strategies on how to link the fragments. One was to join fragments with predefined linkers such as spacer skeletons [6] or fragments from a database [16,25]. Another was to generate a lattice and perform either depth-first search or breadth-first search along the lattice from one fragment to the other to generate linker fragments. Regular diamond lattices [8], irregular lattices from docked fragments [7], or random lattices [19] were tried for this strategy. Other programs employed an iterative buildup procedure, similar to the *grow* strategy, until all site points were connected. FlexNovo [26] uses a k-greedy search for its buildup procedure. LigBuilder [27] used an EA to guide the buildup procedure.

The *grow* strategy starts with a seed point or fragment and builds up a molecule. Two of the seminal programs employing the *grow* strategy were the GROW [28] program and LEGEND [22]. GROW generated peptides from amino acid fragments in multiple conformations using a tree search pruned by predicted binding affinity at each step to guide the growth. LEGEND took the opposite tack and used an atom-based growth strategy with random selection at every decision point (i.e., selection of growth point, selection of next atom, and selection of join type) to guide the search process. Many other approaches have been tried to efficiently search combinatorial space during the buildup procedure including random selection combined with depth-first search [15,29], Metropolis Monte Carlo [30], and various tree-search strategies [9,20,31].

The last category for structure generation can be termed "sampling approaches," which use sampling and optimization processes to control molecule generation, rather than using site points to direct them in a specific direction such as to grow outward or to link fragments. Several strategies of this type have been tried including molecular dynamics [32], Monte Carlo [33], simulated annealing [21,34], particle-swarm [35], and EAs [11,24,36–39], which is the most common algorithm in recent ligand-based programs. Ligand-based programs that lack site points, such as those with a template molecule or QSAR as the primary constraint, all use a sampling-based method to generate structures.

Each of these strategies requires a connection scheme to join building blocks. With atoms, the rules are usually defined by the individual atom valences. Some atom-based programs have linear chains in growing molecules or links between fragments, and look for rings either on the fly [29] during structure generation, by opening, closing, expanding, and contracting rings during sampling [40], or as a postprocessing step after structure generation to search for rings [41]. With fragment-based methods, building blocks can be joined together using a single bond, rings can be fused or spirojoined. Recently, reaction-based connection rules have been used [24,38] as a heuristic to incorporate synthetic accessibility into the structure generation stage. Programs that use molecular templates as building graphs have an additional search step after generation of a molecular skeleton to replace vertices with atom-type identities to match chemical constraints such as hydrophobicity and electrostatics [9,21].

10.2.4 Structure Evaluation

Receptor-based *de novo* programs use an estimation of binding energy for primary structure evaluation. However, predicting binding affinity accurately continues to be one of the biggest hurdles with *de novo* design programs. Early programs focused mainly on steric constraints and hydrogen bonding [5,7,8]. LEGEND [22] was the first to use a molecular mechanics force field for scoring. Force-field scores have many shortcomings due to the approximations in the force field in applying it to ligands, and most notably in neglecting desolvation and entropy terms, and can be computationally demanding. LUDI [42,43] developed the first empirical scoring function by defining a set of ligand–receptor interaction types such as hydrogen bonding electrostatic and lipophilic interactions, as well as penalty terms such as the number of rotatable bonds. It derived weightings for these terms from a least squares regression on a series of ligands with known binding constants and crystal structures. The challenges here were the small size of the available data set at that time, which limits accuracy to proteins and ligands similar to those used in the regression set. Knowledge-based scoring, first implemented in SMoG [44,45], uses atom-based ligand–receptor interaction terms with weights derived from a large statistical study of ligand–receptor complexes and the frequencies of various ligand–receptor pairs in these complexes. The advantage of this approach is that there are a larger number of ligand–receptor complexes than those with known binding energies, and so more diversity went into the set, resulting in a less biased scoring function. A common problem with all receptor-based scoring schemes is that they only take into account a static protein. Skelgen is the first program to handle receptor flexibility [46,47], which was shown to improve the diversity of results in conformational and chemical space, and activity of designed ligands. Many programs that used a receptor-based scoring function also had features to score ligands based on the 3D pharmacophore model [9,10,48,49] either by deriving receptor-based constraints from the model directly or by scoring by similarity to the model.

Ligand-based *de novo* design programs that do not use a pharmacophore model have fundamentally different scoring functions than above. One approach is to derive a scoring function from a QSAR model [11,40,50]. This has the disadvantage that the scoring parameters have to be re-input for every receptor target. Another common approach is to use the similarity to an active template [24,37,51,52] as the scoring function. This is easier to code up, but reduces the diversity of the resulting molecules.

10.2.5 Synthetic Accessibility and ADMET

Synthetic accessibility continues to be the second major hurdle with *de novo* drug design programs. It is evaluated along with prediction of ADMET properties as part of the secondary scoring. Initial *de novo* design programs performed this evaluation on the final set of structures. CAESA [17] was the first program developed to predict synthetic accessibility and was based on retrospective analysis. SEEDS compares core substructures both to reaction databases for synthesis pathways and compound databases to identify derivatives [53]. More recent approaches are based on similarity to available reactants and heuristics for molecular complexity [54]. A recent survey showed that these two latter approaches have superior success in predicting synthetic accessibility [55].

Some programs include synthetic accessibility and ADMET as heuristics during structure generation. One way to do this is to include a user interaction step where an organic chemist evaluates structures during the buildup process, for example, evaluating the initial fragments prior to linking [7,19,56], or as a scoring function during an EA [7,19,56]. Another approach was to generate building blocks from fragmenting known drug molecules. This has the heuristic that the building blocks are "drug-like" [24,26,37,38]. In addition, if the fragmentation is based on retrosynthetic analysis and regenerated using reaction-based joining rules, then this can also serve as a heuristic for synthetic accessibility, such as in TOPAS [12] and FLUX [37]. Similarly, SYNOPSIS [38] chose available molecules (i.e., reactants) as fragments and used a buildup based on synthetic reactions. Another type of heuristic is to use a substructure lookup during structure generation to filter out substructures that are not drug-like or are synthetically intractable.

Finally, some programs include ADMET predictions in a scoring function. For algorithms that build up a structure, this score is usually performed after the set of structures has been generated. For algorithms that sample the chemical space of full-size structures, such as the EAs, it can be included in the scoring function during structure generation. This score can range from simple filters using Lipinski's Rule of Five [57] for drug-like compounds, to more complicated of physicochemical properties, or predictions of hERG activity [58].

Several other drug-design methodologies have their roots in *de novo* design. For example, fragment-based design approaches are similar to the fragment-link *de novo* strategies, except that these take the extra step of validating the fragment positions experimentally prior to linking. The first combinatorial library design programs started from variations in *de novo* programs—PRO_SELECT [31] evolved from PRO_LIGAND [10] and CombiBUILD [59] from BUILDER [19] (Section 8.4).

10.3 COMMON ALGORITHMS IN *DE NOVO* STRUCTURE GENERATION

De novo design algorithms are usually classified according to their structure generation strategy. The three main strategies are (1) grow, (2) fragment-link, and (3) sampling (Figure 10.1).

10.3.1 Grow

The *grow* strategy grows a molecule to complement the target receptor (or pharmacophore model) in a sequential buildup procedure. It starts by identifying site interaction points in the target receptor. A site point is chosen as a seed point to start the structure generation. An initial building block is placed in the site to complement its chemical functionality (i.e., electrostatic properties, H-bonding, and lipophilicity). Growth points are identified on the initial building block. From here, the molecule "grows" through a cycle of adding a building block to the growth points at the end of the partial structure according to connection rules, followed by scoring to evaluate whether to retain the new building block. Growth continues until termination conditions are reached, such as if the molecule extends to all site points or exceeds maximum size. How building blocks are added depends on the search strategy.

a. In the Metropolis Monte Carlo strategy, the acceptance of new building blocks to the growing molecule is biased based on predicted binding affinity according to Boltzmann statistics. A growth point and a building block are randomly selected. The new building block is scored by a measure of predicted binding affinity. The Boltzmann factor BF=exp(–affinity_score/RT) is calculated and a random number is generated. If BF is greater than the random number, the building block is retained and growth points are updated to the newest building block; otherwise it is removed and a new growth point and a building block are randomly selected. Note that the BF for building blocks with scores ≤ 0 is always ≥ 1 (i.e., always retained). This continues until termination conditions exist. The procedure is rerun from the starting seed until the desired number of structures has been generated.

ALGORITHM 10.1: PSEUDOCODE FOR GROW STRATEGY WITH A METROPOLIS MONTE CARLO SEARCH

```
Input: receptor or pharmacophore, building block library
       assign sitepoints
```

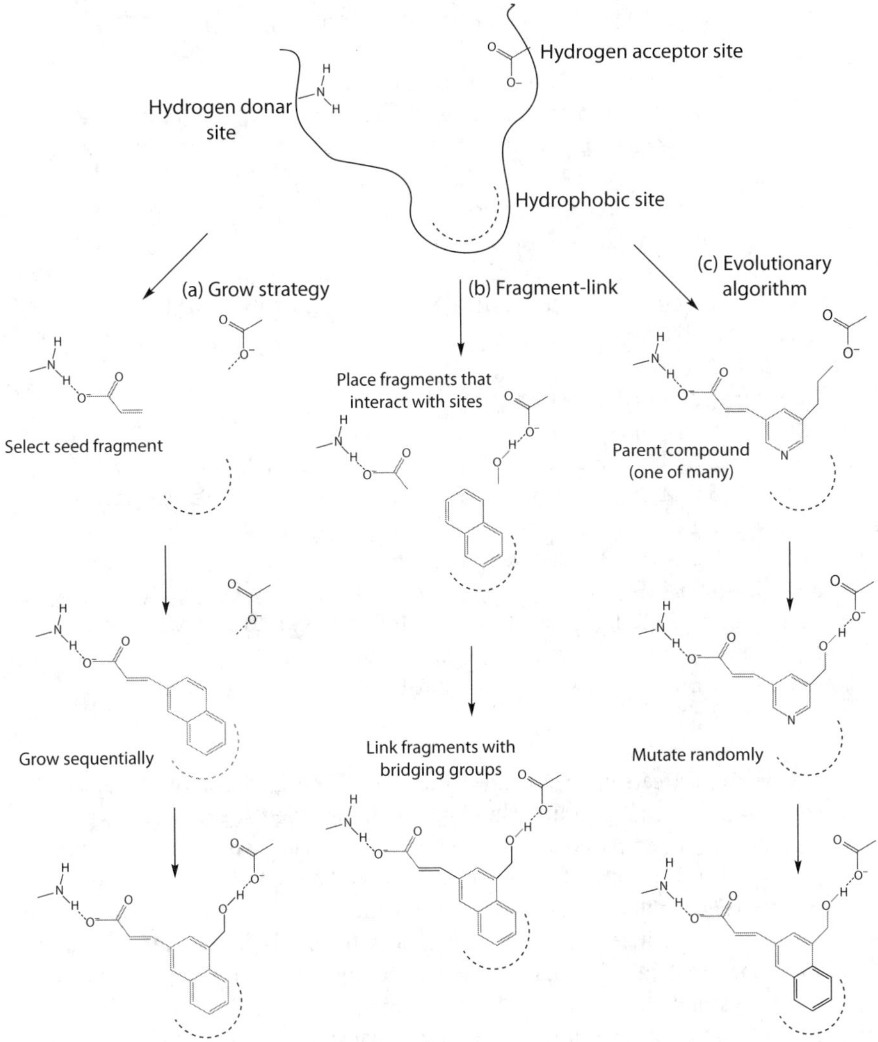

FIGURE 10.1 Comparison of three different categories of structure generation algorithms. All start with identifying site points in the target receptor. (a) Grow strategy—an initial fragment is placed at one site point and grown sequentially. (b) Fragments are placed in all site points and linked together. (c) Complete initial structure is mutated in random locations.

```
WHILE (large number of structures to grow)DO
      ##generate a structure
      place seed building block (b) in starting
       sitepoint
      assign growth points to building block
      WHILE (NOT (End=(all sitepoints fit? or >max#
           atoms?))) DO
```

```
        randomly select a growth point (g) partial
           structure
        randomly select a building block (b) and
          add to growth point using connection rules.
        Prune by primary and secondary constraints
        IF (pruned)THEN CONTINUE
        calculate binding affinity score S for (b,g)
        select or discard according to metropolis
          search criteria
        IF (selected) THEN update growth points to
          g2 on (g,b)
     END DO
     save structure(s) to list
END DO
evaluate final structures for predicted binding affinity
evaluate final structures for synthetic accessibility
   and ADMET
prioritize final structures
Metropolis criteria
    calculate Boltzman Factor BF=exp(-affinity_score/RT)
    generate random number from 0 to 1
    retain building block= TRUE IF BF > random
```

b. In the various tree-search strategies, the inner WHILE loop above is replaced with a tree-search algorithm. Building blocks are tried at each growth point, scored, and added as nodes on a search graph. The nodes may be pruned using primary constraints (e.g., boundary violations) and secondary constraints (e.g., matching to a list of synthetically intractable substructures) and may be prioritized by a scheme such as score or distance to an interaction "hot spot." In depth-first search, the top-scoring node in the graph is selected for expansion (i.e., to examine for growing), whereas in breadth-first search, all nodes at each level are selected for expansion before going to the next level. In this way, depth-first search completes a solution (i.e., generates a structure) before examining the next partial solution, whereas breadth-first search expands all the partial solutions simultaneously until all solutions are found. Depth-first search may find a single solution faster, but may not be the best overall solution, whereas breath-first exhaustively searches for solutions. Functions can be added to estimate the costs of continuing along a partial solution to prioritize nodes in "best-first" searches such as A*.

10.3.1.1 Programs in Current Use that Implement Grow

The *grow* algorithm is implemented in several *de novo* design programs in current use and that have had success in identifying lead compounds in prospective studies (see Table 10.1.a). AlleGrow, the successor of GrowMol [60], uses

TABLE 10.1
***De Novo* Design Programs with Recent Results**

Name	Ligand or Receptor Sites	Building Blocks	Search Type[a]	Scoring Type[a]	Prospective Studies
a. Grow Strategy					
AlleGrow (GrowMol [30])	Receptor	Atoms/fragments	MC	Empirical	Aspartic protease [60] xWNT8 & hWNT8 [70]
Legend [71]	Receptor	Atom	Random	FF	CDK4 inhibitors [53] Aldose reductase [72]
Sprout [9,20,54]	Receptor	Skeleton/fragments	A* search algorithm or exhaustive [54]	Empirical	Dihydroorotate dehydrogenase [73,74] Nk(2) antagonists [73,74]
b. Fragment-Link					
Ludi [16,43]	Receptor	Fragments		Empirical	CYP51(w/ MCSS) [75] Leucine aminopeptidase [76,77]
MCSS [61]	Receptor	Fragments	MD	FF	CYP51(w/LUDI) [75] PPARγ (w/LeapFrog) [62]
c. Sampling					
LEA3D [11,67]	Receptor/ligand	Fragments	EA	Empirical	Thymidine monophosphate kinase [67]
TOPAS [12,24]	Ligand	Fragments	EA	Similarity	Cannabinoid-1 receptor [78,79] Kv1.5 [12]
SkelGen [80]	Receptor/ligand	Fragments	Simulated annealing	Empirical	Cannabinoid-1 receptor [78,79] Kv1.5 [12]
Flux1/Flux2 [37,81]	Ligand	Frag/recap	EA	Similarity	TAR RNA [82]
LeapFrog [48]	Both	Fragments	EA	Empirical	PPARγ (w/MCSS) [62] Link-function only
SYNOPSIS [38]	Receptor	Frag	EA	Target-specific	HIV protease [38]

[a] BFS: breadth-first search, EA: evolutionary algorithm, FF: force field, MC: Monte Carlo.

Note: See review articles [68,69] for a more comprehensive list of *de novo* design programs using (a) Grow strategy, (b) fragment-link, and (c) sampling.

the Metropolis Monte Carlo selection criteria. It is available commercially at http://bostondenovo.com/Allegrow.htm. Legend [22] uses random selection at ever choice point. SPROUT [9] takes the other approach and uses a tree-search algorithm, which can be run in a modified "best-search" algorithm or to completion. It directs growth by prioritizing growth points based on closeness to unsatisfied site points and pruning templates that prevent reaching site points by being too close but not satisfying site point. SPROUT is commercially available at SimBioSys, Inc. (http://www.simbiosys.ca/sprout/index.html). FlexNovo [26] uses FlexX to dock initial fragments and a buildup procedure based on a k-greedy algorithm. It is commercially available at BioSolveIt (http://www.biosolveit.de/FlexNovo/).

10.3.1.2 Advantages, Limitations, and Computational Complexity?

The tree searches are deterministic algorithms. Run to completion, they will find all solutions. The time complexity for most tree searches is $O(b^d)$, where b is the branching factor and d is the depth, although proper heuristics in best-first search can greatly reduce this. The branching factor in this case is a product of the number of attachment points times the number of building blocks, whereas depth is the number of building blocks in a final structure. A quick back-of-the-hand comparison of atom-based versus fragment-based approaches would have b for atom-based methods as ~12 atoms/functional groups times 2 attachment points on average (3 for sp3 atoms, 2 for sp2, 1 for sp) and d (~50) leading to $b^d = 24^{50} \approx 10^{70}$, or roughly all of chemical space. For fragment-based approach with a small fragment library b is 30 fragments times 4 average attachment points each (larger since rings included) and depth approximately 8 is $b^d = 120^8 \approx 10^{16}$. We can see why smaller fragment libraries are usually chosen for tree-search methods, whereas Monte Carlo is chosen for both atom-based and fragment-based methods. This also shows the advantage of using generic templates in tree-search approaches such as SPROUT, which reduces the complexity by greatly reducing size of the template library. Note that the diversity covered in this approach is far greater than 10^{16} because atom types are placed into the generic fragments, but it does not approach the full diversity from an atom-based approach.

Overall, the *grow* algorithms have been successful in finding new drug candidates. However, they tend to behave poorly in situations where the receptor site consists of two or more subpockets separated by a large gap, whereas the fragment-link (next section) performs better in these situations.

10.3.2 Fragment-Link

The fragment-link strategy also starts out by identifying site points in a target receptor or pharmacophore model. In this case, complementary fragments are placed in all of these "hot spots" to maximize interaction. Results at this point can be pruned by visual inspection. Linking groups are then generated or chosen from a link library and fitted to the fragments. Linking groups that do not match the primary constraints (shape & chemistry) or make substructures that violate secondary constraints can be discarded. The final structures are evaluated by predicted binding affinity and secondary scoring characteristics such as synthetic accessibility and ADMET, and prioritized (*cf.* Figure 10.1b and Table 10.1.b).

ALGORITHM 10.2: PSEUDOCODE FOR FRAGMENT-LINK STRATEGY

```
Input: receptor or pharmacophore, library of initial
       building blocks, library of bridging building
       blocks

assign sitepoints
place building block(s) in sitepoint(s) according
 to rules
prune by criteria (visual inspection and/or score)
WHILE (NOT all fragments joined) DO
       identify 2 closest fragments
       identify link points between fragments (closest
        atoms)
       place bridging group(s) to join at these
        points by matching distances and angles to
        bridge library.

END DO
evaluate final structures for binding affinity
evaluate final structures for synthetic accessibility
 and ADMET
prioritize final structures
```

10.3.2.1 Programs in Current Use that Implement Fragment-Link

Several programs have successfully applied the *fragment-link* algorithm to identify lead compounds (see Table 10.1.b). Best-known is the Ludi [14,16] program, available commercially at Accelrys (http://accelrys.com/). The MCSS [61] program has been combined with several others for the bridging step including HOOK [25], Leapfrog [62], LUDI, and by visual inspection. Ligbuilder is a hybrid algorithm that includes *grow* and *fragment-link*, both using an EA to generate structures. Ligbuilder is freely available at (http://www.chem.ac.ru/Chemistry/Soft/LIGBUILD.en.html).

10.3.2.2 Advantages, Limitations, and Computational Complexity?

This approach has the advantage in that it maximizes interactions in the key interaction sites in the target protein. It has the computational advantage that the search for bridge points is an O(n) lookup through a fragment database. The challenge is identifying linking groups of the proper chemistry and geometry that do not greatly alter the orientation of the fragments binding to these sites, and which do not have artificially strained bonds, angles, and torsions. In terms of amount of chemical space sampled, it covers roughly the same chemical space as other fragment-based methods using the *grow* strategy.

10.3.3 Sampling Strategies with EAs

Sampling strategies differ from *grow* and *fragment-link* in that they sample structures without directing generation in a particular direction (outward or explicitly linking interaction sites). EAs are the most common chosen for this purpose. There are many types of EAs: genetic algorithms (GAs) that encode the molecular structure in a "chromosome" of fixed length that is operated on and transformed into the molecular structure for fitness evaluations; genetic programming, where the chromosomes are trees to allow them to have variable length; and evolutionary strategies that operate directly on the phenotype, which is the molecular structure. A basic evolutionary strategy (μ, λ) is shown below [63], where λ is the size of the child population and μ is the size of the parent population in each generation.

The algorithm starts with a population of λ chemical structures (the initial child population) generated from putting a random selection of building blocks together according to the building rules for the building blocks. Each structure in this population is evaluated for "fitness." The fitness is the scoring function that can combine primary and secondary constraints. In receptor-based *de novo* methods, the primary score may be the interaction score such as from a docking calculation [64], minus any boundary violations. For ligand-based employing of a single template ligand, the primary score may be a similarity score. Secondary scoring considerations, such as requirements for Lipinski's [57] rules, or other ADMET considerations, can be added to the fitness function here, along with other molecule properties such as surface area or radius of gyration. The most fit μ structures are selected to be parents for the next generation. Mutation and crossover operators can be performed on the parents. Mutation in this case is to take a parent structure and remove a building block and replace it according to the joining rules. Crossover is to remove building blocks from each parent and swap them again according to joining rules. Some algorithms have only mutation [24] or only crossover [52]. A total of λ child structures are generated. This cycle is repeated until it reaches a maximum generation of children or a termination condition is reached, such as convergence. One feature found with this algorithm was that since building blocks could vary greatly in size, the parent p could grow and shrink as well, while still retaining the same number of building blocks.

ALGORITHM 10.3: PSEUDOCODE FOR BASIC EVOUTIONARY STRATEGY (μ, λ)

```
## μ is the size of parent population
## λ is the size of the child population
Generate λ random structures Sc
DO
   Evaluate fitness Fc of each structure Sc in
      population
   Choose μ most fit (Fc) structures as parents Sp
   Mutate and Crossover of parents Sp to generate
      population λ children Sc
UNTIL (> maximum generations or termination condition)
```

10.3.3.1 Programs in Current Use that Implement Sampling Strategies

EAs are common especially in ligand-based design programs, although several receptor-based programs also employ this approach. See Table 10.1.c for some examples of lead compounds identified using the EA. One successful implementation of this algorithm is in the TOPAS [24] ligand-based *de novo* program, which uses pairwise similarity to a molecular template as the fitness function. It sets $\lambda = 100$ and $\mu = 1$ (i.e., 1 parent) with no crossover operation, so all variations are through mutation. It uses 25,000 fragments from the WDI using 11 retrosynthetic pathways. The variance in each new child structure can be controlled by how similar a new building block is to the original building block being mutated, and is controlled by a parameter ("step-size") that is a Gaussian distribution of random numbers, resulting in a child population that is bell-shaped distribution of variations with the parent at the center. It was found that 100 generations were sufficient to explore chemical space in this program. TOPAS is at Hoffmann-La Roche and is not generally available.

Flux [37] was developed based on TOPAS. It finds optimal results with a 50:50 ratio between crossover and mutation, and typically sets maximum generations to 75 (50 generations were found to converge in most cases). The other main difference from TOPAS is a modified similarity descriptor that is weighted. It is being used at the Goethe University in Germany but is currently not generally available.

LEA3D uses fragments as building blocks generating from fragmenting "drug-like" database of over 8,000 fragments into single rings, fused rings, and acyclic parts. It allows both 1 and 2 point crossover and mutation. It is not generally available, but an in-house version is in use at the Centre De Biochimie Structurale, Montepelier, France.

De novo programs incorporating EAs that are commercially available include AutoLudi and LeapFrog [48] and the Molecule Evoluator [56,65]. AutoLudi is an extension of LUDI that uses an EA to modify an existing lead compound by adding on small fragments. LeapFrog commercially available at Tripos (http://www.tripos.com) evolves a population of molecules in an atom-based method. The Molecular Evoluator [56] uses an unusual fitness function—the user working interactively with the program. It is available at Cidrux Pharmaceuticals [66].

10.3.3.2 Advantages, Limitations, and Computational Complexity?

A general challenge for EAs is the molecule representation. SMILES strings such as in LEA [11] have the problem that invalid molecules result during crossover and mutation, and also more steps are required to build up a molecule. TreeSmiles, a variation of SMILES with all hydrogens explicitly shown, helps avoid unreasonable structures [56]. The LEA3D successor of LEA uses fragments instead of atoms as the building block, with numbers representing fragments for genetic operations [67]. Other approaches operate directly on the 3D structures leading to additional translational and rotational operators [36].

The theoretical chemical space available for a fragment-based approach is $(nb)^{ns}$, where nb is the number of building blocks in the fragment library, and ns is the average number of building blocks is the final structure. For TOPAS that has 25,000 building blocks and approximately four building blocks in a final structure, the total is $(25{,}000)^4 \approx 10^{18}$. For an atom-based method, this would approach all of chemical space ($\approx 10^{60}$). However, the number of structures actually evaluated in an EA run is much smaller as it is given by the function λ ng, where λ represents the population of each generation and ng the number of generations. For example, TOPAS has population size 100×100 generations $\approx 10^4$. Similarly, LEA3D has a a population size of 40×100 generations, that is, 4000. In practice, this seems to be sufficient to generate enough reasonable solutions to find interesting leads.

Compared to the *grow* and *fragment-link*, EA algorithms have the advantage that, since they do not target interaction site points, the output structures are not strained (i.e., have low intramolecular energies). The corresponding disadvantage is that they may not bind to known important interaction sites.

10.4 SUMMARY

In examining all *grow*, *fragment-link*, and sampling-based algorithms, one aspect in common is the use of a random operator of some sort during the structure generation process. This is important for two reasons: first because the scoring functions are not perfect; the best-scoring atom or fragment may not represent the best binder. Second, and more importantly, because the path to construct a *de novo* structure is not a linear function of the scoring function (i.e., higher scoring final structures are not a linear result of the highest scoring precursors; a structure often needs to go through a lower energy construction pathway to get to the final structure).

Ligand-based programs that use similarity to a molecular template or QSAR for scoring require a sampling approach, and the EA is the most commonly chosen one for these programs for its simplicity to program up and its effectiveness in these cases. With receptor-based approaches, the *grow* and *fragment-link* algorithms, which include pharmacophore data in the form of site points, are historically favored. Pure sampling approaches are most commonly seen as lead-optimization once a core has been designed using a *grow* or *fragment-link* approach.

The major hurdles for *de novo* design to overcome to be an effective tool in drug discovery are the same today as when the field began: how to accurately predict receptor-based affinity and predict synthetic accessibility. Without these, it could be a costly effort to synthesize complex molecules, which may not even bind to the target receptor. Newer heuristics for synthetic accessibility, using reaction-based fragment libraries, and heuristics based on molecular complexity have improved the quality of structures resulting from *de novo* design. Several *de novo* design strategies have now been shown to be successful in prospective studies.

REFERENCES

1. Bohacek, R. S., McMartin, C., and Guida, W. C., The art and practice of structure-based drug design: A molecular modeling perspective. *Med. Res. Rev.* 1996, 16(1), 3–50.
2. Anonymous, The numbers game. *Nat. Rev. Drug Discov.* 2002, 1(12), 929.
3. Bolton, E. E., Wang, Y., Thiessen, P. A., and Bryant, S. H., PubChem: Integrated platform of small molecules and biological activities, *Annual Reports in Computational Chemistry*, 2008, p. 4.

4. Danziger, D. J. and Dean, P. M., Automated site-directed drug design: A general algorithm for knowledge acquisition about hydrogen-bonding regions at protein surfaces. *Proc. R. Soc. Lond. B Biol. Sci.* 1989, 236(1283), 101–113.
5. Lewis, R. A. and Dean, P. M., Automated site-directed drug design: The formation of molecular templates in primary structure generation. *Proc. R. Soc. Lond. B. Biol. Sci.* 1989, 236(1283), 141–162.
6. Lewis, R. A. and Dean, P. M., Automated site-directed drug design: The concept of spacer skeletons for primary structure generation. *Proc. R. Soc. Lond. B Biol. Sci.* 1989, 236(1283), 125–140.
7. Lewis, R. A., Roe, D. C., Huang, C., Ferrin, T. E., Langridge, R., and Kuntz, I. D., Automated site-directed drug design using molecular lattices. *J. Mol. Graph.* 1992, 10(2), 66–78, 106.
8. Lewis, R. A., Automated site-directed drug design: Approaches to the formation of 3D molecular graphs. *J. Comput. Aided Mol. Des.* 1990, 4(2), 205–210.
9. Gillet, V. J., Newell, W., Mata, P., Myatt, G., Sike, S., Zsoldos, Z., and Johnson, A. P., SPROUT: Recent developments in the *de novo* design of molecules. *J. Chem. Inf. Comput. Sci.* 1994, 34(1), 207–217.
10. Clark, D. E., Frenkel, D., Levy, S. A., Li, J., Murray, C. W., Robson, B., Waszkowycz, B., and Westhead, D. R., PRO-LIGAND: An approach to *de novo* molecular design. 1. Application to the design of organic molecules. *J. Comput. Aided Mol. Des.* 1995, 9(1), 13–32.
11. Douguet, D., Thoreau, E., and Grassy, G., A genetic algorithm for the automated generation of small organic molecules: Drug design using an evolutionary algorithm. *J. Comput. Aided Mol. Des.* 2000, 14(5), 449–466.
12. Schneider, G., Clement-Chomienne, O., Hilfiger, L., Schneider, P., Kirsch, S., Bohm, H. J., and Neidhart, W., Virtual screening for bioactive molecules by evolutionary *de novo* design. *Angew Chem. Int. Ed. Engl.* 2000, 39(22), 4130–4133.
13. Goodford, P. J., A computational procedure for determining energetically favorable binding sites on biologically important macromolecules. *J. Med. Chem.* 1985, 28(7), 849–857.
14. Bohm, H. J., The computer program LUDI: A new method for the *de novo* design of enzyme inhibitors. *J. Comput. Aided Mol. Des.* 1992, 6(1), 61–78.
15. Rotstein, S. H. and Murcko, M. A., GroupBuild: A fragment-based method for *de novo* drug design. *J. Med. Chem.* 1993, 36(12), 1700–1710.
16. Bohm, H. J., LUDI: Rule-based automatic design of new substituents for enzyme inhibitor leads. *J. Comput. Aided Mol. Des.* 1992, 6(6), 593–606.
17. Gillet, V., Myatt, G., Zsoldos, Z., and Johnson, A., SPROUT, HIPPO and CAESA: Tools for *de novo* structure generation and estimation of synthetic accessibility. *Perspect. Drug Discov. Des.* 1995, 3(1), 34–50.
18. Miranker, A. and Karplus, M., Functionality maps of binding sites: A multiple copy simultaneous search method. *Proteins* 1991, 11(1), 29–34.
19. Roe, D. C. and Kuntz, I. D., BUILDER v.2: Improving the chemistry of a *de novo* design strategy. *J. Comput. Aided Mol. Des.* 1995, 9(3), 269–282.
20. Gillet, V., Johnson, A. P., Mata, P., Sike, S., and Williams, P., SPROUT: A program for structure generation. *J. Comput. Aided Mol. Des.* 1993, 7(2), 127–153.
21. Todorov, N. P. and Dean, P. M., A branch-and-bound method for optimal atom-type assignment in *de novo* ligand design. *J. Comput. Aided Mol. Des.* 1998, 12(4), 335–410.
22. Nishibata, Y. and Itai, A., Automatic creation of drug candidate structures based on receptor structure. Starting point for artificial lead generation. *Tetrahedron* 1991, 47(43), 8985–8990.
23. Lewell, X. Q., Judd, D. B., Watson, S. P., and Hann, M. M., RECAP—retrosynthetic combinatorial analysis procedure: A powerful new technique for identifying privileged

molecular fragments with useful applications in combinatorial chemistry. *J. Chem. Inf. Comput. Sci.* 1998, 38(3), 511–522.
24. Schneider, G., Lee, M. L., Stahl, M., and Schneider, P., *De novo* design of molecular architectures by evolutionary assembly of drug-derived building blocks. *J. Comput. Aided Mol. Des.* 2000, 14(5), 487–494.
25. Eisen, M. B., Wiley, D. C., Karplus, M., and Hubbard, R. E., HOOK: A program for finding novel molecular architectures that satisfy the chemical and steric requirements of a macromolecule binding site. *Proteins* 1994, 19(3), 199–221.
26. Degen, J. and Rarey, M., FlexNovo: Structure-based searching in large fragment spaces. *Chem. Med. Chem.* 2006, 1(8), 854–868.
27. Wang, R., Gao, Y., and Lai, L., LigBuilder: A multi-purpose program for structure-based drug design. *J. Mol. Model.* 2000, 6(7), 498–516.
28. Moon, J. B. and Howe, W. J., Computer design of bioactive molecules: A method for receptor-based *de novo* ligand design. *Proteins* 1991, 11(4), 314–328.
29. Rotstein, S. H. and Murcko, M. A., GenStar: A method for *de novo* drug design. *J. Comput. Aided Mol. Des.* 1993, 7(1), 23–43.
30. Bohacek, R. S. and McMartin, C., Multiple highly diverse structures complementary to enzyme binding sites: Results of extensive application of a *de novo* design method incorporating combinatorial growth. *J. Am. Chem. Soc.* 1994, 116(13), 5560–5571.
31. Murray, C. W., Clark, D. E., Auton, T. R., Firth, M. A., Li, J., Sykes, R. A., Waszkowycz, B., Westhead, D. R., and Young, S. C., PRO_SELECT: Combining structure-based drug design and combinatorial chemistry for rapid lead discovery. 1. Technology. *J. Comput. Aided Mol. Des.* 1997, 11(2), 193–207.
32. David, A. and Pearlman, M. A. M., CONCEPTS: New dynamic algorithm for *de novo* drug suggestion. *J. Comput. Chem.* 1993, 14(10), 1184–1193.
33. Gehlhaar, D. K., Moerder, K. E., Zichi, D., Sherman, C. J., Ogden, R. C., and Freer, S. T., *De novo* design of enzyme inhibitors by Monte Carlo ligand generation. *J. Med. Chem.* 1995, 38(3), 466–472.
34. Todorov, N. P. and Dean, P. M., Evaluation of a method for controlling molecular scaffold diversity in *de novo* ligand design. *J. Comput. Aided Mol. Des.* 1997, 11(2), 175–192.
35. Hartenfeller, M., Proschak, E., Schuller, A., and Schneider, G., Concept of combinatorial *de novo* design of drug-like molecules by particle swarm optimization. *Chem. Biol. Drug Des.* 2008, 72(1), 16–26.
36. Glen, R. C. and Payne, A. W., A genetic algorithm for the automated generation of molecules within constraints. *J. Comput. Aided Mol. Des.* 1995, 9(2), 181–202.
37. Fechner, U. and Schneider, G., Flux (2): Comparison of molecular mutation and crossover operators for ligand-based *de novo* design. *J. Chem. Inf. Model.* 2007, 47(2), 656–667.
38. Vinkers, H. M., de Jonge, M. R., Daeyaert, F. F., Heeres, J., Koymans, L. M., van Lenthe, J. H., Lewi, P. J., Timmerman, H., Van Aken, K., and Janssen, P. A., SYNOPSIS: Synthesize and optimize system *in silico*. *J. Med. Chem.* 2003, 46(13), 2765–2773.
39. Pierce, A. C., Rao, G., Bemis, G. W., BREED: Generating novel inhibitors through hybridization of known ligands. Application to CDK2, p38, and HIV protease. *J. Med. Chem.* 2004, 47(11), 2768–2775.
40. Nachbar, R. B., Molecular evolution: Automated manipulation of hierarchical chemical topology and its application to average molecular structures. *Genet. Prog. Evolv. Mach.* 2000, 1(1–2), 57–94.
41. Andrew, R. and Leach, R. A. L., A ring-bracing approach to computer-assisted ligand design. *J. Comput. Chem.* 1994, 15(2), 233–240.

42. Bohm, H. J., Prediction of binding constants of protein ligands: A fast method for the prioritization of hits obtained from *de novo* design or 3D database search programs. *J. Comput. Aided Mol. Des.* 1998, 12(4), 309–323.
43. Bohm, H. J., The development of a simple empirical scoring function to estimate the binding constant for a protein–ligand complex of known three-dimensional structure. *J. Comput. Aided Mol. Des.* 1994, 8(3), 243–256.
44. Ishchenko, A. V. and Shakhnovich, E. I., SMall molecule growth 2001 (SMoG2001): An improved knowledge-based scoring function for protein–ligand interactions. *J. Med. Chem.* 2002, 45(13), 2770–2780.
45. DeWitte, R. S. and Shakhnovich, E. I., SMoG: *De novo* design method based on simple, fast, and accurate free energy estimates. 1. Methodology and supporting evidence. *J. Am. Chem. Soc.* 1996, 118(47), 11733–11744.
46. Todorov, N. P., Buenemann, C. L., and Alberts, I. L., *De novo* ligand design to an ensemble of protein structures. *Proteins* 2006, 64(1), 43–510.
47. Alberts, I. L., Todorov, N. P., and Dean, P. M., Receptor flexibility in *de novo* ligand design and docking. *J. Med. Chem.* 2005, 48(21), 6585–6596.
48. *Leapfrog, SYBYL 7.1*, TRIPOS: St. Louis, MO.
49. Waszkowycz, B., Clark, D. E., Frenkel, D., Li, J., Murray, C. W., Robson, B., and Westhead, D. R., PRO_LIGAND: An approach to *de novo* molecular design. 2. Design of novel molecules from molecular field analysis (MFA) models and pharmacophores. *J. Med. Chem.* 1994, 37(23), 3994–4002.
50. Pellegrini, E. and Field, M. J., Development and testing of a *de novo* drug-design algorithm. *J. Comput. Aided Mol. Des.* 2003, 17(10), 621–641.
51. Brown, N., McKay, B., Gilardoni, F., and Gasteiger, J., A graph-based genetic algorithm and its application to the multiobjective evolution of median molecules. *J. Chem. Inf. Comput. Sci.* 2004, 44(3), 1079–1087.
52. Globus, A., Lawton, J., and Wipke, T., Automatic molecular design using evolutionary techniques. *Nanotechnology* 1999, 10(3), 290–299.
53. Honma, T., Hayashi, K., Aoyama, T., Hashimoto, N., Machida, T., Fukasawa, K., Iwama, T., et al., Structure-based generation of a new class of potent Cdk4 inhibitors: New *de novo* design strategy and library design. *J. Med. Chem.* 2001, 44(26), 4615–4627.
54. Boda, K. and Johnson, A. P., Molecular complexity analysis of *de novo* designed ligands. *J. Med. Chem.* 2006, 49(20), 5869–5879.
55. Gasteiger, J., *De novo* design and synthetic accessibility. *J. Comput. Aided Mol. Des.* 2007, 21(6), 307–310.
56. Lameijer, E. W., Kok, J. N., Back, T., and Ijzerman, A. P., The molecule evoluator. An interactive evolutionary algorithm for the design of drug-like molecules. *J. Chem. Inf. Model.* 2006, 46(2), 545–552.
57. Lipinski, C. A., Lombardo, F., Dominy, B. W., and Feeney, P. J., Experimental and computational approaches to estimate solubility and permeability in drug discovery and development settings. *Adv. Drug Deliv. Rev.* 2001, 46(1–3), 3–26.
58. Aronov, A. M., Predictive *in silico* modeling for hERG channel blockers. *Drug Discov. Today* 2005, 10(2), 149–155.
59. Kick, E. K., Roe, D. C., Skillman, A. G., Liu, G., Ewing, T. J., Sun, Y., Kuntz, I. D., and Ellman, J. A., Structure-based design and combinatorial chemistry yield low nanomolar inhibitors of cathepsin D. *Chem. Biol.* 1997, 4(4), 297–307.
60. Ripka, A. S., Satyshur, K. A., Bohacek, R. S., and Rich, D. H., Aspartic protease inhibitors designed from computer-generated templates bind as predicted. *Org. Lett.* 2001, 3(15), 2309–2312.

61. Caflisch, A., Miranker, A., and Karplus, M., Multiple copy simultaneous search and construction of ligands in binding sites: Application to inhibitors of HIV-1 aspartic proteinase. *J. Med. Chem.* 1993, 36(15), 2142–2167.
62. Dong, X., Zhang, Z., Wen, R., Shen, J., Shen, X., and Jiang, H., Structure-based *de novo* design, synthesis, and biological evaluation of the indole-based PPARgamma ligands (I). *Bioorg. Med. Chem. Lett.* 2006, 16(22), 5913–5916.
63. Schwefel, H. P., Deep insight from simple models of evolution. *Biosystems* 2002, 64(1–3), 189–198.
64. Pegg, S. C., Haresco, J. J., and Kuntz, I. D., A genetic algorithm for structure-based *de novo* design. *J. Comput. Aided Mol. Des.* 2001, 15(10), 911–933.
65. Lameijer, E. W., Tromp, R. A., Spanjersberg, R. F., Brussee, J., and Ijzerman, A. P., Designing active template molecules by combining computational *de novo* design and human chemist's expertise. *J. Med. Chem.* 2007, 50(8), 1925–1932.
66. Molecular Evoluator [computer software] CIDRUX Pharminformatics: Haarlem, The Netherlands. www.cidrux.com
67. Douguet, D., Munier-Lehmann, H., Labesse, G., and Pochet, S., LEA3D: A computer-aided ligand design for structure-based drug design. *J. Med. Chem.* 2005, 48(7), 2457–2468.
68. Schneider, G. and Fechner, U., Computer-based *de novo* design of drug-like molecules. *Nat. Rev. Drug Discov.* 2005, 4(8), 649–663.
69. Mauser, H. and Guba, W., Recent developments in *de novo* design and scaffold hopping. *Curr. Opin. Drug Discov. Devel.* 2008, 11(3), 365–374.
70. Voronkov, A. E., Baskin, I. I., Palyulin, V. A., and Zefirov, N. S., Molecular modeling of modified peptides, potent inhibitors of the xWNT8 and hWNT8 proteins. *J. Mol. Graph. Model.* 2008, 26(7), 1179–1187.
71. Nishibata, Y. and Itai, A., Confirmation of usefulness of a structure construction program based on three-dimensional receptor structure for rational lead generation. *J. Med. Chem.* 1993, 36(20), 2921–2928.
72. Iwata, Y., Naito, S., Itai, A., and Miyamoto, S., Protein structure-based *de novo* design and synthesis of aldose reductase inhibitors. *Drug Des. Discov.* 2001, 17(4), 349–510.
73. Heikkila, T., Thirumalairajan, S., Davies, M., Parsons, M. R., McConkey, A. G., Fishwick, C. W., and Johnson, A. P., The first *de novo* designed inhibitors of *Plasmodium falciparum* dihydroorotate dehydrogenase. *Bioorg. Med. Chem. Lett.* 2006, 16(1), 88–92.
74. Ali, M. A., Bhogal, N., Findlay, J. B., and Fishwick, C. W., The first *de novo*-designed antagonists of the human NK(2) receptor. *J. Med. Chem.* 2005, 48(18), 5655–5658.
75. Ji, H., Zhang, W., Zhang, M., Kudo, M., Aoyama, Y., Yoshida, Y., Sheng, C., et al., Structure-based *de novo* design, synthesis, and biological evaluation of non-azole inhibitors specific for lanosterol 14alpha-demethylase of fungi. *J. Med. Chem.* 2003, 46(4), 474–485.
76. Grembecka, J., Sokalski, W. A., and Kafarski, P., Computer-aided design and activity prediction of leucine aminopeptidase inhibitors. *J. Comput. Aided. Mol. Des.* 2000, 14(6), 531–544.
77. Grembecka, J., Mucha, A., Cierpicki, T., and Kafarski, P., The most potent organophosphorus inhibitors of leucine aminopeptidase. Structure-based design, chemistry, and activity. *J. Med. Chem.* 2003, 46(13), 2641–2655.
78. Alig, L., Alsenz, J., Andjelkovic, M., Bendels, S., Benardeau, A., Bleicher, K., Bourson, A., et al., Benzodioxoles: Novel cannabinoid-1 receptor inverse agonists for the treatment of obesity. *J. Med. Chem.* 2008, 51(7), 2115–2127.

79. Rogers-Evans, M., Alanine, A. I., Bleicher, K. H., Kube, D., and Schneider, G., Identification of novel cannabinoid receptor ligands *via* evolutionary *de novo* design and rapid parallel synthesis. *QSAR Comb. Sci.* 2004, 23(6), 426–430.
80. Stahl, M., Todorov, N. P., James, T., Mauser, H., Boehm, H. J., and Dean, P. M., A validation study on the practical use of automated *de novo* design. *J. Comput. Aided Mol. Des.* 2002, 16(7), 459–478.
81. Fechner, U. and Schneider, G., Flux (1): A virtual synthesis scheme for fragment-based *de novo* design. *J. Chem. Inf. Model.* 2006, 46(2), 699–707.
82. Schuller, A., Suhartono, M., Fechner, U., Tanrikulu, Y., Breitung, S., Scheffer, U., Gobel, M. W., and Schneider, G., The concept of template-based *de novo* design from drug-derived molecular fragments and its application to TAR RNA. *J. Comput. Aided Mol. Des.* 2008, 22(2), 59–68.

11 Reaction Network Generation

Jean-Loup Faulon and Pablo Carbonell

CONTENTS

11.1 INTRODUCTION

Designing new drugs or chemicals, understanding the kinetics of combustion and petroleum refining, studying the dynamics of metabolic networks, and using metabolism to synthesize compounds in microorganisms are applications that require us to generate reaction networks. As the networks may be quite large, there is a need to automate the procedure. The purpose of this chapter is to present algorithms that have been developed to handle that task. Section 11.2 introduces the problem and the challenges associated with network generation. Section 11.3 gives various methods used to represent chemical reactions and reaction networks. Section 11.4 introduces techniques to simulate the dynamics of reaction networks and, in particular, approaches to estimate rate constants of the reactions. Section 11.5 presents deterministic techniques to generate networks. It is not always possible to deterministically produce reaction networks due to the combinatorial explosion of the number of species involved; stochastic network generation methods are presented in Section 11.6.

11.2 THE CHALLENGES OF GENERATING NETWORKS

Provided an ensemble of possible chemical reactions, the network generation process essentially consists of finding all species that can be synthesized from an initial set of reactants. The process is recursive as reactions can be applied to generated species as well. The process ends depending on the goal of the study, for instance, a specific compound has been produced, or all compounds with predefined physical, chemical, or biological properties have been generated. The same process can be used to search for all species producing a given target when reversing the reactions. This reverse process is used in retrosynthesis analysis where reversed reactions are named transforms, and species generated are named synthons [1]. The network generation algorithms given in this chapter apply to both synthesis and retrosynthesis as it is just a matter of defining the direction of the reactions. As reviewed by Ugi et al. [2], three main approaches have been developed for network generation: empirical approaches, whose strategies are based on data from reaction libraries, semiformal approaches based on heuristic algorithms, where reactions are derived from a few mechanistic elementary reactions, and formal techniques, based on graph theory. This chapter focuses on the last approach. The formal approach has spurred most of the network generation algorithms, which historically started with the work of Corey et al. [3] and the retrosynthesis code Lhasa (http://www.lhasalimited.org). Ugi et al. argue that formal techniques are general enough to be applicable to any type of reacting system in organic or inorganic chemistry; furthermore, formal techniques are the only methods capable of generating new reaction mechanisms and therefore elucidating unresolved chemical processes such as those found with synthesis planning.

While formal techniques are robust, their computational scaling limits their applicability to real reacting systems. Indeed, as has been shown by several authors [4–9], for many processes related to retrosynthesis, combustion, and petroleum refining, the number of reactions and intermediate species that are created generally scales exponentially with the number of atoms of the reactants. With metabolic and signaling networks, the number of possible species also grows prohibitively [10,11] when one considers all the possible states a species can fall into (phosphorylation at several sites, and complex formation with other proteins and ligands). As a matter of fact it has been shown that even simple models of the epidermal growth factor (EGF) receptor signaling network can generate more than 10^{23} different species [12].

With all these systems, not only the time taken to generate networks may scale exponentially, but simulating the dynamics of large networks may also become computationally prohibitive. The most common approach to study the dynamics of a reaction network is to calculate species concentration over time by integrating a system of ordinary differential equations (ODEs). The computational cost of integrating ODEs depends nonlinearly on N, the number of chemical species. For stiff ODEs (cf. definition in Section 11.4), that cost scales N^3, and thus simulating a system for which N is larger than $10^4 - 10^5$ becomes impractical.

An alternative to palliate the exponential scaling problem of formal techniques is the reduction of the reaction mechanism. The question raised when the reducing mechanism is used is how to choose a reaction subset that describes correctly the dynamics of the entire system. Reduction strategies in the area of combustion modeling have been reviewed by Frenklach [6]; these are quite general and applicable to other fields. According to Frenklach, there are five types of reduction strategies: (1) global reduction, (2) response modeling, (3) chemical lumping, (4) statistical lumping, and (5) detailed reduction. Global modeling techniques transform a complete reaction scheme into a small number of global reaction steps. Global techniques comprise ad hoc methods such as empirical fitting, reduction by approximations, and lumping. All global techniques are specific to a particular problem and cannot be generalized. Response modeling techniques consist of mapping model responses and model variables through functional relationships. Generally, model responses are species concentrations, and model variables are the initial boundary conditions of the reacting mixture and the model parameters, such as rate coefficients, and transport properties. Usually, model responses are expressed as simple algebraic functions (such as polynomials) in terms of model variables. As with global techniques, response modeling solutions are problem specific since they require data to build algebraic functions. Chemical lumping was first developed for polymerization-type reactions. Chemical lumping models are used when a polymer grows by reaction between the polymer and monomer species. The lumping strategy is guided by similarity in chemical structure or chemical reactivity of the reacting species. The main assumption of chemical lumping is that the reactions describing the polymer growth are essentially the same and the associated thermochemical and rate parameters exhibit only a weak dependence on the polymer size. Finally, the detailed reduction technique consists of identifying and removing noncontributing reactions. An effective reduction strategy is to compare the individual reaction rates with the rate of a chosen reference reaction. The reference reaction is, for instance, the rate-limiting step or the fastest reaction. The detailed reaction reduction approach is a general technique and is *a priori* applicable to any reacting system.

The goal of this chapter is to present reaction network generation techniques that are computationally tractable and general enough to be applicable to many processes, such as synthesis planning, combustion, petroleum refining, and signaling and metabolic network inferences. Thus, network generation algorithms are presented focusing on the formal generation approach since it is the only approach that is general enough and is applicable to different processes. As mentioned above, formal techniques scale exponentially with the problem size and therefore reduction methods need to be applied. The only reduction technique applicable to general systems is the detailed reduction method. Because the networks generated can be exponentially large, network generation and reduction cannot be performed in sequence but simultaneously. Section 11.6 presents algorithms where the detailed reduction technique is applied "on the fly" while generating networks.

11.3 REPRESENTATION OF CHEMICAL REACTIONS

There are several reasons why one wants to codify chemical reactions. Reactions need to be stored and retrieved from databases and these databases often require us to link information from different sources. There is also a need for automatic procedures for analyzing, performing correlations, and classifying reactions in databases, a prominent example being the classification of enzymatic reactions [13]. Finally, reactions need to be coded for the construction of knowledge bases for reaction predictions with the purpose of network generation or synthesis design. The reaction codifications systems presented next are focusing on this latter application.

11.3.1 Representation Based on Reaction Centers

The traditional method to represent a chemical reaction is based on reaction centers. A reaction center is constituted of the atoms and bonds directly involved in the bond and electron rearrangement process. The reaction centers are thus composed of the bonds broken and formed when the reaction takes place as well as the atoms attached to these bonds. It is easy to detect these reaction centers from the reaction graphs presented in Chapter 1, and an automated procedure to perform this task has been worked out for some time. Among the published solutions are the ClassCode Infochem system (http://infochem.de/en/products/software/classify.html), HORACE [14], and the reaction classification numbers used in the KEGG database [13]. In the solution adopted with KEGG, reaction centers are searched between all possible pairs (reactant × product) involved in a reaction. An example of detection of reaction center is depicted in Figure 11.1.

To find the reaction center between the elements of a given pair, one first searches the maximum common substructure between the elements. Next, each element (reactant and product) is decomposed into a common part and a nonoverlapping part (dashed boxes in Figure 11.1). The reaction center is the atom (circled in Figure 11.1) belonging to the common part and adjacent to the nonoverlapping part. The process is repeated between all possible pairs and as illustrated in Figure 11.2, a given reaction is characterized by several reaction centers, common parts, and nonoverlapping parts. All reaction centers (R), nonoverlapping (D) parts, and common parts (M) of the KEGG database have been numbered uniquely; thus each reactant × product pair is characterized by three numbers R, D and M, also named the RC numbers. For instance, for the glutamate × N-acetyl-glutamate, $R = 173$, $D = 349$ and $M = 14$; the RC number for the pair is thus 173.349.14. As illustrated in Figure 11.3, when several reactant × product are found for a reaction, the code for the reaction is the compilation of the RC numbers of all pairs.

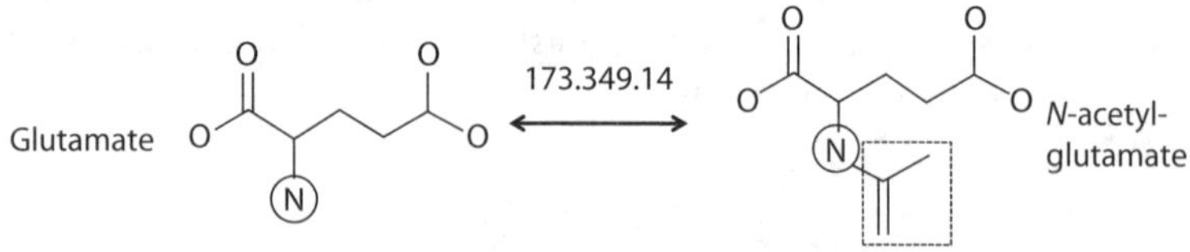

FIGURE 11.1 Reaction center between glutamate and N-acetyl-glutamine. The reaction centers are circles, the nonoverlapping part is in a dashed box, and the remaining parts of the structures are the common parts. The reaction center (circled) is the atom in the common part adjacent to the nonoverlapping part.

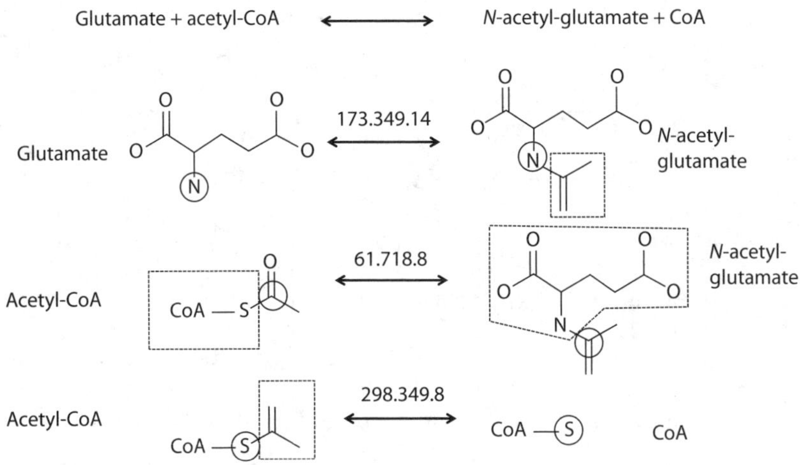

FIGURE 11.2 Reaction classification for amino acid *N*-acetyltransferase (EC 2.3.1.1) between all pair reactants (glutamate, acetyl-Co Enzyme A) × products (*N*-acetyl-glutamate, Co Enzyme A). The figure depicts reaction centers, nonoverlapping parts, and common parts. All parts are depicted using the notation of Figure 11.1.

11.3.2 Bond–Electron Matrices

Bond–electron matrices were first introduced by Ugi and coworkers [15,16]. In this representation, compounds and reactions are coded by bond–electron (*be*-) and reaction (*r*-) matrices (cf. Figure 11.4). In a given *be*-matrix representing a compound, the *i*th row (and column) is assigned to the *i*th atom of the compound. The entry b_{ij}, $(i \neq j)$ of a *be*-matrix is the bond order of the covalent bond between atoms i and j. The diagonal entry b_{ii} is the number of free valence electrons for atom i. It is worth noting that the adjacency and connectivity matrices (cf. Chapter 1) differ from the *be*-matrices in their diagonal entries. The redistribution of valence electrons by

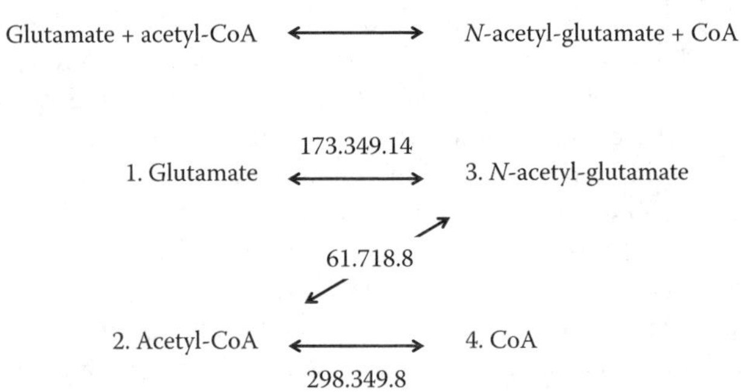

FIGURE 11.3 Codification classification for amino acid *N*-acetyltransferase (EC 2.3.1.1). The code reflects all mapping between reactant and products. The mappings are detailed in Figure 11.2, and the resulting code is 1_3(173.349.14) + 2_3(61.718.8) + 2_4(298.349.8).

	C^1	O^2	H^3	H^4	C^5	N^6
C^1	0	2	1	1	0	0
O^2	2	0	0	0	0	0
H^3	1	0	0	0	0	0
H^4	1	0	0	0	0	0
C^5	0	0	0	0	1	3
N^6	0	0	0	0	3	0

B matrix

	C^1	O^2	H^3	H^4	C^5	N^6
C^1	0	−1	0	0	1	0
O^2	−1	1	0	0	0	0
H^3	0	0	0	0	0	0
H^4	0	0	0	0	0	0
C^5	1	0	0	0	−1	0
N^6	0	0	0	0	0	0

R matrix

	C^1	O^2	H^3	H^4	C^5	N^6
C^1	0	1	1	1	1	0
O^2	1	1	0	0	0	0
H^3	1	0	0	0	0	0
H^4	1	0	0	0	0	0
C^5	1	0	0	0	0	0
N^6	0	0	0	0	3	0

E matrix

FIGURE 11.4 Example of bond–electron (*be*-) matrices and reaction (*r*−) matrices. Negative numbers in the reaction matrix represent broken bonds while positive numbers represent bonds that are created as the reaction proceeds. The sum of a row (column) of *be*-matrices equals the valence of the corresponding atoms. The sum of a row (column) of *r*-matrices equals zero.

which the starting materials of chemical reactions are converted into their products is represented by *r*-matrices.

The fundamental equation of the Dugundji–Ugi model for any given reaction $\alpha_1 + \alpha_2 + \cdots + \alpha_n \rightarrow \beta_1 + \beta_2 + \cdots + \beta_n$ is $B + R = E$, where B (Begin) and E (End) are the *be*-matrices of reactants and products and R is the reaction matrix. The addition of matrices proceeds entry by entry, that is, $b_{ij} + r_{ij} = e_{ij}$. Since there are no formal negative bond orders or negative numbers of valence electrons, the negative entries of R must be matched by positive entries of B of at least equal values.

11.3.3 Representation Based on Fingerprints

Several reaction codification systems are based on molecular fingerprints: examples include the Daylight reaction fingerprints (http://www.daylight.com/dayhtml/doc/theory/theory.finger.html), the reaction signature [5,17], and fingerprints based on atom types [18]. These codification systems have in common that the fingerprint of the reaction is the difference between the fingerprints of the products and the reactants. Taking the example of signature, let us recall from Chapters 3 and 9 that the signature of a molecule (reactant or product) is the sum of the signatures of all atoms of that molecule. The signature of an atom x, ${}^h\sigma_G(x)$, in a molecule G is a subgraph containing the neighborhood of that atom up to a certain distance, named signature height. Examples of atom and molecular signatures are given in Figure 11.5.

Let B and E be two molecular graphs representing the reactants and products of the reaction $R : B \rightarrow E$. Note that *signatures* can be computed on graphs that are not necessarily connected; hence B and E can both be composed of several molecules, including several copies of the same molecule to reflect stoichiometry. The *h-signature*

(a) Atom signature

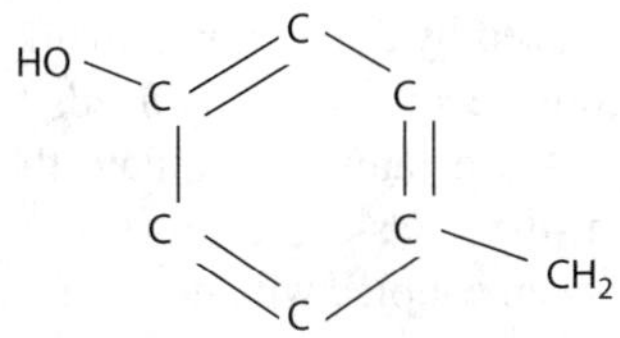

h=0	[C]		
h=1	[C] ([C]	= [C]	[O])
h=2	[C] ([C] (= [C])	= [C] ([C])	
[O] ([H]))			
h=3	[C] ([C] (= [C] ([C]))	= [C] ([C] (= [C])) [O] ([H])	

(b) Molecular signature

$^{h=1}\sigma$ (Tyrosine) =

[C] ([C] = [C] [O]) + 4 [C] ([C] = [C]) + [C] ([C] [C] = [C]) + [O] ([C] [H]) + [C] ([C] [H] [H])

FIGURE 11.5 Atom and molecular signatures. (a) The signature is given for the red carbon atom. (b) The molecular signature of tyrosine is derived by summing the signatures of all atoms (in the example, signatures are not computed for hydrogen atoms).

of reaction *R* is given by the equation

$$^{h}\sigma(R) = {^{h}\sigma(E)} - {^{h}\sigma(B)}. \tag{11.1}$$

An example of reaction signature is given in Figure 11.6.

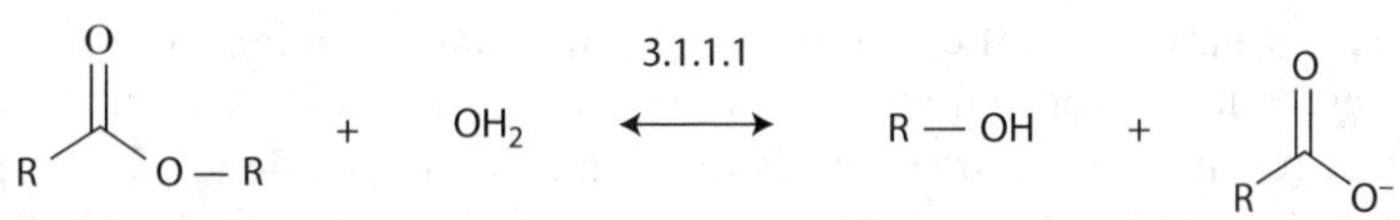

1.0	[R] ([O])	1.0	[O] ([H] [H])	1.0	[R] ([O])	1.0	[R] ([C])
1.0	[R] ([C])			1.0	[O] ([R] [H])	1.0	[O] ([=C])
1.0	[O] ([=C])					1.0	[O–] ([C])
1.0	[O] ([R] [C])					1.0	[C] ([=O] [O–] [R])
1.0	[C] ([=O] [O] [R])						

σ(3.1.1.1)=σ(Alcohol)+σ(Carboxylate)-σ(Carboxylic ester)-σ(H_2O)

−1.0	[O] ([[R] [C])
−1.0	[O] ([H] [H])
1.0	[O–] ([C])
1.0	[O] ([R] [H])
−1.0	[C] ([=O] [O] [R])
1.0	[C] ([=O] [O–] [R])

FIGURE 11.6 Signature for the carboxylesterase enzymatic reaction (EC number 3.1.1.1).

11.4 NETWORK KINETICS

In order to study the kinetics of networks produced by a reaction network generator to, for instance, calculate species concentrations versus time, one needs to estimate reaction rates for each reaction and then use these rates to simulate the kinetics. Methods that have been proposed to perform these tasks are presented next. The presentation is limited to techniques that have been coupled with network generation; these methods are thus constrained by the fact that generated networks may be large; in other words, the techniques must be computational efficient.

11.4.1 ESTIMATING REACTION RATES

In chemical kinetics the rate of a reaction allows one to compute the amount of reactant lost and product formed during an infinitesimal small time interval dt. Considering the chemical reaction $a\mathrm{A} + b\mathrm{B} \rightarrow c\mathrm{C} + d\mathrm{D}$, where a, b, c, and d are the stochiometric coefficients, the concentrations [A], [B], [C], and [D] of the reactants and products are derived from the following equation:

$$-\frac{1}{a}\frac{\mathrm{d[A]}}{\mathrm{d}t} = -\frac{1}{b}\frac{\mathrm{d[B]}}{\mathrm{d}t} = \frac{1}{c}\frac{\mathrm{d[C]}}{\mathrm{d}t} = \frac{1}{d}\frac{\mathrm{d[D]}}{\mathrm{d}t} = r, \tag{11.2}$$

where r is the reaction rate. The reaction rate is in turn calculated with the expression:

$$r = k(T)[\mathrm{A}]^n[\mathrm{B}]^m, \tag{11.3}$$

where $k(T)$ is the rate constant depending on temperature T, and n and m are reaction orders linked to the reaction mechanism. For instance, for first-order reaction of the type $\mathrm{A} \rightarrow \mathrm{C} + \mathrm{D}$, $n = 1$ and $m = 0$; for second-order reaction of the type $\mathrm{A} + \mathrm{A} \rightarrow \mathrm{C} + \mathrm{D}$, $n = 2$ and $m = 0$; and for second-order reaction of the type $\mathrm{A} + \mathrm{B} \rightarrow \mathrm{C} + \mathrm{D}$, $n = 1$ and $m = 1$. After integrating Equation 11.2, one is able to compute the concentrations of the species versus time from the initial concentrations. Assuming that the initial concentrations and the reaction orders are known, the only parameter that remains to be computed is the rate constant. While some rate constants along with other kinetics parameters can be found in databases [for instance, NIST thermodynamics database for chemicals (http://webbook.nist.gov), BRENDA (http://www.brenda-enzymes.org/) for enzyme catalyzed reactions], only a limited number of reactions are stored in these databases, and there is a need to automatically compute rate constants, especially when generating networks. To this end, several approaches have been developed. One approach makes use of an approximation based on the Arrhenius equation:

$$k(T) = A\exp\left(-\frac{E_\mathrm{a}}{RT}\right), \tag{11.4}$$

where R is the gas constant, A is the so-called pre-exponential factor, and E_a is the activation energy. In this approach, which has been utilized for thermal cracking [5,7], the pre-exponential factor is compiled from experimental literature data for five reaction classes (also named elementary transformations) independent of species-specific

			B-matrix			*R*-matrix			*E*-matrix		
Homolysis:	R_1-R_2 -> R_1• + R_2•	R1	0	1		1	−1		0	1	
		R2	1	0		−1	1		0	1	
Recombination:	R_1• + R_2• -> R_1-R_2	R1	1	0		−1	1		0	1	
		R2	0	1		1	−1		1	0	
H-abstraction:	CH + R• -> RH + C•	H	0	1	0	0	−1	1	0	0	1
		C	1	0	0	−1	1	0	0	1	0
		R	0	0	1	1	0	−1	1	0	0
β-scission:	R-C_β-C_α• -> $C_\alpha = C_\beta$ + R•	R	0	0	1	1	0	−1	1	0	0
		C_α	0	1	1	0	−1	1	0	0	2
		C_β	1	1	0	−1	1	0	0	2	0
Addition:	$C_\alpha = C_\beta$ + R• -> R-C_β-C_α•	R	1	0	0	−1	0	1	0	0	1
		C_α	0	0	2	0	1	−1	0	1	1
		C_β	0	2	0	1	−1	0	1	1	0

FIGURE 11.7 The five elementary transformations of hydrocarbon thermal cracking used in Ref. [5] and their associated (*be*-) and (*r*-) matrices.

structures. The elementary transformations are bond homolysis, radical recombination, hydrogen abstraction, b-scission, and b-addition (cf. Section 11.5 and Figure 11.7 for further explanations). In the approach proposed by Susnow et al. [7], the activation energy is species specific and is computed from linear free energy relationships (LFERs) relating the activation energy to the heat of the reaction. The heat of reactions is either retrieved from the NIST thermodynamics database or calculated using the semiempirical molecular orbital package (MOPAC) (http://openmopac.net/).

With the goal of producing more accurate rate constants another approach has been developed based on a group additivity method. In that method the rate constant is related with compound thermodynamics properties (heat of formation and entropy) by the following equation:

$$k(T) = C \exp\left(-\frac{\Delta H}{RT}\right) \exp\left(\frac{\Delta S}{R}\right), \tag{11.5}$$

where C is a constant depending on temperature, molar volume, and reaction path, and ΔH and ΔS are, respectively, the enthalpy and entropy differences between the transition states geometries and the reactants states. Entropy and enthalpy are calculated using group additivity. The main idea of computing thermodynamics properties by summing the properties of individual groups was initially developed by Benson and Buss, as described in [19]. In the Benson and Buss method, the atom properties are compiled for every possible configuration an atom can have with its first neighbors, and the property of the molecule is simply the sum of the properties of each atom. For instance, using the signature notation of Figure 11.5, alkanes (including methane) have only five possible groups [C]([H][H][H][H]), [C]([C][H][H][H]), [C]([C][C][H][H]), [C]([C][C][C][H]), and [C]([C][C][C][C]), and the property of every alkane is simply the scalar product between the molecular signature of the compound and the calculated property of the groups. To compute group properties, quantum calculation can be used. As an example of application of the group additivity method relevant to the chapter, Sumathi et al. [20,21] calculated enthalpies, entropies, and heat capacities for hydrocarbon–hydrogen abstractions using *ab initio* calculations.

11.4.2 Simulating Network Kinetics

Once rates have been estimated, the species concentrations can be calculated integrating reaction equations such as Equation 11.2. There are essentially two main mathematical approaches to integrate a system of reaction equations. In the first approach, the concentrations of chemical species are modeled as continuous variables that change over time and reactions between species are modeled by ODEs. The set of ODEs are often numerically integrated using solvers that can be found in many freeware and commercial packages. In the second approach, chemical species are treated as discrete variables that change over time in a discrete manner. In this approach, reactions are individual events that update the system and can be combined into a chemical master equation [22]. The chemical master equation is often computationally integrated to compute the time evolution of the systems using stochastic simulation algorithm (SSA). Both method (ODEs and SSA) provide exact solutions albeit there are instances where ODEs break down. An example of such a breakdown with ODEs is when an individual reaction event causes a large difference in the likelihood that other reactions will occur, and so the precise order and timing of individual reactions can influence the overall system behavior. ODEs have also difficulties to solve stiff equations, that is, a differential equation for which numerical methods are numerically unstable, unless the step size is taken to be extremely small. For the above reasons, all algorithms presented in the chapter make use of SSAs rather than ODEs.

SSAs were first proposed by Gillespie [22] and are based on Monte Carlo (MC) sampling. The MC–Gillespie technique monitors the number of particles for each species versus time. The initial number of particles of the reactants is computed from their initial concentrations (given by the user) and the initial number, M_p, of particles in the system. In the present chapter, the MC–Gillespie technique is used at constant volume V, which is calculated from the initial number of particles and the particle density (both user inputs). The MC–Gillespie technique is an exact method for numerical integration of the time evolution of any spatially homogeneous mixture of molecular species that interact through a specified set of coupled chemical reaction channels. The technique is based on a fundamental equation giving the probability at time t that the next reaction in V will occur in the differential time interval $[t + \tau, t + \tau + d\tau]$ and will be an r_μ reaction:

$$P(\tau, \mu)d\tau = P_1(\tau)P_2(\mu|\tau), \qquad (11.6)$$

where P_1 is the probability that the next reaction will occur between times $t + \tau$ and $t + \tau + d\tau$:

$$P_1(\tau) = a \exp(-a\tau)d\tau, \qquad (11.7)$$

and P_2 is the probability that the next reaction will be r_μ:

$$P_2(\mu|\tau) = \frac{a_\mu}{a} \tag{11.8}$$

with

$$a = \Sigma_\mu a_\mu. \tag{11.9}$$

In the previous equations, a_μ dτ is the probability, to first order in dτ, that a reaction r_μ will occur in V in the next time interval dτ. The rate constant k_μ is related to a_μ in different ways depending on whether the reaction is monomolecular or bimolecular. For monomolecular reactions

$$a_\mu = [s]k_\mu, \tag{11.10}$$

where $[s]$ is the number of particles of reactant species s. For bimolecular reactions involving two species $s1$ and $s2$,

$$a_\mu = [s_1][s_2]\frac{k_\mu}{V}, \tag{11.11}$$

and for bimolecular reactions involving only one reactant species,

$$a_\mu = [s]([s]-1)\frac{k_\mu}{2V}. \tag{11.12}$$

In order to integrate Equation 11.6, Gillespie proposes the following scheme. First, generate a random value t according to $P_1(t)$ and, second, generate a random integer μ according to $P_2(\mu|\tau)$. The random value τ is computed by simply drawing a random number r_1 from a uniform distribution in the unit interval and taking

$$\tau = \frac{1}{a}\ln\frac{1}{r_1}. \tag{11.13}$$

In turn, the random integer μ is generated by drawing another random number r_2 in the unit interval and by taking μ the smallest integer verifying

$$\sum_{\nu=1}^{\mu} a_\nu > r_2 a. \tag{11.14}$$

In his original paper, Gillespie [22] has proven that expressions 11.13 and 11.14 are indeed correct to simulate stochastically the time evolution of homogeneous reactions. Since this original paper, improvement in SSAs has appeared in the literature, especially regarding computational time; these have recently been reviewed by Gillespie [23].

11.5 REACTION NETWORK GENERATION ALGORITHM

The network generator algorithm presented next is based on the Dugundji–Ugi theory [15] and the computer programs RAIN developed by Fontain and Reitsam [24] and NetGen developed by Broadbelt et al. [4]. The generator is limited to monomolecular and bimolecular reactions but could easily be extended to more complex reactions. Following the Dugundji–Ugi methodology, reactants are represented by *be*-matrices. The reactions are represented in the form of elementary transformations. As any reaction, an elementary transformation is composed of two configurations containing atoms participating in a reaction before and after the reaction has taken place. Unlike full reactions, the configurations with elementary transformations do not take into account the full structure of the species involved, but focuses on the reaction center and its immediate surrounding. In other words, elementary transformations focus on the electronic transformations that atoms undergo when reacting. One notes that the codifications (RC numbers, *be*- and *r*-matrices, and signatures) presented in the Section 11.3 can all be restricted to reaction centers, and thus can all be used to code elementary transformations. Examples of elementary transformations using *be*- and *r*-matrices are given in Figure 11.7 for thermal cracking. Fontain and Reitsan [24] compiled the complete set of elementary transformations that carbon atoms can take in organic reactions. This set is composed of 324 transformations; however, the set can be greatly reduced depending on the studied process. For instance, it is known [25] that hydrocarbon thermal cracking reactions can be generated from the five elementary transformations listed in Figure 11.7. As another example, Susnow et al. [7] proposed 17 transformations to model methane combustion. As a final example, Hatzimanikatis et al. [26] have computed elementary transformation for metabolic reactions. *r*-matrices were generated for all reactions stored in the metabolic KEGG database (http://www.genome.jp/kegg/) resulting in about 250 unique elementary transformations.

The input of the deterministic reaction network generator (Algorithm 11.1) is a list of reactant species, a list of constraints, and a list of elementary transformations. Constraints are user defined; examples of constraints are maximum number of species and reactions generated, maximum species size, maximum number of reactants per reaction, maximum number of lone electrons, and so on. The output of the algorithm is a network composed of all possible species and reactions that can be created from the input. The algorithm described below is illustrated for ethane thermal cracking in Figure 11.8.

The algorithm starts by computing all the possible species ($\mathcal{L}_{s1}$) that can be derived by applying the elementary transformations ($\mathcal{L}_{et}$) to the initial species ($\mathcal{L}_{s0}$) while respecting the constraints. To prohibit duplication of species, $\mathcal{L}_{s1}$ is composed of species that are not already present in $\mathcal{L}_{s0}$. When applying elementary transformations, one has to make the distinction between monomolecular and bimolecular reactions. For each monomolecular elementary transformation, $\mathcal{L}_{s1}$ is composed of all the possible products that can be generated by applying the elementary transformation in all possible ways to the species of $\mathcal{L}_{s0}$. For each bimolecular reaction, $\mathcal{L}_{s1}$ is composed of the products derived by applying the elementary transformations to all possible pairs of species in $\mathcal{L}_{s0}$. The list of reactions $\mathcal{L}_{r1}$ is updated each time an elementary

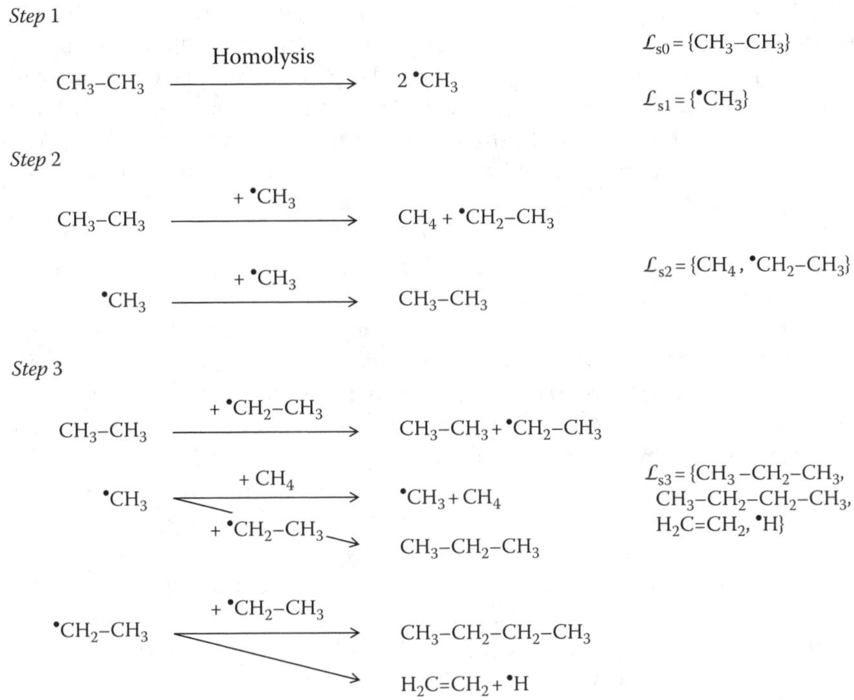

FIGURE 11.8 The first three steps of ethane thermal cracking using the elementary transformations listed in Figure 11.7. These matrices comprise only atoms for which the configuration is changed. The lists of species ($\mathcal{L}_{s0}$, $\mathcal{L}_{s1}$, $\mathcal{L}_{s2}$, and $\mathcal{L}_{s3}$) are given on the right side of the figure.

transformation applies to a species in $\mathcal{L}_{s0}$. Each reaction is represented by n-tuples composed of the elementary transformation (from $\mathcal{L}_{\text{et}}$), the reacting species (from $\mathcal{L}_{s0}$), the product species (from $\mathcal{L}_{s1}$ or $\mathcal{L}_{s0}$), and the rate constant of the reaction. Once the list $\mathcal{L}_{s1}$ has been computed, the algorithm proceeds and computes recursively $\mathcal{L}_{s1}, \mathcal{L}_{s2}, \ldots, \mathcal{L}_{si}$. Note that in order to compute $\mathcal{L}_{si}$ $(i > 1)$ one has to consider monomolecular reactions only from the set $\mathcal{L}_{si-1}$ since all monomolecular reactions from the sets $\mathcal{L}_{si-2}, \mathcal{L}_{si-3}, \ldots$ have already been computed in the previous steps. For the same reason, the products of bimolecular reactions in $\mathcal{L}_{si}$ are generated for every possible pair of species so far generated having at least one element in $\mathcal{L}_{si-1}$. To avoid redundancies of species in the lists $\mathcal{L}_{s0}, \mathcal{L}_{s1}, \ldots, \mathcal{L}_{si}$ at any step $i > 1$, a new species is added to $\mathcal{L}_{si}$ only if the species is not in $\mathcal{L}_{si-1} + \mathcal{L}_{si-2} + \cdots + \mathcal{L}_{s0}$.

The process halts at any step i, where the corresponding lists of species $\mathcal{L}_{si}$ and reaction $\mathcal{L}_{ri}$ are empty. Since bimolecular reactions can potentially create products of infinite molecular weight, in order to keep a finite network size one has to set a limit for the maximum species size; let n be this limit. Note that with the exception of polymerization reactions it is reasonable to assume that all species in a given network will be limited in size. In turn, note that if the species have a limited size, then the network generation algorithm is guaranteed to converge.

Indeed, the maximum number of species, N, that can be formed is bounded by the total number of molecular structures having a number of atoms raging from 1 to the specified value of n. This latter number is equal to the sum of the numbers of isomers containing $1, 2, \ldots, n$ atoms and scales exponentially with n even for simple compounds such as hydrocarbons. Finally, note that the number of reactions is also finite and is bounded by N^4, which is an upper bound on the number of ways of selecting four species (i.e., two reactants and two products) among N species.

The network generator Algorithm 11.1 uses specific data structures to describe species and reactions. A molecular species s is represented by a molecular graph $G(s)$. This graph can in turn be represented by any codification system presented in Section 11.3, *be*-matrix for instance. An elementary transformation (et) is represented by two molecular graphs, B(et) the et of the surrounding atoms before the reaction has taken place, and E(et) the et after the reaction has taken place. These molecular graphs can be represented by any of the codification systems presented in Section 11.3. A reaction, r, is an n-tuple composed of an elementary transformation, a list of reactant species $\mathcal{L}_b(r)$, a list of products $\mathcal{L}_e(r)$, and a rate constant $k(r)$. $\mathcal{L}_b(r)$ is composed of either one or two species for monomolecular and bimolecular reactions, respectively. Additionally, the algorithm maintains several lists of species and reactions: the list of reactants ($\mathcal{L}_{s0}$), the list of species ($\mathcal{L}_{si}$) and reactions ($\mathcal{L}_{ri}$) created at the current step i, and the list of species generated at the previous step ($\mathcal{L}_{si-1}$). As already mentioned, species are added to lists when not already present; since species are represented by molecular graphs, the addition of species requires to check for graph isomorphism between the species to be inserted and the species already in the list. Molecular graph isomorphism algorithms can be found in Chapter 2. In Algorithm 11.1, the function "species-constraints" verifies the user constraints and the function "rate-constant" computes the rate constant of a given reaction using techniques described in Section 11.4.

ALGORITHM 11.1 PSEUDOCODE FOR DETERMINISTIC NETWORK GENERATOR

```
deterministic-network-generator(Ls0,Let,Ls,Lr)
input:   - Lys0: list of initial species
         - Let: list of elementary transformations
output:  - Ls: list of all species in network
         - Lr: list of all reactions in network
   Lys = Ls0, Lr = Ø
   Lsi-1 = Ls0 # Lsi-1 is the list of species at previous
   step i-1
   do
     (Lsi, Lri) = generate-species-reactions (Lsi-1,Ls,Let)
       remove from Lsi all species present in Ls
       remove from Lri all reactions present in Lr
     Ls = Ls + Lsi, Lr = Lr + Lri
     Lsi-1 = Lsi
```

```
   until (Lsi = Ø and Lri = Ø)
generate-species-reactions(Ls1,Ls2,Let)
input:  - Ls1,Ls2: list of species
        - Let: list of elementary transformations
output: - Lsi: list of species created at step i
        - Lri: list of reactions created at step i
  For all tranformations et in Let do
     For all species s1 in Ls1 and s2 in Ls2 do
          # s2 = Ø for monomolecular reactions
          LGe = generate-product(G(s1) + G(s2), et)
        (Lsi,Lri) = update-species-reactions (et,s1,s2,LGe)
     done
   done
   return (Lsi,Lri)
generate-product(Gb,et)
input: - Gb: molecular graph of the reactant(s)
       - et: elementary transformation
output: - LGe: list of products
    LGe = Ø
    B(et), E(et) = Graphs of the reactants and products
     of et
    For all subgraph b of Gb s.t. b is isomorphic
     to B(et) do
       e = Gb - b + E(et) # b is replaced in Gb by E(et)
       if species-constraints(e) and e is not in LGe
       then LGe = LGe + e
    done
    return LGe

update-species-reactions(et,s1,s2,LGe)
input: - LGe: list of graphs returned by
  generate-product
       - et: elementary transformation
       - s1,s2: reactant species
output: - Lsi: list of species created at step i
        - Lri: list of reactions created at step i
   Lsi = Ø, Lri = Ø
   For all graphs Ge in LGe do
      compute Le the list of connected components of Ge
         remove from Le all element already present
            in Lsi
      Lsi = Lsi + Le
      k = rate-constant(et, s1, s2, Le)
      Lri = Lri + (et, s1, s2, Le, k)
   done
   return (Lsi,Lri)
```

Deterministic algorithms similar to the one described above have been applied to processes such as pyrolysis, metabolism, and signal transduction. Broadbelt et al. [4] developed a network generator named NetGen and applied it to hydrocarbon pyrolysis (ethane and cyclohexane). Another pyrolysis network generation process can be found in Ref. [5] for the thermal cracking of butane. With both applications, the elementary transformations and their associated *be*- and *r*-matrices are given in Figure 11.7. All studies generated networks with reactions rates. In Ref. [4], the rates were estimated using linear free energy, and quantitative structure–reactivity relationships as in Faulon and Sault [5]. Finally, in both cases, the numbers of species and reactions were found to scale exponentially with the number of atoms of the initial species.

Beyond pyrolysis, it has been proposed to use techniques similar to NetGen to generate reaction networks to predict toxicity of complex reaction mixtures [27] focusing on degradation of chemicals by cytochrome P450. The deterministic generator proposed by Faulon and Sault has also been used to generate and analyze complex reaction networks in interstellar chemistry [28].

Broadbelt et al. adapted NetGen with a new software named BNICE to explore the diversity of complex metabolic networks [26]. With BNICE, elementary transformations were computed for the KEGG database resulting in about 250 unique elementary transformations. The transformations were then applied to the biosynthesis pathways of aromatic amino acids, phenylalanine, tyrosine, and tryptophan. The three amino acids are synthesized from chrorismate and the cofactors and cosubstrates—glutamate, gutamine, serine, NAD^+/NAD, and 5-phospho-α-D-ribose-1-diphsopahe (PRPP). The native pathways from chorismate to phenylalanine and tyrosine comprise three reactions between less than 10 compounds, and the native pathway leading to tryptophan is composed of five reactions between 19 compounds. The 250 elementary transformations were applied to the initial substrate, cosubstrate, and cofactors, producing 246 compounds with the phenylalanine pathway and 289 and 58 for the tyrosine and tryptophan pathways. These compounds fall into three categories: the compounds that are part of the native pathways (i.e., compounds in KEGG), compounds found in the CAS database (http://www.cas.org/) but not in KEGG, and novel compounds. The main outcome of the study was the *in silico* discovery of novel alternative biosynthesis pathways to aromatic amino acid biosynthesis, which remain to be experimentally verified through enzyme and pathway engineering. BNICE was also applied to polyketide biosynthesis [29]. While about 10,000 polyketides structures have been discovered experimentally, BNICE raised this number over 7 millions.

To generate reaction networks for product biocatalysis and biodegradation, a semi-automated system (UM-PPS) with a database (UM-BBD) has been developed at the University of Minnesota [30] (http://umbbd.msi.umn.edu/predict/). In this system, reaction rules (i.e., elementary transformation) are applied to an initial compound entered by the user. Because several reaction rules applied to an initial compound may result in different products, the user has to select the next product and the next rule to apply. The process is iterated until no more rules can be applied, the user selecting next products and rules at each step. At the time of writing (March 2009) the database contained 259 biotransformation rules. Rules generally transform one functional group into another, for instance, a cyano group can be hydrated with one water molecule to form an amide or hydrolyzed by two water molecules to form carboxylic acid and ammonia. Functional groups are searched through SMARTS matching and transformations are carried using an algorithm whose outcome is identical to the generate-product routine of Algorithm 11.1.

Deterministic network generators have also been used to create models characterizing the dynamics of signal transduction networks. In particular, BioNetGen (http://bionetgen.org) uses an algorithm similar to Algorithm 11.1 with reaction rules instead of elementary transformations [31]. As seen in Section 11.3, elementary transformations depict graph modifications of reactant species undergoing a reaction; similarly, reaction rules code for modifications between reacting biological species (ligand, protein, and complexes). BioNetGen and reaction rules have been used for many biological applications and further information is given in Chapter 15.

As already mentioned, one of the main drawbacks of deterministic network generation is computational complexity due to the fact that an exponentially large number of species may be generated. Techniques that overcome this drawback are discussed in the next section.

11.6 REACTION NETWORK SAMPLING ALGORITHM

As discussed in Algorithm 11.1, the only reduction technique applicable to general systems is the detailed reduction method. The caveat of the method, however, is that the entire network must be known prior to reduction. Although the size of the network generated by the deterministic algorithm is finite due to the limited size of the species, as already mentioned the number of species in the network grows exponentially with n, the number of atoms of the largest species. To overcome the combinatorial explosion issue, network generation and reduction cannot be processed in sequence if the goal is to derive a computationally tractable technique. In this section, three methods are outlined where network generation and reduction are performed simultaneously, concentration-sampling-network-generator (CSNG), MC-sampling-network-generator (MCNG), and single-molecule simulator (SMG). These algorithms are stochastic in nature and are efficient. The idea of the CSNG algorithm was first published by Susnow et al. [7]. The MCNG algorithm was first published by Faulon and Sault [5]. The SMG algorithm simulates the dynamics of the species of an initial molecular graph comprising all reactants, without actually generating a network. The idea of this algorithm was first published by Morton-Firth and Bray [32], for predicting temporal fluctuations in an intracellular signaling pathway. All the three algorithms require on-the-fly assignment of the rate constants of the reactions generated; these computations are detailed in Section 11.4.

11.6.1 CONCENTRATION-SAMPLING NETWORK-GENERATOR ALGORITHM

With the concentration-sampling algorithm, the reduction of the network is based on the species concentrations. The main assumption of this method is that if a species created at any given step i has low concentration values over the reaction time, that species will generate products with concentrations at the most equal to twice that low value (in the particular case the species dissociates into two identical new species).

Therefore, removing low-concentration species from the list $\mathcal{L}_{si}$ should have a negligible effect on the final product distribution. Hence, in the present case, the total number of network species is not limited, but the number of species generated at each step is bounded by a predefined value M_s. Note that this assumption is valid only if the species are consumed during the reactions and do not act as catalysts. Indeed, if a species is a catalyst even with low concentration, this species could have a relatively large impact on the final product distribution. Hence, catalytic species must be identified prior to using this scheme.

Using the above concentration assumption, the algorithm works as follows. At each step i, the deterministic routine generate-species-reaction is run to compute the list of new species. At this point the network is composed of all species in $\mathcal{L}_{s0} + \cdots + \mathcal{L}_{si}$ and associated reactions. Although the network may not be complete, since the reaction rates are calculated on-the-fly, the time evolution of the species concentrations can be computed by solving the system of differential equations associated with the partial network. The algorithm presented next uses the SSA MC–Gillespie algorithm [22] (MC–Gillespie, cf. Section 11.4 for further details) to solve the system. The MC–Gillespie integration (routine MC–Gillespie-step) is called M_c times. That routine updates species concentration according to Equations 11.13 and 11.14. Precisely, if r is the selected reaction, then the numbers of particles of all species in the list of reactants for r ($\mathcal{L}_b(r)$) are decremented by 1 and the numbers of particles of all species in the list of products for r ($\mathcal{L}_e(r)$) are incremented by 1. During the MC–Gillespie integration, the CSNG algorithm retains for each species present in the network the maximum concentration (i.e., maximum number of particles) reached over the time period the system is integrated. This operation requires to maintain two lists: $\mathcal{L}_{[s]}$, the list of species concentrations calculated for a given time step t, and $\mathcal{L}_{[s\max]}$, the list of species maximum concentrations calculated for all M_c integration time steps. Species are then sorted by decreasing concentration in $\mathcal{L}_{[s\max]}$, the first M_s elements of the sorted list are kept in $\mathcal{L}_s$, while the others are eliminated. The numbers M_c and M_s are user input. The algorithm is given in Algorithm 11.2, and the routine generate-species-reaction is given in Algorithm 11.1. The algorithm 11.2 takes as input the list of reactants ($\mathcal{L}_{s0}$), reactant concentration ($\mathcal{L}_{[s0]}$), and elementary transformations ($\mathcal{L}_{et}$); it returns the list of species ($\mathcal{L}_s$) and reactions ($\mathcal{L}_r$) of the generated network.

ALGORITHM 11.2 PSEUDOCODE FOR CSNG

```
concentration-sampling-network-generator(Ls0,L[s0],Let,
  Ls,Lr)
input:   - Ls0: list of initial species (reactants)
         - L[s0]: list of initial species concentrations
         - Let: list of elementary transformations
output:  - Ls: list of all species in network
         - Lr: list of all reactions in network
Ls = Ls0, Lr = Ø
Lsi-1 = Ls0 # Lsi-1 is the list of species at previous
  step i-1
do forever
   (Lsi,Lri) = generate-species-reactions (Lsi-1,Ls,Let)
```

```
      remove from Lsi all species present in Ls
      remove from Lri all reactions present in Lr
   if (Lsi = Ø and Lri = Ø) then end
   Ls = Ls + Lsi, Lr = Lr + Lri
   (Ls, Lsi) = reduce-mechanism-concentration(Ls, L[s0], Lsi)
   Lsi-1 = Lsi
done

reduce-mechanism-concentration(Ls, L[s0], Lsi)
input: - Ls: list of species
       - L[s0]: list of initial species (reactants)
          concentrations
       - Lsi: list of species created at step i
output: - Ls: reduced list of species
        - Lsi: reduced list of species created at step i
  L[s] = L[s0]
  L[smax] = list of species maximum concentration
   over time
  initialized to Ø
  t = 0
  For Mc steps do
     (L[s], t) = MC-Gillespie-step(Ls, L[s], Lr, t)
        L[smax] = MAX(L[smax], L[s])
  done
  While (|Lsi| > Ms) do
        find s in Lsi having the lowest value in L[smax]
        Ls = Ls - s,  Lsi = Lsi - s
  done
  return (Ls, Lsi)

MC-Gillespie-step(Ls, L[s], Lr, t)
input: - Ls: list of species
       - L[s]: list of species concentration
       - Lr: list of reactions
       - t: time
output: - L[s]: updated list of species concentration
        - τ: time after event occurs
   τ = time of next event using eq. (11.13)
   r = selected reaction in Lr occurring at time
      t + τ using eq. (11.14)
   t = t + τ
#Lb(r) and Le(r) are the lists of reactants
  and products for r
   For all s in Lb(r) do [s] = [s] - 1 done
   For all s in Le(r) do [s] = [s] + 1 done
   return (t, L[s])
```

The above algorithm was used by Faulon and Sault for thermal cracking of butane [5]. The results given by the CSNG algorithm are identical to those of the deterministic DNG algorithm when limiting the number of particles created at each step to no more than 8 (for a full deterministic network comprising about 100 species and 1000 reactions).

A different implementation of the CSNG algorithm was developed by Susnow et al. [7]. In this implementation, species concentrations are computed solving ODEs rather than SSAs, and the selection of the species to keep when reducing the network is based not on concentration but on rate of formation. In other words, in Algorithm 11.2, new species are kept in $\mathcal{L}_s$ when their rate of formation is greater than a user-defined threshold R_{min}. Susnow et al. applied their method for the pyrolysis of ethane and butane. The network generation process was iterated until the conversion of the initial reactants was below a user-defined threshold. By changing the R_{min} value, Susnow et al. were able to reduce the number of reacted species and differential equations by an order of magnitude while maintaining suitable complete mechanism.

11.6.2 MC-Sampling-Network-Generator Algorithm

The idea of the MC-sampling algorithm is to perform at the same time both the MC integration and the network generation. The advantage of this technique is that there are no assumptions regarding catalyst species. As in the previous case, one starts with an initial reactant concentration ($\mathcal{L}_{[s0]}$) given in the form of numbers of particles. At each step, the set of new species is computed using the generate-species-reactions routine of Algorithm 11.1, but in the present case, these species are generated by applying the ets only for species having nonzero concentration (i.e., set $\mathcal{L}_{s*}$ in Algorithm 11.3). The concentrations of the new species are set to zero and updated using the MC–Gillespie-step integration routine given in Algorithm 11.2. The process is iterated until the number of steps exceeds a predefined M_c value.

ALGORITHM 11.3 PSEUDOCODE FOR MCNG

```
MC-sampling-network-generator(Ls0, L[s0], Let, Ls, Lr)
input: - Ls0: list of initial species (reactants)
        - L[s0]: list of initial species concentrations
        - Let: list of elementary transformations
output: - Ls : list of all species in network
         - Lr : list of all reactions in network
  Ls = Lsi = Ls0, Lr = Ø, L[s] = L[s0]
  t = 0
  For Mc steps do
  Ls* = species in Ls with non-zero concentration
  Lsi* = species in Lsi with non-zero concentration
  (Lsi, Lri) = generate − species − reactions
   (Lsi*, Ls*, Let)
     remove from Lsi all species present in Ls
     remove from Lri all reactions present in Lr
```

```
Ls = Ls + Lsi, Lr = Lr + Lri
For all species s in Lsi do [s]=0 done
(L[s], t) = MC-Gillespie-step(Ls, L[s], Lr, t)
done
```

Faulon and Sault applied the above scheme to generate the reaction network of butane pyrolysis [5]. When comparing the results with the DNG algorithm, good agreement was found as long as the number of initial particles M_p was around or above 2000; furthermore, the same threshold was found for all alkanes up to octane.

MCNG has been implemented and used to model signaling and protein interaction networks. Faeder et al. [31] implemented MCMG within BioNetGen for signaling cascades such as those found with Toll-like receptors. Lok and Brent [11] developed Moleculizer (http://sourceforge.net/projects/moleculizer/), a software implementing a procedure very much like the MCNG algorithm [33]. Moleculizer has been used to model the receptor–G-protein complex and the MAP kinase cascade and scaffold complex.

11.6.3 SMS Algorithm

This approach, which was first proposed by Monthon-Firth and Bray [32], led to the development of a computer code named StochSim (http://www.ebi.ac.uk/~lenov/stochsim.html). So far it appears that this algorithm has been developed and used exclusively to model biological network dynamics. Below, a general algorithm is proposed, which is not necessarily optimized, but provides some ideas on how the SMS approach could be applied to chemicals. In the StochSim method, the dynamics of the reaction network is simulated without actually compiling species and species concentrations. Instead, the algorithm considers each molecule as a separate entity, all molecules have thus the same concentration (e.g., one particle), and reactions occurs one molecule at a time. Starting with an initial simulation graph ($\mathcal{G}$) comprising all reactants, each reactant being duplicated a number of times equal to its initial concentration, the main task of the algorithm is to apply elementary transformations following an SSA procedure. Precisely, at each time step, one compiles all the reactions ($\mathcal{L}_r$) that can occur according to the list of elementary transformations ($\mathcal{L}_{et}$) and the simulation graph $\mathcal{G}$. In their original paper, Monthon-Firth and Bray [32] provide equations where the time increment (t) is fixed; since there is no guarantee that reactions can occur in a fixed time interval, the dynamics may encompass null events (no reactions occur between t and $t + \tau$). In the implementation given below, Equations 11.13 and 11.14 are used to select a time delay t and a reaction r to apply to $\mathcal{G}$. Let us note than when solving these equations, all concentrations are equal to 1, and that Equation 11.12 does not apply as each molecule is considered to be a separate entity. In Algorithm 11.4 given below, the reaction selected by equation 11.14 is fired at any subgraph (b) of $\mathcal{G}$ matching (e.g., isomorphic to) the et of the reactants (B(et)) of an elementary transformation. The subgraphs are compiled for all elementary transformations by the routine generate-reactions.

ALGORITHM 11.4 PSEUDOCODE FOR SMS

```
single-molecule-simulator(G, Let)
input: - G: simulation graph
       - Let : list of elementary transformations
output: - Printout of the G vs. time
   t = 0
   For Mc steps do
         print t,G
      Lr = generate-reactions(G, Let)
      τ = time of next event using eq. (11.13)
      r = selected reaction in Lr occurring at time t + τ
         using eq. (11.14)
         B(r) = graph of the reactants of reaction r
      E(r) = graph of the products of reaction r
          G = G − B(r) + E(r)
          t = t + τ
   done
generate-reactions(G, Let)
input: - G: simulation graph
       - Let : list of elementary transformations
output: - Lr: list of reactions
   For all et in Let do
      For all subgraph b of G s.t. b is isomorphic to
        B(et) do
         e = G − b + E(et)
         if species-constraints(e) then
            # connected components (full species)
         # of subgraphs b and e are required to
         # compute reaction rates
             Lb = list of connected components of b
             Le = list of connected components of e
         k = rate-constant(et, Lb, Le)
      Lr = Lr + (et, b, e, k)
         fi
   done
done
return Lr
```

One notes that contrary to previous techniques, Algorithm 11.4 does not require us to monitor species concentrations. Nonetheless, species and reactions between species can be retrieved by a postprocess that reads the graphs printed by the algorithm. One advantage of Algorithm 11.4 is that species concentration can be computed only when needed; for instance, with a synthesis planning application, concentration may be computed for only one species (the target product) and a single time step (the final reaction time).

As already mentioned, the SMS algorithm has been used to study the dynamics of biological networks. Morton-Firth and Bray developed the first version of the algorithm (StochSim; http://www.ebi.ac.uk/~lenov/stochsim.html) and used it to predict the fluctuations in numbers of molecules in a chemotactic signaling pathway of coliform bacteria. In particular, they examined the temporal changes in the number of molecules of CheYp, a cytoplasmic protein known to influence the direction of rotation of the flagellar motor of the bacteria.

Colvin et al. [34] developed a similar algorithm (DYNSTOC; http://public.tgen.org/dynstoc) and demonstrated the algorithm with an idealized rule-based model of two systems. A system in which autophosphorylation of a receptor tyrosine kinase can generate a multitude of receptor phosphoforms and phosphorylation-dependent adapter-bound receptor states, and a system in which multivalent ligand–receptor binding can generate a multitude of ligand-induced receptor aggregates.

As a last application example, Kosuri et al. [35] proposed another implementation of the SMS algorithm to model cellular transcription and translation. The software named Tabasco (http://openwetware.org/wiki/TABASCO) directly represents the position and activity of individual molecules on DNA and can be used to study the effect of detailed molecular processes on systemwide gene expression. Tabasco has been demonstrated by simulating the entirety of gene expression during bacteriophage T7 infection.

11.7 CONCLUDING REMARKS

Initially developed for synthesis planning and retrosynthesis, reaction network generators have been used in combustion and fuel processing. For these initial applications, freeware and commercial products exist (for instance, LASHAA, http://lhasa.harvard.edu/, from Harvard University and, WODCA, http://www.molecular-networks.com/software/wodca/index.html, from Molecular Networks). With recent developments in systems biology, one is witnessing novel uses of network generation in biology, in particular for inferring and studying the dynamics of signaling, and metabolic and transcriptional networks. With the even more recent activities in synthetic biology, one may foresee new applications for network generation in particular with enzymatically controlled retrosynthesis, where the synthesis is performed by microorganisms.

REFERENCES

1. Corey, E. J., General methods for the construction of complex molecules. *Pure Appl. Chem.* 1967, 14, 19–37.
2. Ugi, I. B., J., Bley, K., Dengler, A., Dietz A., Fontain, E., Gruber, B., Herges, R. K. M., Reitsam, K., and Stein, N., Computer-assisted solution of chemical problems—the historical development and the present state of the art of a new discipline of chemistry. *Angew. Chem., Int. Ed. Engl.* 1993, 32, 201–227.
3. Corey, E. J., Cramer III, R. D., and Howe, J., Computer-assisted synthetic analysis for complex molecules. Methods and procedures for machine generation of synthetic intermediates. *J. Am. Chem. Soc.* 1971, 94, 440–459.

4. Clymans, P. J. and Froment, G. F., Computer-generation of reaction paths and rates equations in the thermal cracking of normal and branched paraffins. *Comput. Chem. Eng.* 1984, 83, 137–142.
5. Broadbelt, L. J., Stark, S. M., and Klein, M. T., Computer generated pyrolysis modeling: On-the-fly generation of species, reactions, and rates. *Ind. Eng. Chem. Res.* 1994, 33, 790–799.
6. Faulon, J. L. and Sault, A. G., Stochastic generator of chemical structure. 3. Reaction network generation. *J. Chem. Inf. Comput. Sci.* 2001, 41(4), 894–908.
7. Frenklach, M., Modeling of large reaction systems. In: *Complex Chemical Reaction Systems, Mathematical Modelling and Simulation*, J. Warnatz and W. Jäger (eds). Springer: Berlin, 1987; Vol. 47, pp. 2–16.
8. Susnow, R. G., Dean, A. M., Green, W. H., Peczak, P., and Broadbelt, L. J., Rate-based construction of kinetic models for complex systems. *J. Phys. Chem. A* 1997, 101, 3731–3740.
9. Ugi, I., Fontain, E., and Bauer, J., Transparent formal methods for reducing the combinatorial abundance of conceivable solutions to a chemical problem. Computer-assisted elucidation of complex mechanism. *Anal. Chim. Acta* 1990, 235, 155–161.
10. Faeder, J. R., Blinov, M. L., Goldstein, B., and Hlavacek, W. S., Combinatorial complexity and dynamical restriction of network flows in signal transduction. *Syst. Biol. (Stevenage)* 2005, 2(1), 5–15.
11. Lok, L. and Brent, R., Automatic generation of cellular reaction networks with Moleculizer 1.0. *Nat. Biotechnol.* 2005, 23(1), 131–136.
12. Danos, V., Feret, J., Fontana, W., and Krivine, J., Scalable simulation of cellular signaling networks. *Lect. Notes Comput. Sci.* 2007, 4807, 139–157.
13. Kotera, M., Okuno, Y., Hattori, M., Goto, S., and Kanehisa, M., Computational assignment of the EC numbers for genomic-scale analysis of enzymatic reactions. *J. Am. Chem. Soc.* 2004, 126(50), 16487–16498.
14. Rose, J. R. and Gasteiger, J., HORACE: An automatic system for the hierarchical classification of chemical reactions. *J. Chem. Inf. Comput. Sci.* 1994, 34, 74–90.
15. Dugundji, J. and Ugi, I., Theory of the *be*- and *r*-matrices. *Top. Curr. Chem.* 1973, 39, 19–29.
16. Dugundji, J., Gillespie, P., Marquarding, D., and Ugi, I., Metric space and graphs representing the logical structure of chemistry. In: *Chemical Applications of Graph Theory*, A. T. Balaban (ed.). Academic Press: London, 1976.
17. Faulon, J.-L., Stochastic generator of chemical structure: 1. Application to the structure elucidation of large molecules. *J. Chem. Inf. Comput. Sci.* 1994, 34, 1204–1218.
18. Ridder, L. and Wagener, M., SyGMa: Combining expert knowledge and empirical scoring in the prediction of metabolites. *Chem. Med. Chem.* 2008, 3(5), 821–832.
19. Benson, S. W., *Thermochemical Kinetics*. Wiley-Interscience: New York, 1976.
20. Sumathi, R., Carstensen, H.-H., and Green, W. H., Jr., Reaction rate prediction via group additivity, Part 2: H-abstraction from alkenes, alkynes, alcohols, aldehydes, and acids by H atoms. *J. Phys. Chem. A* 2001, 105, 8969–8984.
21. Sumathi, R., Carstensen, H.-H., and Green, W. H., Jr., Reaction rate prediction via group additivity, Part 1: H abstraction from alkanes by H and CH_3. *J. Phys. Chem. A* 2001, 105, 6910–6925.
22. Gillespie, D. T., A general method for numerically simulating the stochastic time evolution of coupled chemical reactions. *J. Comput. Phys.* 1976, 22, 403–434.
23. Gillespie, D. T., Stochastic simulation of chemical kinetics. *Annu. Rev. Phys. Chem.* 2007, 58, 35–55.

24. Fontain, E. and Reitsam, K., The generation of reaction network with RAIN. 1. The reaction generator. *J. Chem. Inf. Comput. Sci.* 1991, 31, 96–101.
25. Nigam, A. and Klein, M. T., A mechanism-oriented lumping strategy for heavy hydrocarbon pyrolysis: Imposition of quantitative structure–reactivity relationship for pure components. *IEC Res.* 1993, 32, 1297–1303.
26. Hatzimanikatis, V., Li, C., Ionita, J. A., Henry, C. S., Jankowski, M. D., and Broadbelt, L. J., Exploring the diversity of complex metabolic networks. *Bioinformatics* 2005, 21(8), 1603–1609.
27. Liao, K. H., Dobrev, I. D., Dennison, J. E., Jr., Andersen, M. E., Reisfeld, B., Reardon, K. F., Campain, A., et al., Application of biologically based computer modeling to simple or complex mixtures. *Environ. Health Perspect.* 2002, 110(Suppl 6), 957–963.
28. Lenaerts, T. and Bersini, H., A synthon approach to artificial chemistry. *Artif. Life* 2009, 15(1), 89–103.
29. Gonzalez-Lergier, J., Broadbelt, L. J., and Hatzimanikatis, V., Theoretical considerations and computational analysis of the complexity in polyketide synthesis pathways. *J. Am. Chem. Soc.* 2005, 127(27), 9930–9938.
30. Hou, B. K., Wackett, L. P., and Ellis, L. B., Microbial pathway prediction: A functional group approach. *J. Chem. Inf. Comput. Sci.* 2003, 43(3), 1051–1057.
31. Faeder, J. R., Blinov, M. L., Goldstein, B., and Hlavacek, W. S., Rule-based modeling of biochemical networks. *Complexity* 2005, 10, 22–41.
32. Morton-Firth, C. J. and Bray, D., Predicting temporal fluctuations in an intracellular signalling pathway. *J. Theor. Biol.* 1998, 192(1), 117–128.
33. Blinov, M. L., Faeder, J. R., Yang, J., Goldstein, B., and Hlavacek, W. S., 'On-the-fly' or 'generate-first' modeling? *Nat. Biotechnol.* 2005, 23(11), 1344–1345.
34. Colvin, J., Monine, M. I., Faeder, J. R., Hlavacek, W. S., Von Hoff, D. D., and Posner, R. G., Simulation of large-scale rule-based models. *Bioinformatics* 2009, 25(7), 910–917.
35. Kosuri, S., Kelly, J. R., and Endy, D., TABASCO: A single molecule, base-pair resolved gene expression simulator. *BMC Bioinform.* 2007, 8, 480.

12 Open Source Chemoinformatics Software and Database Technologies

Rajarshi Guha

CONTENTS

12.1 INTRODUCTION

Algorithm development is integral to the study of chemoinformatics. It is an implicit fact that many of the tools will incorporate fundamental chemoinformatics algorithms (such as canonicalization and aromaticity perception). Although any software dealing with chemoinformatics will need to implement a number of these algorithms, this chapter will focus on two specific classes of chemoinformatics software. Firstly, we discuss toolkit libraries which provide various (low- and high-level) chemoinformatics methods but which are not designed as standalone applications. Secondly, we discuss relational databases and the solutions available for incorporating chemoinformatics within the database environment. We also briefly touch upon workflow tools as a means of aggregating chemoinformatics functionality. Finally, we note that although there are many vendors for each type of software considered in this chapter, we exclusively consider Open Source offerings.

12.2 WHY OPEN SOURCE?

Although the idea of free and shared software has been in existence since the 1960s, such software has not been generally available in chemoinformatics. In contrast, the field of bioinformatics has produced a number of core algorithms (such as BLAST) as freely available software. Part of the reason for this is the preponderance of chemoinformatics research in industry, but equally important is the fact that the majority of chemical information has been proprietary, in contrast to biological data (such as gene sequences and protein structures), which has traditionally been freely exchanged. Recently, there has been increasing interest in Open Source chemoinformatics software, both in academia and industry [1]. It is also interesting to note that this has coincided, to some extent, with increasing amounts of public chemical information (such as structures and assay results). Our focus here is not the business case for Open Source in software (although issues such as cost and accountability are certainly important for both business and nonbusiness users); rather, we would like to stress the transparency of Open Source software in this context. This is especially important when we realize that the bulk of chemoinformatics software depends on a collection of core algorithms. The correctness and validity of such software are directly dependent on that of the underlying algorithms. Open Source implementations of the core algorithms thus allow the user to *verify* that the algorithm is implemented correctly, which is usually not true for closed source implementations. Certainly, with the use of unit tests, one can provide a suite of checks and this is applicable to both closed source and Open Source software. However, in the case of the latter, one is still at the mercy of the vendor who may or may not provide the results of a test suite. And even then, it is not possible to guarantee that an algorithm has been implemented as described. In one sense, Open Source software follows the principles of good scientific conduct: The "experiment" (i.e., the implementation) is freely accessible for everybody to examine, repeat, and verify. Furthermore, if advances are made in algorithms, one need not develop an implementation from scratch. Instead, one can start from a preexisting Open Source implementation. Another important aspect of Open Source software is the ability to fix errors. Given that software invariably contains bugs, it is useful not only to identify them, but also to fix them. The ability to access the source code can allow identification and correction of bugs in a rapid manner. It should be noted that for chemoinformatics software, this is not always the case, given the niche nature of the field and the fact that such fixes do require a certain level of expertise in programming as well as in chemoinformatics. However, the principle still holds. Probably more important than any of the factors described here is the ability to freely reuse software (within the limits of the license) and develop novel applications. No doubt a number of closed source toolkits and applications do provide free licenses for a number of scenarios (e.g., no cost academic licenses); however, one cannot usually redistribute the underlying software with the newly developed tool. As a result, new applications require that other users have access to the underlying toolkit that can be a hindrance to the spread of the higher-level application. The use of Open Source software in such situations avoids this problem.

Although the advent of Open Source software in the chemoinformatics field is relatively recent, there is a growing community that focuses specifically on Open Source and Open Access issues relating to software as well as data. The latter is important since many chemoinformatics algorithms depend on data (such as van der Waals radii) and, consequently, the validity and verifiability of the data are important. One of the more prominent groups in the area is the Blue Obelisk movement [2], which is an amalgamation of Open Source projects. The group maintains an Open Access, Open Source data repository at http://bodr.sourceforge.net/. A recent development under the umbrella of the Blue Obelisk movement is the OpenSMILES project. The SMILES language was designed in the 1980s, and although there are a number of publications on this standard [3,4], the specification contains a number of ambiguities. Given that the original Daylight implementation was proprietary, other vendors (both commercial and noncommercial) have added extensions or provided their own interpretations. The goal of the OpenSMILES project is to explicitly define the SMILES language in a public manner. The end result is expected to be a formal grammar for the language along with reference implementations. Furthermore, the specification will be completely free. Given the core nature of the SMILES language in chemoinformatics, open discussion and resolution of ambiguities are vital. The OpenSMILES project is an excellent example of the transparency afforded by Open Source approaches.

12.3 THE CHEMOINFORMATICS SOFTWARE STACK

The preceding section raises the notion of developing chemoinformatics applications on top of toolkits. From this point of view, we can look at chemoinformatics software development in terms of the chemoinformatics "stack," as shown in Figure 12.1. The main feature of the stack is the increase in user-oriented functionality as we move from the bottom to the top.

At the lowest level are the toolkits and databases, which are primarily the focus of programmers and software developers. It is at this level that most fundamental

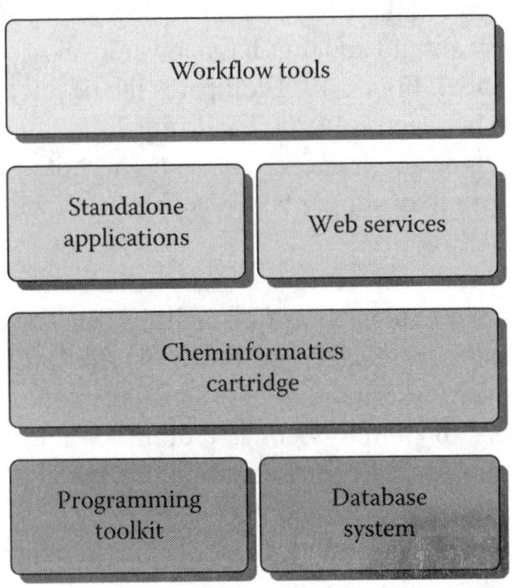

FIGURE 12.1 A schematic representation of an Open Source chemoinformatics software stack.

chemoinformatics algorithms will be implemented. Building on top of the toolkit and database, chemoinformatics cartridges allow one to manipulate chemical information within the database. With this functionality one is in a position to then develop focused applications, which may be stand-alone or deployed in a variety of manners (such as web pages or web services) for end users. Finally, at the top we have workflow tools, which are able to aggregate high- and low-level applications and present them in an easy-to-use manner. It should be noted that Figure 12.1 does not imply that databases and cartridges are employed by every chemoinformatics application. Nor does it imply a rigid hierarchy. However, it does highlight the general role of individual chemoinformatics components. It should be noted that certain components such as web services will employ software such as web servers, application containers, and so on. These components can be commercial or Open Source, the latter accounting for some of the most popular examples (such as Apache and Tomcat).

12.4 TOOLKITS

Given that the focus of the book is chemoinformatics algorithms, it is important to realize that the field is not purely theoretical. In other words, one must be able to employ the algorithms to process real-world data. Although one can build monolithic chemoinformatics applications, good software engineering practice suggests that low-level and frequently used functionality be reused. This issue is addressed by the use of chemoinformatics toolkits that package chemoinformatics functionality into reusable libraries. A number of toolkit libraries are available for various languages and we focus on three Open Source libraries that are under active development, namely, the CDK, OpenBabel, and RDKit. Table 12.1 provides a broad comparison of the three toolkits in terms of supported features.

Before describing the features of the various toolkits, we provide a brief overview of the projects themselves. All three have an active development community, although RDKit is a relatively new entrant and thus has a lower level of participation compared to the CDK and OpenBabel. Figure 12.2 compares the development activity in terms of commits to the trunk between July 2007 and July 2008. It is important to note that the graph of activity includes both maintenance and new development. In that sense, a project with low activity may simply be in a stable state, with only bug fixes being committed.

All three are hosted on Sourceforge, which provides mailing lists and version control. All projects make extensive use of mailing lists for discussions related to both developmental issues and usage questions. In addition to mailing lists, Wikis are also employed to host descriptions, tutorials, and so on. Both OpenBabel and RDKit provide a variety of online documentation in the form of help pages, API documentation, and examples. In contrast, although there is a variety of information available for the CDK, it is not collected in one place.

We next consider the chemoinformatics functionality provided by the three toolkits. It should be noted that these descriptions are not exhaustive and we advise that the reader explore the web sites of each project to get a more exhaustive and up-to-date list of features.

TABLE 12.1
A Broad Comparison of Chemoinformatics Features Provided by the Three Toolkits Discussed

Feature	CDK	OpenBabel	RDKit
License	LGPL	GPL	New BSD
Language	Java	C++	C++/Python
SLOC[a]	188,554	194,358	173,219
Fingerprints			
Hashed	√√√	√√√	√√√
Substructure	√√√	√√√	√√√
File format support	√√	√√√	√
Aromaticity models	√	√	√
Stereochemistry	√	√√	√√√
Canonicalization	√√√	√√√	√√√
Descriptors	√√√	√	√√√
2D coordinate generation	√√√	x	√√√
3D coordinate generation	√	√√√	√√√
2D depictions	√√√	x	√√√
Conformer generation	x	√	√
Rigid alignment	√√√	√√√	√√√
SMARTS searching	√√√	√√√	√√√
Pharmacophore searching	√√	x	√√√

[a] Source Lines of Code as measured by the tool *sloccount* (http://www.dwheeler.com/sloccount/). The count includes all source files, in any language, for the project. In addition, only the trunk for each project was considered.

12.4.1 Chemistry Development Kit

The chemistry development kit (CDK) [5,6] (http://cdk.sourceforge.net) is a chemoinformatics toolkit library written in Java and licensed under the LGPL. Development on the library was initiated in 2000 and since then it has been actively developed. Currently, the project has approximately 20 active developers who address a variety of projects ranging from implementing new functionality, bug fixes to documentation and code quality metrics. The library sees widespread usage as evidenced by 60 citations to the original publications. Although the library is written in Java, it can be accessed from a variety of other languages including Python (via Jython), Ruby, and R [7]. In addition, a variety of CDK functionalities are available as components within the KNIME workflow environment.

The CDK provides both low- and high-level functionality. Firstly, the library provides a set of low-level classes for representing molecular concepts such as atoms, bonds, molecules, and so on. These classes provide a variety of methods to get and set various properties of these objects. Related to core representational classes, the library also provides various low-level operations that are commonly performed in chemoinformatics. This includes atom typing, ring perception, aromaticity perception, and substructure searching (either by SMILES or by SMARTS). The second level

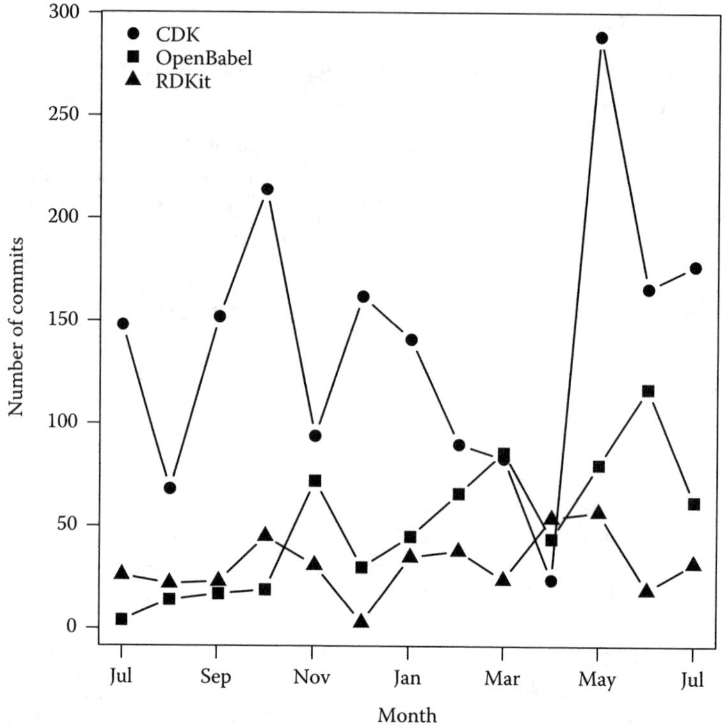

FIGURE 12.2 Number of commits per month (between July 2007 and July 2008) for the three projects discussed here. Note that this graph only considers commits to the trunk of the development trees.

of functionality is oriented toward specific chemoinformatics tasks. Examples include fingerprints, descriptors, and pharmacophore searching. The CDK implements a number of descriptors for atoms, bonds, and molecules. Table 12.2 summarizes the currently available molecular descriptors.

The descriptors are oriented toward QSAR modeling, and Open Source tools are available that provide user-friendly interfaces on top of this functionality (Bioclipse and CDKDescUI). In the area of pharmacophore searching, the CDK provides support for the representation of pharmacophore queries in terms of groups (defined by SMARTS) and geometric constraints. There is no limit on the size of the query (in terms of the number of pharmacophore groups or constraints). However, constraints are currently limited to distance or angle constraints, and more advanced features such as dihedral constraints and excluded volumes remain to be implemented.

Apart from the chemoinformatics functionality, the project also addresses a number of more general aspects of software development. For example, development of the CDK is test-driven. That is, each class is meant to be associated with a unit test that ensures that the class works as expected. This leads to two useful features. Firstly, any modifications to the code base should not lead to errors in working functionality and unit tests can check for this. Secondly, unit tests provide simple usage examples

TABLE 12.2
A Summary of the Various Molecular Descriptors Implemented in the CDK

Class	Descriptor	Reference
Constitutional	Atom and bond counts, molecular weight	
	Aromatic atom and bond counts	
	Hydrogen bond donor/acceptor counts	
	Rotatable bond count	
	X log *P*, *A* log *P*	[8,9]
	E-state fragment counts	[10]
Topological	χ (ordinary and valence) indices	[11]
	κ indices	[11]
	Wiener index	[12]
	Zagreb index	[13]
	Vertex adjacency	
	Petitjean indices	[14]
Geometric	Gravitational indices	[15]
	Moment of inertia	
Electronic	σ electronegativity	
	Partial charges	[16]
	Hybrid BCUT	[17]
	WHIM	[18]
	Topological Surface Area (TPSA)	[19]
	Charged Partial Surface Area (CPSA)	[20]

of the classes and methods they aim to test. Although there are currently 7287 tests, all classes and methods are not yet tested and implementation of new unit tests is ongoing. The project also provides various types of documentation. The main source of this is the API Javadocs, which are augmented to include references, links to source code, and so on. In addition, the project releases a newsletter, *CDK News*, on an approximately quarterly basis that contains articles on new features and examples of products using the CDK. The CDK puts an effort into maintaining good code and documentation quality by using static code and documentation analysis tools [21,22]. Finally, the project also provides a nightly build site, where one can download the latest sources or JAR files and view documentation, code and documentation quality reports, and testing results.

12.4.2 OpenBabel

The OpenBabel project (http://openbabel.org) is a C++ chemoinformatics library released under the GPL. It is derived from the original OELib library from OpenEye, although significant amounts of the code have been rewritten. The project has been existing since 2001 and originally focused on file format conversion (currently supporting 97 different formats). Since then it has expanded into a fully fledged chemoinformatics toolkit. As with the CDK, it provides commonly used, core chemoinformatics functionality such as representations of molecules, bonds, and atoms, aromaticity perception, chirality detection, and substructure searching facilities. Higher-level functionality includes Gasteiger–Marsili partial charges, molecular superimpositions, force fields, 3D coordinate generation, and descriptors. The project employs a Perl-based unit testing framework and currently it has approximately 100,000 unit tests. A variety of documentation is also available on the OpenBabel wiki, ranging from API-level documentation (generated from the source code using Doxygen) to various how-to's and tutorials.

A useful feature of OpenBabel is the ability to develop functionality as plugins. Such plugins exist as dynamic libraries that can be loaded by OpenBabel at runtime. As a result, they can be developed independently of the OpenBabel project. Examples of current plugins include fingerprints and force fields. Currently, OpenBabel supports the UFF, MMFF94, and MMFF94s force fields. In addition to force fields, 3D coordinate generation and conformer generation (using a Monte Carlo algorithm) are available. Descriptors are also provided by way of plugins, although, currently, only a few descriptors (H-bond donor, acceptor counts, TPSA, Lipinski's Rule of Five, and molecular weight) are available.

The project also provides a number of ready-to-use command line tools. Examples of these tools include babel (the file format converter), obconformer (a conformer generation tool), and so on. Table 12.3 summarizes the currently available utilities.

Given that the library is implemented in C++, it is relatively easy to generate bindings for the library in a number of other languages, using SWIG. Currently, bindings are available for Perl, Python, Ruby, C#, and Java. In addition to the SWIG bindings, the Pybel [23] project provides a more Pythonic interface to the C++ library.

12.4.3 RDKit

RDKit (http://www.rdkit.org/) is a relatively new entrant on the Open Source chemoinformatics scene. The toolkit was originally developed at Rational Discovery LLC and was released under the new BSD license in May 2006. The project provides a core C++ library along with a set of higher-level Python functions that make use of the

TABLE 12.3
A Summary of the Command Line Tools Provided by the OpenBabel Project

Tool	Function
Babel	File format conversion
Obchiral	Prints chirality information
Obconformer	Generate conformers
Obenergy	Evaluate the energy of a molecule using different force fields
Obfit	Superimpose two molecules based on SMARTS
Obgen	Generate 3D coordinates
Obgrep	SMARTS-based substructure searches
Obprobe	Generate electrostatic grids using MMFF94

core routines. In addition, a number of GUI tools, command line scripts, and database extensions are also provided.

As with the CDK and OpenBabel, RDKit provides a variety of low- and high-level chemoinformatics functionality including support for multiple file formats, canonicalization, SMARTS substructure searching, 2D and 3D pharmacophores, 2D depiction, and 3D coordination generation. The library also supports SMARTS-based molecular transformations (such as RECAP [24]). In contrast to the CDK and OpenBabel, RDKit provides good support for chirality by providing methods to assign CIP codes to atoms (R/S) and bonds (E/Z). The project also provides a wide variety of molecular descriptors, summarized in Table 12.4.

In the area of database integration, RDkit provides a set of extensions that allow certain methods to be called from within a PostgreSQL database. This aspect is discussed in more detail in Section 12.3.1.

An interesting aspect of the RDKit project is that it provides a variety of machine learning routines. Examples include clustering, decision trees, Naïve Bayes, and random forests. In addition to these methods, various utility methods such as data splitting, serialization of models, and enrichment plots are available. Although external statistical environments are available (such as R), it can be useful to have access to library methods that implement machine learning algorithms. It should be noted that these algorithms are not necessarily highly optimized. The project also provides several GUIs both for chemoinformatics and machine learning. The former is exemplified by a molecule browser and similarity calculator and the latter by interfaces

TABLE 12.4
A Summary of the Molecular Descriptors Provided by RDKit

Class	Descriptor	Reference
Constitutional	Atom and group counts	
	Ring counts, molecular weight	
	log *P*	[9]
	Molar refractivity	
	Topliss-like fragment counts	
Topological	χ (ordinary and valence) indices	[11]
	κ indices	[11]
	Balabans J	[25]
	EState indices	[10]
	Atom pairs	[26]
	Topological torsions	[27]
	Kier and Hall shape indices	[28]
Hybrid	Topological polar surface area (TPSA)	[19]
	Labute ASA	
VSA	PEOE	[29]
	SMR	
	S log *P*	
	Estate	

for clustering and visualization of dendrograms. Note that the GUI components are licensed under the GPL.

The project also provides a nightly build script to ensure that builds are not broken. In addition, unit testing is performed using the Python testing framework. Documentation for the source code is provided for both the C++ and Python components using Doxygen and ePyDoc, respectively. In addition to API documentation, the project also provides user-oriented help pages such as how-to's.

12.5 DATABASE TECHNOLOGIES

Database technologies range from simple flat files with no special formatting to complex relational databases (RDBMS) and object-oriented databases. RDBMSs have been used extensively within the pharmaceutical industry for the purposes of storing information related to compound collections, assay data, and so on. Traditionally, these usage scenarios have employed commercial databases coupled with some form of chemoinformatics intelligence in the form of plugins. Such plugins are also known as "cartridges." Examples include the Torus cartridge from Digital Chemistry and DayCart from Daylight CIS. Note that in this section we do not describe any specific database but rather focus on database technologies. For a review of public databases, the reader is referred to Ref. [30].

With the increase in availability of large amounts of public chemical information via public databases (PubChem and ChemSpider) as well as high-throughput experimental resources such as the Molecular Libraries Screening Network (MLSCN), it is possible for individual researchers both in industry and academia to build large compound collections and associate them with other arbitrary data sources. As noted in Section 12.1.1, Open Source solutions to this data management problem provide a low barrier to entry. Furthermore, depending on the needs of the user, Open Source database technologies provide a cost-effective solution.

Before discussing current options for Open Source chemical information databases, it is useful to consider what is the type of data that such systems are expected to handle. First and foremost is chemical structure information. Structures can be represented in a variety of formats ranging from connection tables and plain text formats such as SMILES and SDF to binary representations of molecules. Depending on the usage scenario of a database, one or more different representations may be stored. Although most chemoinformatics systems can process a wide variety of representations, some forms may lead to more efficient processing than others. Apart from chemical structure, the other types of information will tend to be textual or numeric in nature.

12.5.1 CARTRIDGES

A cartridge is simply a set of extensions to a database, such that certain domain functionalities are available within the database system itself. Such cartridges usually provide support for domain-specific data types and indexing methods. In addition, cartridges will usually provide new SQL functions to support domain-specific queries. In the field of chemoinformatics there are a number of offerings that allow one to access chemoinformatics data types and methods within a database. Commercial examples include DayCart (Daylight CIS), CHORD (gNova Scientific Software), and JChem Cartridge (ChemAxon). Note that, of those mentioned here, only CHORD can be used with the PostgreSQL (an Open Source DBMS), the others being designed for Oracle.

On the Open Source side, there are three main offerings: Tigress (for PostgreSQL), Mychem (for MySQL), and RDKit (for PostgreSQL). The first two cartridges use OpenBabel to provide the underlying chemoinformatics functionality. Regarding substructure searching, Mychem makes use of the underlying SMARTS matching capabilities of OpenBabel. Although Tigress also provides a function similar to that of Mychem, it also employs the checkmol/matchmol suite [31] to detect functional groups and perform substructure searches using this information or else use it as a prefilter for full SMARTS-based substructure searching. The cartridges are written in C and must be compiled and loaded into the database before usage. Given that Mychem and Tigress are both based on OpenBabel, it is natural to expect that they will expose similar functionality using a common API. The ChemiSQL project (http://sourceforge.net/projects/chemdb/) has been recently started and aims to provide a single source for chemoinformatics cartridges for a variety of Open Source databases and toolkits. It should be noted that both Tigress and Mychem do not provide any of their own chemoinformatics functionality being dependent on the OpenBabel project, whereas RDKit, as described in Section 12.2.3, represents a complete chemoinformatics toolkit. Table 12.5 compares the functionality of the three Open Source cartridges.

TABLE 12.5
A Summary of the Functionality Provided by OpenSource Chemoinformatics Cartridges

Cartridge	License	Representation Support	Methods
Tigress	GPL, LGPL	Binary, SMILES, MOL, InChI	Exact and substructure searches, similarity (Tanimoto), property calculation (molecular weight, charge, bond count), fingerprint calculation, salt removal, format conversion
Mychem	GPL	SMILES, MOL, InChI, CML	Exact and substructure searching, similarity (Tanimoto), property calculation (molecular weight, charge, bond count), fingerprints, salt removal, format conversion
RDKit	New BSD	SMILES	Exact and substructure searches,canonicalization, similarity (Dice, Tanimoto, Cosine), property calculation (molecular weight, log *P*), fingerprint calculation

Exposing the chemoinformatics functionality of the underlying toolkit within a database is not particularly difficult and simply requires that one conform to the prescribed database API. On the other hand, the representation used to store molecules can significantly affect query efficiency. Thus, for example, one can store a molecule as a SMILES string—indeed this is probably the most platform-independent way of storing it. However, during a query, each SMILES string must be parsed and an internal molecule object must be created. Invariably this will not be retained for future queries. To alleviate this, one can generate binary representations of a molecule and store them in a column of appropriate type (such as bytea in PostgreSQL). In this scenario, the binary form may simply represent a serialized form of an internal molecule object. Given that deserialization can be much faster than parsing, this representation can lead to improvements in query efficiency. Both Tigress and RDkit currently support binary molecular representations.

The use of chemoinformatics cartridges can result in cleaner and more efficient chemical information infrastructures by virtue of moving complexity away from frontends (or clients) into the backend database.

12.5.2 Indexing Chemical Information

As noted above, chemical information databases will hold chemical structures in addition to traditional data types (text, numeric, dates, etc.). Furthermore, the traditional data types will usually represent some properties of the molecules. Examples might include molecular descriptors, fingerprints, assay readouts, and so on. Given these varied data types, efficient indexing plays an important role in allowing fast queries. One of the key issues that face choice of indexing scheme is the intended query. Thus, for example, if one were simply retrieving records based on a textual compound ID, a single B-tree [32] index on the relevant column would provide a time complexity of $O(\log n)$ for searches. On the other hand, similarity searches require an indexing scheme that is capable of performing efficient near-neighbor searches, in possibly highly multidimensional spaces.

We first consider how one might employ indexing to provide efficient query times when searching for chemical structures. Ignoring the trivial case of retrieving structures based on some textual ID, we focus on how structure and substructure searches can be improved by an indexing scheme. Searching for exact matches to a query molecule can benefit from standard hash indexes. Depending on the nature of the structure representation, this may require some form of canonicalization of the query molecule (as well as for the stored molecules, possibly at registration time). Thus, for example, one can store the molecules in a text field using a SMILES representation. Assuming that they are appropriately canonicalized, one can then identify entries that exactly match a query molecule by performing a string equality search. If this field is indexed by a B-tree index, this will be very fast. Given that canonicalization methods are specific to a given toolkit, a more generalized solution that is independent of any specific toolkit is to employ InChIs for structure representation. Since this is a plain text format, this provides the same advantages as SMILES. But in addition, InChIs for two forms of the same molecule will always be the same since there is only one implementation of the algorithm.

Although exact matches can be useful, a more common task is to perform substructure searches. Substructure searching is performed using graph isomorphism algorithms such as the Ullman algorithm [33]. Given that such algorithms require a full comparison between the query and target structures, there is no way, *a priori*, to store a reduced representation that would allow one to directly answer the question of whether a query is a substructure of the target. Naively, one might expect that a substructure search within a database will boil down to a linear scan over an entire table. However, all is not lost. One way to speed up substructure searches is to employ a binary fingerprint filter. Thus, the molecules in the database will have their fingerprints precomputed and stored in a binary field. Then, given a query, its fingerprint would be evaluated. Next, one would perform a linear scan over the database, but for each row one would check whether the bits of the query fingerprint are also set in the target fingerprint. Only if this is true would one then apply the subgraph isomorphism test to the query and target. This filtering process is summarized in Algorithm 12.1. Since comparison of binary fields is very fast, one avoids having to perform a large fraction of the more expensive isomorphism tests (depending on the nature of the query).

ALGORITHM 12.1 THE USE OF A BINARY FINGERPRINT FILTER TO SPEED UP SUBSTRUCTURE SEARCHES

```
q ← query molecule
Fq ← get fingerprint(q)
R ← { }
for row in TABLE do
t ← target molecule
Ft ← get fingerprint(t)
if Ft contains Fq then
if is subgraph(t,q) then
append t to R
end if
end if
end for
```

The drawback of this approach is that it can be applied only to "well-formed" queries. That is, if the query can be represented as a SMILES string, this approach allows us to speed up the matching process. On the other hand, if the query is a SMARTS pattern, the above procedure fails since it is not ordinarily possible to generate a fingerprint from an arbitrary SMARTS pattern. Sayle [34] has described a number of strategies that can be employed to make SMARTS-based substructure searching more efficient. First one generates a fragment fingerprint, such as for c:c which will be contained in the fingerprint for a molecule such as benzene (c1ccccc1). The next step involves the enumeration of a SMARTS pattern. Taking the example from Ref. [34], the pattern $C = [N, P]$ only allows us to fingerprint the C. However, one can expand the pattern into $C = N$ AND $C = P$, in which case, each term can be fingerprinted. Sayle describes five specific procedures to allow one to optimize a SMARTS pattern for the purposes of substructure searching. Currently, such optimizations are not available in any of the Open Source database cartridges. As a result, chemical database systems based on these cartridges are forced to perform linear scans over the entire table when carrying out SMARTS substructure searches.

We next consider the problem of similarity searching within chemical databases. Although a common task, there are many variations of it, depending on the nature of the chemical representation being employed. We first consider the use of a real-valued descriptor representation of a molecule. Such representations can be used to define arbitrary chemical spaces (such as for QSAR applications) or more realistic properties such as molecular shape [35,36]. In such a representation, a molecule is denoted by a real-valued vector of length N. The similarity between two molecules is then defined in terms of the reciprocal of the Euclidean or Manhattan distance between their descriptor vectors. In these scenarios, similarity searching is equivalent to identifying the nearest neighbors (NNs) of a query molecule in the defined chemical space. Indeed, spatial indices can be profitably used for diversity analysis methods based on NNs [37,38]. A number of algorithms have been described for efficient NN searches such as k-d trees [39] and locality-sensitive hashing [40]. Most database systems employ the R-tree [41]. Traditionally, this type of index has been used in spatial databases, designed for Geographic Information Systems (GIS), and it exhibits good performance for the 2D data types commonly employed in that field. However, it can be used for higher-dimensional data types, such as those representing molecular descriptor spaces. The drawback of using this index for chemoinformatics applications is that performance degrades with increasing dimensionality [42] of the chemical space, which is usually high dimensional. A variety of spatial indexing schemes have been proposed that aim to support efficient NN searches in high-dimensional spaces such as the PK-tree [43] and M-tree [44].

In terms of support for spatial indexing in Open Source databases, both PostreSQL and MySQL support R-trees. Support for this type of indexing in these databases is geared toward GIS applications, although as noted this does not preclude their use in chemoinformatics applications. A distinguishing feature of the PostgreSQL support for spatial indexing is the availability of Generalized Inverse Search Trees (GiST) [45]. This is a generalized data structure that allows one to develop indexing schemes for arbitrary data types (ranging from point data to BLAST sequences and graphs). An implementation of a GiST index requires that one simply implement four fundamental operations:

- Consistent—Given a query q and a key p, returns false if q and p cannot be true for a given item
- Union—Given a set of entries returns a key p that is true for all entries
- Penalty—returns the cost of inserting a new item under a subtree. Items get inserted down the path of lowest cost in the tree
- Picksplit—Decides when and which items in a page go to a new page or stay on the old page

As a result, the indexing scheme is independent of data type. Furthermore, the GiST index can also be used to perform NN searches directly as well as statistical approximations over large datasets. In fact, the implementation of R-trees in PostgreSQL is simply a special case of a GiST index. The PostgreSQL implementation of

GiST indexes also supports extensions such as variable length keys and concurrency control.

Although the GiST implementation of the R-tree in the index does involve an abstraction layer, performance is still quite high. As an example, we have created a database containing 3D structures of 17 M molecules from PubChem. The shape of each molecule is characterized by a 12-vector for a random collection of query compounds. The queries were performed on a machine with 2 GB RAM and a dual core Intel Xeon 2.4 GHz. The database was assigned 1 GB of shared memory [35]. This vector field was indexed using an R-tree. We performed a series of queries where, given the 12-vector for a query molecule, we identified molecules lying within a radius *R* of the query molecule. Figure 12.3 shows the distribution of query times for two different neighborhoods (defined in terms of radii) using a slightly outdated version of the database containing 10M compounds.

The speed of these queries allows us to apply the R–NN method [37] to characterize the density of chemical space of any compound in the context of the 10 M compound collection. This involves performing NN lookups at varying radii and hence requires efficient spatial indexing schemes.

Given that the goal of GiST is to allow efficient indexing for arbitrary data types (assuming that the four fundamental GiST operators can be defined for the data type in question), one might develop a GiST index for molecules. Such an index is currently being developed by the Tigress project [46]. In this context, the molecule data type is the binary representation of a molecule. A fundamental task for a GiST index is to compare items (i.e., molecules). In this context, the consistent function will employ 1024 bit fingerprints and the XOR operator for equality. For the special case of leaf nodes (i.e., individual molecules), equality is determined by comparing MD5 hashes of the InChI representation of the molecule. The penalty function employs the Tanimoto distance between two fingerprints to provide a cost value. Finally, the penalty

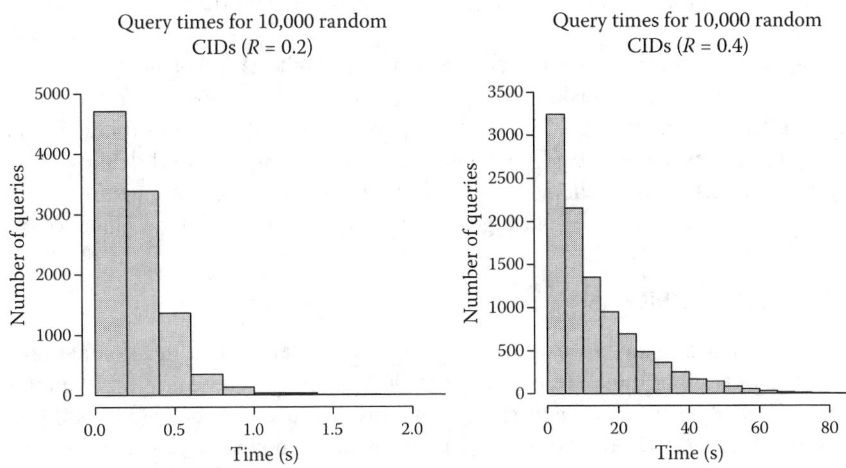

FIGURE 12.3 A summary of the query times in a 3D structure database for two different neighborhoods.

function is based on Guttman quadratic split algorithm [41], using the Tanimoto distance.

Another common scenario in chemoinformatics applications is the use of similarity searches with binary fingerprints. In this case, similarity between two molecules is defined by their Tanimoto index, which can be efficiently computed using bit operations. Currently, there is no Open Source implementation of an index that will support similarity searching on binary string (necessitating linear scans), although the work being performed on GiST indexes for molecules described above is applicable to this problem.

12.6 WORKFLOW ENVIRONMENTS

Workflow (or pipelining) tools have grown in popularity as a means of allowing nonexperts to perform tasks composed of sequential units, without having to write actual programs (although such tools can be enhanced with user-written scripts). Note that, by definition, a workflow tool is not tied to any specific domain. By providing domain-specific tasks (components), one can provide support for any specific domain such as bioinformatics or chemoinformatics.

A variety of commercial and Open Source workflow tools for chemoinformatics are available, and Warr [47] provides a broad overview. Two of the most popular Open Source tools are Taverna [48] and KNIME (http://www.knime.org), both written in Java and supporting chemoinformatics via third-party plugins. In the former case, the CDK is used to provide chemoinformatics support. In the latter case, chemoinformatics support is provided by plugins from Tripos, Schrodinger, as well as the CDK. Note that the plugin functionality can range from simple operations such as fingerprint generation, format conversion, and 2D depictions to much more complex tasks such as pharmacophore searching or docking. Given the rise in chemoinformatics web services, it is useful for workflow tools to be able to handle them. Taverna provides support for SOAP-based web services, although KNIME currently does not support them.

In one sense, workflow tools can be considered the highest level of an Open Source chemoinformatics stack—making using of toolkits and databases and providing an easy-to-use interface on top of these items. Note that workflow environments do not necessarily represent a fixed-goal application. Indeed their flexibility allows one to mix and match various chemoinformatics tools. At the same time, workflows to perform specific tasks can be "packaged" and thus presented as a stand-alone tool.

12.7 CONCLUSIONS

Given the practical nature of chemoinformatics, it is essential that implementations of fundamental algorithms are easily accessible. Although one can always implement specific algorithms as stand-alone programs, it is useful to be able to build on top of previous work. In this sense, chemoinformatics toolkits provide a convenient platform on which to build a variety of applications. By their nature, such toolkits will provide core data models for chemical concepts and implement a number of core chemoinformatics algorithms such as ring perception and canonicalization. As a result, most toolkits exhibit similar core functionalities, although differences in their completeness (such as the handling of chirality) do occur. Toolkits may also provide higher-level functionality that, while not representing a full standalone program, is sufficiently common in various applications. Examples include fingerprint generation, similarity calculations, and so on. The three toolkits discussed here provide a variety of high-level functionality, which, as in the case of RDKit, may extend beyond traditional chemoinformatics.

The choice of toolkit is very dependent on language, features available, support, documentation, and of course personal preference. Indeed, a recent Open Source project called Cinfony (http://code.google.com/p/cinfony/) provides a uniform API to the three toolkits discussed here, allowing one to mix and match functionality from any of them. With the increased availability of publicly accessible data and cheap computing power, the ability to carry out chemoinformatics research and develop applications in a *redistributable* fashion is becoming increasingly feasible. In such a scenario, the license associated with a toolkit can play a major role in choosing which one to use. In this context, Open Source toolkits provide a number of advantages as described in Section 12.1.1. Of course, Open Source toolkits are not always as polished as their commercial counterparts.

Given that some commercial vendors do provide no-cost licenses for certain groups, they can be an attractive alternative when developing applications. The downside is that distribution of such applications is dependent on access to the toolkit.

In this context, it is clear that the viability of an Open Source chemoinformatics stack (Figure 12.1) is very much dependent on the use of Open Source toolkits. Given the free availability of high-performance database systems and web servers, a significant part of the computational infrastructure for large chemoinformatics applications exists in an Open Source fashion and toolkits represent the core domain-specific functionality. With the rise of Grid and Cloud computing, the ability to freely distribute the stack across hundreds or thousands of machines can be severely limited by commercial licenses. In such a scenario, Open Source software provides an attractive approach to making use of emerging computing technologies in chemoinformatics applications.

In conclusion, while Open Source chemoinformatics software may suffer from some disadvantages compared to commercial offerings, they provide a number of significant advantages in terms of transparency (leading to verifiability), reuse, and cost. A number of Open Source toolkits are available and this chapter has focused on three projects that are most active. In addition to toolkits, we have provided a brief discussion on database and pipelining technologies as they relate to chemoinformatics applications.

REFERENCES

1. Delano, W., The case for cpen source software in drug discovery. *Drug Discov. Today* 2005, 10, 213–217.
2. Guha, R., Howard, M. T., Hutchison, G. R., Murray-Rust, P., Rzepa, H., Steinbeck, C., Wegner, J., and Willighagen, E. L., The blue obelisk–interoperability in chemical informatics. *J. Chem. Inf. Model.* 2006, 46, 991–998.

3. Weininger, D., SMILES, a chemical language and information system. 1. Introduction to methodology and encoding rules. *J. Chem. Inf. Comput. Sci.* 1988, 28, 31–36.
4. Weininger, D., Weininger, A., and Weininger, J., SMILES. 2. Algorithm for generation of unique SMILES notation. *J. Chem. Inf. Comput. Sci.* 1989, 29, 97–101.
5. Steinbeck, C., Han, Y. Q., Kuhn, S., Horlacher, O., Luttmann, E., and Willighagen, E., The Chemistry Development Kit (CDK): An open-source java library for chemo- and bioinformatics. *J. Chem. Inf. Comput. Sci.* 2003, 43, 493–500.
6. Steinbeck, C., Hoppe, C., Kuhn, S., Floris, M., Guha, R., and Willighagen, E., Recent developments of the Chemistry Development Kit (CDK)—an open-source java library for chemo- and bioinformatics. *Curr. Pharm. Des.* 2006, 12, 2110–2120.
7. Guha, R., Chemical informatics functionality in R. *J. Stat. Soft.* 2007, 18 [online].
8. Wang, R. and Lai, L., A new atom-additive method for calculating partition coefficients. *J. Chem. Inf. Comput. Sci.* 1997, 37, 615–621.
9. Ghose, A. and Crippen, G., Atomic physicochemical parameters for three-dimensional-structure-directed quantitative structure–activity relationships. 2. Modeling dispersive and hydrophobic interactions. *J. Chem. Inf. Comput. Sci.* 1987, 27, 21–35.
10. Hall, L. and Kier, L., Electrotopological state indices for atom types: A novel combination of electronic, topological, and valence state information. *J. Chem. Inf. Comput. Sci.* 1995, 35, 1039–1045.
11. Hall, L., and Kier, L., The molecular connectivity χ indices and κ shape indices in structure–property modeling.. In: *Reviews of Computational Chemistry*, Vol. 2, K. Lipkowitz and D. Boyd (eds), VCH Publishers: New York, 1991.
12. Wiener, H., Structural determination of paraffin boiling points. *J. Am. Chem. Soc.* 1947, 69, 2636–2638.
13. Gutman, I., Ruscic, B., Trinajstic, N., and Wilcox, Jr., C., Graph theory and molecular orbitals. XII. Acyclic polyenes. *J. Chem. Phys.* 1975, 62, 3399–3405.
14. Petitjean, M., Applications of the radius diameter diagram to the classification of topological and geometrical shapes of chemical compounds. *J. Chem. Inf. Comput. Sci.* 1992, 32, 331–337.
15. Katritzky, A., Mu, L., Lobanov, V., and Karelson, M., Correlation of boiling points with molecular structure. 1. A training set of 298 diverse organics and a test set of 9 simple inorganics. *J. Phys. Chem.* 1996, 100, 10400–10407.
16. Gasteiger, J. and Marsili, M., Iterative partial equalization of orbital electronegativity—a rapid access to atomic charges. *Tetrahedron* 1980, 36, 3219–3228.
17. Pearlman, R. S. and Smith, K. M., Metric validation and the receptor-relevant subspace concept. *J. Chem. Inf. Comput. Sci.* 1999, 39, 28–35.
18. Todeschini, R. and Gramatica, P., 3D modelling and prediction by WHIM descriptors. Part 5. Theory, development and chemical meaning of WHIM descriptors. *Quant. Struct. Act. Relat.* 1997, 16, 113–119.
19. Ertl, P., Rohde, B., and Selzer, P., Fast calculation of molecular polar surface area as a sum of fragment based contributions and its application to the prediction of drug transport properties. *J. Med. Chem.* 2000, 43, 3714–3717.
20. Stanton, D. and Jurs, P., Development and use of charged partial surface area structural descriptors in computer assisted quantitative structure property relationship studies. *Anal. Chem.* 1990, 62, 2323–2329.
21. Copeland, T., *PMD Applied*, Centennial Books: Alexandria, VA, 2005.
22. Sun Microsystems, "DocCheck," http://java.sun.com/j2se/javadoc/doccheck/, last accessed July 2008.
23. O'Boyle, N., Morely, C., and Hutchison, G., Pybel: A python wrapper for the openBabel chemoinformatics toolkit. *Chem. Central J.* 2008, 2, 5.

24. Lewell, X., Judd, D., Watson, S., and Hann, M., RECAP—retrosynthetic combinatorial analysis procedure: A powerful new technique for identifying privileged molecular fragments with useful applications in combinatorial chemistry. *J. Chem. Inf. Comput. Sci.* 1998, 38, 511–522.
25. Balaban, A., Highly discriminating distance based topological index. *Chem. Phys. Lett.* 1982, 89, 399–404.
26. Carhart, R., Smith, D., and Venkataraghavan, R., Atom pairs as molecular features in structure–activity studies: Definition and applications. *J. Chem. Inf. Comput. Sci.* 1985, 25, 64–73.
27. Nilakantan, R., Bauman, N., Dixon, J., and Venkataraghavan, R., Topological torsion: A new molecular descriptor for SAR applications. Comparison with other descriptors. *J. Chem. Inf. Model.* 1987, 27, 82–85.
28. Kier, L., A shape index from molecular graphs. *Quant. Struct.–Act. Relat. Pharmacol., Chem. Biol.* 1985, 4, 109–116.
29. Labute, P., A widely applicable set of descriptors, http://www.chemcomp.com/journal/vsadesc. htm, last accessed August 2008.
30. Williams, A., A perspective of publicly accessible/open-access chemistry databases. *Drug Discov. Today* 2008, 13, 495–501.
31. Haider, N., "checkmol/matchmol," http://merian.pch.univie.ac.at/~nhaider/cheminf/cmmm. html, last accessed August 2008.
32. Cormen, T., Leiserson, C., and Rivest, R., *Introduction to Algorithms*, MIT Press: Cambridge, MA, 1998.
33. Ullmann, J., An algorithm for subgraph isomorphism. *J. ACM* 1976, 23, 31–42.
34. Sayle, R., Improved SMILES substructure Searching, http://www.daylight.com/meetings/emug00/Sayle/substruct.html, 2000.
35. Ballester, P. and Graham Richards, W., Ultrafast shape recognition to search compound databases for similar molecular shapes. *J. Comp. Chem.* 2007, 28, 1711–1723.
36. Good, A., Ewing, T., Gschwend, D., and Kuntz, I., New molecular shape descriptors—applications in database screening. *J. Comput.-Aided Mol. Des.* 1995, 9, 1–12.
37. Guha, R., Dutta, D., Jurs, P., and Chen, T., R–NN curves: An intuitive approach to outlier detection using a distance based method. *J. Chem. Inf. Model.* 2006, 46, 1713–1722.
38. Xu, H. and Agrafiotis, D., Nearest neighbor search in general metric spaces using a tree data structure with a simple heuristic. *J. Chem. Inf. Comput. Sci.* 2003, 43, 1933–1941.
39. Bentley, J., Multidimensional binary search trees used for associative searching. *Commun. ACM* 1975, 18, 509–517.
40. Gionis, A., Indyk, P., and Motwani, R., Similarity search in high dimensions via hashing. In: *VLDB '99: Proceedings of the 25th International Conference on Very Large Data Bases*, Morgan Kaufmann Publishers Inc.: San Francisco, CA, 1999.
41. Guttman, A., R-trees: A dynamic index structure for spatial searching. In: *SIGMOD Conference*, ACM Press: New York, 1984.
42. Bohm, C., Berchtold, S., and Keim, D., Searching in high-dimensional spaces: Index structures for improving the performance of multimedia databases. *ACM Comput. Surv.* 2001, 33, 322–373.
43. Wang, W., Yang, J., and Muntz, R., Information organization and databases: Foundations of data organization. In: *Kluwer International Series in Engineering and Computer Science Series*, Kluwer Academic Publishers: Norwell, MA, 2000; Chapter PK-tree: A Spatial Index Structure for High Dimensional Point Data, pp. 281–293.
44. Ciaccia, P., Patella, M., and Zezula, M., M-tree: An efficient access method for similarity search in metric spaces. In: *VLDB '97: Proceedings of the 23rd International Conference on Very Large Data Bases*, Morgan Kaufmann Publishers, Inc.: San Francisco, CA, 1997.

45. Hellerstein, J., Naughton, J., and Pfeffer, A., Generalized search trees for database systems. In: *VLDB '95: Proceedings of the 21st International Conference on Very Large Databases*, Morgan Kaufmann Publishers Inc.: San Francisco, CA, 1995.
46. Schmid, E.-G., personal communication, 2008.
47. Warr, W., Workflow and pipelining in chemoinformatics, http://www.qsarworld.com/qsar-workflow1.php, last accessed July 2008.
48. Oinn, T., Addis, M., Ferris, J., Marvin, D., Senger, M., Greenwood, M., Carver, T., Glover, K., Pocock, M., Wipat, A., and Li, P., Taverna: A tool for the composition and enactment of bioinformatics workflows. *Bioinformatics* 2004, 20, 3045–3054.

13 Sequence Alignment Algorithms

Applications to Glycans and Trees and Tree-Like Structures

Tatsuya Akutsu

CONTENTS

13.1 INTRODUCTION

Glycans, which are also known as carbohydrate sugar chains, are important biomolecules. In particular, they are quite vital for the development and functioning of multicellular organisms, and they are generally found on the exterior surface of cells. Some glycans play an important role in cell–cell interactions. For example, tumor cells make some abnormal glycans, which are recognized by some receptors on natural killer cells. Some glycans also play an important role in protein folding cooperating with chaperone proteins.

Despite their importance, few computational methods had been developed for analyzing glycans until the beginning of the twenty-first century. The importance of glycans has been recognized in the field of bioinformatics since the beginning of the twenty first-century and then various studies have been carried out. One of the important studies is the construction of databases of glycans. Based on the early work on the CarbBank database, a new publically available database called KEGG Glycan has been constructed [1]. Along with the construction of the database, it was recognized that there was no tool for similarity search for glycans. Thus, a search tool named KCaM (KEGG Carbohydrate Matcher) has been developed [2] along with alignment algorithms for glycans [3]. Following the development of the search tool, several machine learning or computational methods have been developed, which include development of score matrices for glycan alignment [4], probabilistic models the methods for the classification of glycans [5], support vector machine-based methods for the classification of glycans using newly developed tree kernels [6,7], elucidation of glycan structures using gene expression data [8], and tandem mass spectrometry data [9].

Since the purpose of this section is not to give a comprehensive review but to give a detailed explanation on glycan alignment algorithms, we focus on glycan alignment. In this section, we first review tree edit distance and tree alignment because glycans are usually represented as trees and glycan alignment algorithms are based on these concepts. Next, we briefly review glycan structures. Then, after reviewing some basic algorithms, we give a detailed description of the glycan alignment algorithms (KCaM algorithms) [3] along with pseudocodes and examples. Finally, we briefly review the KCaM search tool [2] and discuss the limitation of the algorithms and possible future development.

13.2 TREE EDIT DISTANCE AND TREE ALIGNMENT

Trees are very common data structures in computer science and are special cases of graphs. The problem of comparing trees arises in various areas such as bioinformatics, chemoinformatics, structured text databases (e.g., XML databases) and image analysis [10]. In particular, a lot of tree-based studies have been carried out for comparison of RNA secondary structures [11]. Although trees are classified into rooted trees and unrooted trees, this section focuses on rooted trees since glycans are usually represented as rooted trees as well as RNA secondary structures.

A *tree* T consists of a set of nodes $V(T)$ and a set of directed edges $E(T)$. Nodes are divided into *internal nodes* and *leaves*, where there exists one special internal node called the *root*. In this section, we only consider *labeled trees* in which each node is assigned a symbol from some finite set Σ.

Each internal node has one or more outgoing edges and each leaf does not have an outgoing edge. For each edge (u, v), u is called a *parent* of v, and v is called a *child* of u. For each node u, $Chd(u)$ denotes the set of children. Thus, $Chd(u) = \{\}$ holds if and only if u is a leaf. Then, we can see that the set of edges is given by

$$E(T) = \{(u, v) | v \in Chd(u),\ u \in V(T)\}.$$

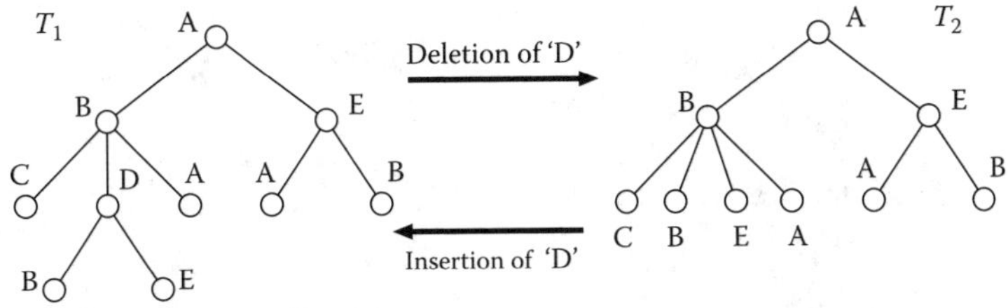

FIGURE 13.1 Deletion and insertion operations. In this example, T_2 is obtained from T_1 by deleting a node labeled 'D'. Conversely, T_1 is obtained from T_2 by inserting a node labeled 'D'.

There are two types of rooted trees: ordered trees and unordered trees. A tree is called *ordered* if a left-to-right order among siblings is given. Otherwise, it is called *unordered*. In ordered trees, this left-to-right order must be preserved among matching nodes.

Although various measures have been proposed for evaluating the similarity between two trees, *tree edit distance* has been extensively studied and applied. Tree edit distance for ordered trees is defined as follows (see Ref. [10] for details), where we only consider the unit cost case (i.e., it takes cost 1 per insertion, deletion, or substitution). Let T be a rooted ordered tree. Let $label(v)$ denote the label of a node v. $|T|$ denotes the size (the number of nodes) of T. An *edit operation on a tree T* is either a *deletion*, an *insertion*, or a *substitution* (see also Figure 13.1).

Deletion: Delete a non-root node v in T with parent u, making the children of v become children of u. The children are inserted in the place of v as a subsequence in the left-to-right order of the children of u.

Insertion: Complement of delete. Insert a node v as a child of u in T making v the parent of a consecutive subsequence of the children of u.

Substitution: Change the label of a node v in T.

The *tree edit distance* between T_1 and T_2 is defined as the minimum number of operations to transform T_1 into T_2.

A close relationship exists between the edit distance and the *ordered edit distance mapping* (or just a *mapping*) [10]. $\mathcal{A} \subseteq V(T_1) \times V(T_2)$ is called a mapping if the following conditions are satisfied for any pair (u_1, v_1), $(u_2, v_2) \in \mathcal{A}$ (see also Figure. 13.2):

i. $u_1 = u_2$ iff. $v_1 = v_2$
ii. u_1 is an ancestor of u_2 iff. v_1 is an ancestor of v_2
iii. u_1 is to the left of u_2 iff. v_1 is to the left of v_2

Let $id(\mathcal{A})$ be the set of pairs having identical labels in $\mathcal{A}$. The mapping $\mathcal{A}$ maximizing $id(\mathcal{A})$ defines the *largest common sub-tree*,* which is the tree obtained by

* In this section, we add a "-" between "sub" and "tree" to distinguish the word from the common notion of a *subtree*.

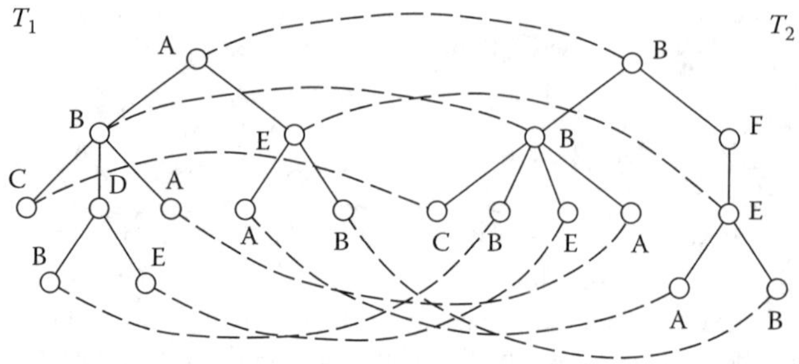

FIGURE 13.2 An example of ordered edit distance mapping. In this example, T_2 is obtained from T_1 by deleting a node labeled 'D', inserting a node labeled 'F' and substituting the label of the root. Thus, the distance between T_1 and T_2 is three. The corresponding ordered edit distance mapping is shown by broken lines.

deleting nodes not appearing in $id(\mathcal{A})$ from T_1 (or T_2). It is well known [10] that the tree edit distance is equal to:

$$|T_1| + |T_2| - |\mathcal{A}| - |id(\mathcal{A})|,$$

which means that the computation of the largest common sub-tree is equivalent to the computation of the tree edit distance. $\mathcal{A}$ can also be regarded as an *alignment* between T_1 and T_2.

Here we review a dynamic programming procedure to compute the tree edit distance [10]. For a forest F_1, let $T_1(u)$ and u be the rightmost tree of F_1 and its root, respectively, where a *forest* is a set of rooted trees and we assume that trees in F_1 are ordered from left to right. Then, $T_1(u) - u$, $F_1 - u$, and $F_1 - T_1(u)$ denote the forests obtained by deleting u from $T_1(u)$, by deleting u from F_1, and by deleting $T_1(u)$ from F_1, respectively. For a forest F_2, v, $T_2(v)$, $T_2(v) - v$, $F_2 - v$, and $F_2 - T_2(v)$ are defined in an analogous way. Then, the tree edit distance is computed by the following dynamic programming procedure (see also Figure 13.3):

$$D(\epsilon, \epsilon) = 0,$$

$$D(F_1, \epsilon) = D(F_1 - u, \epsilon) + 1,$$

$$D(\epsilon, F_2) = D(\epsilon, F_2 - v) + 1,$$

$$D(F_1, F_2) = \min$$

$$\begin{cases} D(F_1 - u, F_2) + 1, \\ D(F_1, F_2 - v) + 1, \\ D(F_1 - T_1(u), F_2 - T_2(v)) + D(T_1(u) - u, T_2(v) - v) \\ \quad + \delta(label(u), label(v)), \end{cases}$$

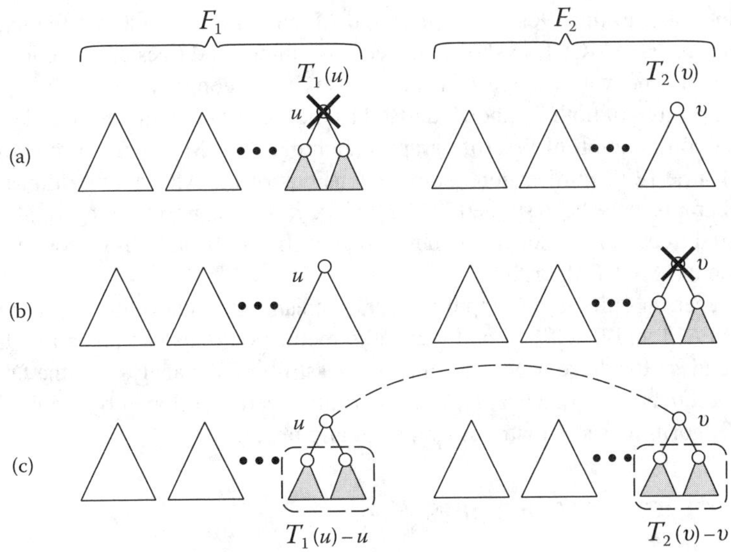

FIGURE 13.3 Explanation of the dynamic programming algorithm for tree edit distance.

where ϵ denotes an empty forest, and $\delta(x, y) = 1$ if $x = y$, otherwise $\delta(x, y) = 0$. The meaning of the recursion can be seen from Figure 13.3. This algorithm can be extended for the general cost function.

The above presented algorithm works in $O(n^4)$ time, where $n = \max(|T_1|, |T_2|)$. There is a long history of efficient algorithms for tree edit distance. Tai-first defined the notion of tree edit distance and proposed an $O(n^6)$ time algorithm [12]. Zhang and Shasha [13] improved it to $O(n^4)$ time. Klein [14] further improved it to $O(n^3 \log n)$ time. Demaine et al. [15] finally developed an $O(n^3)$ time algorithm and showed that it is optimal under a reasonable computation model.

Tree edit distance for unordered trees is defined in the same way as above except that we need not preserve the ordering among siblings. However, it is known that computation of edit distance between unordered trees is NP-hard [16]. Therefore, it is very difficult to compute the edit distance or the largest common sub-tree efficiently for unordered trees. Horesh et al. [17] developed an A* algorithm to compute the largest common sub-tree for unordered trees and they showed that it is possible to compute an optimal solution for trees consisting of dozens of nodes.

Because of NP-hardness of edit distance for unordered trees, some other measures are proposed. Jiang et al. [18] proposed *alignment of trees*. In alignment of trees, a smallest common supertree T is computed from two input trees T_1 and T_2. That is, T is a smallest tree such that both T_1 and T_2 become subtrees of T with allowing substitution of labels paying some cost. It is shown in Ref. [18] that alignment for unordered trees can be computed in polynomial time if the maximum number of children is bounded by a constant, otherwise it is NP-hard. However, alignment of trees is less flexible than tree edit distance.

Going back to the 1960s, Edmonds and Matula [19] showed that the *maximum common subtree* (MCST, in short) between two unordered trees can be computed in polynomial time, where a tree T is called a maximum common subtree of T_1 and T_2 if T is a subtree of both T_1 and T_2 and T has the maximum number of nodes. Using efficient computation of bipartite graph matching, their MCST algorithm works in $O(n^{2.5})$ time [20]. Furthermore, some extensions of the MCST algorithm to more general graphs have been studied [21,22]. Since it is often useful to regard glycans as unordered trees, it is reasonable to develop algorithms for the comparison of glycans based on the MCST algorithm.

However, in order to develop biologically meaningful algorithms, score matrices such as PAM and BLOSUM and local alignment should also be taken into account because these two have been used quite successfully in the analysis of the DNA and protein sequences [23]. Thus, KCaM algorithms were developed by combining the MCST algorithm, score matrices, and local alignment.

13.3 GLYCAN STRUCTURES

Glycans are special kinds of chemical compounds. The basic component of glycans is the monosaccharide unit, or sugar, of which a handful are most common in higher animal oligosaccharides (see Table 13.1). Each unit is linked to one or more other monosaccharides by various types of linkages, depending on the anomer (i.e., α or β) and the hydroxyl group numbers to which they are attached on the monosaccharides. Most glycans have rooted tree structures, where nodes correspond to monosaccharides and edges correspond to linkages between monosaccharides (see Figure 13.4). Although glycans might be regarded as ordered trees, it is better to regard them as unordered trees for providing more flexible matching methods.

There are several classes of glycan based on certain basic patterns mainly in the core structure (i.e., a substructure near to the root), which include *N*-glycan, *O*-glycan, glycoside, and sphingolipid [1]. Parts of these structures are recognized by various

TABLE 13.1
Common Monosaccharide Names, their Abbreviations, and their Symbols [4]

Sugar Name	Abbreviations	Symbols
Glucose	Glc	▲
Galactose	Gla	●
Mannose	Man	○
N-acetyl neuraminic/sialic acid	NeuNAc	◆
N-acetylglucosamine	GlcNAc	■
N-acetylgalactosamine	GalNAc	□
Fucose	Fuc	△
Xylose	Xyl	▽

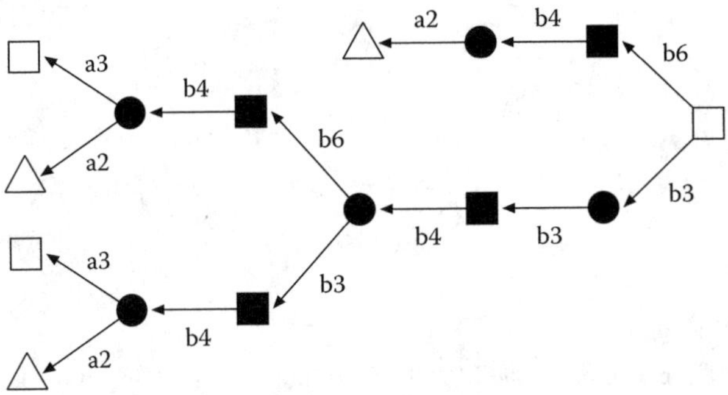

FIGURE 13.4 An example of glycan structure (KEGG Glycan ID: G02178). A label of each edge denotes a pair of ID numbers of atoms connected by the corresponding chemical bond.

agents such as pathogens and proteins and, thus, are closely related to their functions. Thus, analysis of tree patterns of glycan structures is important for understanding and predicting functions of glycans.

13.4 BASIC ALGORITHMS

Before describing KCaM algorithms, we review the MCST algorithm [19,20] and the global and local sequence alignment algorithms because KCaM algorithms are based on them.

13.4.1 MCST ALGORITHM

Different from tree edit, we hereafter consider unordered trees. A *subtree* of tree T is a tree whose nodes and edges are subsets of those of T. Two trees T_1 and T_2 are said to be *isomorphic* if and only if there is a bijection M between nodes in T_1 and T_2 such that the following conditions are satisfied:

- $\forall(v_1, v_2) \in M, label(u) = label(v)$
- $(u_1, v_1) \in E(T_1)$ if and only if $(u_2, v_2) \in E(T_2)$

Then, the MCST problem is defined as a problem of, given two unordered, labeled, rooted trees T_1 and T_2, finding the tree T_c that is isomorphic to a subtree of both T_1 and T_2 and whose number of nodes is the maximum among all such possible trees.

The MCST algorithm is based on dynamic programming. For a node v in tree T, $T(v)$ denotes the subtree of T induced by v and its descendants. For each pair $(u, v) \in V(T_1) \times V(T_2)$, the MCST algorithm computes the size (i.e., the number of nodes) of the MCST between $T_1(u)$ and $T_2(v)$, which is denoted by $R[u, v]$. $\mathcal{M}(u, v)$ denotes the set of one-to-one mappings between $Chd(u)$ and $Chd(v)$. Then, $R[u, v]$ can be computed by using the following dynamic programming procedure (see also

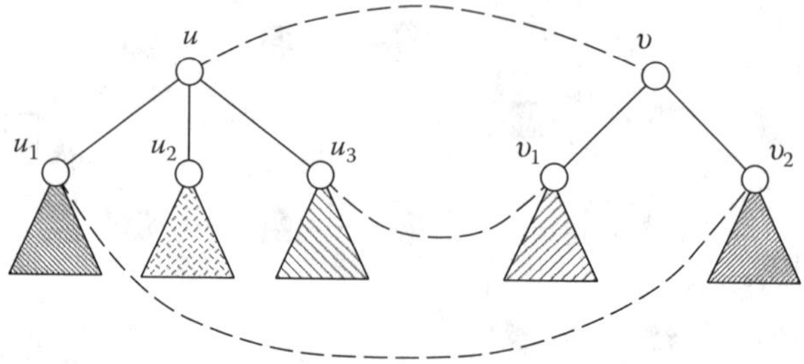

FIGURE 13.5 Explanation of the MCST algorithm. In order to compute score $R[u, v]$, all possible mappings between $\{u_1, u_2, u_3\}$ and $\{v_1, v_2\}$ are examined. In this example, $M = \{(u_1, v_2), (u_3, v_1)\}$ is selected as a mapping between $Chd(u)$ and $Chd(v)$ with the maximum score, where triangles with similar patterns mean similar subtrees.

Figure 13.5):

$$R[u,v] = \begin{cases} 0, & \text{if } u = \epsilon, v = \epsilon \text{ or } label(u) \neq label(v), \\ 1 + \max\limits_{M \in \mathcal{M}(u,v)} \left\{ \sum\limits_{(u_i,v_j) \in M} R[u_i, v_j] \right\}, & \text{otherwise,} \end{cases}$$

where ϵ denotes an empty tree. It is to be noted that the left-to-right ordering of siblings is not taken into account in the above (since there is no restriction on mappings M). Computation of $\max_{M \in \mathcal{M}(u,v)} \{\cdots\}$ can be carried out in polynomial time by using the maximum bipartite graph matching [19,20]. The resulting MCST can be retrieved by using a *traceback* procedure, where the traceback is a standard technique in dynamic programming.

13.4.2 Global and Local Sequence Alignment

Sequence alignment is one of the most basic and important algorithms in bioinformatics. Since sequence alignment is explained in detail in another chapter, we briefly review the algorithms.

Let $s = s_1 s_2 \ldots s_m$ and $t = t_1 t_2 \ldots t_n$ be sequences of DNAs or proteins. *Global sequence alignment* with *linear gap cost* is computed by the following dynamic programming procedure:

$$S[i,0] = d \cdot i,$$

$$S[0,j] = d \cdot j,$$

$$S[i,j] = \max \begin{cases} S[i,j-1] + d, \\ S[i-1,j] + d, \\ S[i-1,j-1] + w(s_i,t_j), \end{cases}$$

where $d < 0$ is a penalty for a gap and $w(x, y)$ is the score between nucleotides or amino acids x and y. $S[i,j]$ gives the score of an optimal global alignment between

$s_1s_2\ldots s_i$ and $t_1t_2\ldots t_j$ and thus the score of an optimal global alignment between s and t is given by $S[m,n]$.

Local sequence alignment is defined as a problem of finding a global sequence alignment with the maximum score between s' and t' where s' and t' are any consecutive subsequences of s and t, respectively. Local sequence alignment with linear gap cost is computed by the following dynamic programming procedure:

$$S[i,0] = 0,$$

$$S[0,j] = 0,$$

$$S[i,j] = \max\begin{cases} 0, \\ S[i,j-1]+d, \\ S[i-1,j]+d, \\ S[i-1,j-1]+w(s_i,t_j). \end{cases}$$

In this case, $S[i,j]$ denotes the score of an optimal local alignment ending at (s_i,t_j) and thus the score of an optimal local alignment between s and t is given by $\max_{i,j} S[i,j]$.

13.5 KCaM ALGORITHMS

KCaM algorithms are combinations of the MCST algorithm and the global/local sequence alignment algorithms. There are basically two versions, *global glycan alignment* and *local glycan alignment*, where the former corresponds to global sequence alignment and the latter corresponds to local sequence alignment. Furthermore, we have variants of these glycan alignment algorithms in which gaps are not allowed. The original versions are called *approximate matching* algorithms and the variants are called *exact matching* algorithms. In this section, we begin with global glycan alignment, then explain local glycan alignment, and finally briefly explain exact matching algorithms.

13.5.1 GLOBAL GLYCAN ALIGNMENT

The global glycan alignment algorithm is a simple combination of the MCST algorithm and the global sequence alignment algorithm. More precisely, it is obtained from the MCST algorithm by allowing deletion of a middle node. However, the deletion operation is different from that of tree edit. In tree edit, all children of the deleted node u become children of the parent of u. However, in glycan alignment, only one child can be a child of the parent of u and the other children are deleted along with their descendants (see Figure. 13.6).

Since it would be a bit complicated to define appropriate edit operations, we do not give edit operations for glycan alignment. Instead, we directly give the dynamic

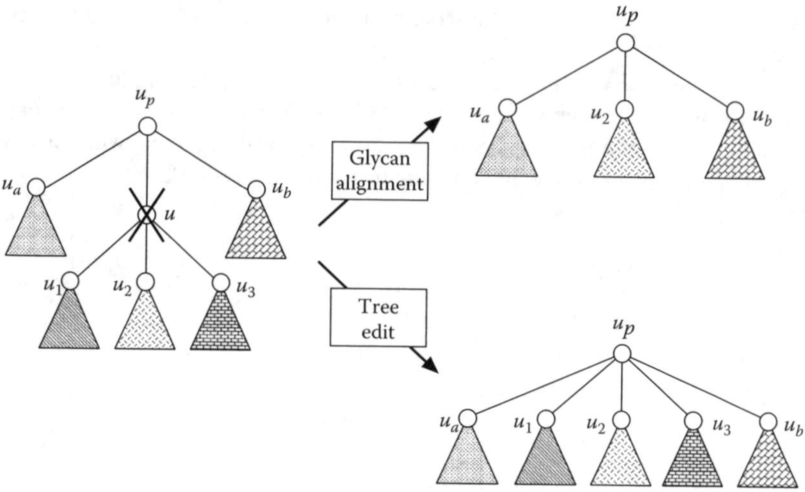

FIGURE 13.6 Difference of deletion operations between glycan alignment and tree edit. If u is deleted, all the children of u become children of the parent of u (denoted by u_p) in tree edit, whereas only one child of u can become a child of u_p in glycan alignment.

programming procedure as follows:

$$Q[u,0] = \sum_{u_i \in T_1(u)} d(u_i),$$

$$Q[0,v] = \sum_{v_i \in T_2(v)} d(v_i),$$

$$Q[u,v] = \max \begin{cases} \max_{v_i \in Chd(v)} \left\{ Q[u,v_i] + d(v) + \sum_{v_j \in Chd(v)-\{v_i\}} Q[0,v_j] \right\}, \\ \max_{u_i \in Chd(u)} \left\{ Q[u_i,v] + d(u) + \sum_{u_j \in Chd(u)-\{u_i\}} Q[u_j,0] \right\}, \\ w(u,v) + \max_{M \in \mathcal{M}(u,v)} \begin{Bmatrix} \sum_{(u_i,v_j) \in M} Q[u_i,v_j] + \\ \sum_{u_k \in Chd(u)-M_1} Q[u_k,0] + \\ \sum_{v_k \in Chd(v)-M_2} Q[0,v_k] \end{Bmatrix}. \end{cases}$$

In the above, $d(u)$ denotes the cost for deleting a node u (corresponding to gap penalty in sequence alignment), $w(u,v)$ represents the similarity between nodes u and v, and M_i denotes the set of nodes of T_i appearing in M. The meanings of three terms (three terms in outer max) appearing in the right-hand side of the third recursion is as follows. The first term corresponds to the deletion of v. The second term corresponds to the deletion of u (see Figure 13.6). The third term corresponds to the matching between u and v. In order to compute the third term, a maximum score matching between the children of u and the children of v is computed as in the MCST algorithm (see Figure. 13.5). However, in this case, the deletion costs for non-matching nodes and their descendants (represented by $Q[u_k,0]$ and $Q[0,v_k]$) are taken into account.

As in the case of tree edit, we can relate the score with a kind of mapping or alignment. $\mathcal{A} \subseteq V(T_1) \times V(T_2)$ is called a *glycan alignment* if the following conditions are satisfied for any pair $(u_1, v_1), (u_2, v_2) \in \mathcal{A}$:

i. $u_1 = u_2$ iff. $v_1 = v_2$
ii. u_1 is an ancestor of u_2 iff. v_1 is an ancestor of v_2
iii. If $u_1 \in V(T_1)$ (respectively $v_1 \in V(T_2)$) does not appear in $\mathcal{A}$, at most one child of u_1 (respectively v_1) and its descendants can appear in $\mathcal{A}$

It is to be noted that condition (iii) is not included in tree edit mapping. Instead, the condition on sibling orderings is not included here. We define the score of an alignment $\mathcal{A}$ by

$$\sum_{(u,v)\in\mathcal{A}} w(u,v) + \sum_{u\in V(T_1)-\mathcal{A}_1} d(u) + \sum_{v\in V(T_2)-\mathcal{A}_2} d(v),$$

where $\mathcal{A}_i$ denotes the set of nodes of T_i appearing in $\mathcal{A}$. Then, the score of an optimal glycan alignment $\mathcal{A}$ is equal to $Q[r_1, r_2]$ where r_1 and r_2 are the roots of T_1 and T_2, respectively.

In the above, the similarities of edges are not taken into account. However, edges have important biological meanings. Thus, the similarities of edges should be taken into account. For that purpose, we define $w(u, v)$ as follows:

$$w(u,v) = \max \begin{cases} 0, \\ \alpha \cdot \delta(label(u), label(v)) \\ \quad -\beta \cdot (1 - \delta(ulabel(p(u),u), ulabel(p(v),v))) \\ \quad -\beta \cdot (1 - \delta(dlabel(p(u),u), dlabel(p(v),v))), \end{cases}$$

where $p(u)$ denotes the parent of u, $label(u)$ denotes the name of the monosaccharide unit or sugar, and $ulabel(p(u), u)$ (resp. $dlabel(p(u), u)$) indicates the carbon number (id) of sugar $p(u)$ (respectively sugar u) to which the edge $(p(u), u)$ is connected. In the current implementation, we use $\alpha = 100.0$ and $\beta = 25.0$ based on several trials.

13.5.2 Local Glycan Alignment

As in sequence alignment, we can develop a *local* version of glycan alignment. The local glycan alignment problem is defined as a problem of finding a global glycan alignment with the maximum score between T_1' and T_2' where T_1' and T_2' are any subtrees of T_1 and T_2, respectively.

The algorithm for local glycan alignment is obtained by combining the local sequence alignment algorithm and the global glycan alignment algorithm. The following is its dynamic programming procedure:

$$Q[u, 0] = 0,$$

$$Q[0, v] = 0,$$

$$Q[u, v] = \max \begin{cases} 0, \\ \max_{v_i \in Chd(v)} \left\{ Q[u, v_i] + d(v) + \sum_{v_j \in Chd(v) - \{v_i\}} Q[0, v_j] \right\}, \\ \max_{u_i \in Chd(u)} \left\{ Q[u_i, v] + d(u) + \sum_{u_j \in Chd(u) - \{u_i\}} Q[u_j, 0] \right\}, \\ w(u, v) + \max_{M \in \mathcal{M}(u,v)} \left\{ \sum_{(u_i, v_j) \in M} Q[u_i, v_j] \right\}. \end{cases}$$

13.5.3 Exact Matching Algorithms

There exists another variant of the above mentioned glycan alignment algorithms. In this variant, gaps are not allowed. Furthermore, the degree of specificity can be specified by selecting either to match just monosaccharide names (i.e., sugar names) or both names and linkage information (i.e., sugar names and bond types). As in approximate matching, there exist two versions: exact global matching and exact local matching. Since the algorithms are almost the same as the ones for approximate matching, we do not give algorithms or pseudocodes of the exact algorithms. See Ref. [3] the details of the exact matching algorithms.

13.6 PSEUDOCODE

Although outlines (i.e., recursions) of the glycan alignment algorithms have been provided above, some details are unclear. In particular, how to compute maximum matchings is unclear. Therefore, we here provide pseudocodes of the algorithms. In these codes, each node is given an ID number, where the numbers are given in the postorder traversal: the ID number of a parent is larger than the ID numbers of its children. The node with ID number i (respectively j) in T_1 (respectively T_2) is denoted by u_i (respectively v_j). For each node with ID number i in T_1, $Chd(i)$ denotes the set of ID numbers of the children of u_i, and $ChdID(u_i, k)$ denotes the ID number of the kth child of v_i. $Chd(j)$ and $ChdID(v_j, k)$ are defined in the same way for T_2. For a tree T and a set of vertices W, $T(W)$ denotes a subtree induced by W.

In both cases, we give codes only for computing the score of an optimal alignment and do not give codes for retrieving an optimal alignment itself because an optimal alignment can be obtained by using the standard traceback technique.

13.6.1 Code for Global Glycan Alignment

We start with the code for global glycan alignment. In this code, *GlobalGlycan Alignment*(T_1, T_2) is the main procedure. $F[i, j]$ stores the score (i.e., $Q[u_i, v_j]$) for $T_1(u_i)$ and $T_2(v_j)$. It is to be noted that $F[i, 0]$ (respectively $F[0, j]$) corresponds to the score for deleting $T_1(u_i)$ (respectively $T_2(v_2)$). $Match(i, j)$ is a subroutine to compute the score of a maximum matching between the children of u_i and the children of v_j.

Since the number of children is small (less than four in most cases), we did not employ an efficient maximum bipartite matching algorithm. Instead, we employed a simple exhaustive procedure $MatchSub(i_1, deg0, CurScore)$. In this procedure, we examine all complete bipartite matchings between $deg0$ children of u_i and $deg0$ children of v_j, where $deg0 = \max\{deg(u_i), deg(v_j)\}$ and the score between the i_1th child of u_i and j_1th child of v_j is stored in $M[i_1, j_1]$. It is to be noted that if $i_1 > deg(u_i)$ (respectively $j_1 > deg(v_j)$), the i_1th (respectively j_1th) child is regarded as a null child and $M[i_1, j_1]$ denotes the score for deletion of $T_2(v_{j_1})$ (respectively $T_1(u_{i_1})$).

$MatachSub(i_1, deg0, CurScore)$ computes the scores of all possible matchings in a recursive manner. It starts with the null matching and extends partial matching gradually by adding pairs of nodes (including pairs with null nodes) one-by-one. When $MatchSub(i_1, deg0, CurScore)$ is called, an assignment for up to the $(i_1 - 1)$th children of u_i was already computed and $CurScore$ keeps the score for such a partial assignment. When i exceeds $deg0$, a complete matching is obtained and the maximum score is updated if the current score is greater than the previous maximum score.

```
Procedure GlobalGlycanAlign(T1,T2)
begin
  Let 1,···,n1 be the id numbers of nodes in T1
  in postorder traversal;
  Let 1,···,n2 be the id numbers of nodes in T2
  in postorder traversal;
  for i=1 to n1 do
    F[i,0] ← d(ui) + Σ_{k∈Chd(i)} F[k,0];
  for j=1 to n2 do
    F[0,j] ← d(vj) + Σ_{k∈Chd(j)} F[0,k];
  for i=1  to n1 do
    for j=1 to n2 do
    begin
      x1 ← Match(i,j) + w(ui,vj);
      x2 ← min_{k∈Chd(j)}{F[i,k] + d(vj) + Σ_{h≠k,h∈Chd(j)} F[0,h]};
      x3 ← min_{k∈Chd(i)}{F[k,j] + d(ui) + Σ_{h≠k,h∈Chd(i)} F[h,0]};
      F[i,j] ← max{x1,x2,x3}
    end;
  return F[n1,n2]
end
```

```
Procedure Match(i,j)
begin
  deg0 ← max{deg(ui),deg(vj)};
  for i1=1 to deg0 do
    for j1=1 to deg0 do
      if i1 > deg0 then M[i1,j1] ← F[0,ChdId(vj,j1)]
      else if j1 > deg0 then M[i1,j1] ← F[ChdId(ui,i1),0]
      else M[i1,j1] ← F[ChdId(ui,i1),ChdId(vj,j1)];
  for j1=1 to deg0 do  ASSIGNED[j1] ← 0;
```

```
    MatchScore ← −∞;
    MatchSub(1,deg0,0.0);
    return MatchScore
end

Procedure MatchSub(i1, deg0, CurScore)
begin
    if i1 > deg0 then
        if CurScore > MatchScore then MatchScore ← CurScore;
    else
        for j1 = 1 to deg0 do
            if ASSIGNED[j1] = 0 then
            begin
                ASSIGNED[j1] ← 1;
                MatchSub(i1 + 1, deg0, CurScore + M[i1][j1]);
                ASSIGNED[j1] ← 0
            end
end
```

Here we briefly analyze the time complexity of this glycan alignment algorithm. Since the maximum number of children in glycans is bounded by a constant, we can assume that *Match*(i,j) works in constant time. Since the main loop of the main procedure is iterated n_1n_2 times, the total time complexity of the global glycan alignment algorithm is $O(n_1n_2)$. It is easily seen that the space complexity of the algorithm is also $O(n_1n_2)$.

13.6.2 Modification for Local Glycan Alignment

The code for global glycan alignment can be modified for local glycan alignment. The required modification is simple and is a direct consequence of the change of the recursion.

We first modify the initialization part for $F[i,0]$ and $F[0,j]$ as below.

$$\textbf{for } i = 1 \textbf{ to } n_1 \textbf{ do } F[i,0] \leftarrow 0;$$

$$\textbf{for } j = 1 \textbf{ to } n_2 \textbf{ do } F[0,j] \leftarrow 0;$$

Next, we replace "$F[i,j] \leftarrow \max\{x_1,x_2,x_3\}$" with "$F[i,j] \leftarrow \max\{x_1,x_2,x_3,0\}$." Finally, we replace "**return** $F[n_1,n_2]$" with "**return** $\max_{i,j} F[i,j]$." As in the case of global glycan alignment, the time and space complexities of the local glycan alignment algorithm are $O(mn)$.

13.7 ILLUSTRATIVE EXAMPLE

Here we provide an example for illustrating the global glycan alignment algorithm. We employ two glycan data with KEGG Glycan ID numbers G06886 and G05554 (see also Figure 13.7).

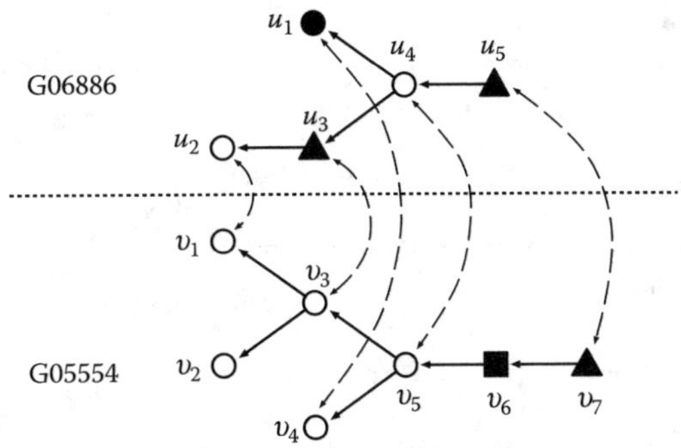

FIGURE 13.7 An example of global glycan alignment.

Recall that nodes are numbered according to the postorder traversal. For the simplicity of presentation, we use the following score function and gap penalty:

$$w(u, v) = \begin{cases} 1, & \text{if } label(u) = label(v), \\ 0, & \text{otherwise,} \end{cases}$$

$$d(v) = d(u) = -1 \text{ for all nodes } u \in V(T_1) \text{ and } v \in V(T_2).$$

The algorithm initializes $F[i, 0]$ and $F[0, j]$ as follows:

$$F[1,0] = -1,\ F[2,0] = -1,\ F[3,0] = -2,\ F[4,0] = -4,\ F[5,0] = -5$$
$$F[0,1] = -1,\ F[0,2] = -1,\ F[0,3] = -3,\ F[0,4] = -1,$$
$$F[0,5] = -5,\ F[0,6] = -6,\ F[0,7] = -7$$

Next, it determines the values of $F[1, j]$ as follows:

$$F[1,1] = 0,\ F[1,2] = 0,\ F[1,3] = -2,\ F[1,4] = 0,\ F[1,5] = -4,$$
$$F[1,6] = -5,\ F[1,7] = -6$$

It is to be noted that the value of $F[1, 3]$ is determined by $F[1, 3] = w(u_1, v_3) + F[0, 1] + F[0, 2]$. Similarly, the algorithm determines the values of $F[2, j]$ as follows:

$$F[2,1] = 1, F[2,2] = 1, F[2,3] = -1, F[2,4] = 1, F[2,5] = -3,$$
$$F[2,6] = -4, F[2,7] = -5$$

Then, it determines the values of $F[3, j]$ as follows:

$$F[3,1] = 0, F[3,2] = 0,\ F[3,3] = 0,\ F[3,4] = 0,\ F[3,5] = -2,$$
$$F[3,6] = -3,\ F[3,7] = -3$$

It is to be noted that the value of $F[3,3]$ is determined by $F[3,3] = w(u_3, v_3) + F[2,1] + F[0,2]$ (or $F[3,3] = w(u_3, v_3) + F[0,1] + F[2,2]$). Then, it determines the values of $F[4,j]$ and $F[5,j]$ as follows:

$$F[4,1] = -2,\ F[4,2] = -2,\ F[4,3] = 1,\ F[4,4] = -2,$$
$$F[4,5] = 1,\ F[4,6] = 0,\ F[4,7] = -1$$
$$F[5,1] = -3,\ F[5,2] = -3,\ F[5,3] = 0,\ F[5,4] = -3,$$
$$F[5,5] = 0,\ F[5,6] = 1,\ F[5,7] = 1$$

Finally, the score of the global glycan alignment is given as $F[5,7] = 1$ and the alignment is obtained as

$$\mathcal{A} = \{(u_1, v_4), (u_2, v_1), (u_3, v_3), (u_4, v_5), (u_5, v_7)\},$$

where this alignment comes from the following:

$$F[5,7] = w(u_5, v_7) + F[4,6],$$
$$F[4,6] = d(v_6) + F[4,5],$$
$$F[4,5] = w(u_4, v_5) + F[1,4] + F[3,3],$$
$$F[1,4] = w(u_1, v_4),$$
$$F[3,3] = w(u_3, v_3) + F[2,1] + F[0,2],$$
$$F[2,1] = w(u_2, v_1).$$

In the case of local glycan alignment, $F[i,0]$ and $F[0,j]$ are initialized to be 0, and the other $F[i,j]$ are determined as follows:

$$F[1,1] = 0, F[1,2] = 0, F[1,3] = 0, F[1,4] = 0, F[1,5] = 0, F[1,6] = 0, F[1,7] = 0,$$
$$F[2,1] = 1, F[2,2] = 1, F[2,3] = 1, F[2,4] = 1, F[2,5] = 1, F[2,6] = 0, F[2,7] = 0,$$
$$F[3,1] = 0, F[3,2] = 0, F[3,3] = 1, F[3,4] = 0, F[3,5] = 1, F[3,6] = 1, F[3,7] = 1,$$
$$F[4,1] = 1, F[4,2] = 1, F[4,3] = 1, F[4,4] = 1, F[4,5] = 2, F[4,6] = 1, F[4,7] = 1,$$
$$F[5,1] = 0, F[5,2] = 0, F[5,3] = 1, F[5,4] = 0, F[5,5] = 1, F[5,6] = 2, F[5,7] = 2.$$

In this case, optimal alignments can be obtained by the traceback procedure beginning from any of $F[4,5]$, $F[5,6]$, $F[5,7]$. Beginning from $F[4,5]$, one of the local alignments will be

$$\mathcal{A} = \{(u_1, v_4), (u_2, v_1), (u_3, v_3), (u_4, v_5)\},$$

which corresponds to a global alignment between $T_1(\{u_1, u_2, u_3, u_4\})$ and $T_2(\{v_1, v_3, v_4, v_5\})$. Beginning from $F[5,7]$, one of the local alignments will be

$$\mathcal{A} = \{(u_2, v_3), (u_3, v_5), (u_4, v_6), (u_5, v_7)\}$$

which corresponds to a global alignment between $T_1(\{u_2, u_3, u_4, u_5\})$ and $T_2(\{v_3, v_5, v_6, v_7\})$.

13.8 KCaM WEB SERVER

The above mentioned algorithms were implemented in the KCaM web server [2]. KCaM is a kind of search tool for the KEGG Glycan database [1]. Though the codes are not available in public, users can use the algorithms through the web server. As in the Blast tool for sequence homology search, KCaM requires each user to specify a query glycan structure and then searches glycan structures similar to the given query structure in the KEGG Glycan database or the CarbBank database.

There exist three ways for inputting a query glycan structure. One way is to specify the KEGG Glycan ID number. Each glycan data stored in the KEGG Glycan database has its own ID number like G00001. Thus, users can use this ID number to specify a query structure. Another way is to create a KCF (KEGG Chemical Function) file where KCF is a special format for describing chemical structures including glycans. The other way is to use the glycan structure editor in the KEGG Glycan database. It is a user-friendly graphical interface for entering any glycan structure via the web. Users can also input a KCF format file and modify it.

After specifying a query glycan structure, users can choose search algorithms, where KCaM provide the following search algorithms.

Gapped & Global: Combination of approximate (i.e., gapped) and global glycan alignment

Gapped & Local: Combination of approximate and local glycan alignment

Ungapped & Global: Combination of exact (i.e., ungapped) and global glycan alignment

Ungapped & Local: Combination of exact and local glycan alignment

Furthermore, users can change optional parameters such as gap penalty and weight of matches and can choose the target database (the KEGG Glycan database or the CarbBank database).

Once the search algorithm has been selected, KCaM is invoked and structures in the database are listed from the highest score to the lowest score. Of course, users can specify the number of structures shown per page. Users can also specify whether or not structures are shown in the list. A snapshot of a KCaM search result is shown in Figure 13.8.

13.9 CONCLUDING REMARKS

We have explained algorithms for glycan alignment. The algorithms are simple and fast enough for practical use. The algorithms were implemented in the KCaM web server, on which search against more than 10,000 glycan structures can be done in several or several tens of seconds. Furthermore, users can adjust several parameters (e.g., gap penalty, weight of matches) used in the algorithms via the web.

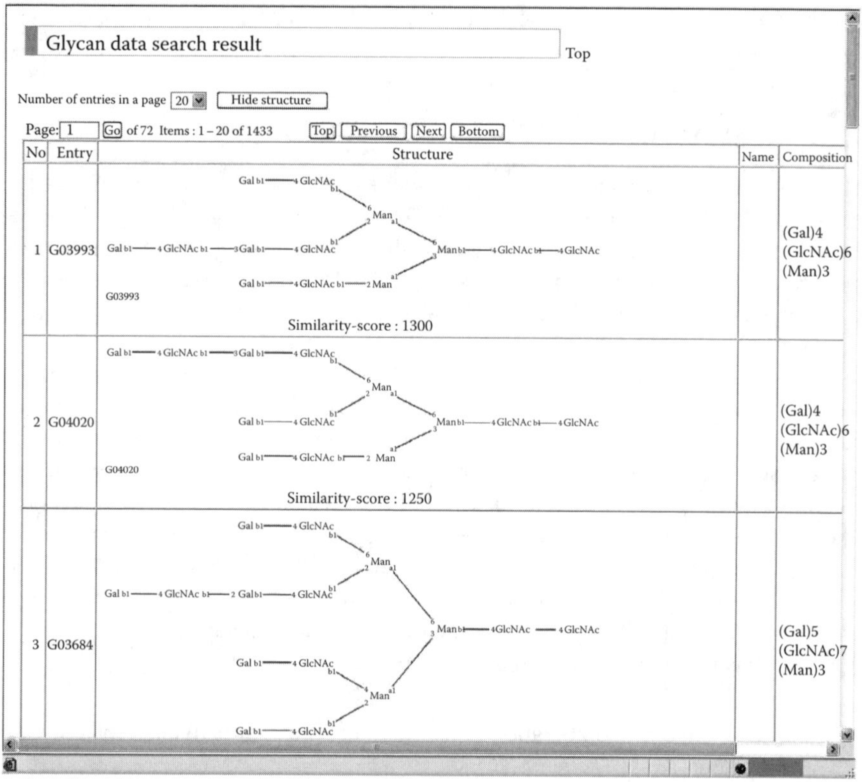

FIGURE 13.8 Snapshot of KCaM search result.

The most crucial drawback of the current glycan alignment algorithms lies in the deletion operation. As seen in Figure 13.6, if a node is deleted, only one subtree branching from the node can survive and the other subtrees are deleted. On the other hand, under the standard definition of tree edit distance, if a node is deleted, all of subtrees branching from the node can survive and become descendants of the parent of the deleted node. If we could use this definition of the deletion operation, more flexible matching would be possible. However, it is known that the tree edit distance problem is NP-hard for unordered trees [16], though it is polynomial time solvable for ordered trees. It suggests that it is almost impossible to develop polynomial time algorithms for glycan alignment with standard deletion operations if glycan structures are regarded as unordered trees. However, some practical algorithm was developed for tree edit distance for unordered trees using the A* algorithm [17]. Based on that algorithm, more flexible algorithms for glycan alignment might be developed.

REFERENCES

1. Hashimoto, K., Goto, S., Kawano, S., Aoki-Kinoshita, K. F., Ueda, N., Hamajima, M., Kawasaki, T., and Kanehisa, M., KEGG as a glycome informatics resource. *Glycobiology* 2006, *16*, 63R–70R.

2. Aoki, K. F., Yamaguchi, A., Ueda, N., Akutsu, T., Mamitsuka, H., Goto, S., and Kanehisa, M., KCaM (KEGG carbohydrate matcher): A software tool for analyzing the structure of carbohydrate sugar chains. *Nucleic Acids Res.* 2004, *32*, W267–W272.
3. Aoki, K. F., Yamaguchi, A., Okuno, Y., Akutsu, T., Ueda, N., Kanehisa, M., and Mamitsuka, H., Efficient tree-matching methods for accurate carbohydrate database queries. *Genome Inform.* 2003, *14*, 134–143.
4. Aoki, K. F., Mamitsuka, H., Akutsu, T., and Kanehisa, M., A score matrix to reveal the hidden links in glycans. *Bioinformatics* 2005, *21*, 1457–1463.
5. Hashimoto, K., Aoki-Kinoshita, K. F., Ueda, N., Kanehisa, M., and Mamitsuka, H., A new efficient probabilistic model for mining labeled ordered trees applied to glycobiology. *ACM Trans. Knowl. Discov. Data* 2008, *2*, Article No. 6.
6. Kuboyama, T., Hirata, K., and Aoki-Kinoshita, K. F., Efficient unordered tree kernel and its application to glycan classification. *Lect. Notes Comput. Sci.* 2008, *5012*, 184–195.
7. Yamanishi, Y., Bach, F., and Vert, J-P., Glycan classification with tree kernels. *Bioinformatics* 2007, *23*, 1211–1216.
8. Kawano, S., Hashimoto, K., Miyama, T., Goto, S., and Kanehisa, M., Prediction of glycan structures from gene expression data based on glycosyltransferase reactions. *Bioinformatics* 2005, *21*, 3976–3982.
9. Shan, B., Ma, B., Zhang, K., and Lajoie, G., Complexities and algorithms for glycan sequencing using tandem mass spectrometry. *J. Bioinfor. Comput. Biol.* 2008, *6*, 77–91.
10. Bille, P., A survey on tree edit distance and related problem. *Theoret. Comput. Sci.* 2005, *337*, 217–239.
11. Zhang, K., RNA structure comparison and alignment. In J. T-L. Wang, et al. (Eds.) *Data Mining in Bioinformatics*; Springer: Heidelberg, 2005, pp. 59–81.
12. Tai, K-C., The tree-to-tree correction problem. *J. ACM* 1979, *26*, 422–433.
13. Zhang, K. and Shasha, D., Simple fast algorithms for the editing distance between trees and related problems. *SIAM J. Comput.* 1989, *18*, 1245–1262.
14. Klein, P. N., Computing the edit-distance between unrooted ordered trees. *Lect. Notes Comput. Sci.* 1998, *1461*, 91–102.
15. Demaine, E., Mozes, S., Rossman, B., and Weimann, O., An optimal decomposition algorithm for tree edit distance. *Lect. Notes Comput. Sci.* 2007, *4596*, 146–157.
16. Zhang, K. and Jiang, T., Some MAX SNP-hard results concerning unordered labeled trees. *Inf. Proc. Lett.* 1994, *49*, 249–254.
17. Horesh, Y., Mehr, R. and Unger, R., Designing an A* algorithm for calculating edit distance between rooted-unordered trees. *J. Comput. Biol.* 2006, *13*, 1165–1176.
18. Jiang, T., Wang, L., and Zhang, K., Alignment of trees—an alternative to tree edit. *Theoret. Comput. Sci.* 1995, *143*, 137–148.
19. Edmonds, J. and Matula, D., An algorithm for subtree identification. *SIAM Rev.* 1968, *10*, 273–274.
20. Akutsu, T., An RNC algorithm for finding a largest common subtree of two trees. *IEICE Trans. Inf. Syst.* 1992, *E75-D*, 95–101.
21. Akutsu, T., A Polynomial time algorithm for finding a largest common subgraph of almost trees of bounded degree. *IEICE Trans. Fundam.* 1993, *E76-A*, 1488–1493.
22. Yamaguchi, A., Aoki, K. F., and Mamitsuka, H., Finding the maximum common subgraph of a partial k-tree and a graph with a polynomially bounded number of spanning trees. *Inf. Proc. Lett.* 2004, *92*, 57–63.
23. Durbin, R., Eddy. S., Krogh, A., and Mitchison, G., *Biological sequence analysis: Probabilistic models of proteins and nucleic acids.* Cambridge University Press: Cambridge, UK, 1998.

14 Machine Learning–Based Bioinformatics Algorithms
Application to Chemicals

Shawn Martin

CONTENTS

14.1 APPLICATIONS OF CLUSTERING

Clustering refers to an array of data analysis techniques that can be used to partition a dataset into groups [1,2]. Clustering is generally used as a first pass through a dataset about which little is known in an effort to organize the information for further analysis. Once this information is organized in some way, the user will often make additional hypotheses and/or perform additional analysis.

Clustering techniques are often divided into two categories: hierarchical and partitional. In hierarchical clustering (for background see Ref. [1]), a dataset is organized into a tree structure known as a dendrogram. The tree is typically drawn upside down with the trunk at the top of the page and the leaves at the bottom. In this drawing, the vertical axis is proportional to the similarity of the nodes in the tree so that clusters can be obtained by drawing a horizontal line through the tree and taking clusters to be subtrees below that line. An example is shown in Figure 14.1.

There are two main approaches to hierarchical clustering: agglomerative and divisive. In agglomerative clustering, the dendrogram is formed from the bottom (leaves) up by merging nodes to form clusters. In divisive clustering, the dendrogram is formed

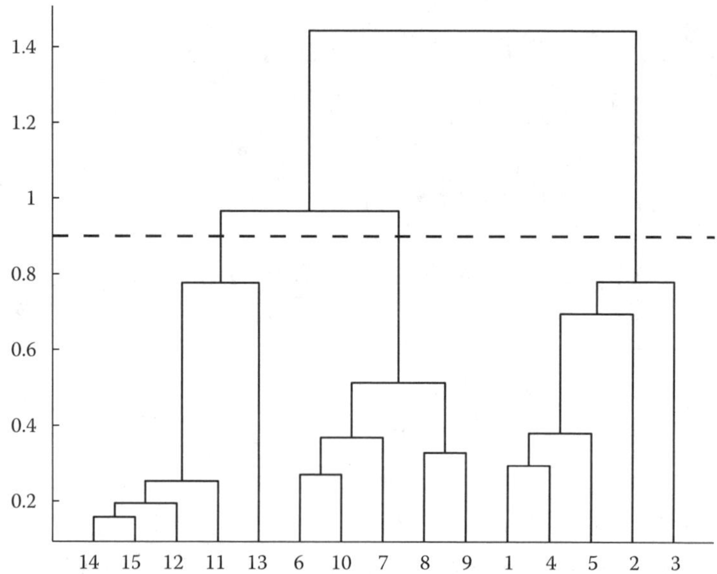

FIGURE 14.1 A dendrogram produced by hierarchical clustering, with clusters taken to be subtrees beneath the horizontal line. This dendrogram was produced using 15 artificially generated data points consisting of three clusters arranged linearly with approximately one unit between them. The x-axis shows the data point indices and the y-axis shows the single-link Euclidean distance between the clusters.

from the top (trunk) down by dividing the dataset into smaller and smaller clusters. Agglomerative clustering is the more common approach.

Within agglomerative methods, there are several variations available for deciding when to merge two clusters. The more common methods are single-link, average-link, and complete-link clustering. Single-link clustering merges two clusters based on the closest two data points within the two clusters; average-link clustering merges two clusters based on the centroids of the two clusters; and complete-link clustering merges clusters based on the maximum distance between two data points within the two clusters. Agglomerative hierarchical clustering proceeds in the following manner:

ALGORITHM 14.1 AGGLOMERATIVE HIERARCHICAL CLUSTERING

1. Compute a pair-wise distance matrix for all available samples. Distance here can be any of a number of possibilities, including Euclidean, Mahalanobis, and Hamming.
2. Search the distance matrix for the most similar pair of clusters, as measured using single-link, complete-link, average-link, or other method. Merge the most similar clusters.
3. Update the distance matrix to reflect distances between clusters.
4. Repeat steps 2 and 3 until there is only one cluster left.

Actual code for hierarchical clustering is widely available. Commercial applications include MATLAB® (http://www.mathworks.com) and SAS (http://www.sas.com). There are also open source codes available [3].

In chemoinformatics, the most common method of hierarchical clustering is agglomerative using Ward's method for merging clusters [4]. Ward's method uses square error, also known as within-cluster-variance, to measure distance between clusters. Square error is defined as

$$e^2 = \sum_{r=1}^{s} \|\mathbf{x}_r - \mathbf{x}_c\|^2, \tag{14.1}$$

where $\boldsymbol{x}_c$ is a cluster centroid and r ranges over the points in the cluster. For the case of cluster-to-cluster distance, r ranges over the points in a neighboring cluster. Ward's method has the advantage of minimizing total cluster variance (it is also known as the minimum variance method).

Hierarchical clustering is a favorite method in bioinformatics, owing to the fact that the dataset is not only partitioned, but visualization is also provided. Further, multiple partitions can be obtained by using a different horizontal threshold to cut the dendrogram. However, hierarchical clustering suffers from a computational limitation. Hierarchical clustering is memory intensive $O(n^2)$ and time intensive $O(n^2 \log n)$. There are, of course, numerous variations on hierarchical clustering that boast better performance (see, e.g., Refs. [5,6]).

Partitional clustering algorithms more strictly follow the definition of clustering previously given, in that partitional methods supply only a partition of a dataset, with no effort at visualization or different possible partitions. The most common partitional method is probably k-means. In chemoinformatics, however, a method known as Jarvis–Patrick clustering [7] is generally preferred. The popularity of the Jarvis–Patrick method in chemoinformatics has resulted from early publications by Willett et al. [8–10] comparing different clustering techniques. The method itself works by comparing data points with neighboring data points to form clusters. The algorithm proceeds in two steps:

ALGORITHM 14.2 JARVIS–PATRICK CLUSTERING

1. Generate a list of top k neighbors for each data point, using (for example) Euclidean distance or the Tanimoto coefficient [11]
2. Cluster points according to the following criterion (points that fail a, b, or c below are not in the same cluster)
   a. Data point i is in the top k neighbors of data point j
   b. Data point j is in the top k neighbors of data point i
   c. Data points i and j have at least k_{min} of their top k neighbors in common

The Jarvis–Patrick algorithm is $O(n^2)$ in time and $O(n)$ in space, making it more applicable to large datasets than hierarchical clustering. The Jarvis–Patrick method is readily available from Daylight software (http://www.daylight.com).

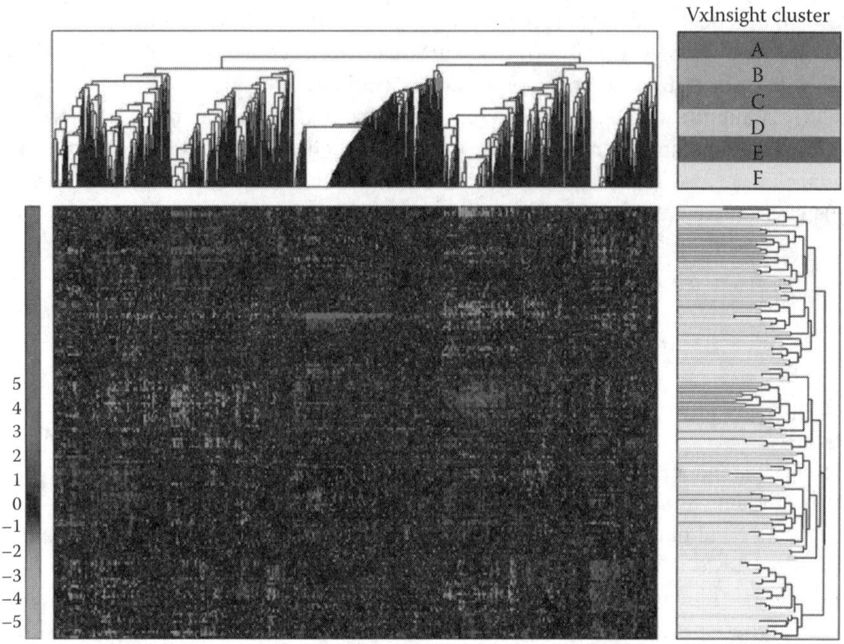

FIGURE 14.2 A prototypical heat map/dendrogram combination plot produced using hierarchical clustering of gene expression data. This plot originally appeared in Wilson et al. [16] and was produced using the clustering algorithms available in MATLAB (http://www.mathworks.com).

14.1.1 Applications in Bioinformatics

Most of the clustering applications in bioinformatics have been driven by the availability of postgenomic data, and in particular, the advent of microarray gene expression data. Gene expression data are taken from a high-throughput experimental method using microarrays [12,13]. A microarray consists of a glass slide with a printed array of microscopic dots, each containing cDNA. When an experiment is performed, gene expression is measured via the amount of mRNA that binds to the cDNA on each spot. Flourophores are used to detect when mRNA binds to a given spot.

Perhaps the earliest application of clustering in bioinformatics was using gene expression data. In the work of Eisen et al. [14], the yeast cell cycle was analyzed using microarray data from [15] and hierarchical clustering. Genes were grouped together according to correlation through time using average link hierarchical clustering. The results were displayed using a heat map for gene expression combined with a dendrogram tree for grouping genes. Eisen's heat maps are now a standard in bioinformatics. An example is shown in Figure 14.2 using data taken from Wilson et al. [16].

In addition to hierarchical clustering of gene expression data, numerous other methods have been applied and/or developed [2,17]. These methods include standard algorithms such as k-means [18] and self-organizing maps (SOMs) [19], as well as algorithms developed specifically for microarray data. Microarray-specific algorithms include graph theoretic algorithms such as cluster identification via connectivity kernels (CLICK) [20] and the cluster affinity search technique (CAST) [21]. Other algorithms specialized to microarray data relate to the simultaneous clustering of genes (rows) and samples (columns) in a dataset. These algorithms include biclustering [22] and coupled two-way clustering (CTWC) [23].

Microarray data are by far the most commonly clustered data in bioinformatics. However, if we expand our definition of clustering just slightly we encounter two other data types that have received a lot of attention. The first is protein data. Protein data are often grouped according to methods such as BLAST to infer function [24]. These types of methods cluster proteins together according to similarity of sequence (homology) and then make predictions about function and interaction.

There has also been substantial effort exerted toward grouping and organizing scientific papers published in medicine and biology and tracked by MEDLINE and PubMed (http://www.pubmed.gov). By analyzing these papers, practitioners hope to make information more readily available to researchers and (potentially) combine information across disciplines to achieve greater insight [25,26].

14.1.2 Applications in Chemoinformatics

Drug discovery is one of the central motivations for the use of clustering in chemoinformatics [4]. Recently, there has also been interest in materials design [27]. Hence, most applications are to combinatorial chemistry data, high-throughput screening (HTS) data [28] and quantitative structure–activity relationships (QSARs)/quantitative structure–property relationships (QSPRs) [4,29,30].

In combinatorial chemistry, libraries of chemicals are simultaneously created. These libraries can then be subjected to HTS. In HTS, (up to) hundreds of thousands of chemicals are arranged on grids so that they may be tested simultaneously for various properties. These datasets are similar in philosophy to gene expression arrays; hence, it is natural to apply clustering algorithms HTS data. Unlike gene expression data, however, the goal of HTS analysis is to select a heterogeneous subset of chemicals that still represent the entire library [4]. This subset is then used for further study or calculation.

Applications in chemoinformatics tend to favor two clustering algorithms. The favorite method is the Jarvis–Patrick algorithm. One of the first applications of the Jarvis–Patrick algorithm to a large chemical dataset (containing ~240,000 compounds) was performed using two-dimensional fragment descriptors [31]. Jarvis–Patrick variants are also commonly used to cluster HTS data [32–35] and improve QSAR analysis by selecting clusters containing representative chemicals [36]. k-means have also been used in combination with QSAR [37].

The next most commonly used method in chemoinformatics is agglomerative hierarchical clustering using the Ward criterion for merging clusters. Like Jarvis–Patrick clustering, Ward hierarchical clustering has also been used to provide better sampling of a compound dataset [38,39] as well as to improve QSAR performance [40].

14.1.3 Comparisons

In comparing the usage of clustering in bioinformatics and chemoinformatics, we can conclude that bioinformatics uses a much wider variety of algorithms, but that chemoinformatics uses a much broader set of data types. Bioinformatics applications use a huge variety of algorithms but are generally restricted to microarray (numerical) and sequence (string) data. Chemoinformatics applications generally use two to three clustering algorithms (Jarvis–Patrick, Ward clustering, and k-means) but use very large datasets and a huge number of chemical descriptors (see Chapters 2–4) as their input.

Clustering in both bioinformatics and chemoinformatics is fairly mature. Both fields have standard approaches that work well for their data types and applications. Cross-fertilization, however, is certain to benefit both areas of study. Bioinformatics, for example, has spurred the development of many new alternative clustering techniques, some of which work on large datasets. After adaptation to data types from chemoinformatics, it seems likely that many of these algorithms will be useful. Conversely, the diversity of data types in chemoinformatics and the general interest of drug design in bioinformatics have spurred the development of new techniques, including graph kernels in support vector machines (SVMs) (see Section 14.2.2) and the inference of protein–chemical interaction networks (Section 14.2.3).

14.2 APPLICATIONS OF CLASSIFICATION AND REGRESSION

In this section, we consider applications of methods in classification and regression to problems in bioinformatics and chemoinformatics. In the previous section (Equation 14.1), we saw that the choice of clustering algorithm varied widely between the two fields. This is due to historical reasons, as well as dataset size and type differences between biology and chemistry. In contrast, the choice of algorithm in classification and regression is fairly uniform between bioinformatics and chemoinformatics. Instead, most of the difference lies in the choice of data description. In bioinformatics, descriptors are often based on sequence (largely due to the vast amount of sequence data available), while in chemoinformatics, descriptors are typically tied more directly to structure.

Classification and regression are both methods for correlating elements in a dataset with properties of those elements. In classification, elements are correlated with discrete properties, as in the case of the classification of a molecule according to its activity (e.g., active or inactive). In regression, elements are correlated with continuous properties, as in the case of relating molecular structure to boiling point or log P.

There are many available algorithms for use in both classification and regression. In classification, popular algorithms include k-nearest neighbors [41], neural networks [42], SVMs [43], and numerous statistical methods [44,45]. In regression, popular algorithms include linear and multiple regression, principal component regression, and partial least squares [46,47]. In addition, there are several algorithms that can be used for both classification and regression, including neural networks and SVMs.

Since techniques in classification and regression are applied fairly uniformly in both bioinformatics and chemoinformatics, we focus our discussion on SVMs. SVMs can be used for classification [43,48], regression [49], and even clustering [50]. We highlight data-type differences between bioinformatics and chemoinformatics and how these differences can be accommodated using SVMs.

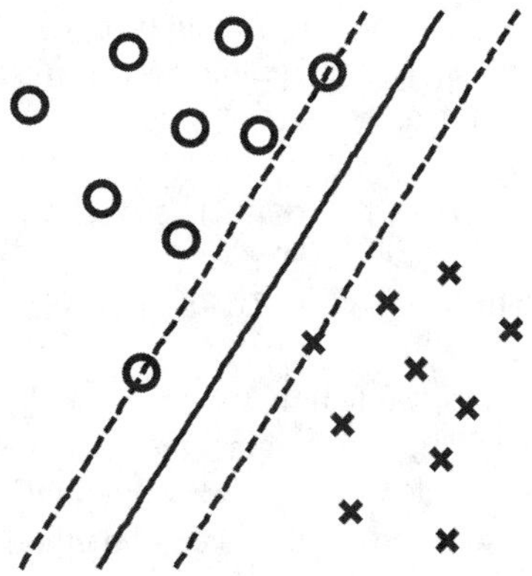

FIGURE 14.3 A linear SVM. The SVM decision boundary is defined by the support vectors, which are the examples falling on the dotted lines. The distance between the dotted lines is known as the margin and is maximized to obtain the SVM decision boundary, shown as the solid line separating the circles (class 1) from the x marks (class −1).

At its core, an SVM is a linear binary classifier. Suppose that we have a dataset $\{\mathbf{x}_i\} \subseteq R^n$ and that each point $\mathbf{x}_i$ in our dataset has a corresponding class label $y_i \in \{\pm 1\}$. Our goal is to separate the points in our dataset according to their class label. Since there are two classes, this is known as binary classification. An SVM attempts this classification by using a linear hyperplane $\mathbf{w}^T\mathbf{x} + b, \mathbf{w} \neq 0$, as shown in Figure 14.3.

Assuming that our dataset is in fact linearly separable, there will in general be many possible hyperplanes that can achieve the separation. An SVM uses an optimal separating hyperplane known as the maximal margin hyperplane. The hyperplane margin is twice the distance from the separating hyperplane to the nearest point in one (or the other) of the two classes. In Figure 14.3, this is the distance between the two dotted lines. The SVM hyperplane is found by solving the quadratic programming problem [51,52]:

$$\begin{aligned} &\max_\alpha \frac{1}{2}\sum_{i,j} y_i y_j \alpha_i \alpha_j \mathbf{x}_i^T \mathbf{x}_j - \sum_i \alpha_i \\ &\text{s.t.} \sum_i y_i \alpha_i = 0, \quad 0 \leq \alpha_i \leq C, \end{aligned} \tag{14.2}$$

where $\mathbf{w} = \sum_i y_i \alpha_i \mathbf{x}_i$ is the normal to the SVM hyperplane. Using $\mathbf{w}$ we form the SVM decision function $f(\mathbf{x}) = \text{sign}(\mathbf{w}^T\mathbf{x} + b)$, where b is obtained implicitly [51,52]. We note that $\alpha_i \neq 0$ only when $\mathbf{x}_i$ is a support vector.

The SVM problem given in Equation 14.2 only applies to datasets $\{\mathbf{x}_i\} \subseteq R^n$. Often, however, we want to use an SVM on a dataset that is not a subset of R^n. This occurs in the case of biology and chemistry problems when we are likely to use amino acid sequences or chemical structures to describe our data. Fortunately, there is a ready solution to this problem, formalized in the use of kernel functions.

Suppose that our data $\{\mathbf{x}_i\} \subseteq S$, where S might be the set of all finite length protein sequences or all finite diameter chemical graphs. We can then define a kernel function as a map $k : S \times S \rightarrow R$ such that

$$k(\mathbf{x}_i, \mathbf{x}_j) = \Phi(\mathbf{x}_i)^T \Phi(\mathbf{x}_j), \tag{14.3}$$

where $\Phi : S \rightarrow F$ is a map from our original data space S into a space F with a defined dot product such as R^N.

Once we have defined a kernel function, we simply replace the dot product $\mathbf{x}_i^T \mathbf{x}_j$ in Equation 14.2 with the kernel $k(\mathbf{x}_i, \mathbf{x}_j)$ to obtain the full SVM quadratic programming problem. A similar procedure can be used for any method that is written in terms of dot products, thus giving rise to the moniker kernel methods.

SVMs have a number of advantages over competing methods, including a unique solution of a fairly straightforward quadratic programming problem (compared to a nonunique solution for a neural network), ability to employ nonlinearity by choice of kernel, and widespread availability of software [53,54]. On the other hand, SVMs can be memory intensive $O(n^2)$ and the flexibility in choice of kernel can make SVMs difficult for beginners. Neither of these disadvantages, however, has slowed the widespread use of SVMs in both bioinformatics and chemoinformatics.

14.2.1 Applications in Bioinformatics

The central applications of classification and regression methods in bioinformatics are to protein function prediction [55] and protein interaction prediction [56]. These two topics are not unrelated and are often approached using the same basic methods. In particular, both protein function and protein interaction prediction have benefited from the application of classification and regression methods.

The success of function and interaction prediction often depends more on the formulation of the problem than on the algorithm used for the classification or regression. For proteins, this means the use of either sequence or structure and the particular assumptions used when the data are put into the algorithm. For sequence data, predictions are often made by comparing proteins using sequence similarity (homology) across organisms [24,57], using subsequence motifs or identifying motifs and using them to predict either function or interaction [58,59] and by using amino acid residue-specific features such as van der Waals volume, polarity, hydrophobicity, charge, and so on to represent a sequence as a vector [60,61]. For structural data, predictions can be made by comparing proteins based on geometric considerations such as amino acid proximity and similarity of 3D ball protein models [62].

In terms of SVMs, data types used to predict function include gene expression data [63], motif-based subsequences [58], and feature vectors based on sequence [61] and structure [62]. Data used to predict interactions include generalized subsequences [64], feature vectors [60], and combinations of sequence and structure [65,66].

The success of classification and regression methods has been widespread in bioinformatics, due largely to the availability of vast amounts of sequence data. Although many different algorithms have been applied in the field, SVMs have been a favorite for a variety of data types. One reason that SVMs are popular is that SVMs encourage

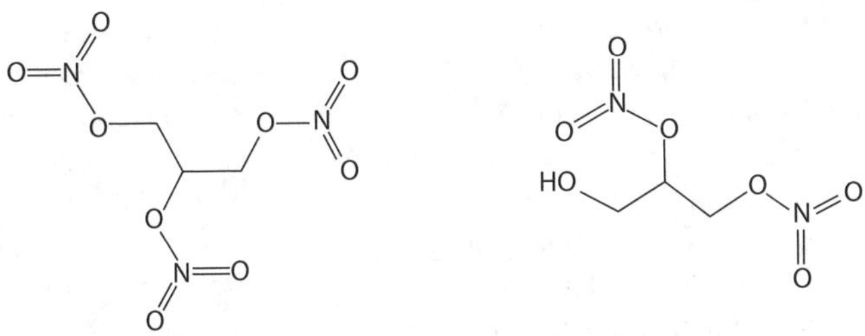

Nitroglycerine	1,2-Dinitroglycerine
3 O(NC) $\leftrightarrow$ $\mathbf{z}_1$	2 O(NC) $\leftrightarrow$ $\mathbf{z}_1$
6 O(=N) $\leftrightarrow$ $\mathbf{z}_2$	1 O(HC) $\leftrightarrow$ $\mathbf{z}_7$
3 N(O=O=O) $\leftrightarrow$ $\mathbf{z}_3$	4 O(=N) $\leftrightarrow$ $\mathbf{z}_2$
5 H(C) $\leftrightarrow$ $\mathbf{z}_4$	2 N(O=O=O) $\leftrightarrow$ $\mathbf{z}_3$
2 C(OHHC) $\leftrightarrow$ $\mathbf{z}_5$	1 H(O) $\leftrightarrow$ $\mathbf{z}_8$
1 C(OHCC) $\leftrightarrow$ $\mathbf{z}_6$	5 H(C) $\leftrightarrow$ $\mathbf{z}_4$
	2 C(OHHC) $\leftrightarrow$ $\mathbf{z}_5$
	1 C(OHCC) $\leftrightarrow$ $\mathbf{z}_6$
$\updownarrow$	$\updownarrow$
$(3,6,3,5,2,1,0,0)^T$	$(2,4,2,5,2,1,1,1)^T$

FIGURE 14.5 Height-1 signature representations of nitroglycerine (left) and 1,2-dinitroglycerine (right). Proceeding from the top to the bottom we show (top row) the molecular graph representations of the two molecules, (middle row) the number of occurrences of height-1 atomic signatures and (bottom row) the signature vector representation of the two molecules. The atomic signature occurrences give the number of times a given chemical fragment occurs in the molecules. In the case of nitroglycerine, we have three oxygen atoms bonded to a nitrogen atom and a carbon atom [shown as 3 O(NC)], six oxygen atoms double bonded to a nitrogen atom [shown as 6 O(=N)], three nitrogen atoms bonded to an oxygen atom and double bonded to two other oxygen atoms [shown as 3 N(O=O=O)], and so on.

To provide an example, consider the two molecules shown in Figure 14.5. Both molecules, nitroglycerine and 1,2-dinitroglycerine, are represented as undirected edge- and vertex-labeled molecular graphs (carbons and hydrogens are implicit). To obtain vectors from these graphs, we first visit each node in each graph and record the subgraph formed by that node and its neighbors. This is known as a height 1 atomic signature. If we wanted to compute height 2 signature, we would have to visit the neighbors of the neighbors.

The atomic signatures are recorded as strings $A_1(b_2A_2b_3A_3\ldots)$, where A_1 is the vertex type of the root node (the node we are visiting), A_2 is a neighbor of A_1 with bond type b_2, A_3 is a neighbor of A_1 with bond type b_3, and so on. If $b_2, b_3, \ldots$ are single bonds, then they are omitted. This representation is canonical if we alphabetize the list $(b_2A_2, b_3A_3, \ldots)$ of bonds and atoms [81]. In Figure 14.5, we have written oxygen bonded to nitrogen and carbon as O(NC) and oxygen double bonded to nitrogen as $O(=N)$.

After visiting each node of each molecular graph, we obtain a list of atomic signature string representations. This list is identified with a set $z_1, z_2, \ldots$ of basis vectors.

In Figure 14.5, we have identified O(NC) with z_1 and $O(=N)$ with z_2. Using these basis vectors, we record the number of times a given atomic signature subgraph occurs in a molecular graph to obtain our molecular signature vector representation. Since O(NC) occurs three times in nitroglycerine and $O(=N)$ occurs six times, the first two entries of our signature vector for nitroglycerine are 3 and 6. The full vector is $(3, 6, 3, 5, 2, 1, 0, 0, 0)^T$. This analysis is also performed on 1,2-dinitroglycerine to get the signature vector $(2, 4, 2, 5, 2, 1, 1, 1)^T$.

Once we have molecular signature vectors for the various molecular graphs in our dataset, it is a simple matter to compute kernel similarities by taking dot products of signature vectors. Using the signature vectors for nitroglycerine and 1,2-dinitroglycerine, we compute a similarity of $(3, 6, 3, 5, 2, 1, 0, 0)^T \cdot (2, 4, 2, 5, 2, 1, 1, 1)^T = 66$.

14.2.3 Comparisons

The use of classification and regression are fairly mature in both bioinformatics and chemoinformatics. The two fields, however, are again benefiting from cross-fertilization, this time in terms of problem formulation and molecular description. A motivating example is drug design, where we are interested in predicting protein–chemical interaction [78,83].

One of the standard chemoinformatic approaches to drug design is the use of virtual screening, where a library of chemicals is examined (e.g., using QSAR) for a certain activity. In this approach, the corresponding protein is usually fixed and all attention is focused on the drug. In bioinformatics, it is more common to consider a network of interactions, as in the case of a protein–protein interaction network [56]. If we combine both types of approaches, we arrive at the prospect of predicting a protein–chemical interaction network [78,83]. [Although the general area of large-scale (informatics-based) protein–chemical interaction prediction is relatively unexplored, both approaches mentioned here use SVMs.]

REFERENCES

1. Jain, A., Murthy, M. N., and Flynn, P., Cluster analysis: A review. *ACM Comput. Surv.* 1999, 31(3), 264–323.
2. Xu, R., and Wunsch, D., Survey of clustering algorithms. *IEEE Trans. Neural Netw.* 2005, 16(3), 645–678.
3. de Hoon, M. J., Imoto, S., Nolan, J., and Miyano, S., Open source clustering software. *Bioinformatics* 2004, 20(9), 1453–1454.
4. Downs, G. M. and Barnard, J. M., Clustering methods and their uses in computational chemistry. In: *Reviews in Computational Chemistry*, K. B. Lipkowitz and D. B. Boyd (eds), Wiley, Hoboken, NJ, 2002; Vol. 18.
5. Krause, A., Stoye, J., and Vingron, M., Large scale hierarchical clustering of protein sequences. *BMC Bioinform.* 2005, 6, 15.
6. Loewenstein, Y., Portugaly, E., Fromer, M., and Linial, M., Efficient algorithms for accurate hierarchical clustering of huge datasets: Tackling the entire protein space. *Bioinformatics* 2008, 24(13), i41– i49.

7. Jarvis, R. A. and Patrick, E. A., Clustering using a similarity measure based on shared near neighbors. *IEEE Trans. Comput.* 1973, C-22(11), 1025–1034.
8. Rubin, V. and Willett, P., A comparison of some hierarchical agglomerative clustering algorithms for structure–property correlation. *Anal. Chim. Acta* 1983, 151, 161–166.
9. Willett, P., A comparison of some hierarchical agglomerative clustering algorithms for structure–property correlation. *Anal. Chim. Acta* 1982, 136, 29–37.
10. Willett, P., Evaluation of relocation clustering algorithms for the automatic classification of chemical structures. *J. Chem. Inf. Comput. Sci.* 1984, 24(1), 29–33.
11. Willett, P., Barnard, J. M., and Downs, G. M., Chemical similarity searching. *J. Chem. Inf. Comput. Sci.* 1998, 38(6), 983–996.
12. Schena, M., Shalon, D., Davis, R. W., and Brown, P. O., Quantitative monitoring of gene expression patterns with a complementary DNA microarray. *Science* 1995, 270(5235), 467–470.
13. Shalon, D., Smith, S. J., and Brown, P. O., A DNA microarray system for analyzing complex DNA samples using two-color fluorescent probe hybridization. *Genome Res.* 1996, 6(7), 639–645.
14. Eisen, M. B., Spellman, P. T., Brown, P. O., and Botstein, D., Cluster analysis and display of genome-wide expression patterns. *Proc. Natl. Acad. Sci. USA* 1998, 95(25), 14863–14868.
15. Spellman, P. T., Sherlock, G., Zhang, M. Q., Iyer, V. R., Anders, K., Eisen, M. B., Brown, P. O., Botstein, D., and Futcher, B., Comprehensive identification of cell cycle-regulated genes of the yeast *Saccharomyces cerevisiae* by microarray hybridization. *Mol. Biol. Cell* 1998, 9(12), 3273–3297.
16. Wilson, C. S., Davidson, G. S., Martin, S. B., Andries, E., Potter, J., Harvey, R., Ar, K., et al., Gene expression profiling of adult acute myeloid leukemia identifies novel biologic clusters for risk classification and outcome prediction. *Blood* 2006, 108(2), 685–696.
17. Jiang, D., Tang, C., and Zhang, A., Cluster analysis for gene expression data: A survey. *IEEE Trans. Knowl. Data Eng.* 2004, 16(11), 1370–1386.
18. Tavazoie, S., Hughes, J. D., Campbell, M. J., Cho, R. J., and Church, G. M., Systematic determination of genetic network architecture. *Nat. Genet.* 1999, 22(3), 281–285.
19. Tamayo, P., Slonim, D., Mesirov, J., Zhu, Q., Kitareewan, S., Dmitrovsky, E., Lander, E. S., and Golub, T. R., Interpreting patterns of gene expression with self-organizing maps: Methods and application to hematopoietic differentiation. *Proc. Natl. Acad. Sci. USA* 1999, 96(6), 2907–2912.
20. Shamir, R. and Sharan, R., Click: A clustering algorithm for gene expression analysis. In: *Proceedings of the 8th International Conference on Intelligent Systems for Molecular Biology (ISMB)*, AAAI Press, San Diego, CA, 2000; Vol. 8, pp. 307–316.
21. Ben-Dor, A., Shamir, R., and Yakhini, Z., Clustering gene expression patterns. *J. Comput. Biol.* 1999, 6(3–4), 281–297.
22. Cheng, Y. and Church, G. M., Biclustering of expression data. In: *Proceedings of the 8th International Conference on Intelligent Systems for Molecular Biology (ISMB)*, AAAI Press, San Diego, CA, 2000; Vol. 8, pp. 93–103.
23. Getz, G., Levine, E., and Domany, E., Coupled two-way clustering analysis of gene microarray data. *Proc. Natl. Acad. Sci. USA* 2000, 97(22), 12079–12084.
24. Altschul, S. F., Gish, W., Miller, W., Myers, E. W., and Lipman, D. J., Basic local alignment search tool. *J. Mol. Biol.* 1990, 215(3), 403–410.
25. Cohen, A. M. and Hersh, W. R., A survey of current work in biomedical text mining. *Brief Bioinform.* 2005, 6(1), 57–71.
26. Janssens, F., Glanzel, W., and DeMoor, B., Dynamic hybrid clustering of bioinformatics by incorporating text mining and citation analysis. In: *Proceedings of the 13th ACM SIGKDD*

International Conference on Knowledge Discovery and Data Mining, San Jose, CA, 2007; pp. 360–369.

27. Rodgers, J. R. and Cebon, D., Materials informatics. *Mater. Res. Soc. Bull.* 2006, 31.
28. Mayr, L. M. and Fuerst, P., The future of high-throughput screening. *J. Biomol. Screen* 2008, 13(6), 443–448.
29. Kubinyi, H., From narcosis to hyperspace: The history of QSAR. *QSAR* 2002, 21(4), 348–356.
30. Selassie, C. D., *History of Quantitative Structure–Activity Relationships* (6th edition), Wiley, New York, NY, 2003.
31. Willett, P., Winterman, V., and Bawden, D., Implementation of nonhierarchic cluster analysis methods in chemical information systems: Selection of compounds for biological testing and clustering of substructure search output. *J. Chem. Inf. Comput. Sci.* 1986, 26, 109–118.
32. Doman, T. N., Cibulskis, J. M., Cibulskis, M. J., McCray, P. D., and Spangler, D. P., Algorithm5: A technique for fuzzy similarity clustering of chemical inventories. *J. Chem. Inf. Comput. Sci.* 1996, 3(6), 1195–1204.
33. McGregor, M. J. and Pallai, P. V., Clustering of large databases of compounds: Using the MDL "keys" as structural descriptors. *J. Chem. Inf. Comput. Sci.* 1997, 37(3), 443–448.
34. Menard, P. R., Lewis, R. A., and Mason, J. S., Rational screening set design and compound selection: Cascaded clustering. *J. Chem. Inf. Comput. Sci.* 1998, 38(3), 497–505.
35. Shemetulskis, N. E., Dunbar, J. B., Jr., Dunbar, B. W., Moreland, D. W., and Humblet, C., Enhancing the diversity of a corporate database using chemical database clustering and analysis. *J. Comput.-Aided Mol. Des.* 1995, 9(5), 407–416.
36. Nouwen, J. and Hansen, B., An investigation of clustering as a tool in quantitative structure–activity relationships (QSARs). *SAR QSAR Environ. Res.* 1995, 4(1), 1–10.
37. Lawson, R. G. and Jurs, P. C., Cluster analysis of acrylates to guide sampling for toxicity testing. *J. Chem. Inf. Comput. Sci.* 1990, 30(2), 137–144.
38. van Geerestein, V. J., Hamersma, H., and van Helden, S. P., Exploiting molecular diversity: Pharmacophore searching and compound clustering. In: *Computer-Assisted Lead Finding and Optimization*, D. H. van de Waterbeemd, P. B. Testa, and P. D. G. Folkers (eds), Wiley-VCH Basel, Switzerland, 2007; pp. 157–178.
39. Wild, D. J. and Blankley, C. J., VisualiSAR: A web-based application for clustering, structure browsing, and structure–activity relationship study. *J. Mol. Graph. Model* 1999, 17(2), 85–89, 120–125.
40. Engels, M. F., Thielemans, T., Verbinnen, D., Tollenaere, J. P., and Verbeeck, R., CerBeruS: A system supporting the sequential screening process. *J. Chem. Inf. Comput. Sci.* 2000, 40(2), 241–245.
41. Dasarathy, B., *Nearest Neighbor Pattern Classification Techniques*. IEEE Computer Society, Los Alamitos, CA, 1991.
42. Gurney, K., *An Introduction to Neural Networks*. CRC Press, London, UK, 1997.
43. Shawe-Taylor, J. and Cristianini, N., *Support Vector Machines and other Kernel-Based Learning Methods*. Cambridge University Press: Cambridge, 2000.
44. Fukunaga, K., *Introduction to Statistical Pattern Recognition*. Academic Press, San Francisco, CA, 1990.
45. Theodoridis, S. and Koutroumbas, K., *Pattern Recognition*. Academic Press, San Diego, CA, 2003.
46. Frank, I. E. and Friedman, J. H., A statistical view of some chemometrics regression tools. *Technometrics* 1993, 35(2), 109–135.
47. Montgomery, D. and Peck, E., *Introduction to Linear Regression Analysis*. Wiley, New York, 2001.

48. Vapnik, V., *Statistical Learning Theory*. Wiley: New York, 1998.
49. Smola, A. and Scholkopf, B., *A Tutorial on Support Vector Regression*; NeuroCOLT NC-TR-98-030, University of London, UK, 1998.
50. Ben-Hur, A., Horn, D., Siegelmann, H. T., and Vapnik, V., Support vector clustering. *J. Mach. Learn. Res.* 2001, 2, 125–137.
51. Bennett, K. and Campbell, C., Support vector machines: Hype or hallelujah? *SIGKDD Explor.* 2000, 2(2), 1–14.
52. Burges, C., A tutorial on support vector machines for pattern recognition. *Data Min. Knowl. Discov.* 1998, 2, 121–167.
53. Chang, C.-C. and Lin, C.-J., LibSVM: A library for support vector machines, http://www.csie.ntu.edu.tw/~cjlin/libsvm, 2001.
54. Joachims, T., Making large-scale SVM learning practical. In: *Advances in Kernel Methods—Support Vector Learning*, B. Scholkopf, C. Burges, and A. Smola, (eds). MIT Press, Cambridge, MA, 1999.
55. Pandey, G., Kumar, V., and Steinbach, M., *Computational Approaches for Protein Function Prediction*. Wiley Book Series on Bioinformatics, Hoboken, NJ, 2008.
56. Shoemaker, B. A. and Panchenko, A. R., Deciphering protein–protein interactions. Part II. Computational methods to predict protein and domain interaction partners. *PLoS Comput. Biol.* 2007, 3(4), e43.
57. Pellegrini, M., Marcotte, E. M., Thompson, M. J., Eisenberg, D., and Yeates, T. O., Assigning protein functions by comparative genome analysis: Protein phylogenetic profiles. *Proc. Natl. Acad. Sci. USA* 1999, 96(8), 4285–4288.
58. Ben-Hur, A. and Brutlag, D., *Protein Sequence Motifs: Highly Predictive Features of Protein Function*. Springer: Berlin, 2006.
59. Sprinzak, E. and Margalit, H., Correlated sequence-signatures as markers of protein–protein interaction. *J. Mol. Biol.* 2001, 311(4), 681–692.
60. Bock, J. R. and Gough, D. A., Predicting protein–protein interactions from primary structure. *Bioinformatics* 2001, 17(5), 455–460.
61. Cai, C. Z., Han, L. Y., Ji, Z. L., Chen, X., and Chen, Y. Z., SVM-Prot: Web-based support vector machine software for functional classification of a protein from its primary sequence. *Nucleic Acids Res.* 2003, 31(13), 3692–3697.
62. Wang, C. and Scott, S. D. New kernels for protein structural motif discovery and function classification, In: *Proceedings of the 22nd International Conference on Machine Learning (ICML)*, AAAI press, Bonn, Germany, 2005; pp. 940–947.
63. Brown, M. P., Grundy, W. N., Lin, D., Cristianini, N., Sugnet, C. W., Furey, T. S., Ares, M., Jr., and Haussler, D., Knowledge-based analysis of microarray gene expression data by using support vector machines. *Proc. Natl. Acad. Sci. USA* 2000, 97(1), 262–267.
64. Martin, S., Roe, D., and Faulon, J. L., Predicting protein–protein interactions using signature products. *Bioinformatics* 2005, 21(2), 218–226.
65. Bradford, J. R. and Westhead, D. R., Improved prediction of protein–protein binding sites using a support vector machines approach. *Bioinformatics* 2005, 21(8), 1487–1494.
66. Koike, A. and Takagi, T., Prediction of protein–protein interaction sites using support vector machines. *Protein Eng. Des. Sel.* 2004, 17(2), 165–173.
67. Hammett, L. P., The effect of structure upon the reactions of organic compounds. *J. Am. Chem. Soc.* 1937, 59(1), 96–103.
68. Taft, R. W., Polar and steric substituent constants for aliphatic and *o*-benzoate groups from rates of esterification and hydrolysis of esters. *J. Am. Chem. Soc.* 1952, 74, 3120–3128.
69. Hansch, C. and Fujita, T., A method for the correlation of biological activity and chemical structure. *J. Am. Chem. Soc.* 194, 86, 1616–1626.

70. Free, S. M. and Wilson, J. W., A mathematical contribution to structure–activity studies. *J. Med. Chem.* 1964, 7, 395–399.
71. Liu, H. X., Xue, C. X., Zhang, R. S., Yao, X. J., Liu, M. C., Hu, Z. D., and Fan, B. T., Quantitative prediction of logk of peptides in high-performance liquid chromatography based on molecular descriptors by using the heuristic method and support vector machine. *J. Chem. Inf. Comput. Sci.* 2004, 44(6), 1979–1986.
72. Tugcu, N., Song, M., Breneman, C. M., Sukumar, N., Bennett, K. P., and Cramer, S. M., Prediction of the effect of mobile-phase salt type on protein retention and selectivity in anion exchange systems. *Anal. Chem.* 2003, 75(14), 3563–3572.
73. Xue, C. X., Zhang, R. S., Liu, H. X., Liu, M. C., Hu, Z. D., and Fan, B. T., Support vector machines-based quantitative structure–property relationship for the prediction of heat capacity. *J. Chem. Inf. Comput. Sci.* 2004, 44(4), 1267–1274.
74. Burbidge, R., Trotter, M., Buxton, B., and Holden, S., Drug design by machine learning: Support vector machines for pharmaceutical data analysis. *Comput. Chem.* 2001, 26(1), 5–14.
75. Warmuth, M. K., Liao, J., Ratsch, G., Mathieson, M., Putta, S., and Lemmen, C., Active learning with support vector machines in the drug discovery process. *J. Chem. Inf. Comput. Sci.* 2003, 43(2), 667–673.
76. Kashima, H., Tsuda, K., and Inokuchi, A., Marginalized kernels between labeled graphs, In: *20th International Conference on Machine Learning (ICML)*, T. Fawcett and N. Mishra (eds), AAAI Press, Washington, DC, 2003; pp. 321–328.
77. Mahe, P., Ueda, N., Akutsu, T., Perret, J. L., and Vert, J. P., Graph kernels for molecular structure–activity relationship analysis with support vector machines. *J. Chem. Inf. Model.* 2005, 45(4), 939–951.
78. Faulon, J. L., Misra, M., Martin, S., Sale, K., and Sapra, R., Genome scale enzyme–metabolite and drug–target interaction predictions using the signature molecular descriptor. *Bioinformatics* 2008, 24(2), 225–233.
79. Martin, S., Brown, W. M., and Faulon, J. L., Using product kernels to predict protein interactions. *Adv. Biochem. Eng. Biotechnol.* 2007, 110, 215–245.
80. Faulon, J. L., Churchwell, C. J., and Visco, D. P., Jr., The signature molecular descriptor. 2. Enumerating molecules from their extended valence sequences. *J. Chem. Inf. Comput. Sci.* 2003, 43(3), 721–734.
81. Faulon, J. L., Collins, M. J., and Carr, R. D., The signature molecular descriptor. 4. Canonizing molecules using extended valence sequences. *J. Chem. Inf. Comput. Sci.* 2004, 44(2), 427–436.
82. Faulon, J. L., Visco, D. P., Jr., and Pophale, R. S., The signature molecular descriptor. 1. Using extended valence sequences in QSAR and QSPR studies. *J. Chem. Inf. Comput. Sci.* 2003, 43(3), 707–720.
83. Nagamine, N. and Sakakibara, Y., Statistical prediction of protein chemical interactions based on chemical structure and mass spectrometry data. *Bioinformatics* 2007, 23(15), 2004–2012.

15 Using Systems Biology Techniques to Determine Metabolic Fluxes and Metabolite Pool Sizes

Fangping Mu, Amy L. Bauer, James R. Faeder, and William S. Hlavacek

CONTENTS

15.1 INTRODUCTION

The metabolic network of an organism consists of spontaneous and enzyme-catalyzed reactions that transform small molecules, or metabolites, into others to produce energy and building blocks for essential macromolecules. Large-scale metabolic networks have been reconstructed from genome sequence and literature data for a number of organisms, including *Escherichia coli*, yeast, and humans [1–3]. These networks shed light on the metabolic capabilities of these organisms and their differences. Now with the ability to identify metabolic networks from sequence data and the vast accumulated knowledge of metabolism, we are challenged to characterize how these networks function. A metabolic pathway defines the static sequence of feasible and observable biochemical reaction steps involved in the conversion of inputs into this pathway into a product. Metabolic fluxes are the rates at which material is processed through the steps of a pathway. Flux distributions indicate patterns of network utilization. The complete analysis of how a cell functions will include not only a description of its metabolic network, but also an understanding of flux distribution in different environments.

Flux balance analysis, or FBA, is a computational technique used to quantitatively determine a steady-state flux distribution in a biochemical reaction network. Constraints imposed by a network's structure are used in FBA to define limits on the steady-state metabolic fluxes in a network. Constraints on fluxes arise because the rate at which a metabolite accumulates must equal the sum of fluxes that produce the metabolite minus the sum of fluxes that consume the metabolite. At steady state, these two sums are equal, which leads to linear balance constraints on fluxes. The dynamics of a metabolic reaction network can be represented as a system of linear equations for the material balances on the reactions between metabolites, that is, $\mathbf{N} \cdot \mathbf{v} = 0$, where $\mathbf{N}$ denotes the stoichiometric matrix and $\mathbf{v}$ represents the fluxes. For a reaction network with n reactions and m metabolites, the stoichiometric matrix will have n columns and m rows. The coefficients of the stoichiometric matrix define the relationship between the metabolites and the reactions, that is, these coefficients express to what extent each metabolite is produced or consumed in a reaction. The expression $\mathbf{N} \cdot \mathbf{v}$ combines the stoichiometric matrix with a rate vector to produce a system of equations that describes the rate of change of the metabolites. This system of linear equations forms a set of mass-balance equations describing the dynamics of the metabolite pools. The mass-balance equations can be further constrained using measurements of the rates of substrate uptake, rates of metabolite secretion into the growth medium, and rates of biomass formation to derive a map of net metabolic fluxes. These extra constraints provide necessary restrictions on the fluxes; however, these are still not enough to fully determine the flux distributions in a metabolic network. To fully determine a steady-state flux distribution, the most common additional assumptions are to assume that the system either maximizes biomass production or minimizes nutrient utilization.

FBA can also be used to study the effects of gene deletions and other types of perturbations on the system. Gene deletion studies can be performed by setting the reaction fluxes corresponding to the genes, and therefore, of their corresponding enzymes, to zero. It has also been argued that the assumption of maximization of biomass production for a wild-type bacterium may be justifiable, although the assumption may not be valid for genetically engineered knockouts or other bacterial strains that were not exposed to long-term evolutionary pressures. Minimization of metabolic adjustment (MOMA), whereby knockout metabolic fluxes undergo a minimal redistribution with respect to the flux configuration of the wild type, has been proposed to predict fluxes of knockout strains [4].

Metabolic flux analysis (MFA) is an experimental technique to determine the flux distribution in a biochemical reaction network. The stoichiometric equations used in MFA are the same as those employed in FBA, but instead of using optimization to constrain fluxes, MFA uses experimental results. Information about metabolic fluxes can be obtained from isotopic labeling experiments, where a cell population is fed with labeled nutrients, such as glucose containing ^{13}C atoms [5,6]. These labeled atoms are then transferred to internal metabolites through biochemical reactions. The relative abundances of different labeling patterns in internal metabolites depend on the atom mapping of the reactions from reactants to products and on the fluxes of the reactions producing them. Measurements of these labeling patterns provide additional constraints on metabolic fluxes. Estimating metabolic fluxes from these labeling patterns is the goal of ^{13}C MFA. MFA based on stable isotope labeling experiments is a powerful quantitative method for metabolite flux determination. This method requires multiple techniques (1) to determine the fate of carbon atoms in metabolic reactions, (2) to generate isotopomer, or isotope isomer, (mass-)balance equations, and (3) to perform complex parameter fitting for flux determination. In this chapter, we provide a description of each of these steps and give special attention to the role of chemoinformatics in ^{13}C flux analysis.

15.2 ISOTOPOMER MODELING METHODS

15.2.1 Isotopomer Mapping Matrices

Several different methods have been proposed to derive isotopomer mass-balance equations, which balance the isotopomers in each metabolite pool. To derive the isotopomer mass-balance equations, we need to know the fate of each atom, more specifically each carbon in this case, as it moves from reactants to products for each metabolic reaction. This information is captured in an atom fate map. For a metabolite with n carbons, there are 2^n possible isotopomers because each carbon can be either labeled or unlabeled. Atom mapping matrix methods [7,8] allow these 2^n isotopomer mass-balance equations to be written compactly as a single matrix equation. For each metabolite in a given reaction network, the isotope distribution can be represented by an isotopomer distribution vector (IDV), which has 2^n state variables. The entries of the IDV specify the mole fraction of the metabolite for every possible ^{13}C labeling pattern. With the IDVs determined, an isotopomer mapping matrix (IMM) for each reactant–product pair of a metabolic reaction can be formulated. For a given metabolic reaction, the reactions between reactants with n carbons and products with m carbons can be summarized in a $2^m \times 2^n$ IMM. Columns of the IMM represent the distinct labeling patterns of each reactant molecule and rows correspond to the product isotopomer labeling patterns. Therefore, the IMM defines all possible combinations of product isotopomers that can arise from reactant isotopomers. For instance, given a reaction where A → B, the isotopomer balance equations can be derived from the following matrix equation:

$$V_{\mathrm{B}} \frac{\mathrm{d}\mathbf{I}_{\mathrm{B}}}{\mathrm{d}t} = r \cdot (\mathbf{IMM}_{\mathrm{A>B}} \otimes \mathbf{I}_{\mathrm{A}}), \tag{15.1}$$

where V_{B} is the metabolite pool size, r specifies the reaction flux, $\mathbf{I}_{\mathrm{A}}$ and $\mathbf{I}_{\mathrm{B}}$ are the IDVs for substrates A and B, respectively, and the operator $\otimes$ is the elementwise multiplication of two equally long column vectors. For reactions involving multiple reactants and products, such as A + B → C + D, the picture is more complicated.

The equations describing the isotopomer dynamics for metabolite C in this reaction, for example, are given by

$$V_C \frac{d\mathbf{I}_C}{dt} = r \cdot (\mathbf{IMM}_{A>C} \cdot \mathbf{I}_A) \otimes (\mathbf{IMM}_{B>C} \cdot \mathbf{I}_B). \qquad (15.2)$$

An example of flux determination using the IMM method is presented in detail in a subsequent section. For more details about the IMM approach, the interested reader should refer to Refs. [7–11].

15.2.2 BioNetGen Language (BNGL)-Encoded Atom Fate Maps

Isotopomer mass-balance equations are essential for estimating reaction fluxes from measurements of the relative abundances of isotopomers [5,6]. To specify these equations, the fates of individual atoms from substrates to products must be traced [7–9,12,13]. In tracer-based flux profiling studies, atom fate maps are usually assembled *ad hoc* and are highly problem specific. Moreover, the data are stored in different formats across studies. Atom fate maps are not included in the most widely used metabolic databases, such as the Kyoto Encyclopedia of Genes and Genomes (KEGG) database [14], MetaCyc [15], BRENDA [16], or the biochemical pathways wall chart of Roche Applied Science [17]. Recently, two large-scale collections of fate maps have been assembled, one by Arita [18] and another by Mu et al. [19]. Arita subsequently used these maps to investigate the connectivity of the *Escherichia coli* metabolic network [20]. These maps account for 2764 reactions involving 2100 metabolites and rely on MDL® Mol files downloaded from the KEGG COMPOUND database. In comparison, the collection of Mu et al. [19] includes a greater number of reactions, almost double that of Arita [18]. In the collection of Mu et al. [19], specific chemoinformatics considerations are applied to facilitate fate map display and interpretation (1) by using InChI™, a systematic structure-based system for naming and referencing metabolites and their carbon atoms, (2) by accounting for prochiral carbon atoms, which are present in 17% of the metabolites considered, and (3) by encoding the fate maps in a formal model-specification language, the BNGL [21–24]. As a result of the latter feature, maps can be automatically interpreted by the BioNetGen software tool [21] to obtain mass-balance equations and simulate stationary and dynamic labeling patterns. Compound and atom identification is systematically considered in the design of this database [19], which is available at http://cellsignaling.lanl.gov/FateMaps/. A comprehensive review of the procedure for molecular canonization, such as the InChI™ method, is provided in a separate chapter of this book, and thus we forego further discussion of it here and refer the reader to that chapter.

BioNetGen represents atom fate maps using the executable model-specification language BNGL. Table 15.1 shows the reaction catalyzed by sedoheptulose-7-phosphate: D-glyceraldehyde-3-phosphate glycolaldehyde transferase in the database. The metabolite name is represented as an InChI™ string in quotation marks. For example, the first substrate, D-sedoheptulose 7-phosphate, is represented as "InChI=1/C7H15O10P/c8-1-3(9)5(11)7(13)6(12)4(10)2-17-18(14,15)16/h4-8,

TABLE 15.1
An Entry for a Reaction from the Database [19]

Field	Value
1	Sedoheptulose-7-phosphate: D-glyceraldehyde-3-phosphate glycolaldehyde transferase
2	2.2.1.1
3	“InChI=1/C7H15O10P/c8-1-3(9)5(11)7(13)6(12)4(10)2-17-18(14,15)16/h4-8,10-13H,1-2H2,(H2,14,15,16)/t4-,5-,6,7+/m1/s1”(C1%1,C2%2,c3%3, C4%4,c5%5, C6%6,C7%7)+ “InChI=1/C3H7O6P/c4-1-3(5)2-9-10(6,7)8/h1,3,5H,2H2,(H2,6,7,8)/t3/m0/s1”(c1%8,C2%9,C3%10) < − > “InChI=1/C5H11O8P/c6-3-2(1-12-14(9,10)11)13-5(8)4(3)7/h2-8H,1H2,(H2,9,10,11)/t2-,3-,4,5?/m1/s1”(C1%2,C2%4,C3%6,C4%7,c5%5)+ “InChI=1/C5H11O8P/c6-1-3(7)5(9)4(8)2-13-14(10,11)12/h4-6,8-9H,1-2H2,(H2,10,11,12)/t4-,5-/m1/s1”(C1%1,C2%9,c3%3,C4%10,c5%8)
5	R01641
6	http://www.genome.jp/dbget-bin/www_bget?rn:R01641

Note: Figure 15.1 illustrates a FateMapViewer visualization of the carbon fate map defined for this reaction in Field 3. The three metabolites participating in this reaction are indicated by the InChI™ strings included in the map.

10-13H,1-2H2,(H2,14,15,16)/t4-,5-,6-,7+/m1/s1.” The InChI™ identifier of the metabolite includes up to six layers of information. The first layer defines the chemical formula, and the second layer defines the atom connectivity of the metabolite. For each atom in a metabolite, its InChI™ identifier provides a unique index. Since atoms represented in an InChI™ string are indexed in Hill order (i.e., carbon atoms first), the carbon atoms in a metabolite are referenced by integer indices 1 to n, where n is the number of carbon atoms. For D-sedoheptulose 7-phosphate, we know it has seven carbons from its formula, and the indices of these carbons are assigned from 1 to 7 (Figure 15.1). In BioNetGen, the carbon fate map for a reaction $A + B \leftrightarrow C + D$

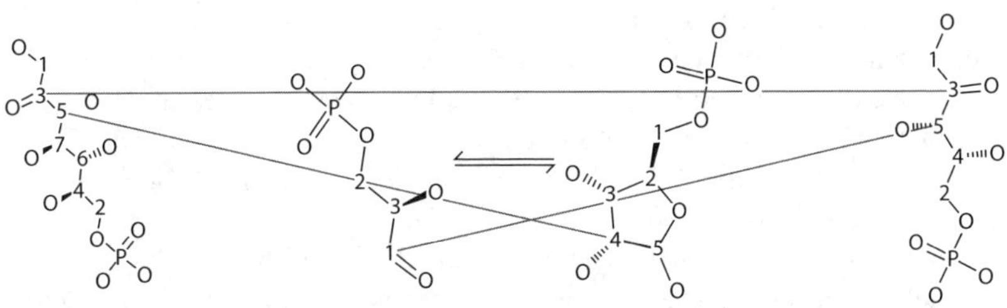

FIGURE 15.1 Illustration using FateMapViewer [25] of the carbon fate map in Table 15.1. Lines indicate the fates of carbon atoms in reaction centers. Mapping of other carbon atoms is suppressed for clarity.

is represented as follows:

$$\begin{aligned} s_{\mathrm{A}}(\mathrm{c}1\%x_1, \ldots, \mathrm{c}m\%x_m) + s_{\mathrm{B}}(\mathrm{c}1\%x_{m+1}, \ldots, \mathrm{c}n\%x_{m+n}) < - > \\ s_{\mathrm{C}}(\mathrm{c}1\%y_1, \ldots, \mathrm{c}p\%y_p) + s_{\mathrm{D}}(\mathrm{c}1\%y_{p+1}, \ldots, \mathrm{c}q\%y_{p+q}), \end{aligned} \tag{15.3}$$

where s_{A}, s_{B}, s_{C}, and s_{D} are quote-delimited InChI™ strings for metabolites; $c \in \{\mathrm{c,C}\}$ is a single character that precedes an atom index, which is uniquely specified by the InChI™ string. A lowercase character indicates that a carbon atom is in a reaction center. The number of carbon atoms in A, B, C, and D is given by m, n, p, and q such that $m + n = p + q$, and x_i, $y_i \in [1, \ldots, m + n]$ are integer indices that indicate how the carbon atoms of reactants map to the carbon atoms of products. Atoms that share the same index map to each other. This representation specifies detailed information on the carbon maps of each reaction. Software tools can be developed to parse and visualize the carbon fate maps. The database of Mu et al. [19] includes carbon atom fate maps for 4605 reactions involving 3713 metabolites. This database is available online. Carbon fate maps for each metabolic reaction define the structural correspondence between reactants and products at the atomic scale. For metabolic reactions, these mappings are consistent in each reaction, and there is no probabilistic arbitrariness. To define the carbon fate maps, we first use a maximum common subgraph algorithm (SIMCOMP, http://web.kuicr.kyoto-u.ac.jpa/simcomp/) to identify the maximum common substructures of all substrate and product pairs within each metabolic reaction. The algorithm output identifies the matching atoms in the common substructures of the two input chemical structures. An initial carbon fate map is defined based on these matching atoms for each reaction of interest. The carbon fate maps are written in BNGL. Initial carbon fate maps are then refined manually through text editing at the level of BNGL with help from FateMapViewer for visualization. FateMapViewer is a specific software tool, which we wrote to visualize fate maps [25]. Figure 15.1 is a graphic display of the carbon fate map shown in Table 15.1 using FateMapViewer.

Using the carbon atom fate maps described by Mu et al. [19], BioNetGen translates a set of fate maps into isotopomer mass-balance equations. The BNGL/BioNetGen-based method for MFA is represented in flowchart form in Figure 15.2. BioNetGen translates the carbon fate maps into reaction rules, mapping carbons from reactants to products. Molecules are represented as structured objects and molecular interactions as rules for transforming the attributes of these objects. These reaction rules serve as generators of a list of reactions. The core component of BioNetGen, BNG2.pl, which has a command-line interface, processes BioNetGen input files to generate two kinds of outputs: a chemical reaction network derived by processing rules and/or the results of simulating a model. BioNetGen is open source freely available software [25]. For additional information about BNGL and BioNetGen, we refer the interested reader to Refs. [19,21,24,26].

In contrast with the IMM-based approach, the BNGL/BioNetGen-based approach has several advantages. First, we can distinguish between the labeling patterns of symmetric pairs and we can handle reversible reactions without added complexity. Second, even though an IMM for each reaction only needs to be generated once, when the number of carbons in a molecule is large, for example, succinate-CoA, the matrices become quite large and pose a computational challenge themselves. Another feature of

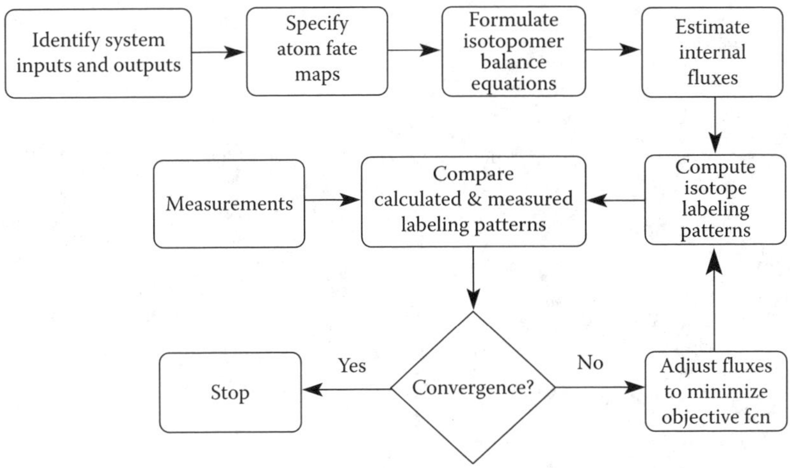

FIGURE 15.2 Carbon transition diagram for a simple metabolic network with four metabolites (S, $M1$, $M2$, and P), four external fluxes (v_S, v_{M1}, v_{M2}, and v_P), and four internal fluxes (v_1, v_2, v_3, and v_4). The external fluxes can be either controlled or measured in an experiment. These fluxes include one source and three sinks.

BioNetGen over IMM is that it automatically generates the balance equations relevant for a given experiment directly from fate maps. The possible number of isotopes for a metabolite with n carbons is 2^n and there is a balance equation for each one of them. In the IMM approach, all balance equations are generated and used to make predictions, whereas the BNGL/BioNetGen-based approach generates only the relevant balance equations. For example, if only one carbon of a metabolite is labeled, then many of its isotopomers will not be populated. IMM would generate balance equations for all possible isotopomers, whereas BioNetGen only generates the relevant balance equations. This can potentially result in greater computational efficiency.

15.3 A SIMPLE EXAMPLE

The most widely used isotope tracer method detects the steady-state ^{13}C labeling patterns in proteinogenic amino acids. In Figure 15.3, we consider a simple example to illustrate the concept of ^{13}C labeling for metabolic flux determination. A microbial culture is fed with a ^{13}C-labeled substrate, S, whose chemical structure is known. For this example, we assume that the system is fed with ^{13}C labeled at carbon position 1, denoted C1. The substrate is then metabolized to intermediate metabolites $M1$ and $M2$ and then to product P. After the system has reached a steady state, the labeling patterns of P can be measured to obtain valuable information that provides additional constraints on the internal fluxes.

The input and output fluxes, v_S, v_{M1}, v_{M2}, and v_P, of the metabolic pathways are estimated using quantitative physiological parameters from the experimental setup, including cell density, chemostat nutrient feed and efflux, and the biomass of exiting cells. Flux analysis is then employed to determine the internal fluxes v_1, v_2, v_3, and

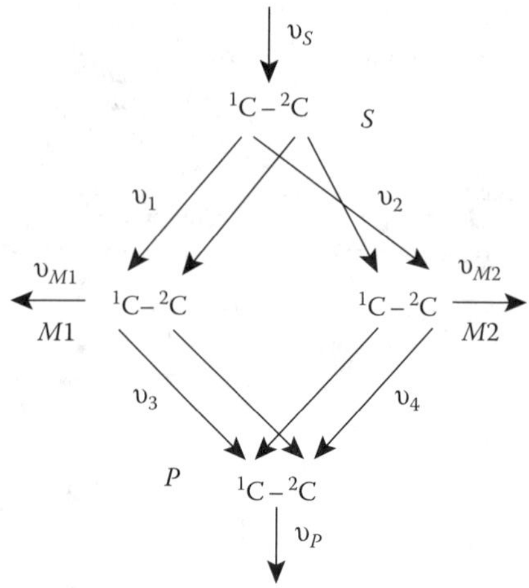

FIGURE 15.3 Framework for ^{13}C MFA. For a given system, physiological parameters from the experimental setup are used to derive system input and output fluxes. The isotopomer balance equations are derived using the BNGL or IMM approach, intermediate fluxes are estimated, and isotope patterns are simulated. Optimization methods, such as gradient optimization or simulated annealing, provide new estimates for the intracellular fluxes. Based on the revised flux estimates, new isotope labeling patterns are generated and again compared with measurements. This process repeats until the patterns converge.

v_4. From Figure 15.3, at steady state, $v_3 = v_1 - v_{M1}$. Thus, $v_1 = v_{M1} + v_3$. Applying a similar logic, we get the following stoichiometric balance equations:

$$v_S = v_1 + v_2, \tag{15.4}$$

$$v_1 = v_{M1} + v_3, \tag{15.5}$$

$$v_2 = v_{M2} + v_4, \tag{15.6}$$

$$v_P = v_3 + v_4. \tag{15.7}$$

These balance equations can be represented in matrix form, that is, $\mathbf{N} \cdot \mathbf{v} = 0$, where $\mathbf{v}$ contains the fluxes and $\mathbf{N}$ denotes the stoichiometric matrix:

$$N = \begin{matrix} \\ S \\ M1 \\ M2 \\ P \end{matrix} \begin{pmatrix} v_S & v_1 & v_2 & v_{M1} & v_{M2} & v_3 & v_4 & v_P \\ -1 & 1 & 1 & 0 & 0 & 0 & 0 & 0 \\ 0 & -1 & 0 & 1 & 0 & 1 & 0 & 0 \\ 0 & 0 & -1 & 0 & 1 & 0 & 1 & 0 \\ 0 & 0 & 0 & 0 & 0 & -1 & -1 & 1 \end{pmatrix}. \tag{15.8}$$

These balance equations alone, however, do not uniquely determine the internal fluxes. As shown in Figure 15.3, the fates of the two carbons from S to P through $M1$ or $M2$ are different. Balancing of isotopomers can provide extra constraints. For the simple example shown in Figure 15.3, four different isotopomers exist for S and we represent the fraction of isotopomers in S as S_{00}, S_{01}, S_{10}, and S_{11}, where the first (second) digit in the subscript corresponds to the atomic mass of the carbon C1 (C2) in S, a 1 indicates ^{13}C and a 0 means ^{12}C. Based on this definition, we have that

$$S_{00} + S_{01} + S_{10} + S_{11} = 1. \tag{15.9}$$

For each isotopomer, we can write down a mass-balance equation. For mass balance, the difference between the input rate and the output rate is the accumulation rate. For the $M1$ isotopomer with both carbons unlabeled, that is, ^{12}C, the accumulation rate can be written as $V_{M1}(\mathrm{d}M1_{00}/\mathrm{d}t)$, where V_{M1} is the pool size of $M1$. Because the only source of this isotopomer is from S with both carbons ^{12}C, the input rate is $v_1 S_{00}$. There are two possible outputs and the output rate is given by $(v_3 + v_{m1})M1_{00}$. Therefore, we can write down the balance equation for this isotopomer as

$$V_{M1}\frac{\mathrm{d}M1_{00}}{\mathrm{d}t} = v_1 S_{00} - (v_3 + v_{m1})M1_{00}. \tag{15.10}$$

At isotope steady state, that is, when the isotopomer fractions have stopped changing in time, we have

$$v_1 S_{00} - (v_3 + v_{m1})M1_{00} = 0. \tag{15.11}$$

Using similar reasoning, we can write down the isotopomer balance equations for the other isotopomers. These isotopomer balance equations define extra constraints used to solve for the internal fluxes. In this example, all constraints are linear because we are dealing with a reaction A → B. However, for reactions with multiple reactants, for example, A + B → C + D, the isotopomer balance equations are nonlinear equations of isotopomer fractions, as shown in Equation 15.2.

For this simple example, we can also use intuitive reasoning. For product P, ^{13}C labeled at carbon position 1 and position 2 comes from v_3 and v_4, respectively. If we define p_1 as the fraction of ^{13}C at C1 in P and p_2 as the fraction of ^{13}C at C2 in P at isotope steady state, then

$$\frac{v_3}{v_4} = \frac{p_1}{p_2}. \tag{15.12}$$

With this extra constraint, we can now solve the balance equations to get the internal fluxes.

For complex networks, we need systematic representations of carbon atom fate maps to derive mass-balance equations and isotopomer balance equations from these carbon fate maps. In the next sections, we review two methods: IMM-based methods and BNGL/BioNetGen-based methods.

15.3.1 Carbon Fate Maps and Isotopomer Balance Equations

Figure 15.3 illustrates how isotope labeling data can be used in an iterative procedure to estimate metabolic fluxes from observed isotopomer distributions. The goal is to identify the unknown intracellular fluxes. For this analysis, all relevant reaction steps and the fate of carbon atoms within each step must be known *a priori*. The system's input and output fluxes are determined by quantitative physiological parameters obtained from the experiment, including nutrient feed, cell growth rates and densities in the chemostat, and the efflux. The stoichiometric constraints on all internal fluxes are formulated using the principle of mass conservation, that is, $\mathbf{N} \cdot \mathbf{v} = 0$. The isotopomer balance equations for the internal reactions are then derived using either the BNGL/BioNetGen-based method or the IMM-based method. The underlying mathematical model for ^{13}C MFA is a set of isotopomer balance equations as shown below in Equations 15.19 through 15.34. Given an initial guess for the intermediate fluxes, the balance equations, in algebraic or ordinary differential equation (ODE) form, are used to compute the steady-state or dynamic labeling patterns. The mathematical model for the dynamic ^{13}C labeling experiments is generally a high-dimensional coupled system of differential algebraic equations, which in fact constitutes an inverse problem for the unknown intracellular fluxes. Thus, predictions for the internal fluxes are obtained by minimizing the difference between simulated and measured isotope patterns. Essentially, this is a complex parameter fitting problem, which can be solved using a variety of techniques. For steady-state ^{13}C labeling experiments, the model is a set of algebraic equations, and the algorithms employed for numerical flux estimation are primarily gradient-based local optimization [27] or gradient-free global optimization [10, 28–30] techniques, such as simulated annealing or genetic algorithms. In addition, a hybrid technique of global-local optimization has been applied [31]. For dynamic ^{13}C labeling analysis, the system is described by a set of algebraic–differential equations. Although analysis of dynamic ^{13}C labeling data has been proposed by a number of groups and used to estimate fluxes in small metabolic subnetworks [32–34], to the best of our knowledge, this type of analysis is yet to be used to estimate fluxes in large-scale networks. A barrier to estimating large numbers of fluxes from this type of data is the large number of isotopomer balance equations that must be specified. This problem is solved by the BNGL/BioNetGen-based approach we review here.

Minimizing the objective function [31], that is, the difference between the simulated and measured labeling patterns using, for example, simulated annealing, provides new estimates for the internal fluxes. The objective function is given by

$$\text{Obj} = \sum_{i=1}^{N} \left(\frac{M_i(t) - E_i(\vec{V}, \vec{v}, t)}{\delta_i} \right)^2, \tag{15.13}$$

where M_i are the N individual labeling measurements and E_i their corresponding simulated values. The quantity δ_i is the confidence value of the ith measurement. For dynamic isotopomer experiments, M_i is a function of time, and E_i is a function of time, V represents metabolite pool sizes, and $\mathbf{v}$ represents flux distributions. For steady-state isotopomer experiments, E_i is a function of flux distributions ($\mathbf{v}$) only. Using these revised fluxes, new isotope labeling patterns are generated and again compared with the observed labeling patterns. This process continues until some measure of convergence is achieved, for example, the intracellular fluxes do not significantly change between iterations or some predefined number of iterations is performed.

15.3.2 IMM-Based Approach

We now apply the IMM method to the four-flux problem posed in Figure 15.3 to determine the metabolic fluxes. For a metabolite S with two carbons, the IDV is given by

$$\vec{I}_S = \begin{pmatrix} I_S(00) \\ I_S(01) \\ I_S(10) \\ I_S(11) \end{pmatrix}, \tag{15.14}$$

where the binary representation $(\cdot)$ corresponds to the state (C1 C2), and 0 and 1 indicate ^{12}C and ^{13}C, respectively. $0 \leq I_S(j) \leq 1$ is the fraction of S molecules that show a labeling pattern corresponding to the binary representation of j. We can derive the IMM for each of the internal reactions directly from the carbon atom fate map provided in Figure 15.3. $\mathbf{IMM}_{S>M1}$ denotes the IMM for the reaction between S and $M1$. We obtain

$$\mathbf{IMM}_{S>M1} = \begin{pmatrix} 1 & 0 & 0 & 0 \\ 0 & 1 & 0 & 0 \\ 0 & 0 & 1 & 0 \\ 0 & 0 & 0 & 1 \end{pmatrix}, \tag{15.15}$$

$$\mathbf{IMM}_{S>M2} = \begin{pmatrix} 1 & 0 & 0 & 0 \\ 0 & 0 & 1 & 0 \\ 0 & 1 & 0 & 0 \\ 0 & 0 & 0 & 1 \end{pmatrix}, \tag{15.16}$$

$$\mathbf{IMM}_{M1>P} = \begin{pmatrix} 1 & 0 & 0 & 0 \\ 0 & 1 & 0 & 0 \\ 0 & 0 & 1 & 0 \\ 0 & 0 & 0 & 1 \end{pmatrix}, \tag{15.17}$$

$$\mathbf{IMM}_{M2>P} = \begin{pmatrix} 1 & 0 & 0 & 0 \\ 0 & 1 & 0 & 0 \\ 0 & 0 & 1 & 0 \\ 0 & 0 & 0 & 1 \end{pmatrix}. \tag{15.18}$$

Each column of the IMM indicates how the substrate isotopomer evolves to the corresponding isotopomers in the product, which is represented by each row. For example, in $\mathbf{IMM}_{S>M1}$, the first column indicates that the isotopomer S_{00} can only be transferred to a product $M1_{00}$.

Again, for a reaction A→B, the isotopomer balance equations can be derived from Equation 15.1. The corresponding isotopomer balance equations for our four-flux example are therefore

$$V_S \frac{dS_{00}}{dt} = v_S S_{00} - (v_1 + v_2) S_{00}, \tag{15.19}$$

$$V_S \frac{dS_{01}}{dt} = v_S S_{01} - (v_1 + v_2) S_{01}, \tag{15.20}$$

$$V_S \frac{dS_{10}}{dt} = v_S S_{10} - (v_1 + v_2) S_{10}, \tag{15.21}$$

$$V_S \frac{dS_{11}}{dt} = v_S S_{11} - (v_1 + v_2) S_{11}, \tag{15.22}$$

$$V_{M1} \frac{dM1_{00}}{dt} = v_1 S_{00} - (v_3 + v_{M1}) M1_{00}, \tag{15.23}$$

$$V_{M1} \frac{dM1_{01}}{dt} = v_1 S_{01} - (v_3 + v_{M1}) M1_{01}, \tag{15.24}$$

$$V_{M1} \frac{dM1_{10}}{dt} = v_1 S_{10} - (v_3 + v_{M1}) M1_{10}, \tag{15.25}$$

$$V_{M1} \frac{dM1_{11}}{dt} = v_1 S_{11} - (v_3 + v_{M1}) M1_{11} \tag{15.26}$$

$$V_{M2} \frac{dM2_{00}}{dt} = v_2 S_{00} - (v_4 + v_{M2}) M2_{00}, \tag{15.27}$$

$$V_{M2} \frac{dM2_{01}}{dt} = v_2 S_{01} - (v_4 + v_{M2}) M2_{01}, \tag{15.28}$$

$$V_{M2} \frac{dM2_{10}}{dt} = v_2 S_{10} - (v_4 + v_{M2}) M2_{10}, \tag{15.29}$$

$$V_{M2} \frac{dM2_{11}}{dt} = v_2 S_{11} - (v_4 + v_{M2}) M2_{11}, \tag{15.30}$$

$$V_P \frac{dP_{00}}{dt} = v_3 M1_{00} + v_4 M2_{00} - v_P P_{00}, \tag{15.31}$$

$$V_P \frac{dP_{01}}{dt} = v_3 M1_{01} + v_4 M2_{01} - v_P P_{01}, \tag{15.32}$$

$$V_P \frac{dP_{10}}{dt} = v_3 M1_{10} + v_4 M2_{10} - v_P P_{10} \tag{15.33}$$

$$V_P \frac{dP_{11}}{dt} = v_3 M1_{11} + v_4 M2_{11} - v_P P_{11}, \tag{15.34}$$

and are subject to the following algebraic constraints:

$$S_{00} + S_{01} + S_{10} + S_{11} = 1, \tag{15.35}$$

$$M1_{00} + M1_{01} + M1_{10} + M1_{11} = 1, \tag{15.36}$$

$$M2_{00} + M2_{01} + M2_{10} + M2_{11} = 1, \quad (15.37)$$

$$P_{00} + P_{01} + P_{10} + P_{11} = 1. \quad (15.38)$$

These equations arise due to the definition of the isotopomer fractions, that is, mass conservation dictates that these fractions must sum to one. The isotopomer balance equations coupled with the algebraic constraints comprise a system of differential–algebraic equations that needs to be solved to determine the dynamic isotopomer labeling patterns.

15.3.3 BNGL/BioNetGen-Based Approach

As discussed previously, BioNetGen derives the isotopomer balance equations relevant for a given experiment directly from fate maps and generates the dynamic distribution of observable metabolites given the input parameters. Here we simulate the forward problem shown in Figure 15.3 using BioNetGen and generate the isotope labeling patterns given the assumed flux patterns. The BioNetGen input file for the four-flux example is provided below. It includes several blocks: parameters, species, reaction rules, and observables. In the first block, the metabolite pool sizes and the flux estimates, which are the input parameters, are specified. The second block identifies the species, which are the metabolites, and their initial concentrations. For each differently labeled initial metabolite, define a precursor species with fixed concentration (defined by prepending the "$" character) corresponding to the fraction in that labeling state. The block entitled "reaction rules" defines the carbon atom fate maps. Finally, the observables, or model outputs, are given in the fourth block. In this input file, the observables are the metabolite isotopomer fractions for P. For example, the observable pattern P(C1~1, C2~0) represents the isotopomer P_{10}. The remaining sections are commands for generating the isotopomer balance equations and simulating the dynamic labeling patterns. Specifically, the next block of this input file contains commands that generate the detailed isotopomer reactions from the specified reaction rules and write BioNetGen and MATLAB® output files. The next several sections simulate the dynamic labeling patterns given different pool sizes and/or fluxes. The command setParameter() defines the pool sizes or the fluxes. The system is perturbed by feeding the system with a labeled substrate. In this case, we have labeled 10% at the first carbon position. The command simulate_ode() runs a simulation of the dynamic trajectories of the isotope labeling patterns.

```
begin parameters
# Pools
V_S          1.0
V_M1         1.0
V_M2         1.0
V_P          1.0
```

```
# Fluxes
v_v1          0.5
v_v2          0.5
v_v3          0.4
v_v4          0.4
v_vm1         0.1
v_vm2         0.1
v_vp          0.8
v_vs          1.0


# Labeled fraction
f_label        0.1
end parameters


begin species
$proS(C1~0,C2~0)          1-f_label
$proS(C1~1,C2~0)          f_label
S(C1~0,C2~0)              V_S
M1(C1~0,C2~0)             V_M1
M2(C1~0,C2~0)             V_M2
P(C1~0,C2~0)              V_P
NULL                      0
end parameters


begin reaction rules
pros(C1%1,C2%2) -> M1(C1%1,C2%2)          v_v1/V_S
S(C1%1,C2%2) -> M1(C1%1,C2%2)             v_v1/V_S
S(C1%1,C2%2) -> M2(C1%2,C2%1)             v_v2/V_S
M1(C1%1,C2%2) -> P(C1%1,C2%2)             v_v3/V_M1
M2(C1%1,C2%2) -> P(C1%1,C2%2)             v_v4/V_M2
M1() -> NULL                              v_vm1/V_M1
M2() -> NULL                              v_vm2/V_M2
P() -> NULL                               v_vp/V_P
end reaction rules


begin observables
Molecules P00         P(C1~0,C2~0)/V_P
Molecules P01         P(C1~0,C2~1)/V_P
Molecules P10         P(C1~1,C2~0)/V_P
Molecules P11         P(C1~1,C2~1)/V_P
end observables
```

```
generate_network({overwrite=>1});
writeSBML();
writeMfile();

saveConcentrations();
imulate_ode({t_end=>20, n_steps=>40});

resetConcentrations();
setParameter(V_yM2, 0.5);
simulate_yode({suffix=>"M2low", t_end=>20,
n_steps=>40});

resetConcentrations();
setParameter(V_yM2, 2.0);
simulate_ode({suffix=>"M2high", t_end=>20,
n{_}ysteps=>40});

resetConcentrations();
setParameter(V_M2, 1.0);
setParameter(v_v1, 0.6);
setParameter(v_v2, 0.4);
setParameter(v_v3, 0.5);
setParameter(v_v4, 0.3);
simulate_ode({suffix=>"dottedline", t_end=>20,
n_steps=>40});
```

The raw experimental measurements of isotopomer distributions contain the contributions due to naturally occurring isotopes of its elements, such as hydrogen, carbon, nitrogen, oxygen, and so on. Correction of the isotopomer distribution is required to take into account differences in the relative natural abundance distribution of each mass isotopomer (skewing) as described in Ref. [35]. A software tool for this correction is reported in Ref. [36].

Solving the system of differential algebraic equations presented earlier gives us a dynamic trajectory of labeling patterns. Figure 15.4 depicts the temporal dynamics of the reaction product P_{01} from a simulation of the four-flux network. P_{01} indicates that the second carbon is ^{13}C labeled. At time zero, the input feed changes from $S_{00} = 1.0$, $S_{01} = S_{10} = S_{11} = 0$ to $S_{00} = 0.9$, $S_{01} = 0$, $S_{10} = 0.1$, and $S_{11} = 0$, indicating that 10% of the feed is ^{13}C labeled at carbon position 1. We assume the following values for the fluxes: $v_S = 1.0$, $v_1 = v_2 = 0.5$, $v_3 = v_4 = 0.4$, $v_{M1} = v_{M2} = 0.1$, and $v_P = 0.8$. In Figure 15.4, we study the effect of metabolite pool size on product dynamics. The pool sizes of the metabolites S, $M1$, and P are set to 1, and the pool size of $M2$ is varied. We see that the pool size has no effect on the steady-state distribution. Interestingly, however, Figure 15.4 shows that the dynamic labeling patterns depend, not only on the fluxes, but also on the metabolite pool sizes. Consequently, valuable information about the relative metabolite pool sizes can potentially be inferred from plots of the temporal dynamics.

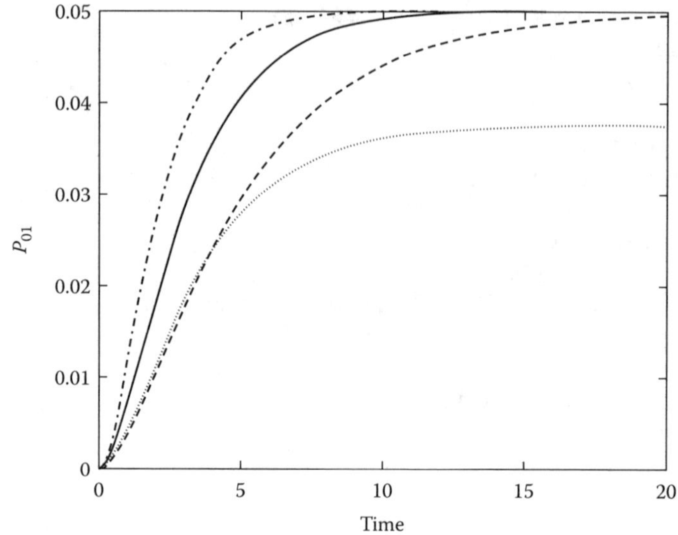

FIGURE 15.4 Dynamic simulation of the four-flux problem. At time zero, the input feed changes from $S_{00} = 1.0, S_{01} = S_{10} = S_{11} = 0$ to $S_{00} = 0.9, S_{01} = 0, S_{10} = 0.1$, and $S_{11} = 0$. Time evolution of the reaction product with ^{13}C at carbon position 2 (P_{01}) for different pool sizes is plotted. Flux patterns are assumed as follows: $v_S = 1.0, v_1 = v_2 = 0.5, v_3 = v_4 = 0.4$, $v_{M1} = v_{M2} = 0.1$, and $v_P = 0.8$. The pool sizes of metabolites S, $M1$, and P are set to 1, and the pool size of $M2$ is varied. $M2 = 1$ (solid line), $M2 = 2$ (dashed line), and $M2 = 0.5$ (dash-dot line). For the dotted line, flux patterns are assumed as follows: $v_S = 1.0, v_1 = 0.6, v_2 = 0.4$, $v_3 = 0.5, v_4 = 0.3, v_{M1} = 0.1, v_{M2} = 0.1$, and $v_P = 0.8$, and the pool sizes of metabolites S, $M1$, $M2$ and P are set to 1. This plot reveals that the dynamics depend on the pool size in addition to the flux distribution and provide valuable information on the relative pool sizes that can be used for dynamic labeling experiments. The pool size does not affect the steady-state distribution.

15.4 DETAILED EXAMPLE OF THE BNGL/BioNetGen-BASED APPROACH FOR THE CALVIN CYCLE

The Calvin cycle is a series of biochemical reactions required for carbon fixation. The carbon source for this cycle is CO_2. Five metabolites D-ribulose-1,5-bisphosphate (R5P), D-erythrose 4-phosphate (E4P), 3-phospho-D-glycerate (G3P), (2R)-2-hydroxy-3-(phosphonooxy)-propanal (T3P), and D-fructose 6-phosphate (F6P) link this pathway with central metabolic pathways. Figure 15.5 provides a network rendition of the Calvin cycle. The program we developed to visualize chemical structures, such as that shown in Figure 15.5, uses routines available in the Chemistry Development Kit [37]. The catalysts involved in the Calvin cycle are functionally equivalent to those used in other pathways, including the pentose phosphate pathway and in gluconeogenesis, and therefore, this example may be of interest to a broad community. The Calvin cycle is of particular interest because this network involves only a single-carbon input metabolite, and therefore steady-state labeling

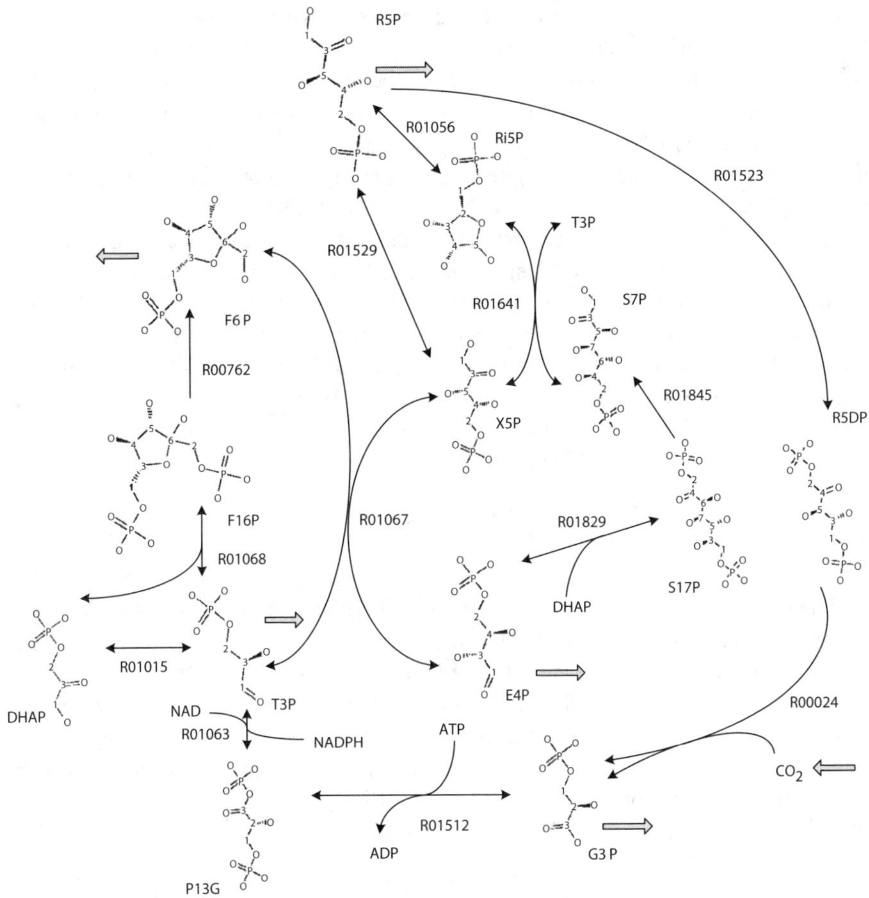

FIGURE 15.5 The Calvin cycle. The following abbreviations are used for metabolites in the network: R5P for D-ribulose 1,5-bisphosphate, G3P for 3-phospho-D-glycerate, P13G for 3-phospho-D-glyceroyl phosphate, and T3P for (2R)-2-hydroxy-3-(phosphonooxy)-propanal, DHAP for glycerone phosphate, F16P for D-fructose 1,6-bisphosphate, F6P for D-fructose 6-phosphate, E4P for D-erythrose 4-phosphate, X5P for D-xylulose 5-phosphate, S17P for D-sedoheptulose 1,7-bisphosphate, S7P for D-sedoheptulose 7-phosphate, Ri5P for D-ribose 5-phosphate, and R5P for D-ribulose 5-phosphate. The KEGG ID is used for reaction identification. There is an input flux feeding the pool of CO_2 and five flux sinks from pools of R5P, E4P, G3P, T3P, and F6P. For each metabolite, the carbon index shown in the figure is based on InChI™ naming. The indices are used to define the carbon atom fate maps as shown in the BioNetGen input file.

of CO_2 is completely uninformative. Analysis of the transient labeling patterns of metabolites is required to learn anything about the internal fluxes from a ^{13}C-labeling experiment. We employ the BNGL/BioNetGen-based approach to determine the intracellular fluxes for the Calvin cycle. The BioNetGen input file shown below specifies a model and simulation for labeling dynamics in the Calvin cycle (Figure 15.5) when a feed of unlabeled CO_2 is replaced with a mixture of unlabeled and ^{13}C-labeled CO_2. The raw carbon fate maps need to be processed before they can be used as BioNetGen input. This preprocessing is described in Ref. [19]. The model tracks 562 isotopomers. In the simulation, the system is taken to be in a steady state. At time $t = 0$, the composition of the CO_2 feed is changed, such that 10% of incoming CO_2 is labeled with ^{13}C. The steady-state fluxes in the system remain unchanged (up to isotope effects, which are expected to be small), but the ^{13}C labeling pattern of each intermediate metabolite changes until eventually all intermediates are fully labeled. BioNetGen is capable of simulating the labeling dynamics. BioNetGen can also generate a specification of the same model and simulation in M-file and Systems Biology Markup Language (SBML) formats, which can then be interpreted by other software tools capable of simulating ODEs, such as MATLAB®. For reversible reactions, we need to specify two fluxes: a forward flux and a backward flux. Instead of using forward and backward fluxes, reversible reactions can also be modeled using a net flux and an exchange flux [27]. That is,

$$v_{\text{net}} = v_{\text{forward}} - v_{\text{backward}} \quad \text{and} \quad v_{\text{exchange}} = \min(v_{\text{forward}}, v_{\text{backward}}). \tag{15.39}$$

The exchange flux can be in the range [0, ∞]. The characterization of forward and backward fluxes as net and exchange fluxes provides a measure of the reversibility of a reaction, where an exchange flux of 0 corresponds to irreversible reactions and infinity (∞) indicates fast reaction equilibrium. Isotopomer distributions are not very sensitive to exchange fluxes. Because it has been shown that the exchange flux can only be determined to within an order of magnitude [9], an exchange flux can be transformed into the so-called [0, 1]-exchange flux,

$$v_{\text{exchange}} \in [0,1] = \frac{v_{\text{exchange}}}{(1 - v_{\text{exchange}})}. \tag{15.40}$$

In addition, when the exchange flux is very large, numerical instabilities may arise. In these cases, this range can be specified as, for example, [0.1, 0.9] to remove the possibility that an exchange flux is either too small or too large. As a measure of the efficiency of BioNetGen for a problem with some computational complexity (562 isotopomers), we have calculated the CPU time requirements for the Calvin cycle simulation. On an Intel® Pentium® 4 CPU 3.2 GHz machine, the entire simulation took a total of 33.8 s. It took 14.4 s to generate the isotopomer balance equations from the BNGL-encoded input and 17.2 s to equilibrate the system, which corresponds to the first simulate_ode command. The second simulate_ode command runs another ODE simulation to identify the isotopomer fractions and took 2.2 s. The computational cost of simulating a set of stiff differential equations typically scales as N^3, where N is the number of equations (number of species). In general, N and the number of isotopomer balance equations will tend to increase exponentially with the number of reactions in the metabolic network. However, for a recently reported kinetic Monte Carlo method [38], the cost of simulation is independent of the number of isotopomers and scales with the number of reactions (rules).

```
begin parameters
#Pools
V_R5P                                1.0
V_R5DP                               1.0
V_CO2                                1.0
V_G3P                                1.0
V_P13G                               1.0
V_T3P                                1.0
V_DHAP                               1.0
V_F16P                               1.0
V_F6P                                1.0
V_E4P                                1.0
V_X5P                                1.0
V_Ri5P                               1.0
V_S17P                               1.0
V_S7P                                1.0

# Fluxes
v_R00024                             1.00
v_R01512                             1.95
v_R01063_f                           0.05
v_R01063_b                           2.00
v_R01015_f                           0.86
v_R01015_b                           0.10
v_R01068_f                           0.15
v_R01068_b                           0.58
v_R00762                             0.43
v_R01067_f                           0.50
v_R01067_b                           0.12
v_R01829_f                           0.20
v_R01829_b                           0.53
v_R01845                             0.33
v_R01641_f                           0.43
v_R01641_b                           0.10
v_R01529_f                           0.10
v_R01529_b                           0.82
v_R01056_f                           0.50
v_R01056_b                           0.17
v_R01523                             1.00
vs_T3P                               0.05
vs_F6P                               0.05
vs_R5P                               0.05
vs_E4P                               0.05
vs_G3P                               0.05
v_vi_CO2_unlabeled                   1
v_vi_CO2_labeled                     0
end parameters

begin species
R5P(C1~0,C2~0,C3~0,C4~0,C5~0)                   V_R5P
R5DP(C1~0,C2~0,C3~0,C4~0,C5~0)                  V_R5DP
CO2(C1~0)                                       V_CO2
G3P(C1~0,C2~0,C3~0)                             V_G3P
P13G(C1~0,C2~0,C3~0)                            V_P13G
T3P(C1~0,C2~0,C3~0)                             V_T3P
DHAP(C1~0,C2~0,C3~0)                            V_DHAP
F16P(C1~0,C2~0,C3~0,C4~0,C5~0,C6~0)             V_F16P
F6P(C1~0,C2~0,C3~0,C4~0,C5~0,C6~0)              V_F6P
E4P(C1~0,C2~0,C3~0,C4~0)                        V_E4P
X5P(C1~0,C2~0,C3~0,C4~0,C5~0)                   V_X5P
```

```
Ri5P(C1~0,C2~0,C3~0,C4~0,C5~0)                              V_Ri5P
S17P(C1~0,C2~0,C3~0,C4~0,C5~0,C6~0,C7~0)                    V_S17P
S7P(C1~0,C2~0,C3~0,C4~0,C5~0,C6~0,C7~0)                     V_S7P
I                                                           1
NULL                                                        0
end species

begin reaction rules
R5DP(C1%1,C2%2,C3%3,C4%4,C5%5)+CO2(C1%6) -> \
G3P(C1%2,C2%4,C3%6)+G3P(C1%1,C2%3,C3%5)               v_R00024/(V_R5DP*V_CO2)
G3P(C1%1,C2%2,C3%3) -> P13G(C1%1,C2%2,C3%3)           v_R01512/V_G3P
T3P(C1%1,C2%2,C3%3) <-> P13G(C1%2,C2%3,C3%1)          v_R01063_f/V_T3P,
                                                      v_R01063_b/V_P13G
T3P(C1%1,C2%2,C3%3) <-> DHAP(C1%1,C2%2,C3%3)          v_R01015_f/V_T3P,
                                                      v_R01015_b/V_DHAP
F16P(C1%1,C2%2,C3%3,C4%4,C5%5,C6%6) <-> \
DHAP(C1%5,C2%2,C3%6)+T3P(C1%4,C2%1,C3%3)              v_R01068_f/V_F16P,
                                                      v_R01068_b/(V_DHAP*V_T3P)
F16P(C1%1,C2%2,C3%3,C4%4,C5%5,C6%6) -> \
F6P(C1%1,C2%2,C3%3,C4%4,C5%5,C6%6)                    v_R00762/V_F16P
F6P(C1%1,C2%2,C3%3,C4%4,C5%5,C6%6) + \
T3P(C1%7,C2%8,C3%9)<->E4P(C1%5,C2%1,C3%4,C4%3)+\
X5P(C1%2,C2%8,C3%6,C4%9,C5%7)                         v_R01067_f/(V_F6P*V_T3P),
                                                      v_R01067_b/(V_E4P*V_X5P)
S17P(C1%1,C2%2,C3%3,C4%4,C5%5,C6%6,C7%7) <-> \
DHAP(C1%6,C2%2,C3%4) +
E4P(C1%7,C2%1,C3%5,C4%3)                              v_R01829_f/V_S17P,
                                                      v_R01829_b/(V_DHAP*V_E4P)
S17P(C1%1,C2%2,C3%3,C4%4,C5%5,C6%6,C7%7) -> \
S7P(C1%2,C2%1,C3%4,C4%3,C5%6,C6%5,C7%7)               v_R01845/V_S17P
S7P(C1%1,C2%2,C3%3,C4%4,C5%5,C6%6,C7%7) +\
T3P(C1%8,C2%9,C3%10) <-> \
Ri5P(C1%2,C2%4,C3%6,C4%7,C5%5) +\
X5P(C1%1,C2%9,C3%3,C4%10,C5%8)                        v_R01641_f/(V_S7P*V_T3P),
                                                      v_R01641_b/(V_Ri5P*V_X5P)
R5P(C1%1,C2%2,C3%3,C4%4,C5%5) <-> \
X5P(C1%1,C2%2,C3%3,C4%4,C5%5)                         v_R01529_f/V_R5P,
                                                      v_R01529_b/V_X5P
Ri5P(C1%1,C2%2,C3%3,C4%4,C5%5) <-> \
R5P(C1%5,C2%1,C3%4,C4%2,C5%3)                         v_R01056_f/V_Ri5P,
                                                      v_R01056_b/V_R5P
R5P(C1%11,C2%12,C3%13,C4%14,C5%15) -> \
R5DP(C1%12,C2%11,C3%14,C4%13,C5%15)                   v_R01523/V_R5P
T3P() -> NULL                                         vs_T3P/V_T3P
F6P() -> NULL                                         vs_F6P/V_F6P
R5P() -> NULL                                         vs_R5P/V_R5P
E4P() -> NULL                                         vs_E4P/V_E4P
G3P() -> NULL                                         vs_G3P/V_G3P
I -> I+CO2(C1~0)                                      v_vi_CO2_unlabeled
I -> I+CO2(C1~1)                                      v_vi_CO2_labeled
end reaction rules

begin observables
Molecules T3P_0  T3P(C1~0,C2~0,\
                 C3~0)/V_T3P
Molecules T3P_1  T3P(C1~1,C2~0,\  T3P(C1~0,C2~1,\  T3P(C1~0,C2~0,C3~1)/\
                 C3~0)/V_T3P      C3~0)/V_T3P      V_T3P
Molecules T3P_2  T3P(C1~1,C2~1,\  T3P(C1~1,C2~0,\  T3P(C1~0,C2~1,C3~1)/\
                 C3~0)/V_T3P      C3~1)/V_T3P      V_T3P
Molecules T3P_3  T3P(C1~1,C2~1,\
                 C3~1)/V_T3P
end observables
```

```
generate_network({overwrite=>1});
writeSBML();
writeMfile();
simulate_ode({prefix=>"equil", t_end=>10000, n_steps=>10});
setParameter("v_vi_CO2_unlabeled", 0.9);
setParameter("v_vi_CO2_labeled", 0.1);
simulate_ode({t_end=>200, n_steps=>200});
writeNET();
```

With reference to the Calvin cycle and BioNetGen input file given above, we can now recap the method of setting up a BioNetGen input file to perform a forward simulation of ^{13}C labeling dynamics, which is the core component of any flux estimation procedure.

1. Define the list of molecular species and reactions of interest and specify the atom fate maps for these reactions. In our example, there are 13 internal reactions and six exchange reactions defined in the reaction rules block of the BioNetGen input file.
2. In the parameters block, for each reaction, specify a flux consistent with the stoichiometric constraints. For reversible reactions, we need to specify both the forward and backward fluxes. Metabolite pool sizes are also specified in the parameters block. For this example, we use the default pool size of one.
3. Next, specify the observables. These should be experimentally measurable quantities of interest. We specified T3P based on the mass differences of the isotopomers.
4. Run the forward simulation to obtain the steady state.
5. Change the input feed to isotope-labeled fluxes and simulate the ^{13}C labeling dynamics.

15.5 DISCUSSION AND CONCLUSION

Systems biology is the integration of experimental and computational approaches to achieve the overall goal of explaining and predicting complex cellular behaviors of biological systems. Using different "omics" technologies—genomics, proteomics, and metabolomics—a wealth of information is becoming available about many different cellular components and, to some extent, on the interaction between some of these components. These approaches are providing a static view of biological systems. However, it is of growing importance to augment these analyses with a higher-level understanding of cellular dynamics. Fluxes are an integrated functional output of a metabolic network, which are highly relevant for understanding how the network operates as a system and for applications that focus on manipulating metabolic behaviors, as in metabolic engineering. In recent years, MFA has become an important tool for quantifying metabolic pathways.

MFA allows the detailed quantification of all intracellular fluxes in the metabolism of a biological system. Among the developed tools, ^{13}C MFA, which uses ^{13}C labeling patterns of metabolic products that result from feeding ^{13}C-labeled substrates, is a powerful quantitative method for intracellular pathway flux determination within metabolic networks. We have provided an algorithmic overview of the procedure of ^{13}C MFA. ^{13}C-based MFA requires the knowledge of fate of carbons in metabolic reactions, and the carbon atom fate information is transferred into isotopomer balance equations, which describe the mathematical relationship between unknown fluxes and the available measurement dataset. Using these equations, fluxes can be computed from the measurements through nonlinear parameter fitting.

In recent years, with the development of experimental techniques and computational methods, steady-state ^{13}C MFA has reached a state of relative maturity. The experiments themselves have become a routine procedure, the measurement techniques are well established, and sophisticated mathematical evaluation algorithms are available. However, ^{13}C MFA is still a time-consuming and low-throughput process. Typically, modeling carbon isotopomer networks, such as central carbon pathways, involves an equation system containing a few to several thousand variables to balance reactions between isotopomers. This problem is addressed by the BNGL/BioNetGen-based approach to ^{13}C MFA reviewed here. The BNGL/BioNetGen-based approach allows isotopomer balance equations to be constructed quickly from a database of carbon fate maps and then simulated.

Metabolic steady state is a precondition for current flux analysis methods. The routine methods for flux analysis are based on ^{13}C steady state using isotopic labeling patterns of proteinogenic amino acids, which is reflective of their 12 precursors in central metabolism. Detecting the labeling patterns of free intracellular metabolic intermediates can dramatically shorten the labeling experiments because these intermediates can reach ^{13}C steady state much faster than proteinogenic amino acids. The labeling patterns can also go beyond the 12 precursors. Dynamic ^{13}C labeling patterns of intracellular metabolic intermediates contain more information than the ^{13}C steady-state isotopic labeling patterns. For example, intracellular metabolite pool sizes affect the dynamic labeling patterns, whereas they have no effect on steady-state labeling patterns.

ACKNOWLEDGMENTS

This work was supported in part by the NIH, under grants GM080216, CA132629, and GM076570, and by the DOE, under contract DE-AC52-06NA25396. We thank P.J. Unkefer, C.J. Unkefer, and R.F. Williams for helpful discussions.

REFERENCES

1. Feist A. M., Henry C. S., Reed J. L., Krummenacker M., Joyce A. R., Karp P. D., Broadbelt L. J., Hatzimanikatis V., and Palsson B. O., A genome-scale metabolic reconstruction for Escherichia coli K-12 MG1655 that accounts for 1260 ORFs and thermodynamic information, *Molecular Systems Biology* 2007, 3: 121.
2. Forster J., Famili I., Fu P., Palsson B. O., and Nielsen J., Genome-scale reconstruction of the *Saccharomyces cerevisiae metabolic* network, *Genome Research* 2003, 13: 244–253.
3. Duarte N. C., Becker S. A., Jamshidi N., Thiele I., Mo M. L., Vo T. D., Srivas R., and Palsson B. O., Global reconstruction of the human metabolic network based on genomic and bibliomic data, *Proceedings of the National Academy of Sciences* 2007, 104: 1777–1782.

4. Segre D., Vitkup D., and Church G. M., Analysis of optimality in natural and perturbed metabolic networks, *Proceedings of the National Academy of Sciences* 2002, 99: 15112–15117.
5. Sauer U., High-throughput phenomics: Experimental methods for mapping fluxomes, *Current Opinion in Biotechnology* 2004, 15: 58–63.
6. Sauer U., Metabolic networks in motion: 13C-based flux analysis, *Molecular Systems Biology* 2006, 2:62.
7. Schmidt K., Carlsen M., Nielsen J., and Villadsen J., Modeling isotopomer distributions in biochemical networks using isotopomer mapping matrices, *Biotechnology and Bioengineering* 1997, 55: 831–840.
8. Zupke C. and Stephanopoulos G., Modeling of isotope distribution and intracellular fluxes in metabolic networks using atom mapping matrices, *Biotechnology Progress* 1994, 10: 489–498.
9. Wiechert W., 13C Metabolic flux analysis, *Metabolic Engineering* 2001, 3: 195–206.
10. Arauzo-Bravo M. J. and Shimizu K., An improved method for statistical analysis of metabolic flux analysis using isotopomer mapping matrices with analytical expressions, *Journal of Biotechnology* 2003, 105: 117–133.
11. Wiechert W., Mollney M., Petersen S., and de Graaf A., A universal framework for 13C metabolic flux analysis, *Metabolic Engineering* 2001, 3: 265–283.
12. Szyperski T., Biosynthetically directed fractional 13C-labeling of proteinogenic amino acids. An efficient analytical tool to investigate intermediary metabolism, *European Journal of Biochemistry* 1995, 232: 433–448.
13. Szyperski T., 13C-NMR, MS and metabolic flux balancing in biotechnology research, *Quarterly Reviews of Biophysics* 1998, 31: 41–106.
14. Goto S., Okuno Y., Hattori M., Nishioka T., and Kanehisa M., LIGAND: Database of chemical compounds and reactions in biological pathways, *Nucleic Acids Research* 2002, 30: 402–404.
15. Krieger C. J., Zhang P., Mueller L. A., Wang A., Paley S., Arnaud M., Pick J., Rhee S. Y., and Karp P. D., MetaCyc: Recent enhancements to a database of metabolic pathways and enzymes in microorganisms and plants, *Nucleic Acids Research* 2004, 32: D438–D442.
16. Schomburg I., Chang A., Ebeling C., Gremse M., Heldt C., Huhn G., and Schomburg D., BRENDA, the enzyme database: updates and major new developments, *Nucleic Acids Research* 2004, 32: D431–D433.
17. Michal G., *Biochemical Pathways: An Atlas of Biochemistry and Molecular Biology*. Wiley: New York, Heidelberg, Germany, 1999.
18. Arita M., In silico atomic tracing by substrate-product relationships in Escherichia coli intermediary metabolism, *Genome Research* 2003, 13: 2455–2466.
19. Mu F., Williams R. F., Unkefer C. J., Unkefer P. J., Faeder J. R., and Hlavacek W.S., Carbon fate maps for metabolic reactions, *Bioinformatics* (Oxford) 2007, 23: 3193–3199.
20. Arita M., The metabolic world of Escherichia coli is not small, *Proceedings of the National Academy of Sciences* 2004, 101: 1543–1547.
21. http://bionetgen.org.
22. Faeder J. R., Blinov M. L., Goldstein B., and Hlavacek W. S., Rule-based modeling of biochemical networks, *Complexity* 2005, 10: 22–41.
23. Blinov M. L., Yang J., Faeder J. R., and Hlavacek W. S., Graph theory for rule-based modeling of biochemical networks. *Transactions on Computational Systems Biology VII*, 2006, 89–106.
24. Hlavacek W. S., Faeder J. R., Blinov M. L., Posner R. G., Hucka M., and Fontana W., Rules for modeling signal-transduction systems, *Science's STKE* 2006, p. re6.
25. FateMapViewer Software http://cellsignaling.lanl.gov/FateMaps

26. Faeder J. R., Blinov M. L., and Hlavacek W. S., Rule-based modeling of biochemical systems with BioNetGen, *Methods in Molecular Biology* 2009, 500: 113–169.
27. Wiechert W. and de Graaf A. A., Bidirectional reaction steps in metabolic networks: I. Modeling and simulation of carbon isotope labeling experiments, *Biotechnology and Bioengineering* 1997, 55: 101–117.
28. Dauner M., Bailey J. E., and Sauer U., Metabolic flux analysis with a comprehensive isotopomer model in Bacillus subtilis, *Biotechnology and Bioengineering* 2001, 76: 144–156.
29. Forbes N. S., Clark D. S., and Blanch H. W., Using isotopomer path tracing to quantify metabolic fluxes in pathway models containing reversible reactions, *Biotechnology and Bioengineering* 2001, 74: 196–211.
30. Schmidt K., Nielsen J., and Villadsen J., Quantitative analysis of metabolic fluxes in Escherichia coli using two-dimensional NMR spectroscopy and complete isotopomer models, *Journal of Biotechnology* 1999, 71: 175–190.
31. Zhao J. and Shimizu K., Metabolic flux analysis of Escherichia coli K12 grown on 13C-labeled acetate and glucose using GC-MS and powerful flux calculation method, *Journal of Biotechnology* 2003, 101: 101–117.
32. Schaub J., Mauch K., and Reuss M., Metabolic flux analysis in Escherichia coli by integrating isotopic dynamic and isotopic stationary 13C labeling data, *Biotechnology and Bioengineering* 2008, 99: 1170–1185.
33. Young J. D., Walther J. L., Antoniewicz M. R., Yoo H., and Stephanopoulos G., An elementary metabolite unit (EMU) based method of isotopically nonstationary flux analysis, *Biotechnology and Bioengineering* 2008, 99: 686–699.
34. Yuan J., Fowler W. U., Kimball E., Lu W., and Rabinowitz J. D., Kinetic flux profiling of nitrogen assimilation in Escherichia coli, *Nature Chemical Biology* 2006, 2: 529–530.
35. Wahl S. A., Dauner M., and Wiechert W., New tools for mass isotopomer data evaluation in (13)C flux analysis: Mass isotope correction, data consistency checking, and precursor relationships, *Biotechnology and Bioengineering* 2004, 85: 259–268.
36. http://vimss.lbl.gov/DvHFlux.
37. Steinbeck C., Hoppe C., Kuhn S., Floris M., Guha R., and Willighagen E. L., Recent developments of the chemistry development kit (CDK)—an open-source java library for chemo- and bioinformatics, *Current Pharmaceutical Design* 2006, 12: 2111–2120.
38. Yang J., Monine M. I., Faeder J. R., and Hlavacek, W. S., Kinetic Monte Carlo method for rule-based modeling of biochemical networks, *Physical Review E* 2008, 78: 031910.

Index

B

C

D

E

F

G

J

K

N

O

R

S

W

Z